REDUCING HEALTH DISPARITIES

Health Communication

Gary L. Kreps, General Editor

Vol. 6

The Health Communication series is part of
the Peter Lang Media and Communication list.
Every volume is peer reviewed and meets
the highest quality standards for content and production.

PETER LANG
New York • Washington, D.C./Baltimore • Bern
Frankfurt • Berlin • Brussels • Vienna • Oxford

REDUCING HEALTH DISPARITIES

Communication Interventions

Edited by Mohan J. Dutta
and Gary L. Kreps

PETER LANG
New York • Washington, D.C./Baltimore • Bern
Frankfurt • Berlin • Brussels • Vienna • Oxford

Library of Congress Cataloging-in-Publication Data

Reducing health disparities: communication interventions /
edited by Mohan J. Dutta, Gary L. Kreps.
p. ; cm. — (Health communication; v. 6)
Includes bibliographical references and index.
1. Healthcare Disparities. 2. Communication Barriers.
3. Health Communication—methods. 4. Health Services Accessibility.
5. Quality of Health Care. I. Dutta, Mohan J. II. Kreps, Gary L.
III. Series: Health communication (New York, N.Y.); v. 6.
W 84.1 362.1—dc23 2012036412
ISBN 978-1-4331-1918-7 (hardcover)
ISBN 978-1-4331-1905-7 (paperback)
ISBN 978-1-4539-0935-5 (e-book)
ISSN 2153-1277

Bibliographic information published by **Die Deutsche Nationalbibliothek**.
Die Deutsche Nationalbibliothek lists this publication in the "Deutsche Nationalbibliografie"; detailed bibliographic data is available on the Internet at http://dnb.d-nb.de/.

29 Broadway, 18th floor, New York, NY 10006
www.peterlang.com

CONTENTS

Acknowledgments

The increasing health disparities across the globe draw attention to the emergent need for communication interventions that are directed at addressing these health disparities. We acknowledge the contributions of the communication scholars who have actively contributed toward the development of theoretical, empirical, and applied communication knowledge directed at addressing health disparities globally at micro-, meso-, and macro- levels.

The ideas and voices articulated in this book draw out the convictions of the underserved who work every day to address the large health inequities they experience locally and globally. Their strength and courage offer the intellectual and theoretical impetus for this project.

Mohan would like to thank Gary Kreps, Collins Airhihenbuwa, Teri Thompson, Eileen Berlin Ray, Barbara Sharf, Heather Zoller, Jeff Peterson, Shiv Ganesh and Kathy Miller for their inspiration and support in developing a nuanced understanding of the various communication issues that play out in the realm of health disparities. I would also like to thank my team of community and activist partners and graduate students in various culture-centered projects across the globe who work tirelessly toward creating a just world. Most importantly, I extend my gratitude to my family for their continued support in my academic and activist projects. I thank my baba, Chanchal Dutta, for teaching me the principles of social activism and social change, and ma, Rama Dutta, for teaching me the value of helping others. My son, Shloke, inspires me with his indomitable spirit. Most importantly, I am thankful to my wife, Debalina, who challenges me to do better, stands by me at every step of my life, and brings so much joy in this journey of wanting to make a difference.

Gary would like to thank Mohan for taking the lead on this important book and always following through on completing endless array of needed tasks. I am grateful for my generous friends and mentors who sensitized me to the role of communication in addressing cultural inequities and health disparities, such as Ev Rogers, Gerry Phillips, Don Cassata, Mara Adelman, David Smith, Thorrel Fest, and Paula Kim. Thanks to the personal inspiration I received from my mother, Rhoda Cohen Kreps, who taught me to support the needs of the vulnerable and disenfranchised, as well as the lessons I learned from my father, Sid Kreps, about building productive collaborative partnerships through personal sincerity, concern, and generosity. Most of all, however, I thank my loving partner and best friend for more than 40 years, Stephanie, for always supporting, encouraging, and inspiring me to do important work to help others.

We would also like to thank Mary Savigar and Sophie Appel at Peter Lang for their extensive editorial support with the project. Sarah Comer at the Center for Culture-Centered Approach to Research and Evaluation (CARE) at the National University of Singapore (NUS) assisted with the formatting of the book.

Finally, this book is dedicated to the memory of Chuck Atkin, who played a key role in the development and growth of health communication as an area of study. Chuck passionately demonstrated in his active program of intervention research that strategic communication could be used to effectively mitigate barriers to support, promote informed health decisions, and reduce health disparities.

Dedicated to the memory of Chuck Atkin

1

Reducing Health Disparities: Communication Interventions

Mohan J. Dutta and Gary L. Kreps

Health disparities, where vulnerable populations are heir to much poorer health outcomes than other, more privileged groups, are clearly evident in every part of the globe, including in both the poorest and wealthiest, and the most technologically impoverished and advanced nations of the world (Betancourt, 2007). Typically, the poorest, least well-educated, and most disenfranchised groups of people (minority populations, immigrants, the elderly, people with physical and mental disabilities) tend to suffer from disparities in health outcomes despite concerted efforts to improve the quality of health care around the globe. This insidious pattern of health inequities is closely related to breakdowns in communication processes and limited access to relevant health information (Kreps, 2006). Those who suffer from health disparities are often uninformed and misinformed about imminent health risks, are unaware of the best strategies for detecting and treating health problems, and have difficulties accessing needed health-care services. Poor access to relevant health information limits the abilities of disenfranchised individuals to make good health decisions. They often have difficulties communicating their health concerns, having their voices heard, and eliciting needed support. This pattern of health communication breakdowns inevitably leads to dangerous exposures to health risks, participation in risky health behaviors, high incidence of avoidable health problems, late-stage diagnoses, missed opportunities for needed care, limited access to the best therapies, and ultimately to high levels of health-related morbidity and mortality. Since pathological health communication patterns contribute so mightily to health disparities, it makes sense that opportune interventions to improve health communication could also help to reduce health inequities (Freimuth & Quinn, 2004; Kreps, 2006). This book examines those opportunities to use enhanced health communication to improve health outcomes and reduce health disparities for vulnerable populations around the world.

The patterns of disparities in health outcomes across the globe demonstrate systematically entrenched gaps, with widening gaps in some instances and gaps continuing to be persistent in other areas of health experiences, experiences with health care, and health outcomes (Airhihenbuwa & Dutta, 2012; Millen & Holtz, 2000; Millen, Irwin, & Kim, 2000). These gaps in the global context also map onto the increasing patterns of global inequities accompanying neoliberal political and economic programs (Dutta, 2011). Within nation states, the inequalities between the haves and have-nots continue to increase, with increasing imbalances in the distribution of resources among the different sectors of society (Dutta, 2008, 2011; Marmot, 2004; Navarro, 1999). These patterns of material differentials in the distribution of resources also play out in systematic differentials in the distribution of communicative infrastructures and communicative opportunities for participation (Dutta, 2008, 2009, 2010, 2011, 2012).

Depending upon how the gaps in health experiences, access to health care services and technologies, and health outcomes are measured and compared, patterns of health disparities are observed along the lines of class, gender, race, geographical neighborhood, community context, and nationality of origin. Also, with the increasing patterns of global flows of labor across national boundaries, new patterns of disparities emerge in the context of migrant health outcomes and access to services and treatment resources in migrant communities as compared to native populations. Other research documents declining patterns of health when comparing the health outcomes of migrants before and after migration. It is in this backdrop that the present book is set up as an effort to understand the underlying dimensions of health disparities and the communicative processes, policies, methodologies, and messages that are deployed toward the goal of addressing these disparities, especially as it relates to increasing access, improving quality, and addressing the underlying causes of health disparities (Kreps, 2006).

Broadly, interventions seeking to address health disparities are categorized on the basis of their level of impact. Whereas a large number of health disparities interventions focus on the individual-level determinants of health disparities, other health interventions in the context of disparities address the structural facets of health disparities, seeking to address the social, economic, and political roots of health disparities. Whereas a large percentage of communication interventions within the domain of directed social change fall within the agendas of the dominant structures of society, other interventions are explicitly activist in nature, and are directed at bringing about structural transformations through changes in politics, policies, and programs. Furthermore, in certain instances, communication interventions utilize the combination of both approaches, and address both individual-level as well as structural features of health disparities, shuffling between the structurally transformative and status quo aspects of health disparities work (Airhihenbuwa & Dutta, 2012; Dutta, 2008, 2011).

Background on Health Disparities

Health disparities are documented on a global scale as well as within nation states. Covering experiences and outcomes of health throughout the lifespan, health disparities scholars document disparities in the sociostructural determinants of health; disparities in access to health-promoting resources and information capacities; disparities in access to preventive resources; as well as disparities in access to and quality of health services, treatments received, and health outcomes (LaVeist, 2005; Marmot, 2004). The research on health disparities is driven by the underlying notion that health inequities reflect deep-seated inequities in distribution of resources, and therefore, reflect the organizing processes of political and economic foundations of societies.

Space

Where a person lives has a great deal of impact on the health outcomes of the individual and her or his family (Barr, 2008). This role of space in the context of health inequities may also be noted at familial, household, neighborhood, and community levels. Research on health effects of neighborhoods point out that specific neighborhood contexts constitute specific relationships of power with reference to the mainstream, and therefore offer frameworks for differential distribution of risks within communities, and simultaneously impact the health outcomes within communities (Kawachi & Berkman, 2003). Explanatory mechanisms on social capital, for instance, document that the neighborhoods with greatest inequities are also the ones that have poor social capital—i.e., poor community networks, trust, and community ties—which in turn result in poorer health outcomes. In other work, community-based research in poor communities documents the overall poor outcomes in low-income contexts, marked by lower access to health-promoting resources as well as a variety of structural barriers that are experienced in these communities. For instance, lower-income communities have lower levels of access to a variety of health services such as hospitals and clinics, which therefore impacts the health-care access of community members.

Social Class

Extant research on health disparities has historically been conducted in the domain of social class, documenting class-based differentials in the distribution of resources and the impact of these differentials on a variety of health-based resources as well as health outcomes (Isaacs & Schroeder, 2004; Marmot, 2004). The research on social class documents class-based inequities, pointing out that access to health resources is dependent upon social class and upon where a person belongs in the class structure. The basic explanation of class-based inequities documents the differential patterns of access to resources that are based upon differential economic capacities. Research documents class-based disparities in a variety of contexts ranging from access to quality education and access to opportunities for growth to access to health-promoting resources, access to employment, and access to health care.

Race

Within the United States, a consistent body of literature on health disparities documents differentials in access to preventive services, quality of care, quality of treatment, and health outcomes by race (Barr, 2008; LaVeist, 2005). For instance, there are systematic patterns of differences in overall morbidity and mortality by race, as well as systematic patterns of differences in disease-specific morbidity and mortality. The effect of race also intersects with the effect of class, and therefore, scholars addressing the intersections of race and class have sought to delineate the intertwined and complex effects of race and class, foregrounding both mediating as well as moderating effects of these variables (Barr, 2008). In spite of controlling for class and access to resources, health scholars report differentials in quality of service and treatment by race. Scholarship on racial differences in health outcomes and experiences of health care document a variety of reasons, including racisms at personal, institutional, and societal levels; cultural patterns; patient attitudes; and systematic patterns of structural inequities at social, economic, and political levels.

Global Policies

At a global level, in seeking to understand the structural causes of health inequities, health scholars point to the underlying political and economic configurations of neoliberalism, which constitutes the overall framework of the growing inequities between the haves and have-nots (Dutta, 2008, 2011; Millen & Holtz, 2000; Navarro, 1999). Neoliberalism, understood in the form of the Bretton Woods configurations that were imposed globally through the agendas of the international financial institutions (IFIs) during the regimes of Ronald Reagan in the US and Margaret Thatcher in Britain, established a series of political and economic configurations globally that privileged privatization of resources, liberalization of trade, minimization of tariffs, and minimization of state expenditure on public resources. Imposed globally through structural adjustment programs (SAPs) pushed by the IFIs, nation states in the Global South implemented liberalization policies in exchange for loans sponsored by IFIs, opening up their markets to transnational corporations (TNCs) and consolidating power in the hands of TNCs. The large-scale adoption of neoliberal policies resulted in the upward redistribution of resources, accompanied by rising unemployment, marginalization of the rural sector, rising food prices, and further marginalization of the poor (Dutta, 2011; Millen & Holtz, 2000). Poverty globally has been constituted in the backdrop of neoliberal policies, resulting in the disenfranchisement of the poor from spaces of political and economic decision making, and accompanied by large-scale displacements of the poor through land grabs, and increasing exposure of the poor to disease risks and to structural and physical forms of violence brought about by changing migratory and demographic patterns.

Communication Interventions Addressing Health Disparities

The various types of communicative interventions that are directed at addressing health disparities may be categorized into (a) message-based persuasion interventions that seek to change the individual-level behaviors of community members from underserved groups; (b) training-based communication interventions that seek to develop communication training directed at providers serving marginalized communities or at patients from marginalized contexts; (c) technology-driven communication interventions that propose to build communication technologies in underserved communities; (d) community-driven interventions that catalyze community structures and decision-making processes in order to address community needs; and (e) policy-driven, activist interventions that seek to transform the inequitable structures that underlie health disparities.

Persuasion-Based Models Targeting Underserved Populations

Given the existing research on health disparities that documents differences in access to preventive resources and individual-level differences in enactment of preventive behaviors, the line of research on prevention-based models focuses on developing health communication interventions that address the cultural characteristics of underserved communities. The core assumptions in this line of work build on the argument that developing preventive behavioral capacities in underserved communities works toward addressing health disparities. In this body of work that is specifically driven by persuasion theories seeking to disseminate healthy behaviors in underserved communities, the emphasis is placed on figuring out the appropriate cultural characteristics of the target culture in order to develop effective health interventions.

Training-Based Communication Interventions

Training-driven communication interventions focus on providing training programs for patients from underserved contexts, as well as on providing training programs for physicians who interact with patients from underserved contexts. Based on the understanding that provider biases impact the nature of physician-patient communication, the treatments and services received by patients, as well as the health outcomes of underserved patients, training-based communication programs that are directed at providers seek to build their communication skills as well as cultural competence. In addition, communication-training programs directed at underserved patients focus on developing communication skill sets and patient efficacy, with the goal of building participatory decision-making capacities in underserved populations of patients. Programs on multicultural training and multicultural competence constitute key components of training-based communication interventions.

Technology-Driven Communication Interventions

Technology-driven communication interventions are driven by the underlying idea that fostering technology-driven access works toward addressing health disparities by building technological capacities in underserved communities. The technology therefore becomes the channel for delivering health information, for building preventive resources, for enhancing social-support resources that are available to underserved communities, as well as for creating entry points for patient empowerment and patient participation in decision making. The technology becomes a conduit for promoting greater participation in decision-making processes, and for fostering greater access to health information. In turn, such increased participation and access are expected to improve the quality of outcomes for the underserved segments of the population. Strategies for addressing technology-based components of the digital divide are seen as ways of creating greater access to health-promoting resources in underserved communities.

Community-Driven Communication Interventions

Community-driven communication interventions focus on utilizing community channels and avenues to develop participatory mechanisms that are directed at addressing health disparities. At the heart of community-driven communication interventions is the notion of the participatory decision-making capacity of the community, fostering spaces of community engagement in the development of specific dissemination mechanisms that are directed at addressing the health-care disparities that are faced by community members. Fostering spaces for communication in communities creates local spaces of mobilizing for community members, through which they can come together to identify the key issues facing underserved communities, and mobilize collective efforts and resources to address these problems through locally meaningful solutions. Community-driven efforts vary widely in terms of the level of participation as well as the nature of participation within communities. Whereas some forms of community-based efforts are primarily expert-driven—where community participation serves as a mechanism for diffusing the expert-driven solutions—other forms of community-driven efforts are shaped by community members, and are driven by the needs and understandings as articulated by community members. Community-based interventions also vary widely, in terms of the types of interventions they propose, ranging from individual behavior modifications to structurally directed programs to programs that are oriented toward bringing about changes in structures through their impact on policies, programming, etc.

Activist Interventions

Interventions of activism are directed toward challenging social structures and the dominant forms of organizing resources in political, economic, and social systems. Therefore, activist interventions address disparities by directly seeking to change the underlying political and economic configurations that drive health disparities. Essential to projects of activism is the explicit emphasis on transformative processes that are directed at changing the inequitable forms of economic and communicative distribution of resources, seeking to disrupt the erasures constituted within dominant spaces of policy making and programming that carry out oppressions on disenfranchised communities through their systematic erasure from discursive spaces. Whereas activist projects on the one hand seek to change the ways in which material resources are inequitably distributed, they do so by seeking to build spaces for recognition and representation for underserved communities so that they may have their voices heard. In seeking to resist dominant structures of organizing, they critically interrogate the relationships of power that are written into these structures, and question the configurations of power distribution in the backdrop of inequities. Maintaining a skeptical stance toward the structural formations that constitute the inequitable distribution of resources, activist interventions seek to bring about transformations in these structures by questioning their taken-for-granted logics and underlying assumptions.

Organization of This Book: Communication Interventions at Various Levels

This book is organized into four different sections, categorized under the four key umbrellas of communication interventions that are directed at addressing health disparities in different contexts. Patient-based communication interventions are specifically directed at addressing physician-patient communication, with an emphasis on the perpetuation of health disparities in the interactions between providers and patients. Community-based communication interventions take communities as entry points for action and social change, with an emphasis on developing community participatory strategies for addressing health issues and developing localized solutions from within the community. Media-based communication interventions are focused on developing preventive communication interventions directed at addressing health disparities through the deployment of mass-media campaigns. Whereas some of these media-based interventions deploy traditional mass media to disseminate the campaign message, other interventions apply new media to address health disparities. Critical perspectives on communication and health disparities focus on interrogating structural inequities in distribution of resources, and seek to foster transformative spaces for the politics of interventions by interrogating the taken-for-granted assumptions constituting the distribution of resources. Critical perspectives interrogate the underlying dimensions of power and control that perpetuate patterns of inequities, and seek to foster change by developing spaces for recognition and representation in the politics of social change globally.

In the first section, in the chapter titled, "Shifting Communication Challenges to Education and Reflective Practice: Communication to Reduce Disparities Toward Patients With Disabilities," Ashley P. Duggan and Ylisabyth S. Bradshaw describe a communication intervention that is directed at training providers in communicating with patients with disabilities. This is followed by a chapter by Elaine Hsieh titled, "Health Literacy and Patient Empowerment: The Role of Medical Interpreters in Bilingual Health Communication," that discusses the role of medical interpreters in fostering patient empowerment in bilingual health communication. The notion of empowerment is further carried forward in the chapter by Melinda Weathers and Gina Castle Bell titled, "Assessing Health Disparities Through Communication Competence, Social Support, and Coping Among Hispanic Lay Caregivers for Loved Ones With Alzheimer's Disease." Christina Jones and Mohan J. Dutta engage the culture-centered approach to depict the structural loci of health disparities in their chapter titled, "Culture-Centered Constructions Among Health Providers in Rural Indiana: Setting Agendas for Interventions."

In the second section of the book, on community-based health communication interventions, Tamar Ginossar discusses a communication leadership program in her chapter titled, "Promoting Women Leadership as a Strategy for Reducing Health and Digital Disparities in Latino/a Immigrant Communities." This theme of localized leadership and ownership of locally situated campaigns is also the focus of Patrick J. Dillon and Ambar Basu's chapter on a culture-centered project, "Preventing HIV/AIDS Through a Culture-Centered Health Campaign: The Sonagachi HIV/AIDS Intervention Program." Similarly, Nurit Guttman, Anat Jaffe, and Seffefe Aycheh discuss localized grassroots efforts to address health disparities in their chapter titled, "Challenges and Dilemmas in Developing Health Communication Programs for Ethiopian Immigrants in Israel—The *Tene Briut* Health Promotion Organization." The theme of localized participation is also reiterated in a chapter by Elisia L. Cohen, Robin C. Vanderpool, Rick Crosby, Seth M. Noar, Wallace Bates, Tom Collins, Katharine J. Head, Margaret McGladrey, and Baretta Casey titled, "1-2-3 Pap: A Campaign to Prevent Cervical Cancer in Eastern Kentucky." Torill Bull, Maurice B. Mittelmark, and Peter Sanbian Gmatieyindu in their chapter titled, "Vulnerable Old Women of Northern Ghana at Risk of Being Ostracized for Witchcraft: How Wise Village Leadership Brings Them Back From the Brink," discuss the powerful role of local leadership in identifying innovative, community-directed solutions. The chapter by Robin Kelly, Al Hannans, and Gary L. Kreps titled, "A Case Study of the Community Liaison Pilot Program: A Culturally Oriented, Participatory Pilot Health Education and Communication Program to Decrease Disparities in Minority HIV Vaccine Trial Participation," builds on the theme of community-based interventions to demonstrate the defining role of participatory strategies in addressing a culturally oriented health education and communication program. Community participation also becomes a theme in the chapter titled, "Evaluation of the Influence of an Urban Community Park Revitalization on African American Youth Physical Activity," by Gary L. Kreps, Melinda M. Villagran, Janey Trowbridge, Paula Baldwin, Yolanda Barbier, Amy Chang, Jung-mi Jun, Megan Tucker, Autumn Saxton-Ross, and Sieglinde Friedman, that depicts the participatory strategies at the local community level in an urban community park revitalization effort. Muhiuddin Haider and Nicole I. Wanty offer us insights into the role of community engagement in a global context in their chapter titled, "Reducing Health Disparities Using Health Communication Interventions and Community Engagement." Finally, Deborah S. Ballard-Reisch, Rick McNary, Pamela O'Neal, Ashley Archiopoli, Bobby Rozzell, Melissa Granville, and Alyssa Ballard-Reisch describe a grassroots, community-based project in their chapter titled, "Numana Feeds Haiti: A Case Study of a Grassroots, Community-Based Social Movement to 'Save the Starving.'"

The third section of the book offers us insights into communication interventions that are directed at addressing health disparities through the dissemination of health messages in the mass media, as well as through new media. The interventions discussed in this section have a spelled-out, mediated component although they also, in multiple instances, have community-driven components. We begin the section with a chapter written by Elisia L. Cohen, Charlene A. Caburnay, Jon Stemmle, María Len-Ríos, Tim Poor, Barbara Powe, Susie Rath, Erin Robinson, Katharine Semenkovich, Glen T. Cameron, Douglas A. Luke, and Matthew W. Kreuter titled, "The Ozioma News Service: Targeting Cancer Communication to African American Communities," that discusses the role of a news service in fostering disparities-oriented stories in African American communities. In their chapter titled, "Reducing Smoking Disparities for Hispanic Adolescents: Empowerment Through Media Literacy," Kathryn Greene and Smita C. Banerjee discuss media literacy strategies in empowering Hispanic adolescents to reduce smoking disparities. Similarly, Sarah Diamond, Leslie Snyder, and Jean J. Schensul describe a multimedia campaign targeting substance use in their chapter titled, "Preventing Substance Use Among Multiethnic Urban Teens Using Education Entertainment, Social Marketing, and Peer Content." Ann Neville Miller, Julie Gathoni Muraya, Ann Muthoni-Thuo, and Leonard Mjomba discuss the role of formative research in communication interventions addressing health disparities through mass media in Kenya in their chapter titled, "The Importance of Formative Research in Mass-Media Campaigns Addressing Health Disparities: Two Kenyan Case Studies." The coverage of mass-media interventions in this section is accompanied by chapters that depict communication interventions that utilize the new media. David B. Buller and W. Gill Woodall describe the role of an Internet intervention in the context of health inequities in their chapter titled, "*5 a Day, the Rio Grande Way*: Internet Intervention and Health Inequities in the Upper Rio Grande Valley." Along these lines, Kevin B. Wright documents the role of computer-mediated support groups in his chapter titled, "Comprehensive Care Center and Computer-Mediated Support Group Approaches to Targeting Health Disparities: Two Case Studies." Fiona McTavish, Helene McDowell, and David Gustafson offer us an overview of the specific communication functions that can be performed by an e-health system seeking to serve the medically underserved in their chapter titled, "How e-Health Systems Can Help the Medically Underserved."

The final section of the book emphasizes critical approaches to addressing health disparities, and draws attention to the questions of power and control that constitute the broader realms of social, political, and economic organizing of societies. Situating discussions of health disparities within the contexts of broader material inequities, the chapters in this section discuss communication interventions that seek to disrupt the status quo and interrogate the relationships of power that are embodied in mainstream structural configurations. Rebecca de Souza opens up the section with her chapter, "The Business of Doing Interventions:, A Paradox for the Participatory Method" that interrogates the interplays of power and control embodied in the relationships of NGOs seeking to serve underserved communities and their key stakeholders. This theme of interrogating the power and control of the dominant structures working with health disparities is carried into the chapter written by Lalatendu Acharya and Mohan J. Dutta titled, "Culture-Centered Deconstructions: In-Depth Interviews With HIV Planners Targeting Tribals in Orissa, India," that foregrounds the dialectical tensions in the narratives of HIV/AIDS program planners working with a highly underserved community of indigenous tribes in the Orissa state of India. Heather Zoller and Lisa Melancon take forward the concepts of power and resistance to the realm of health activism in their chapter titled, "The Good-Neighbor Campaign: A Communication Intervention to Reduce Environmental Health Disparities." Mohan J. Dutta, Sydney Dillard, Rati Kumar, Shaunak Sastry, Christina Jones, Agaptus Anaele, U. Dutta, William Collins, Titilayo Okoror, and Calvin Roberson present the reflexive tensions that are foregrounded in the processes of developing academic-community partnerships in building culturally centered community capacities for evidence-based decision making in their chapter titled, "Relational Tensions in Academic-Community Partnerships in the Culture-Centered Approach (CCA): Negotiating Communication in Creating Spaces for Voices." The culture-centered approach also emerges as a theme in the chapter by Mohan J. Dutta, Christina Jones, Abigail Borron, Agaptus Anaele, Haijuan Gao, and Sirisha Kandukuri titled, "Voices of Hunger: A Culture-Centered Approach to Addressing Food Insecurity." Extending the transformative framework of the culture-centered approach, Mohan J. Dutta writes about the health organizing of grassroots groups seeking to build greater domains of accountability and local participation in "Communication as Health Activism: The Case of the Campaign to Stop the Gardasil Clinical Trials in India." In the final chapter in this section, titled, "Voices of Resistance: The Niyamgiri Movement of the Dongria Kondh to Stop Bauxite Mining," Mohan J. Dutta further extends the framework of communication interventions in the context of health disparities by theorizing displacement as a fundamental health problematic in the margins of development programs of land acquisition, mining, and industrialization under neoliberal governance, and by conceptualizing indigenous activism as transformative interventions.

In summary, the chapters presented throughout the pages of the book seek to offer a multiparadigmatic, multicontextual, multilevel, and multimethod framework for understanding the role of communication interventions addressing health disparities. It is our hope that as you read these chapters, you draw out theoretical, methodological, and application-driven frameworks for understanding health disparities, and for developing communication interventions that are directed toward addressing them.

Bibliography

Airhihenbuwa, C. O., & Dutta, M. J. (2012). New perspectives on global health communication: Affirming spaces for rights, equity, and voices. In R. Obregon & S. Waisbord (Eds.), *The handbook of global health communication* (pp. 34–51). Chichester, UK: Wiley-Blackwell.

Barr, D. A. (2008). *Health disparities in the United States*. Baltimore, MD: Johns Hopkins University Press.

Betancourt, J. R. (2007). Racial and ethnic disparities in health care. In L. Epstein (Ed.), *Culturally appropriate health care by culturally competent health professionals* (pp. 18–32). Caesarea, Israel: Israel National Institute for Health Policy and Health Services Research.

Chomsky, A. (2000). The threat of a good example: Health and revolution in Cuba. In J. Y. Kim, J. V. Millen, A. Irwin, & J. Gershman (Eds.), *Dying for growth: Global inequality and the health of the poor* (pp. 331–357). Monroe, ME: Common Courage Press.

Dutta, M. J. (2008). *Communicating health: A culture-centered approach*. London, UK: Polity Press.

Dutta, M. J. (2009). On Spivak: Theorizing resistance—Applying Gayatri Chakravorty Spivak in public relations. In Ø. Ihlen, B. van Ruler, & M. Fredriksson (Eds.), *Public relations and social theory: Key figures and concepts* (pp. 278–300). New York, NY: Routledge.

Dutta, M. J. (2010). The critical cultural turn in health communication: Reflexivity, solidarity, and praxis. *Health Communication, 25*(6–7), 534–539.

Dutta, M. J. (2011). *Communicating social change: Culture, structure, agency*. New York, NY: Routledge.

Dutta, M. J. (2012). *Voices of resistance: Communication and social change*. West Lafayette, IN: Purdue University Press.

Freimuth, V. S., & Quinn, S. C. (2004). The contributions of health communication to eliminating health disparities. *American Journal of Public Health, 94*(12), 2053–2055.

Isaacs, S. L., & Schroeder, S. A. (2004). Class—The ignored determinant of the nation's health. *New England Journal of Medicine, 351*(11), 1137–1142.

Kawachi, I., & Berkman, L. F. (2003). *Neighborhoods and health*. London, UK: Oxford University Press.

Kreps, G. L. (2006). Communication and racial inequities in health care. *American Behavioral Scientist, 49*(6), 760–774.

LaVeist, T. A. (2005). *Minority populations and health: An introduction to health disparities in the United States*. San Francisco, CA: Jossey-Bass.

Marmot, M. (2004). *The status syndrome: How social standing affects our health and longevity*. New York, NY: Henry Holt.

Millen, J. V., & Holtz, T. (2000). Dying for growth, Part I: Transnational corporations and the health of the poor. In J. Y. Kim, J. V. Millen, A. Irwin, & J. Gershman (Eds.), *Dying for growth: Global inequality and the health of the poor* (pp. 177–223). Monroe, ME: Common Courage Press.

Millen, J. V., Irwin, A., & Kim, J. Y. (2000). Introduction: What is growing? Who is dying? In J. Y. Kim, J. V. Millen, A. Irwin, & J. Gershman (Eds.), *Dying for growth: Global inequality and the health of the poor* (pp. 3–10). Monroe, ME: Common Courage Press.

Navarro, V. (1999). Health and equity in the world in the era of 'globalization.' *International Journal of Health Services, 29*(2), 215–226.

2

Shifting Communication Challenges to Education and Reflective Practice: Communication to Reduce Disparities Toward Patients With Disabilities

Ashley P. Duggan and Ylisabyth S. Bradshaw

People with disabilities represent a large, underserved population (Bryan, 2007; Wilson & Merrill, 2002). While members of most minority groups grow up in a recognized subculture, developing shared norms and expectations, people with chronic diseases and disabilities are not similarly prepared. They are often born with or acquire such conditions within families who neither have these conditions nor associate with others who do (Zola, 1993). Adults who develop disabilities have been isolated traditionally without networks of others with similar conditions. Patients with disabilities face obstacles in interactions with health-care providers and are especially vulnerable to adverse health-care experiences, including time constraints to address complex issues, sensory and cognitive communication barriers, limited financial resources, and physically inaccessible care sites (President's Advisory Commission on Consumer Protection and Quality in the Health Care Industry, 1997). Patients with disabilities describe lack of dignity during examinations, inadequate medical explanations, and feeling rushed and overlooked (President's Advisory Commission on Consumer Protection and Quality in the Health Care Industry, 1997). Disability is considered a risk factor for dissatisfaction with health care, frustration with unmet health-care needs, and lack of health-promotion services (Veltman, Stewart, Tardif, & Branigan, 2001). Patients with disabilities describe apprehension regarding both physician attitudes and physician competence about disability issues (Shapiro & Gottman, 2005). Similarly, physicians report limited experience, and express reluctance in addressing patients with disabilities (Aulagnier et al., 2005).

Over 54 million Americans, spanning all ages, have some disability (U.S. Department of Health and Human Services, 2010), with disability defined as impairments, activity limitations, and participation restrictions. Furthermore, with an aging population the prevalence of disability continues to increase. Many people with early-diagnosed disabilities today live virtually normal life spans (Iezzoni, McCarthy, Davis, Harris-David, & O'Day, 2001), but need responsive health-care providers in order to maximize quality of life. Individuals with disabilities have narrow margins of health, and are at increased risk for preventable secondary conditions (U.S. Census Bureau, 2001), yet access to reliable primary health care remains a major unresolved issue for them (DeJong, 1997)

Primary-care physicians are often the first point of contact for health promotion and treatment for individuals with disabilities (Memel, 1996), but health providers describe lack of training (Jain, 2006) and lack of preparation to effectively work with patients with disabilities (Sabharwal & Fiedler, 2003). This chapter describes ways in which an educational module designed to teach medical students about patients with disabilities has evolved into an education and research program that can be used to reduce health disparities in patients with disabilities.

Disabled as a Student but Enabled as a Doctor

This project illustrates the ways in which examination of interpersonal communication processes between health-care providers and standardized patient educators (SPEs) addresses challenges in providing health care to people with disabilities, reducing disparities through education. The educational module and communication analysis allow for understanding how medical students can be disabled as a student, but enabled as a doctor (language borrowed from Maughan, 2005). We examine ways in which providers traverse boundaries of disability identity; confront disparity-inducing attitudes; recognize patient preferences for disability language; and anticipate frequent, unmet needs of patients with disabilities, such as common overuse injuries.

We examine communication behaviors in educational interactions at Tufts University School of Medicine, where individuals with visually apparent disability serve as standardized patient educators (SPEs). The interactions are designed as nonevaluative opportunities for medical students to learn to communicate about disability. The medical interviewing and feedback/education components were videotaped, transcribed, and analyzed across a full year of family medicine clerkship rotations; measures of health outcomes were collected from medical students and SPEs.

Collaborative efforts allow for translating research into practice by positioning the research within the larger framework of medical education and public-health advocacy. A Communication professor (Ashley Duggan, Boston College) leads the research component, in close collaboration with medical education and public health colleagues (Ylisabyth Bradshaw, DO, MS pain research, education and policy academic director; and Wayne Altman, MD, family medicine clerkship director, both at Tufts University School of Medicine). Collaboration ensures that communication processes associated with health outcomes are well-grounded in the foundations of medical education, disability advocacy, and population-based, public-health goals. An existing educational module allowed us to evaluate communication behavior. The communication project builds on an educational module employing individuals with disabilities as medical educators (for a detailed description of the disability curriculum, see Minihan et al., 2004). The communication focus added videotaping, coding and analysis, and outcome evaluation components to an existing family medicine clerkship curriculum. Following IRB approval, family medicine clerkship students (N = 142) were video-recorded across an academic year of 11 clerkship rotations. Students initially interviewed SPEs with disabilities, then received feedback. Recognizing the ways health providers ask about disability or avoid disability language allows for reducing disparities by responding to patient preferences and recognizing common overuse injuries (Duggan, Bradshaw, & Altman, 2010). Recognizing the ways health providers build reflective capacity allows for reducing disparities by identifying communication obstacles and opportunities of resolution. The immediacy of education based on lived experience delivered by a person with a disability, combined with *evidence-based* communication research and evaluation, allow communication challenges to be shifted to empower providers to deliver improved care to patients with disabilities.

Method

Interactions occurred as part of the medical students' family medicine clerkship—a four-week clinical rotation required of all students in their third or fourth year—and included a twelve- to fifteen-minute, simulated medical interview, followed by a twelve- to fifteen-minute nonevaluative feedback session with a faculty observer, peer observer, and SPE. All family medicine clerkship students completed the interactions and feedback; students opting to add research components of videotaping and self-report measures received a copy of their DVD, and were provided an ongoing opportunity to receive descriptions from project analysis. Interactions were assigned arbitrary ID numbers to preserve anonymity.

SPEs were recruited through flyers seeking educators with apparent disability, which were distributed through independent living centers. SPEs were selected and retained based on articulate communication about navigating the health-care system, affinity for teaching, and willingness to share their experience with disability and to address medical students as novice learners. During the interactions, SPEs role-played "Chris Walker," an individual of their same age, sex, and having their own disability, who reports shoulder pain from tendonitis. SPEs' disabilities included multiple sclerosis, juvenile rheumatoid arthritis, traumatic spinal cord injury, neurodegenerative disorders, cerebral palsy, muscular dystrophy, blindness, paraplegia, quadriparesis, respiratory failure, neurofibromatosis, and mobility constraints. None of the disabilities include cognitive deficit. The range of visible dimensions of SPEs' disabilities included assistance dogs, Canadian crutches, manual wheelchairs, scooters, and motorized wheelchairs. SPEs participated in two weekend-long training sessions and a series of practice interviews prior to meeting with medical students, and received financial compensation for their work. Medical students focused on history-taking and negotiated a treatment plan (without a physical exam) in a case designed to elicit questions about disability.

The case was designed in consultation with an advisory committee composed of people with disabilities, family members of people with disabilities, state disability agency representatives, and medical educators; the case was piloted and refined prior to the current project (see Minihan et al., 2004). At the recommendation of the advisory committee, medical students do not know about the disability before the initial interaction. The case includes multiple opportunities to hear the SPE refer to disability and to provide a biopsychosocial portrayal of patients with disability and their professional, family, and relational identities, as well as multiple competencies. During the medical interview, opportunities for disclosure can be probed further, to provide a comprehensive sense of both biopsychosocial components of shoulder pain, activities of daily living with disability, and language preferences for addressing disability.

Feedback sessions following "patient" interviews were designed to give constructive, descriptive (as opposed to evaluative) comments to medical students regarding their performance (Kahn, 2003). To encourage medical students' reflection about their performance, they begin with the SPE asking the medical student, "How did it go for you?" Strengths and areas for improvement are highlighted regarding the medical student's interviewing skills, treatment plan negotiation, and knowledge of disability, including the SPE's disability, treatment of overuse syndromes, and disability language. In addition, feedback sessions include an opportunity for medical students to ask the SPE any questions they would like about disability, specifically inviting questions about the individual's disability.

Transcription and Analysis

For analysis of asking about disability, all interview portions with utterances about disability were transcribed verbatim by a primary investigator. In addition, four research assistants transcribed feedback sessions in their entirety. Coders used Glaser and Strauss's (1967) constant comparative method to categorize into themes. Coders read and initially categorized all of the interactions independently, and then reviewed transcripts together. Where there was disagreement, coders shifted and reorganized independently and then re-reviewed together until reaching consensus. To ensure that identified themes and framing were consistent with the experiences of participants, a subset of SPEs, medical faculty, and health-communication leaders provided feedback on identified themes.

Asking About Disability or Avoiding Disability Language

Identified themes in asking about disability in medical interviews are distinct in both language and nonverbal communication behavior. Seventy-four percent of the students (n = 105) asked about disability; twenty-six percent of the students did not ask about disability (n = 37) (Duggan, Bradshaw, & Altman, 2010). Themes are identified for all interview interactions; for students who ignored the disability in the interview, themes are also identified in feedback responses.

Students Who Asked About Disability

Within the interactions where medical students ask about disability (n = 105), five primary themes were identified:

1. Ask and integrate disability with shoulder pain
2. Ask indirectly
3. Ask but do *NOT* integrate disability with shoulder pain
4. Pick up cues from SPE-led disclosure
5. Ask about past medical history

Secondary themes within each of the primary themes are related to timing (when in the interaction the medical student asks) and to finer distinctions in medical students' language (Duggan, Bradshaw, & Altman, 2010). A brief summary is included below. For the full paper, please see Duggan, Bradshaw, and Altman (2010).

Ask and integrate disability disclosure. Twenty-six medical students asked about the disability and also addressed biological and psychosocial implications connecting the shoulder pain with the disability, and using the word, *disability* following SPE indication of preferred language choice. These medical students integrated disability information with further questions about shoulder pain and treatment options (i.e., asking about wheelchair design and/or whether the personal care attendant can help with physical therapy exercises). The students' straightforward exploration of presenting symptoms and conditions elicited information from the SPE about disability, and eventually integrated the disability disclosure into recommended treatment (Duggan, Bradshaw, & Altman, 2010).

Ask indirectly. Twenty-three medical students asked indirectly by naming implications of the disability rather than directly using the word, *disability.* These students distinguished between shoulder pain and disability by addressing symptom differences, and they acknowledged the ways assistive devices (i.e., crutches) might be related to shoulder pain. Of these twenty-three medical students who asked indirectly, seventeen asked about the wheelchair, crutches, or assistance dog. Three medical students asked about injury or illness rather than using the word, *disability*. The remaining three students who asked indirectly also referred to the wheelchair or crutches, but not until halfway through the medical interview, and used more limiting, potentially negative language (i.e., "Can you tell me what made you wheelchair-bound?").

Ask but do not integrate. Seventeen medical students asked about the disability, but did not integrate the disability diagnosis, or any of its implications, with information about the presenting shoulder pain. After these students asked about disability, they reverted to biomedical questions, changed topics, or managed the interaction through closed-ended questions. After initially eliciting the disability diagnosis, these students kept questions focused on biomedical aspects of pain rather than psychosocial implications of either acute pain or chronic disability (i.e., following disclosure of disability in the arm, asking about pain in the other joints). In five of these interactions the medical student's language can be interpreted as overaccommodating or using overly positive attributions for what would be normative disclosure if the disability were not a consideration (i.e., responding to disclosure that "I have been in a wheelchair since college" by saying, "It's great you went to college) (Duggan, Bradshaw, Swergold, & Altman, 2011).

Pick up cues from SPE-led disclosure. Fifteen medical students responded to SPE disclosure rather than asking probing questions about disability. These students responded to SPE references to disability and then integrated psychosocial aspects of disability, but without ever naming the disability. For example, these students asked about shoulder pain associated with transferring from a wheelchair or using a personal-care assistant. The pain associated with transferring in these interactions was first mentioned by the SPE.

Ask about previous medical history. Twenty-four students asked about previous medical history as a general question, and did not differentiate the patient's long medical history involving disability with implications for shoulder pain. They avoided specific disability language, also indicating high uncertainty or hesitancy in the ways they asked about other medical problems. Asking about medical "problems" rather than medical "history" was included here. Labeling "problems" does not distinguish between disability as a part of the natural human condition and disability as a disease or problem.

Interactions Where Medical Students DO NOT Ask About Disability

Thirty-seven medical students did not ask about the disability, and did not reciprocate when the SPE disclosed about the disability.

Disability questions absent. Ten medical students did not ask about disability, and the SPE similarly did not name the disability. These students addressed the pain and its diagnosis, moving quickly in a direct inquiry from initial symptoms through prior pain diagnosis. In these interactions, disability disclosure is inhibited by high provider conversation control with closed-ended questions and rapid shifts from topic to topic.

SPE discloses, but medical student does not reciprocate. In twenty-seven interactions, the SPE names the disability or implications of the wheelchair or crutches, but the medical student does not respond to cues about disability. Within five of these interactions, the medical students ask a general question about whether there is anything else to address, which prompts the SPE to name the disability; these medical students then change topics and do not return to disability questions. In twenty of these interactions, SPEs discuss implications of disability, but medical students revert to biomedical questions in response to the SPE talking about broader implications of the disability. In these twenty interactions, the SPE brought up psychosocial issues, including concerns about the pain becoming chronic and further interfering with mobility or independence, but these medical students changed topics or responded to concerns as if the disability had not been part of the disclosure. The other two students who did not reciprocate disability disclosure named the disability as if unrelated, treating the disability disclosure as an aside and then returning to the "primary concern" of the shoulder pain.

Feedback Following Interactions Where Medical Students Do Not Ask About Disability

Analysis of the feedback portion of the interactions provides additional information about the meaning of interpersonal communication in the medical interview. The feedback session addresses strengths and opportunities for growth, and is intended to be descriptive rather than evaluative and to prompt medical student reflection about their communication. As an actor or actress talks about a character he or she played, third-person reflections are presented by the SPE in terms of what the "patient" may have perceived during the interview, for the purpose of providing a first-hand report or direct patient data for the student's benefit in developing their interview skills and knowledge base.

Recognizing limitations. In feedback sessions for the ten interactions where disability was not mentioned by either the medical student or the SPE, the medical student immediately acknowledged his or her attitude in not addressing the disability in five interactions, and in the other five interactions, feedback functioned as a teaching tool. When students recognized their own attitudes in not mentioning disability, self-reflection prompted feedback that guided the development of alternative attitudes and relationships. For example, if medical students talked about not wanting to stigmatize a person with a disability, the SPE shared ways to reciprocate the patient's disclosure. When medical students' initial reflection on "how it went" still avoided disability, comments directed to the student were less discussion-focused, and instead highlighted the importance of recognizing the disability. For example, the faculty observer in these interactions pointed out the ways in which the wheelchair and disability are directly related to the shoulder pain.

Learning to reciprocate patient disclosure. Of the twenty-seven interactions where SPEs disclose disability-related information, but medical students do not respond, feedback indicated that medical students were self-aware without prompting in eleven interactions; feedback was used as a teaching tool in ten interactions; and feedback suggested that disability was addressed nonverbally in interactions, and thus acknowledged without naming the disability in six interactions. Medical students who indicated their self-awareness about avoiding disability in the interview initiated conversation and commented that they may have needed to address disability. In medical interviews where the SPE disclosed disability information, but medical students neither picked up on these cues nor mentioned disability in their own reflections, observers and/or the SPE highlighted the importance of recognizing disability. These feedback sessions paralleled teaching opportunities in cases where disability was not ever mentioned. Medical students who addressed the disability implicitly in the interview, but not directly, reflected on the meaning, rather than the content of their interview experience, in feedback sessions.

Learning and Reflective Practice With Patients With Disabilities

In the analysis of learning and reflective practices from transcripts of the feedback following medical interviews, coders were in clear consensus about four primary learning areas identified in feedback sessions. Three of these identified areas integrate disability and self-reflection, and were published in detail in Duggan, Bradshaw, Carroll, Rattigan, and Altman (2009). These evidence-based areas of integrating disability and reflective practice include (1) learning the ways in which disability impacts the treatment plan; (2) self-reflection and recognizing attitudes about disability; and (3) learning about the practice of medicine generally beyond disability.

These are only briefly highlighted below, as these three learning areas are described in greater detail in Duggan, Bradshaw, Carroll, Rattigan, and Altman (2009). A fourth learning area, learning about disability per se, was also analyzed, and is included in greater detail below as a conceptually distinct communication area.

Disability per se: Learning More About Disability

Medical students learned about disability per se through opportunities in feedback sessions to ask the SPE any questions about the specific disability or experiences with disability. Within communication suggesting desire to learn about disability per se, five themes were identified:

1. Lifestyle
2. Role as SPEs

3. Transportation and mobility
4. Past medical history
5. Approaching disability

Lifestyle. Medical students' reflections about the lifestyle of an individual with a disability include what a typical day is like, life at home, sports, and work, as demonstrated in the following illustrative questions:

> MS #111: I would definitely be curious to know what a typical day *is* like for you, in terms of NF [Neurofibromatosis]—just kind of what living with it is like.

> MS #418: How is your daily routine? How do you like, get your groceries?

Medical students asked about *life at home* and how disability affects the patient's family and living situation, and whether they used PCAs:

> MS #347: Do you have a family? (Yes.) Husband? (Yes.) Kids? (Yes, two daughters.) How have they been—your daughters—living with this?

> MS #112: Can you tell me about the PCAs? Do they help with chores and things like grocery shopping? How do you find them?

Medical students asked about SPEs' participation in *sports* and outdoor activities:

> MS #241: I am surprised by your cross-country skiing and distance hiking. What's that like for you? What does it feel like?

> MS #324: You said you play basketball. Do you use a special chair? Is the ball softer? Do you dribble? Does someone help you with the sports?

Medical students also inquired about SPEs' *work*, including "Do you work?" and "What kind of work do you do?"

Role as SPEs. Many students expressed *questions about being an SPE,* including the SPE's involvement in the project and the extent to which the portrayed case pertained to them:

> MS #221: How did you get involved in your work as a Standardized Patient here?

> MS #246: How much of the story you told as a patient is you versus the character? Have you ever had pain problems connected to the wheelchair before?

Medical students *thanked the SPE* for the memorable experience, and for sharing personal stories and knowledge about disability; they expressed appreciation for the educational exercise and for learning the importance of integrating disability with other health issues.

Transportation and mobility. Means of *assistive technology*, transportation, extent of mobility, and dealing with accessibility were common questions directed about the SPE's experience, and students expressed great interest in the *assistance dog*:

> MS #244: How much do you use your wheelchair? How do you get around in your daily life when you're not using the wheelchair? How do you get around the city? Can you drive? When do you use a motorized wheelchair?

> MS #237: How do you tell the dog which way to go? How many commands do you have for the dog? Are there enough dogs for everyone who would like one? Have there been any accidents with the dog? Does she go with you everywhere?

> MS #336: When you go to work do you have to make sure everything is accessible for you? And when you go to the doctors, how do you use the exam table?

Questions about transportation sometimes led to discussion of recent technology, revealing discoveries new to students, including directing a wheelchair with a headband and using GPS to navigate if blind. Medical students said they wanted to learn about *insurance coverage* for assistive devices.

Past medical history. Medical students asked about personal medical history of the SPE, including asking *time since disability diagnosis, previous experience with doctors*, and *adjustment to disability*. Medical students were particularly interested in previous negative experience with doctors, and probed SPEs' advice for improving the encounter and effectively addressing PCAs. SPEs often shared their stories of doctors who addressed the PCA, rather than the patient, and sometimes compared the experience to a doctor "talking to your mother," sometimes prompting students to reflect on their own communication during the simulated exercise:

> MS #412: And how do you cope with the disability emotionally? What was it like for you to find out? Was there a lot of anger and frustration? How have you adjusted to disability?

> MS #424: I think I only addressed you when I was asking questions. How would you feel if I just asked your PCA? Or is that a no-no completely?

> MS #123: I think I could have asked [the PCA] who she was earlier. How should I address your PCA? Is she always there, or is it appropriate to ask her to leave?

Approaching disability. Medical students asked SPEs about their disability ("What is your disability?" or "What is your condition?"), acknowledged that they should have asked about disability ("I should have asked about why you're in a wheelchair."), or inquired how to ask about disability ("I did not know how to account for the wheelchair."). Similarly, they indicated desire to learn how to greet a patient with a disability. Students who said they missed asking about disability also generally said they lacked experience and exposure to individuals with disabilities:

> MS #3210: I don't have any personal experience with patients with disabilities; I mean my experience is very limited, not nearly where I'd like it to be. I don't think I've ever seen anyone on a forced respirator before. How should I have asked about it?

> MS #1210: If somebody has a disability, do you reach for their hand if the person does not extend it? For me it was a bit uncomfortable because I didn't know how to shake your hand.

Addressing patients with disabilities was a topic of feedback. The feedback sessions also demonstrated three other communications regarding learning, including the impact on the treatment plan, attitudes to disability, and learning about medicine in addition to disability.

Disability Impacts the Treatment Plan

Communication that demonstrated learning how disability impacts the treatment plan included learning to integrate disability with the chief complaint, to address the patient's daily life, and to implement an effective treatment plan (Duggan, Bradshaw, Carroll, Rattigan, & Altman, 2009). By the end of the feedback, medical students who ignored disability disclosure during the interaction made the connection that the disability and shoulder pain were related. Medical students' learning about addressing the patient's daily life was observed in their recognizing ways to better address the patient's typical day, caretaking, work, intimacy, and social support.

Implementing an Effective Treatment Plan

Medical students indicated learning additional disability considerations for treatment, including learning about medications, recognizing ways disabilities may increase reliance on shoulder function for mobility, added barriers and burdens to resting the shoulder, the need for outside services, the role of physical therapy, and timing and disability-related needs in follow-up appointments. Medical students noted that they learned that the patient is not able to stop using the shoulder, and indicated learning to recognize the ways the shoulder is required for mobility, work, and daily activities. Medical students who acknowledged the reliance on the shoulder for mobility indicated learning alternatives to rest, such as physical therapy to build strength. Similarly, medical students learned new ways to consider outside services for patients.

Attitudes About Disability

Students conveyed their attitudes about disability in the ways they described awkward or tense interactions. The way in which students indicated concern during feedback in asking about disability revealed students' uneasiness, particularly in reasons provided for not asking about disability and not inquiring about the previous doctor. Student comments indicated that addressing disability seemed inappropriate or irrelevant, or indicated concerns that it would take too long. The ways in which medical students indicated reflection on their communication about adapting treatment for patients with disabilities and communicating respectfully indicated medical students' learning and self-reflection about their own attitudes. Medical students' self-reflection about their assumptions specific to disability indicated (sometimes inaccurate) beliefs that living with a disability should evoke sympathy, and that individuals with disability are dependent, particularly for mobility and transportation.

The Practice of Medicine Overall

Medical students expressed learning about the practice of medicine overall, particularly for understanding the SPE experience and learning strategies for effective interviewing. Students indicated learning to notice their comfort level and to find ways to calm their anxiety about being observed with the patient. Similarly, they expressed feeling confident about ways to increase comfort in SPE interactions. They recognized sometimes asking two questions at the same time or interrupting the patient, and they practiced explaining the diagnosis to reinforce clarity. Students' reflections indicated learning the importance of balancing the patient's needs and opinions with their medical advice. Addressing the patient's concerns with the previous doctor was a recurring comment with regard to effective communication as well, providing opportunities for medical students to recognize the need to ask the patient why he or she is not returning to the previous doctor, while also maintaining professional integrity.

Discussion

The analysis involved in this project illustrates the ways the current educational module provides experiences for medical students to shift their communication challenges to learning opportunities and to better understand patients with disabilities. The project illustrates the ways in which good intentions may actually interfere with best practices in health care, perhaps increasing disparities toward people with disabilities due to lack of comprehensive disability education. Avoiding asking about disability may be interpreted through a lens of expected politeness norms, as medical students may have been hesitant to offend the "patient" by asking, but instead noticed and responded in turn to disability disclosure. The current project enables physicians to work to decrease disparities by recognizing the ways their language about disability reinforces negative stereotypes about people with disabilities, or in contrast, promotes independence and acknowledges disability as one aspect of a larger context of the life world of the patient. In addition, the deliberate effort in the case's educational construction to provide the medical students with three separate opportunities to hear the SPE refer to the patient's disability (the previous doctor is uncomfortable with the "patient's" disability, disclosure of disability to questions about past medical history or any previous medical problems, and reference to the patient's disability activities around exacerbating the shoulder pain) provided astute medical students with an abundance of clues about highlighted language preferences from this sector of the disability community.

The thorough examination of actual interactions shifts the experience from intuitive knowledge to evidence-based training and practice. In addition, the project illustrates the ways multiple stakeholders, including communication experts, medical educators, disability advocates, and public-health professionals can work together to begin to address communication training in ways that empower future health providers. Although research identifies a quality chasm between perspectives of physicians and individuals with disabilities (Iezzoni, 2006), the current project suggests that innovations in medical education can provide medical students the opportunities to learn how to work toward relationship-centered communication, less hampered by prior assumptions and attitudes about disability.

Incorporating the feedback session information suggests that students who ignored disability-related disclosure may be "paralyzed with uncertainty." Although they were presented with implications of the disability around this recurrent episode of shoulder pain from several angles, they ignored SPE disclosures about work, implications for the relationship with a spouse, or responsibilities with their recently sick mother. These medical students seemed to infer that the disability is the cause of the limitations on the "patient's" life. Paralyzing uncertainty and ignoring the disability altogether may be related to medical students trying to "treat them the same as they would treat anyone else." Thus, reflective practice encouraged in feedback can be a tool for recognizing a need for accommodation and modified communication while also recognizing disability as a natural part of the human condition.

Medical students expressed learning how to ask about disability, how to consider disability implications for patients' everyday lives, and how treatment includes disability-particular services but still addresses the chief complaint. Analysis suggested medical students' awareness of disability conditions, yet their lack of vocabulary to effectively articulate disability implications. Reflections indicated that medical students who had not adequately addressed disability during the interview articulated lessons learned about limitations of ignoring disability disclosure. This analysis of themes and learning occurrences indicated that individuals with disabilities working as medical educators offer an opportunity for medical students to learn social and medical implications of disability and ways to consider disability implications for patients' everyday lives before potentially ignoring disability disclosure in clinical practice. Medical students who neglected to inquire about the patient's everyday life were prompted to reflect upon routine activities that could potentially increase pain, and were encouraged to ask the patient about a typical day. Students' reflections in feedback suggested increased breadth and depth in understanding patients with disabilities as multicompetent and as individuals whose relational roles include workers, caretakers, and support providers. The breadth of understanding can decrease disparities by encouraging integration of the complete life world of individuals with disabilities.

Medical students also learned about disability per se, particularly in opportunities to ask the SPE questions about their lives, including questions about their personal disability and disability generally. Students' questions suggested that they perceived high rapport with SPEs and found the opportunity to ask about disability helpful and affirming. Although content areas addressed to SPEs about disability were similar to questions addressed to "patients," the ways of asking prompted less formal inquiry with SPEs, and a greater familiarity and vulnerability, achieving person-to-person inquiries. Furthermore, reflections about disability suggested that medical students developed a further context for interpreting the SPE's professional roles as educators, navigators of health-care and insurance systems, and disability advocates. Medical students' reflections indicating attitudes about disability suggest that they learned to recognize the ways they interpreted disability disclosure. Particularly poignant examples include describing themselves as feeling tense and awkward in facing disability, as having difficulties asking about disability, and as facing challenges and assumptions about respectful communication. Medical students left this educational experience with a broader sense of disability as a natural part of the human condition, thus reducing the stigma sometimes associated with disability and then perhaps limiting resources allocated to disability.

The educational module and examination of communication behaviors may be particularly important for medical students who do not ask about disability, or whose interactions could be most easily interpreted by describing "what was wrong" with the interaction. This project illustrates the ways in which communication competencies are developed and reflective capacity is encouraged, thus building systems to empower providers to recognize and move past their communication challenges. Analysis of learning in feedback indicates an intense level of learning for medical students, a more valuable experience than initial "success" in medical interviewing. In addition to analysis described in this chapter, medical students also receive a copy of their interactions and are offered opportunities to read reports and results from our ongoing work in this area. Identifying communication obstacles and opportunities opens doors for dispelling myths and nonconstructive attitudes toward disability, for addressing social implications of disability, and for anticipating needs of patients with disabilities. The experience of interviewing a person with a disability provides an opportunity to learn about disability communication in a nonjudgmental setting. Overall, students were better able to describe disability as a natural part of the human condition following the interactions. Communication challenges included distinguishing between asking about disability and integrating patient disclosure within the interactions. Determining the clinical diagnosis of acute shoulder pain does not in itself provide sufficient information to manage an effective treatment plan, nor address the fundamental cause of overuse injuries. It is important to note that directly asking about the disability or the individual's specific clinical diagnosis does not always elicit nuances in information required to address patients' needs or preferences about disability.

This educational intervention illustrates the ways interview experience with people with disabilities enables physicians to overcome communication challenges in treating individuals with disabilities. The feedback and ways to encourage reflective practice can decrease disparities in individuals with disabilities by providing a comprehensive context for understanding physical disability per se. In addition, the project highlights the ways providers learn to recognize their own communication challenges in addressing people with disabilities and shift the communication obstacles to learning and reflective practice.

Bibliography

Aulagnier, M., Verger, P., Ravaud, J.-F., Souville, M., Lussault, P.-Y., Garnier, J.-P., & Paraponaris, A. (2005). General practitioners' attitudes towards patients with disabilities: The need for training and support. *Disability and Rehabilitation, 27*(22), 1343–1352.

Bryan, W. V. (2007). *Multicultural aspects of disabilities: A guide to understanding and assisting minorities in the rehabilitation process* (2nd ed.). Springfield, IL: C. C. Thomas.

DeJong, G. (1997). Primary care for persons with disabilities: An overview of the problem. *American Journal of Physical Medicine and Rehabilitation, 76*(3), S2–S8.

Duggan, A. P., Bradshaw, Y. S., & Altman, W. (2010). How do I ask about your disability? An examination of interpersonal communication processes between medical students and patients with disabilities. *Journal of Health Communication, 15*(3), 334–350.

Duggan, A. P., Bradshaw, Y. S., Carroll, S. E., Rattigan, S. H., & Altman, W. (2009). What can I learn from this interaction? A qualitative analysis of medical student self-reflection and learning in a standardized patient exercise about disability. *Journal of Health Communication, 14*(8), 797–811.

Duggan, A. P., Bradshaw, Y. S., Swergold, N., & Altman, W. (2011). When rapport building extends beyond affiliation: Communication overaccommodation toward patients with disabilities. *The Permanente, 15*(2), 23–30.

Glaser, B. G., & Strauss, A. L. (1967). *The discovery of grounded theory: Strategies for qualitative research.* Chicago, IL: Aldine.

Iezzoni, L. I. (2006). Make no assumptions: Communication between persons with disabilities and clinicians. *Assistive Technology, 18*(2), 212–219.

Iezzoni, L. I., McCarthy, E. P., Davis, R. B., Harris-David, L., & O'Day, B. (2001). Use of screening and preventive services among women with disabilities. *American Journal of Medical Quality, 16*(4), 135–144.

Jain, S. (2006). Care of patients with disabilities: An important and often ignored aspect of family medicine teaching. *Family Medicine, 38*(1), 13–15.

Kahn, P. (2003). Teaching tomorrow's docs. *New Mobility, August*, 45–48.

Maughan, D. (2005). Disabled as a medical student, enabled as a doctor. *British Medical Journal, 330*(7505), 1455.

Memel, D. (1996). Chronic disease or physical disability? The role of the general practitioner. *British Journal of General Practice, 46*(403), 109–113.

Minihan, P. M., Bradshaw, Y. S., Long, L. M., Altman, W., Perduta-Fulginiti, S., Ector, J., Foran, K. L., Johnson, L., Kahn, P., & Sneirson, R. (2004). Teaching about disability: Involving patients with disabilities as medical educators. *Disability Studies Quarterly, 24*(4). Retrieved from http://dsq-sds.org/article/view/883/1058.

President's Advisory Commission on Consumer Protection and Quality in the Health Care Industry. (1997, November 18). *Consumer bill of rights and responsibilities.* Washington, DC: Author.

Sabharwal, S. (2001). Objective assessment and structured teaching of disability etiquette. *Academic Medicine, 76*(5), 509.

Sabharwal, S., & Fiedler, I. G. (2003). Increasing disability awareness of future spinal cord injury physicians. *Journal of Spinal Cord Medicine, 26*(1), 45–47.

Shapiro, A. F., & Gottman, J. M. (2005). Effects on marriage of a psycho communicative-educational intervention with couples undergoing the transition to parenthood, evaluation at 1-year post intervention. *Journal of Family Communication, 5*(1), 1–24.

U.S. Census Bureau. (2001). *Americans with disabilities: Household economic status.* Washington, DC: Author.

U.S. Department of Health and Human Services (2010). *Healthy people 2010.* Washington, DC: Author.

Veltman, A., Stewart, D. E., Tardif, G. S., & Branigan, M. (2001). Perceptions of primary healthcare services among people with physical disabilities—part 1: Access issues. *Medscape General Medicine, 3*(2), 1–4.

Wilson, J. S., & Merrill, A. S. (2002). Teaching students to care for and about people with disabilities. *Nurse Educator, 27*(2), 89–93.

Zola, I. K. (1993). Self, identity, and the naming question: Reflections on the language of disability. *Social Science and Medicine, 36*(2), 167–173.

3

Health Literacy and Patient Empowerment: The Role of Medical Interpreters in Bilingual Health Communication

Elaine Hsieh

In 2000, 4.4 million households were linguistically isolated, meaning that no one in the household aged 14 or over spoke English at least "very well"; up from 2.9 million in 1990 (Shin, 2003). In 2007, nearly 55.4 million Americans spoke a language other than English at home, among which 24.5 million (8.7% of the population) have limited English proficiency (LEP); in 1990, the numbers were 31.8 and 13.9 million (6.6% of the population) respectively (Shin & Kominski, 2010). These demographic changes of the United States present unique challenges to the delivery of quality health care to individuals with LEP.

Researchers have noted that when language barriers exist in provider-patient communication, a patient is likely to receive more diagnostic testing (Hampers, Cha, Gutglass, Binns, & Krug, 1999); is less likely to receive preventive care (Woloshin, Schwartz, Katz, & Welch, 1997) and follow-up appointments after an emergency department visit (Sarver & Baker, 2000); is less likely to understand health-care providers' instructions (Doty, 2003); and is less satisfied with the quality of care (Ngo-Metzger, Sorkin, & Phillips, 2009). Compared to their English-speaking counterparts, patients with LEP often make fewer comments and receive less information, responses, and social support from their provider (Rivadeneyra, Elderkin-Thompson, Silver, & Waitzkin, 2000; Thornton, Pham, Engelberg, Jackson, & Curtis, 2009). As a result, patients with LEP are significantly disadvantaged when interacting with providers.

Low health literacy (e.g., lacking the skills to effectively communicate with providers) is a prevalent problem in the United States, with more than one-third of the English-speaking population and half of the Spanish-speaking population struggling with health literacy (Gazmararian, Curran, Parker, Bernhardt, & DeBuono, 2005). Challenges faced by populations with LEP are not limited to their lack of literacy skills (e.g., inability to read the English materials). Even when education materials are translated into the patients' native language, they may still experience difficulties in understanding health information (Leyva, Sharif, & Ozuah, 2005). Populations with LEP often experience multilevel challenges that may lead to experiences of health disparities. In addition to the common factors that lead to low health literacy (e.g., education, socioeconomic status, and language proficiency), they also face challenges in seeking and providing information in health-care settings due to differences in their social norms and cultural expectations (Hsieh, 2006a, 2007). For example, they are likely to have different cultural expectations and normative beliefs for providers, patients, and patients' family members as they coordinate the management of an illness event. Researchers have argued that low health literacy contributes to health disparities experienced by racial and ethnic minorities (Paasche-Orlow & Wolf, 2010).

Interpreters are in a critical position to assist the patients with LEP to navigate the complex health-care system, which may require them to negotiate provider-patient differences in cultural expectations, social norms, and illness ideology. Two recent reviews found that providing professional interpreter services can improve patients' quality of care, treatment processes, health outcomes, satisfaction, and adherence (Flores, 2005; Karliner, Jacobs, Chen, & Mutha, 2007). In some studies, researchers have found that patients who receive interpreting services perform the same, if not better, than their English-speaking counterparts (e.g., Andrulis, Goodman, & Pryor, 2002; Tocher & Larson, 1996). Researchers have found interpreters to be active participants who systematically adopt purposeful strategies to improve a patient's health literacy (e.g., ability to seek, provide, and process information when communicating with providers), to protect institutional resources (Davidson, 2000), to reduce the cultural gap between the provider and the patient, to reconcile provider-patient conflicts, and to ensure the quality of provider-patient interactions (Hsieh, 2006a). However, as researchers noticed interpreters' active involvement in the communicative process, they also have questioned interpreters' ethics, and raised concerns about how some of their communicative strategies may infringe on providers' authority or patients' autonomy (Hsieh, 2010; Leanza, 2008; Rosenberg, Seller, & Leanza, 2008).

In this chapter, I will critically examine interpreters' communicative strategies, focusing on how they facilitate provider-patient interactions and influence patients' information-seeking and help-seeking skills with providers. Extending from past literature that recognizes interpreters' active involvement in medical encounters (Dysart-Gale, 2005), I will further explore whether the interpreters' strategies may improve or hinder patients' health literacy and autonomy in their illness events, which may have significant implications for the health disparities experienced by the populations with LEP.

Background

The findings presented here are part of a larger study that examines the roles of medical interpreters. Three data sets were presented here. The first data set include a one-year ethnographic study. I recruited two Mandarin Chinese interpreters, four patients, and 12 providers. I shadowed the interpreters in their daily routines and audio-recorded their interactions with the patients and the providers. In total, 12 medical encounters (each lasting 1–1.5 hours) were observed, audiotaped, and transcribed. The other two data sets are in-depth interviews and focus groups with health-care providers and interpreters. I recruited 26 interpreters (from 17 languages), and conducted 14 individual and six dyadic interviews (each lasting 1–1.5 hours). Interpreters included in this study are all considered professional, on-site interpreters. The interviews focused on exploring interpreters' understanding and practice of their roles.

After the initial analysis of the interpreters' interview data, my research team recruited 39 health-care providers from a major health-care facility in the southern United States as a part of NIH-funded research to examine the providers' views on the roles of medical interpreters. We recruited 39 providers from five specialty areas: OB/GYN (n = 8), emergency medicine (n = 7), oncology (n = 11), mental health (n = 7), and nursing (n = 6). In total, the research team conducted 8 specialty-specific focus groups (each lasting 1–1.5 hours) and 14 individual interviews (each lasting 1–1.5 hours). The research questions were designed to examine providers' perceptions, expectations, and evaluations of interpreters' roles and practices. All procedures of the study have been approved by the Institutional Review Boards involved.

Based on these data sets, I have examined interpreters' communicative strategies through their performances of specific roles (Hsieh, 2006a, 2007, 2008) and their corresponding challenges to provider-interpreter collaboration (Hsieh, 2010; Hsieh, Ju, & Kong, 2010). Some discussions here are similar to a study that involved discussions of interpreters' advocate role (cf. Hsieh, 2008); however, instead of focusing on role performance, the discussion here centers on interpreters' influence on patient health literacy and patient empowerment and its corresponding consequences.

In the transcript, health-care providers are denoted as H, interpreters as I, and patients as P. In the following discussion, I also denote interpreters with a superscript I (i.e., ^{I}) and health-care providers with a superscript H (i.e., ^{H}) after their pseudonyms. I *italicize* some texts to highlight my emphases.

Interpreters' Influence on Patient Health Literacy and Patient Empowerment

We found that interpreters adopt a variety of strategies aiming to improve patient's health literacy. Researchers have found that patients' active participation (e.g., asking questions, providing information) in provider-patient interactions can lead to increase in providers' patient-centered communication (Cegala & Post, 2009) and better coordination between providers' and patients' goals (Cegala, Street, & Clinch, 2007). In this chapter, I will highlight two distinctive strategies employed by interpreters to facilitate provider-patient communication: (a) making inexplicit information explicit, and (b) providing the patient with the means of self-advocacy.

Making Inexplicit Information Explicit

Individuals rely on conventional norms and other relevant contexts to understand the speakers' intended meaning (Grice, 1975). Because providers and patients do not share the same conventional norms or contexts in bilingual and/or intercultural health care, providers and patients are likely to experience miscommunication or confusion if an interpreter provides little assistance in helping them to be aware of the relevant contexts.

Interpreters can elaborate on a speaker's comment to improve a patient's ability to request services, to understand medical procedures, and to engage in effective provider-patient interactions.

In Extract 001, Claire[I] elaborated the provider's comment to improve the patient's understanding.

Extract 001

H: Has she ever heard of Equal?

I: 你有沒有聽過 Equal?英文叫 Equal 的這個糖，代糖，他們叫代糖，不是真 正自然生產的糖，他們叫代糖，老美叫代糖。名牌叫 Equal。(Have you heard of Equal? The English term for this sugar, Equal, they call it substitute sugar, it's not a naturally produced sugar. They call it substitute sugar. Americans call it substitute sugar; the brand name is Equal.)

P: 不知道 (No.)

I: No.

Claire's elaboration of the term, *Equal*, helps the patient to better understand the providers' information by providing background information about what substitute sugar is and that Equal is just a brand name. When communicating with one another, people often make cultural-specific inferences and assumptions that are embedded in their language practices. These inferences and assumptions, however, may not be transferable to another language and culture (see also Lee, 2009). Interpreters' familiarity with the social norms and cultural knowledge can be valuable in ensuring providers' and patients' accurate and effective exchange of information.

It is important to note that these cultural differences are not always transparent; as a result, interpreters' ability to detect and clarify the divergent understanding between the provider and the patient can be critical to ensuring the quality of care. The following interaction took place during an education session in which a dietician is educating a patient with gestational diabetes about portion size.

Extract 002

H: One-third of a cup of rice is a serving.

I: 三分之一杯的煮熟的米飯就是，一個分量。(One-third of a cup of cooked rice is a serving.)

Does she have measuring cups at home?

I: 家裡有沒有量杯？(Do you have a measuring cup at home?)

P: 就是米飯的那個杯子。(Just the cup for rice.)
I: Is it the same measuring cup for rice?
H: Hmhm, yes.
P: 都是一樣的。(It's the same.)
I: Is it the same?
H: The—
I: Are they?
H: I don't know what she is referring to. I'll show her what I have.
I: 她有一個。(She has one.)
(H took out a glass measuring cup from the drawer.)
P: 這麼大阿！(That's big!)

Claire[I] adopted several strategies here that are critical to the patient's health literacy. First, she used "cooked rice" (line 102) rather than the term "rice" (line 101) used by the provider. Recognizing that there are significant portion size differences between cooked versus uncooked rice (i.e., one cup of uncooked rice makes about three cups of cooked rice), Claire[I] preemptively changed the provider's term, while keeping the provider's comment accurate, to ensure that the patient does not misunderstand the providers' comments.

Second, when the patient made statements (i.e., "just the cup for rice" [line 107] and "it's the same" [line 111]), Claire[I] modified the statements into direct information-seeking questions. Previous studies in provider-patient communication have found that due to the power differences between providers and patients, patients often (a) do not seek questions even when they desire more information, and (b) when they do, they avoid using direct questions when seeking information from their providers (Cegala, 1997). Instead of treating the patients' assertion (which can be a form of an embedded question) as a statement, Claire[I] changed it to a direct question to seek and clarify information ("Is it the same measuring cup for rice?" [line 109]). Similarly, when the patient acknowledged the provider's confirmation (line 110) by reinstating the implications ("It's the same." [line 111]), Claire[I] treated the patient's utterance as an active information-seeking question (line 113). In fact, Claire[I]'s persistence in seeking and verifying the information exposed a miscommunication that would have been originally ignored by the provider. Claire[I]'s direct question ("Is it the same?"

[line 113]) and its reinstatement ("Are they?" [line 115]) forced the provider to admit potential misunderstanding ("I don't know what she is referring to."), and to resolve the problem ("I'll show her what I have."). The patient's comment on line 121 ("That's big!") showed that there was a real difference in the size of measuring cups that the provider and the patient had in mind, and that there could have been significant clinical consequences had Claire[I] not insisted on pursuing accurate information.

Third, an important feature in this interaction is how closely Claire[I] followed the patient's narrative, but at the same time, changed the nature of the narrative. Claire[I]'s strategies could not be categorized as neutral or faithful; yet, these strategies are essential in facilitating provider-patient communication. By explicitly stating the inexplicit information (e.g., "cooked rice" as opposed to "rice"), Claire[I] also improved the patient's ability to understand the information accurately and effectively. Because of the power differences, a patient may not feel comfortable asking the provider a direct question (Cegala, Post, & McClure, 2001). By reformulating the patient's narratives into active, information-seeking strategies, Claire[I] significantly increased the patient's ability to seek information.

Providing Means of Self-Advocacy

In addition to modifying others' utterances to enhance patients' health literacy, interpreters may enhance a patient's quality of care through *providing means of self-advocacy to the patient.* This is different from the advocate role which requires patients to accept other-advocacy to rectify problematic situations (e.g., inequality or injustice of health-care services) (Roat, Putsch, & Lucero, 1997). In contrast, by providing means of self-advocacy, interpreters provide patients access to resources available in various areas (e.g., medical knowledge and relevant information), and empower patients to act as self-advocates.

Interpreters' familiarity with the topic transitions and information flow allows the patients to communicate in a way that is consistent with providers' expectations and management of the medical encounter. In Extract 003, the patient indicated her concerns in an earlier conversation with the interpreter that her baby was losing weight. After the provider asked the mother about her feeding pattern, Christine[I] initiated the following information exchange sequence.

Extract 003

I: 你是不是要問醫生她今天量的體重比出生的時候還要少？(Didn't you want to ask the doctor why she is weighing less today than when she was born?)

P: 對啊 (Yes.)

I: Another question, before, the nurse weighed her, and she was born 7 pounds 11 oz., then today, the weight is only—It lost about 6 oz.

H: 6 oz. It's normal.

I: 她說這是正常的(She said it's normal.)

The interaction between Christine[I] and the providers in lines 306–309 is consistent with the earlier discussion in which the interpreter issued direct questions while making explicit statements on inexplicit information. Christine[I] not only provided the baby's original birth weight, but also highlighted the amount of weight loss. The information was obtained in a previous conversation with a nurse, but not with the current provider. Christine[I]'s use of meta-communication (i.e., "another question"), and inclusion of the exact amount of weight loss (i.e., 6 oz.), allow the provider to respond to and evaluate the patient's concerns more effectively.

What is particularly interesting in this interaction, however, is that when Christine[I] first initiated the question (line 301), the question was directed to the patient. Christine[I] might feel that the patient's concern was appropriate to be the next topic after the information exchanged about the baby's feeding patterns. Such a judgment requires a speaker's ability to understand the norm of information exchange patterns in provider-patient interactions, which an interpreter may have more experience with than the patient. In the interviews, several interpreters noted that they often remind and prompt patients about topics to discuss with the providers. By directing the question to the patient first (as opposed to directing the question to the provider), the patient is empowered to have control over the provider-patient interactions, and is socialized into the norms of topic transitions. In the future, the patient may be able to initiate the questions without the interpreter's assistance. In other words, the interpreter's strategies in Extract 003 not only allow the patient to have control over the topics discussed in the current interaction, but also educate the patient about the norms of future provider-patient interactions.

An interpreter may also coach a patient about *how* to request proper services or information. In Extract 004, the patient complained that the hospital provided water with ice for her to drink after her delivery, a practice that contradicts a Chinese custom which specifies that women after delivery should only drink hot water.

Extract 004

P: Yeah, 我這天-我兩天在醫院都是喝冰水 (Yeah, these days—I drank chilled water in the hospital.)

I: 其實醫院它有熱開水，你跟護士講，她就會給你熱開水 (Actually, they have hot water in the hospital. If you tell the nurse, she'd give you hot water.)

P: 我說-我說給她聽，我不要 ice，它都不是凍的，她就沒有放 ice (I said—I told her that I don't want ice. So, it's not icy. She didn't put ice in it.)

I: 沒有放冰，對對。你應該跟她說 hot water (No ice. Right. Right. You should tell her, "hot water.")

P: Hot water.

This interaction happened when the patient and Christine[I] were alone in the exam room, waiting for the provider. Christine[I] informed the patient about services available in the hospital (line 404) and the proper way to ask for the services (line 410). Christine[I] used the English term, "hot water" (line 410), instead of saying its equivalent Chinese term, showing that Christine[I] was providing the patient with the tool to obtain the services. The patient's verification of the information and repetition of the English term in line 413 shows that she understood Christine[I]'s communicative goal of empowering the patient.

For patients who are from a different culture and/or society, it is possible that they are not familiar with the kinds of services available, their rights as patients, or the appropriate norms of provider-patient interaction. Because interpreters are familiar with the typical exchanges in medical encounters, and have observed successful and problematic provider-patient interactions, they can be extremely helpful and valuable in assisting patients in navigating the health-care systems. Stacy[I] explained,

> Because [patients] might be worried, but they don't know what they are worried about. So, I ask them, "Would you like to ask if taking that medication has side effects?" So, I help them to understand some of the procedures and foresee something so that they would not worry about it later on.

Sara[I], another interpreter, talked about a situation in which a Spanish-speaking father explicitly told her that he did not know what questions to ask; in response, she coached him to ask the physician to provide clarification of the diagnosis, to discuss alternative treatments, and to explain long-term consequences of the disease. In another case, after witnessing a provider's prejudicial attitude, Colin[I] informed the patient that if he wished to file a complaint, he would be able to take him to the complaint office and interpret for him. In such situations, by providing access to illness-related information and health-care facilities, interpreters significantly enhance the patients' abilities to obtain quality care.

Balancing Benevolence and Social Justice With Patient Autonomy

In the last section, we provided examples of how interpreters' strategies may empower patients and enhance patients' health literacy, not only for the current medical encounter, but also for future provider-patient interactions. It is important, however, to take a critical view on the interpreters' strategies, exploring potential concerns about and consequences of interpreters' efforts in empowering patients.

Patient Empowerment and Patient Autonomy

Recent studies have increasingly emphasized the importance of interpreters' active role in bilingual health care (Fatahi, Hellstrom, Skott, & Mattsson, 2008; Messias, McDowell, & Estrada, 2009); however, it is important to recognize that interpreters' involvement and intervention is not without limits (Hsieh, 2010). Although acting as direct advocates (e.g., acting on behalf of the patient without the patients' or providers' explicit consent) may be efficient in meeting the patients' needs (Hsieh, 2008; Messias et al., 2009; Rosenberg et al., 2008), such strategies may blur the lines between the patients' and the interpreters' agenda and communicative goals (Leanza, Boivin, & Rosenberg, 2010; Rosenberg et al., 2008).

A few interpreters in our study raised concerns about how interpreters' active involvement may infringe on patient autonomy (i.e., patients' right to make decisions without the influences of providers and/or other health-care professionals). Shirley[I], manager of the interpreter office of a major hospital and a trainer of an interpreting program, explained, "By NOT empowering, by giving your opinion, going immediately into an advocacy role as a medical interpreter, I feel that you keep [that patient], that parent or that guardian, or primary caretaker, from becoming empowered." Sharon[I], the director of an interpreting agency, echoed, "They do more harm than good when you try to be more than an interpreter, when you try to be an advocate for the patient." Although patient empowerment is often perceived as the communicative goal of the advocate role (Hsieh, 2008), some interpreters argued that advocating for patients or acting on a patient's behalf can be problematic because it compromises patient empowerment. From this perspective, the best form of patient empowerment is to allow patients to have full control over the medical encounter without the influence of the interpreter (see Hsieh, 2008). Several interpreters discussed their decisions of *not* being an advocate. For example, Vicky[I] explained,

> Who are you to tell the doctor what to do? Because you have patients who are very submissive, very afraid, depending on what they went through. So, the interpreter thinks that he or she has a right to advocate. *But were you asked to do so?*

The emphasis on patient autonomy echoes the traditional emphasis on the conduit role, in which interpreters are only the voices of others (Hsieh, 2009; Kaufert & Putsch, 1997; Messias et al., 2009). In the United States, patient autonomy emerged in the 1960s in the overall atmosphere of antipaternalism and redefined beneficence as medical paternalism (Rothman, 2001). As a result, an ideal patient is an informed patient who is capable and willing to assume all responsibilities in their illness events (Kapp, 2007).

Conceptualizing interpreters' noninterference as respect for patient autonomy, however, is based on problematic presumptions. The interpreter-as-conduit model assumes that: (a) all participants are competent speakers who can communicate effectively and appropriately, (b) it is desirable to maintain the existing structure of relationships and patterns of communication, and (c) there are minimal differences between speakers' cultural knowledge and social practices. However, medical interpreters do not operate on an equal playing field, in which providers and patients are equal participants (Kaufert & Putsch, 1997). Literature on provider-patient communication has long documented that the differences in power and medical knowledge often compromise patients' full participation in provider-patient interactions (Elderkin-Thompson, Silver, & Waitzkin, 2001). Providers do not spend more time interacting with patients with LEP (Thornton et al., 2009), even though they often believe that they do (Tocher & Larson, 1999). Patients with LEP often experienced prejudicial attitudes and cultural differences, which can lead to reduced patient satisfaction and compromised care (Dohan & Levintova, 2007; Hicks, Tovar, Orav, & Johnson, 2008). In addition, patients from different cultures may desire different degrees of patient autonomy in their illness events (Back & Huak, 2005; Blackhall, Murphy, Frank, Michel, & Azen, 1995).

In bilingual health communication, patients and providers are not on equal footing, nor do they share similar cultural, social, or communicative norms. A patient may not know how to voice their concerns. Stacey [I] said that she often asks her patients to "give [her] a word or something that they don't understand" so she could "ask different questions to the doctor on their behalf." In other cultures, it may not be appropriate to disclose one's illness. For example, Yetta[I] explained that in Nigerian culture, even a common illness (e.g., high blood pressure) can be used as a personal attack against an individual. As a result, Nigerian people go to a great extent to hide their symptoms. She explained,

> The secrecy of not exposing what they have...I have to let the patients know that they are here to be treated, "TELL THEM, what's wrong with you. How you are going to get help."...They are not used to revealing what's wrong with them.

A patient who is unwilling to disclose his or her problem certainly cannot have meaningful provider-patient communication. In these situations, if interpreters maintain neutrality by avoiding active intervention (e.g., encouraging patients to voice their concerns, to seek information, or to assert their rights), they ignore their important functions in addressing the social inequality embedded in bilingual health communication (Messias et al., 2009).

Benevolence, Social Justice, and Health Literacy

Patient autonomy is not simply allowing the patient to make their own decisions; rather, providers are obligated to create the conditions necessary for autonomous choice in others (Pantilat, 2008). If the patients do not know what information to seek, what services are available, or what rights they have, they are not in a position to make autonomous decisions. To ensure patient autonomy in bilingual health care, it is essential that medical interpreters (and other providers) actively respond to the situational and contextual demands to ensure that a patient is informed and empowered to make autonomous decisions.

Providers rely on interpreters to offer culturally sensitive care (Kaufert & Putsch, 1997; Schapira et al., 2008). Interpreters are extremely valuable to providers in understanding a patient's distress. Gemma[H] noted, "If [interpreters] can be more than just the verbal communication, if they can help bridge the culture, it's HUGE." Celia[H] commented on a case in which she thought that the patient's mother was overly emotional,

> [Once I noticed] the mother was so horrified about the child losing hair. [The interpreter] explained to me, in this particular culture, hair is regarded as a wedding veil, and you don't cut your hair. That was very helpful to me to understand that reaction.

Because of the cultural, power, and socioeconomic differences between the providers and the patients with LEP, it is likely that patients may not actively express all of their concerns. Providers also may face challenges in interpreting patients' subtle nonverbal cues, which are often cultural-specific. As a result, an interpreter who not only relays all information but also actively monitors (and intervenes in) the communicative process is essential in protecting patient autonomy. For example, an interpreter may choose to explicitly express the emotional tone (e.g., "I'm very nervous because…") or address the patients' concerns (e.g., "The patient looks puzzled, can I ask the patient if he has any questions?"). As interpreters actively incorporate patients' nonverbal behaviors (that are neglected or misunderstood by the providers) into the communicative process, they empower patients to have a full range of expressions and control over the provider-patient interactions.

It is important to note that some of the interpreters' communicative strategies may appear to facilitate provider-patient communication to seek or give "accurate" information, but do not necessarily empower patients or enhance other speakers' communicative competence. For example, Ulysses[I] talked about modifying providers' utterances when they may be considered problematic in his culture (i.e., Muslim). He explained,

> [The doctors] ask about sexual contacts outside of the marriage, which is a really really very bad question. It is very offensive. For Muslim ladies in particular.... I said, "Does *your husband* go to other women?"... Even though they might be practicing [adultery], they do not talk. But there is a way to make it soft, just ask another way.

Ulysses[I] intervened into the diagnostic process by changing the provider's actual question (e.g., "Do *you* have other sexual partners?") to what he considered a more culturally appropriate one. He argued that Muslim women would recognize the doctor's intention in evaluating sex-related health risk, and would provide accurate answers to their situation (e.g., "Yes, my husband goes to other women"), even if the husband is faithful. Although Ulysses[I] was able to avoid a potential provider-patient conflict, his intervention denied the opportunity for the provider to become more culturally sensitive, or for the patient to become familiar with the medical dialogue in the United States.

When interviewing providers, I used the Ulysses[I] scenario to solicit their reflection on his strategies. While recognizing the potential cultural differences and conflicts, providers often emphasize their desire to be informed about the cultural differences. Gloria[H] explained,

> If we don't understand the cultural aspect, then we need to know: That's offensive. Well, you can preface it with "I understand that this might be something that would not be asked in your home country. This is why I need to ask this question."... I absolutely want to know. Because we want the patient to understand, in order to have choices, in order to feel right about it, because it's what her decision is. We want her to feel good about it.

From this perspective, Gloria[H] viewed it as her responsibility to know the cultural differences, as it empowers her to honor patient autonomy. Several providers also voiced concerns about interpreters' strategies in modifying their narratives without consulting with them first, believing that such strategies infringe on their control over the medical dialogue (see Hsieh, 2010). In addition, they argued that cultural knowledge is essential in helping them to interact with future patients appropriately and effectively.

Interpreters in Bilingual Health Care

It is important to note that both researchers and practitioners have long recognized interpreters' tremendous power in influencing provider-patient communication and relationships (Hsieh, 2010; Rosenberg, Leanza, & Seller, 2007; Rosenberg et al., 2008). The traditional response, however, is to limit interpreters' control and influence over bilingual health care by restricting their performance in a limited role and minimizing their presence in health-care settings (Leanza, 2008). This attitude also is reflected in the responses of some interpreters in our study, as they equated advocating for patients with infringing on patient autonomy. It is not uncommon for interpreters to explicitly claim a limited role (Bot, 2005; Hsieh, 2009). However, interpreters' beneficent actions (i.e., behaviors to help prevent or remove harms or to simply improve the situation of others) (Pantilat, 2008) should not be viewed as medical paternalism. They may serve as interventions aiming to ensure patient autonomy and communicative competence in bilingual health care.

Examining interpreters' performance and its *long-term* influence on health literacy and patient empowerment is critical in advancing the theoretical development and practice implications for bilingual health communication. Most studies in bilingual health care do not consider prior interactions between participants. This is consistent with the traditional emphasis on the conduit model, in which interpreters are viewed as language machines (Leanza, 2005). If interpreters perform properly, like good machines do, they would transfer information from one language to another neutrally and faithfully. There should be minimal differences from one interpreter to another or from one session to another. After all, a machine either works or does not. One would expect the computer in the library to behave in exactly the same way as the computer at home. There is no history or prior relationship between the user and the computer that may influence their future interactions. In other words, if interpreters were treated simply as a functional tool to facilitate communication, the interpreters' performance and impact are limited to the specific interaction. As a result, the examination of interpreter-mediated interactions traditionally has been limited to the immediate, turn-by-turn talk. Few studies explore how interpreters' strategies may influence the quality of communication or the interpersonal relationship over time.

However, researchers have argued that the temporal aspect of illness management requires individuals to learn to coordinate with others efficiently and appropriately (Brashers, Goldsmith, & Hsieh, 2002). As participants learn about each other's communicative needs and styles, there may be important changes in their communicative behaviors. For example, a family interpreter may act on behalf of the patient because they also are caregivers and are aware of the patient's concerns (Rosenberg et al., 2008). Professional interpreters have reported that prior relationship with a patient makes the interpreting task less challenging (Hsieh, 2006a). Providers also noted that familiarity with an interpreter's communicative styles allows them to work more efficiently with the interpreter (Fatahi et al., 2008; Hsieh et al., 2010). The history within the provider-patient-interpreter triad influences how they interact with each other.

Interpreting agencies and providers are aware of such possibilities. However, traditionally, their strategies are to limit the interactions between the interpreter and the patient (e.g., avoiding assigning the same interpreter to the patient repeatedly, and prohibiting interpreter-patient interaction in waiting rooms), and to limit interpreters to the conduit role (Hsieh, 2006a; Leanza, 2008). Several recent studies, however, have challenged the usefulness and practicality of such practices (Hsieh, 2010; Hsieh et al., 2010; Leanza et al., 2010). Rather than prescribing a limited role or fixed functions to interpreters, researchers have recommended that health-care institutions and providers recognize the diversity of medical interpreters (Hsieh, 2006b; MacFarlane et al., 2009), be adaptive to the communicative needs and contexts (Hsieh & Hong, 2010; Leanza et al., 2010), and nurture institutional cultures and practices that promote interpreters' involvement in the delivery of care (Schapira et al., 2008).

Medical interpreters also need to be mindful about the complexity of their roles and functions in health-care settings. Their performance in bilingual health care and its corresponding influences entail both immediate and long-term consequences. Interpreters adopt various strategies to meet the multiple, emergent, and often conflicting demands in provider-patient interactions

(Hsieh, 2006a, 2009; White & Laws, 2009). During provider-patient interactions, interpreters need to anticipate providers' and patients' communicative needs and actively address potential issues of social inequality and cultural differences (Hsieh, 2010). For example, when interpreters made a speaker's inexplicit information explicitly expressed in their interpreted text, they drew little attention to the nuances of the potential misunderstanding, while ensuring that the communication continued smoothly. When an interpreter changed an indirect information-seeking statement into a direct information-seeking question, they improved the provider's ability to hear and to respond to the patient's information needs. The success of these strategies, however, relies on the interpreters' ability to anticipate and understand the speakers' communicative goals. If the interpreters misunderstand the speakers' intended meaning, the misinterpreted texts may lead to confusion, if not conflict, or even constitute unethical intervention (Lee, 2009). From this perspective, developing effective communication practices (e.g., meta-communication about the speakers' objectives prior to the medical encounter) to help interpreters anticipate the speakers' intended meaning is critical in ensuring the quality of interpreter-mediated interactions.

On the other hand, interpreters need to recognize that their communicative strategies can influence how patients and providers interact with each other in the future (e.g., their health literacy, cultural sensitivity, and communication competence). When interpreters take the time to provide means of self-efficacy to the patients, the patients are not only empowered to participate in the immediate interaction, but also have the opportunity to extend their new skills and knowledge with other providers in future interactions. In contrast, when interpreters actively change a speaker's utterances without informing the speaker about the changes, they deprive the speaker of the opportunity to develop communicative competence (e.g., to learn about the social norms of the medical discourse and the cultural expectations of the provider or patient role). It is from this perspective that interpreters need to further evaluate the appropriateness of their involvement and intervention in provider-patient interactions. Some strategies may be efficient in the immediate contexts, but do not empower the speakers for future interactions.

Previous studies have suggested that due to the pressure to maintain a neutral performance, interpreters often "hide" their intervention (a) outside of the medical encounter (e.g., coaching patients about their rights or questions to ask when providers are not around) (Hsieh, 2006a), or (b) in the voices of others (e.g., imposing their own agenda while appearing to be interpreting for others) (Leanza et al., 2010; Rosenberg et al., 2008). These strategies allow the interpreters to appear neutral, while manipulating the process and content of provider-patient communication. Although these strategies may be well-intended, they also limit other speakers' abilities to understand the complexity of bilingual health care, as the interpreters' manipulation of the communicative process is hidden from others. As indicated by the providers in our study, they are interested in learning more about the cultural expectations and social norms of their patients (Fatahi et al., 2008; Rosenberg et al., 2007). These knowledge and communicative skills will allow them to be better providers in the future.

As interpreters provide their linguistic services, they are also in a position to inform, educate, and empower other speakers for future interactions. It is important for the providers and health-care organizations to recognize this central aspect of interpreters' functions. Some recent studies have highlighted that health-care providers continue to emphasize interpreters' conduit role (Fatahi et al., 2008; Leanza, 2005). Such attitudes, however, put pressure on interpreters to (a) avoid intervening in the provider-patient communication even when they perceive problematic interactions (Hsieh, 2006a, 2009), or (b) conceal or disguise their intervention to avoid others' scrutiny (Keselman, Cederborg, & Linell, 2010; Leanza et al., 2010). As researchers highlighted that successful interpreter-mediated interaction requires providers, patients, and interpreters to coordinate and negotiate their communicative goals (Hsieh, 2010; Leanza et al., 2010), it is important to incorporate these aspects of interpreters' functions into the communicative practices and organizational cultures in health-care settings. For example, developing organizational cultures that support and value interpreters' clarification and/or elaboration on cultural issues (as opposed to viewing these behaviors as intrusions on providers' time) will allow both providers and patients to have better communicative competence in future interactions.

While research on health disparities traditionally has focused on system-level factors, interpreters are in a unique position to reduce health disparities at both the system and interpersonal levels. Although at a system level, starting from the late 1970s, there have been federal and state legislative efforts to require clinicians to provide interpreters for the populations with LEP in the United States (Youdelman, 2008), recent studies have found that providers still underuse interpreters (Ginde, Sullivan, Corel, Caceres, & Camargo, 2010; Schenker, Pérez-Stable, Nickleach, & Karliner, 2011). In addition, few providers receive training on working with interpreters (Flores et al., 2008) despite significant differences in the communicative practices between providers, interpreters, and patients (Hsieh, 2010). To address the problematic utilization of medical interpreters, researchers have recommended (a) incorporating bilingual health communication and cultural sensitivity training for providers (Tribe & Lane, 2009; Wu, Leventhal, Ortiz, Gonzalez, & Forsyth, 2006); (b) taking advantage of the cultural diversity within the health-care team (e.g., bilingual nurses) while providing clear boundaries of role responsibilities and expectations (Moreno, Otero-Sabogal, & Newman, 2007; Wros, 2009); and (c) developing organizational cultures and practices that encourage equality between members of the health-care team, rather than allowing physicians to have unchallenged authority (Hsieh, 2010; Tribe & Lane, 2009).

When interpreters are reconceptualized as having an active role, providers and interpreters are empowered to address health disparities at an interpersonal level. A recent study indicated that on-site interpreters allow other speakers to effectively address cultural nuances in ways that other types of interpreters (e.g., videoconference and telephone interpreters) do not (Nápoles et al., 2010), which highlights the interpersonal aspects of bilingual health care. Interpreters can serve as an important resource to educate providers about relevant cultural and linguistic issues (Wu et al., 2006), and to address potential discrimination and social inequality (Messias et al., 2009). As interpreters contemplate the variety of strategies available in managing the effectiveness and appropriateness of provider-patient interactions, they also should consider the long-term impact of those strategies. For example, as interpreters educate providers about cultural differences, the providers also develop the needed skills for future interactions. By anticipating patients' communicative needs, they provide the necessary conditions (e.g., health literacy) that enable patients to make autonomous decisions. As researchers highlight the importance of integrating research in health literacy and health disparities (Baur, 2010; Paasche-Orlow & Wolf, 2010), bilingual health communication represents an important field of study. While requiring providers to provide interpreters can be viewed as a legislative intervention to address health disparities at the system and/or organizational level (Bonet, 2009; Hsieh, 2012), interpreters can have a much more profound effect in improving the quality of care. Interpreters' interventions not only facilitate the immediate provider-patient interactions, but also can have long-lasting impacts on providers' and patients' communicative competence, empowering them to address and negotiate their communicative needs and therapeutic objectives in future interactions (with or without the interpreters). It is from this perspective that interpreters are particularly valuable in improving health disparities faced by populations with LEP.

Acknowledgments

This study was partially supported by Grant #1R03MH76205-01-A1, funded by the National Institutes of Health/National Institute of Mental Health. Part of the data has been published in previous studies that examined the roles of medical interpreters.

Bibliography

Andrulis, D., Goodman, N., & Pryor, C. (2002). What a difference an interpreter can make: Health care experiences of uninsured with limited English proficiency. Retrieved from http://www.accessproject.org/downloads/c_LEP reportENG.pdf.

Back, M. F., & Huak, C. Y. (2005). Family centred decision making and non-disclosure of diagnosis in a South East Asian oncology practice. *Psycho-Oncology*, *14*(12), 1052–1059.

Baur, C. (2010). New directions in research on public health and health literacy. *Journal of Health Communication*, *15*(2), 42–50.

Blackhall, L. J., Murphy, S. T., Frank, G., Michel, V., & Azen, S. (1995). Ethnicity and attitudes toward patient autonomy. *Journal of the American Medical Association*, *274*(10), 820–825.

Bonet, J. E. (2009). Full circle: The qualified medical interpreter in the culturally competent healthcare system. In S. Kosoko-Lasaki, C. T. Cook, & R. L. O'Brien (Ed.), *Cultural proficiency in addressing health disparities* (pp. 103–115). Boston, MA: Jones and Bartlett.

Bot, H. (2005). *Dialogue interpreting in mental health.* New York, NY: Rodopi.

Brashers, D. E., Goldsmith, D. J., & Hsieh, E. (2002). Information seeking and avoiding in health contexts. *Human Communication Research, 28*(2), 258–271.

Cegala, D. J. (1997). A study of doctors' and patients' communication during a primary care consultation: Implications for communication training. *Journal of Health Communication, 2*(3), 169–194.

Cegala, D. J., & Post, D. M. (2009). The impact of patients' participation on physicians' patient-centered communication. *Patient Education and Counseling,* 77(2), 202–208.

Cegala, D. J., Post, D. M., & McClure, L. (2001). The effects of patient communication skills training on the discourse of older patients during a primary care interview. *Journal of the American Geriatrics Society, 49*(11), 1505–1511.

Cegala, D. J., Street, R. L., Jr., & Clinch, C. R. (2007). The impact of patient participation on physicians' information provision during a primary care medical interview. *Health Communication, 21*(2), 177–185.

Davidson, B. (2000). The interpreter as institutional gatekeeper: The social-linguistic role of interpreters in Spanish-English medical discourse. *Journal of Sociolinguistics, 4*(3), 379–405.

Dohan, D., & Levintova, M. (2007). Barriers beyond words: Cancer, culture, and translation in a community of Russian speakers. *Journal of General Internal Medicine, 22*(2), S300–S305.

Doty, M. M. (2003, February 1). Hispanic patients' double burden: Lack of health insurance and limited English. The Commonwealth Fund. Retrieved from http://www.commonwealthfund.org/Publications/Fund-Reports/2003/Feb/Hispanic-Patients-Double-Burden--Lack-of-Health-Insurance-and-Limited-English.aspx.

Dysart-Gale, D. (2005). Communication models, professionalization, and the work of medical interpreters. *Health Communication, 17*(1), 91–103.

Elderkin-Thompson, V., Silver, R. C., & Waitzkin, H. (2001). When nurses double as interpreters: A study of Spanish-speaking patients in a US primary care setting. *Social Science & Medicine, 52*(9), 1343–1358.

Fatahi, N., Hellstrom, M., Skott, C., & Mattsson, B. (2008). General practitioners' views on consultations with interpreters: A triad situation with complex issues. *Scandinavian Journal of Primary Health Care, 26*(1), 40–45.

Flores, G. (2005). The impact of medical interpreter services on the quality of health care: A systematic review. *Medical Care Research & Review, 62*(3), 255–299.

Flores, G., Torres, S., Holmes, L. J., Salas-Lopez, D., Youdelman, M. K., & Tomany-Korman, S. C. (2008). Access to hospital interpreter services for limited English proficient patients in New Jersey: A statewide evaluation. *Journal of Health Care for the Poor and Underserved, 19*(2), 391–415.

Gazmararian, J. A., Curran, J. W., Parker, R. M., Bernhardt, J. M., & DeBuono, B. A. (2005). Public health literacy in America: An ethical imperative. *American Journal of Preventive Medicine, 28*(3), 317–322.

Ginde, A. A., Sullivan, A. F., Corel, B., Caceres, J. A., & Camargo, C. A., Jr. (2010). Reevaluation of the effect of mandatory interpreter legislation on use of professional interpreters for ED patients with language barriers. *Patient Education and Counseling, 81*(2), 204–206.

Grice, H. P. (1975). Logic and conversation. In P. Cole & J. L. Morgan (Eds.), *Syntax and semantics: Speech acts* (Vol. 3, pp. 41–58). Cambridge, MA: Harvard University Press.

Hampers, L. C., Cha, S., Gutglass, D. J., Binns, H. J., & Krug, S. E. (1999). Language barriers and resource utilization in a pediatric emergency department. *Pediatrics, 103*(6), 1253–1256.

Hicks, L. S., Tovar, D. A., Orav, E., & Johnson, P. A. (2008). Experiences with hospital care: Perspectives of Black and Hispanic patients. *Journal of General Internal Medicine, 23*(8), 1234–1240.

Hsieh, E. (2006a). Conflicts in how interpreters manage their roles in provider-patient interactions. *Social Science & Medicine, 62*(3), 721–730.

Hsieh, E. (2006b). Understanding medical interpreters: Reconceptualizing bilingual health communication. *Health Communication, 20*(2), 177–186.

Hsieh, E. (2007). Interpreters as co-diagnosticians: Overlapping roles and services between providers and interpreters. *Social Science & Medicine, 64*(4), 924–937.

Hsieh, E. (2008). 'I am not a robot!' Interpreters' views of their roles in health care settings. *Qualitative Health Research, 18*(10), 1367–1383.

Hsieh, E. (2009). Moving beyond a conduit model: Medical interpreters as mediators. In D. E. Brashers & D. J. Goldsmith (Eds.), *Communicating to manage health and illness* (pp. 121–146). New York, NY: Routledge.

Hsieh, E. (2010). Provider-interpreter collaboration in bilingual health care: Competitions of control over interpreter-mediated interactions. *Patient Education and Counseling, 78*(2), 154–159.

Hsieh, E. (2012). Interpreter services. In S. Loue & M. Sajatovic (Eds.), *Encyclopedia of immigrant health* (pp. 936–941). New York, NY: Springer.

Hsieh, E., & Hong, S. J. (2010). Not all are desired: Providers' views on interpreters' emotional support to patients. *Patient Education and Counseling, 81*(2), 192–197.

Hsieh, E., Ju, H., & Kong, H. (2010). Dimensions of trust: The tensions and challenges in provider-interpreter trust. *Qualitative Health Research, 20*(2), 170–181.

Kapp, M. B. (2007). Patient autonomy in the age of consumer-driven health care: Informed consent and informed choice. *Journal of Legal Medicine, 28*(1), 91–117.

Karliner, L. S., Jacobs, E. A., Chen, A. H., & Mutha, S. (2007). Do professional interpreters improve clinical care for patients with limited English proficiency? A systematic review of the literature. *Health Services Research, 42*(2), 727–754.

Kaufert, J. M., & Putsch, R. W. (1997). Communication through interpreters in healthcare: Ethical dilemmas arising from differences in class, culture, language, and power. *The Journal of Clinical Ethics, 8*(1), 71–87.

Keselman, O., Cederborg, A-C., & Linell, P. (2010). 'That is not necessary for you to know!' Negotiation of participation status of unaccompanied children in interpreter-mediated asylum hearings. *Interpreting, 12*(1), 83–104.

Leanza, Y. (2005). Roles of community interpreters in pediatrics as seen by interpreters, physicians and researchers. *Interpreting, 7*(2), 167–192.

Leanza, Y. (2008). Community interpreter's power: The hazards of a disturbing attribute. *Journal of Medical Anthropology, 31*(2–3), 211–220.

Leanza, Y., Boivin, I., & Rosenberg, E. (2010). Interruptions and resistance: A comparison of medical consultations with family and trained interpreters. *Social Science & Medicine, 70*(12), 1888–1895.

Lee, J. (2009). Interpreting inexplicit language during courtroom examination. *Applied Linguistics, 30*(1), 93–114.

Leyva, M., Sharif, I., & Ozuah, P. O. (2005). Health literacy among Spanish-speaking Latino parents with limited English proficiency. *Ambulatory Pediatrics, 5*(1), 56–59.

MacFarlane, A., Dzebisova, Z., Karapish, D., Kovacevic, B., Ogbebor, F., & Okonkwo, E. (2009). Arranging and negotiating the use of informal interpreters in general practice consultations: Experiences of refugees and asylum seekers in the west of Ireland. *Social Science & Medicine, 69*(2), 210–214.

Messias, D. K. H., McDowell, L., & Estrada, R. D. (2009). Language interpreting as social justice work: Perspectives of formal and informal healthcare interpreters. *Advances in Nursing Science, 32*(2), 128–143.

Moreno, M. R., Otero-Sabogal, R., & Newman, J. (2007). Assessing dual-role staff-interpreter linguistic competency in an integrated healthcare system. *Journal of General Internal Medicine, 22*(2), S331–S335.

Nápoles, A. M., Santoyo-Olsson, J., Karliner, L. S., O'Brien, H., Gregorich, S. E., & Pérez-Stable, E. J. (2010). Clinician ratings of interpreter mediated visits in underserved primary care settings with ad hoc, in-person professional, and video conferencing modes. *Journal of Health Care for the Poor and Underserved, 21*(1), 301–317.

Ngo-Metzger, Q., Sorkin, D. H., & Phillips, R. S. (2009). Healthcare experiences of limited English-proficient Asian American patients: A cross-sectional mail survey. *The Patient: Patient-Centered Outcomes Research, 2*(2), 113–120.

Paasche-Orlow, M. K., & Wolf, M. S. (2010). Promoting health literacy research to reduce health disparities. *Journal of Health Communication, 15*(2), 34–41.

Pantilat, S. (2008). Ethics fast fact: Autonomy vs. beneficence. Retrieved from http://missinglink.ucsf.edu/lm/ethics/Content%20Pages/fast_fact_auton_bene.htm.

Rivadeneyra, R., Elderkin-Thompson, V., Silver, R. C., & Waitzkin, H. (2000). Patient centeredness in medical encounters requiring an interpreter. *American Journal of Medicine, 108*(6), 470–474.

Roat, C. E., Putsch, R. W., III, & Lucero, C. (1997). *Bridging the gap over the phone: A basic training for telephone interpreters serving medical settings.* Seattle, WA: Cross Cultural Health Care Program.

Rosenberg, E., Leanza, Y., & Seller, R. (2007). Doctor-patient communication in primary care with an interpreter: Physician perceptions of professional and family interpreters. *Patient Education and Counseling, 67*(3), 286–292.

Rosenberg, E., Seller, R., & Leanza, Y. (2008). Through interpreters' eyes: Comparing roles of professional and family interpreters. *Patient Education and Counseling, 70*(1), 87–93.

Rothman, D. J. (2001). The origins and consequences of patient autonomy: A 25-year retrospective. *Health Care Analysis, 9*(3), 255–264.

Sarver, J., & Baker, D. W. (2000). Effect of language barriers on follow-up appointments after an emergency department visit. *Journal of General Internal Medicine, 15*(4), 256–264.

Schapira, L., Vargas, E., Hidalgo, R., Brier, M., Sanchez, L., Hobrecker, K., ... Chabner, B. (2008). Lost in translation: Integrating medical interpreters into the multidisciplinary team. *Oncologist, 13*(5), 586–592.

Schenker, Y., Pérez-Stable, E., Nickleach, D., & Karliner, L. (2011). Patterns of interpreter use for hospitalized patients with limited English proficiency. *Journal of General Internal Medicine, 26*(7), 712–717.

Shin, H. B. (2003). Language use and English-speaking ability: 2000. Retrieved from http://www.census.gov/prod/2003pubs/c2kbr-29.pdf.

Shin, H. B., & Kominski, R. A. (2010). Language use in the United States. Retrieved from http://www.census.gov/prod/2010pubs/acs-12.pdf.

Thornton, J. D., Pham, K., Engelberg, R. A., Jackson, J. C., & Curtis, J. R. (2009). Families with limited English proficiency receive less information and support in interpreted intensive care unit family conferences. *Critical Care Medicine, 37*(1), 89–95.

Tocher, T. M., & Larson, E. B. (1996). Interpreter use and the impact of the process and outcome of care in type II diabetes. *Journal of General Internal Medicine, 11*(1), 150.

Tocher, T. M., & Larson, E. B. (1999). Do physicians spend more time with non-English-speaking patients? *Journal of General Internal Medicine, 14*(5), 303–309.

Tribe, R., & Lane, P. (2009). Working with interpreters across language and culture in mental health. *Journal of Mental Health, 18*(3), 233–241.

White, K., & Laws, M. (2009). Role exchange in medical interpretation. *Journal of Immigrant and Minority Health, 11*(6), 482–493.

Woloshin, S., Schwartz, L. M., Katz, S. J., & Welch, H. G. (1997). Is language a barrier to the use of preventive services? *Journal of General Internal Medicine, 12*(8), 472–477.

Wros, P. (2009). Giving voice: Incorporating the wisdom of Hispanic RNs into practice. *Journal of Cultural Diversity, 16*(4), 151–157.

Wu, A. C., Leventhal, J. M., Ortiz, J., Gonzalez, E. E., & Forsyth, B. (2006). The interpreter as cultural educator of residents: Improving communication for Latino parents. *Archives of Pediatrics & Adolescent Medicine, 160*(11), 1145–1150.

Youdelman, M. K. (2008). The medical tongue: U.S. laws and policies on language access. *Health Affairs, 27*(2), 424–433.

4

Assessing Health Disparities Through Communication Competence, Social Support, and Coping Among Hispanic Lay Caregivers for Loved Ones With Alzheimer's Disease

Melinda R. Weathers and Gina Castle Bell

Drawing from the United States Census Bureau's (2006) population estimates, more than 37.3 million Americans are aged 65 or older, and over 5.3 million Americans are older than 85. However, what is often not reported in such projections is the rising numbers of minority elders. As Gallagher-Thompson et al. (2003) lamented, "less appreciated is the fact that ethnic and racial minorities over 65 years of age are increasing faster than other segments of the population" (p. 423). By 2030, minorities will comprise well over 25% of the older population, increasing the importance of cultural and ethnic influences on health and aging (U.S. Bureau of the Census, 1996; see also Gallagher-Thompson et al., 2003). Advanced age is also a risk factor in Alzheimer's disease (AD) (Alzheimer's Association, 2010).

There are an estimated 5.3 million people living with AD in the United States (Alzheimer's Association, 2010), and 9.9 million unpaid lay caregivers (Alzheimer's Association, 2009). AD affects one in 10 individuals over the age of 65, nearly half of those over 85, and individuals as young as 30 (Alzheimer's Association, 2010). However, these statistics greatly differ for the Hispanic community, who are almost three times as likely to develop AD as Caucasians (Alzheimer's Association, 2010; Gallagher-Thompson et al., 2003). Hispanics are also more likely to acquire an onset of symptoms of AD five to seven years earlier than their Caucasian counterparts.

In addition to AD being more prevalent in Hispanics, symptoms can be more severe. A study conducted by Sink, Covinsky, Newcomer, and Yaffe (2004) concluded that "Latino community-dwelling patients with moderate to severe dementia have a higher prevalence of dementia-related behaviors than whites" (p. 1277).

Results illustrated that 57% of Latino patients reported having four or more behaviors, compared with 46% of White patients.

It is essential to also consider societal factors that influence patient care and health outcomes when confronting AD, specifically in Hispanic families. As noted previously, Hispanics are at greater risk for contracting AD (Alzheimer's Association, 2010; Gallagher-Thompson et al., 2003). This ethnic group also experiences many financial difficulties. According to 2009 census data, slightly over one quarter (25.3%) of all Hispanics in the United States are classified as living below the poverty level (DeNavas-Walt, Proctor, & Smith, 2010). Poverty can also influence Hispanics' public health-care needs, such as insufficient health-care coverage (Macias & Morales, 2000), often resulting in little or no basic health-care services. Hispanics also constitute the largest group of uninsured individuals in the United States; further limiting their access to care (American Psychiatric Association, 2011). As a result, the health disparities mentioned above (i.e., differences in the quality of health and the access and quality of health care) may have serious implications for Hispanics caring for loved ones with AD.

Caregiving

AD occurs across a vast landscape, affecting not only those afflicted with the disease, but also family members, friends, and society at large. According to Aranda, Villa, Trejo, Ramirez, and Ranney (2003), "between 2.4 and 3.1 million spouses or partners, relatives, and friends take care of people with AD" (p. 259). These lay caregivers are usually female, older in age, of moderate means, unemployed, and in poor health themselves (Query & Flint, 1996). In fact, Hispanic lay caregivers are more likely to be younger, less educated, poorer, and in worse health than Caucasian caregiver counterparts (Aranda et al., 2003). Further, most lay caregivers are not formally trained, and may lack the necessary experience needed to care for chronically ill individuals (Query & Flint, 1996). Families and friends caring for individuals with AD face a multitude of coping, as well as relational, financial, and cultural challenges.

Relational and Financial Challenges

As AD progresses, family members and friends may experience significant losses, as they witness the deterioration of their loved one (Query, Baker Stuart, & Golden, 2008). Learning to cope often poses new and difficult challenges to overcome. Family members and friends may also be confronted with ongoing identity confusion (Query et al., 2008; Query & Flint, 1996). Forgetfulness and bewilderment may be sporadic and seem insignificant in the early stages of AD; as the disease progresses, however, individuals with AD typically lose the ability to recall and recognize those around them. Witnessing a loved one's mind and body deteriorate can be emotionally taxing for lay caregivers and other family members.

Managing the financial well-being of a person with AD often becomes the responsibility of family and friends (Kuhn, 2003). Tasks, such as paying bills on time, tracking investments, and making financial transactions eventually have to be assumed by a guardian duly authorized. According to the Alzheimer's Association (2010), individuals with AD will live an average of eight years, and as many as twenty years or more, from the onset of symptoms. The possible longevity and progressive health decline inevitably lead to changes in relationships (Kuhn, 2003; Query et al., 2008; Query & Flint, 1996). As profound changes occur in the individual with AD, family members and friends must learn to adapt and accept their loved one's declining abilities—which is often daunting, emotionally charged, and highly stressful.

Cultural Challenges

In addition to caring for a loved one with AD and coping with relational and financial challenges, it is essential to consider the cultural factors that influence patient care and health outcomes. According to Cox and Monk (1993), "the Hispanic culture is characterized by six values that shape behaviors and the development of individual identity" (p. 93). A number of specific values have been linked to Hispanic culture, including: familism, simpatía, respeto, fatalism, espiritismo, and machismo (Gallagher-Thompson et al., 2003; Marin & Marin, 1991; Villarreal, Blozis, & Widaman, 2005). These six values are defined and discussed below.

According to Villarreal et al. (2005), "Familism has been studied extensively, with much recent empirical support for the contention that familism is core to Hispanic culture" (p. 410). *Familism* has been defined as "a cultural value that involves individual's strong identification with and attachment to their nuclear and extended families, and strong feelings of loyalty, reciprocity, and solidarity among members of the same family" (Marin &

Marin, 1991, p. 13). Within the Hispanic culture, emphasis is placed on the family unit rather than on the individual (Cox & Monk, 1993; Gallagher-Thompson et al., 2003). Subsequently, individual well-being or success is a reflection on the entire family, not merely on the individual (Cox & Monk, 1993). Hence, not meeting cultural standards with regards to the care of the elderly would reflect negatively on the whole familial group (Cox & Monk, 1993). A further characteristic of familism is to "provide both practical and emotional support to elderly people" (Cox & Monk, 1993, p. 93). Thus, it is very common in the Hispanic culture to privilege the extended family as the primary support for an individual with AD, while formal support services are less likely to be used (Gallagher-Thompson et al., 2003).

With regard to simpatía, respeto, fatalism, espiritismo, and machismo, only a few studies have explored these values as central to Hispanic culture (Villarreal et al., 2005). *Simpatía* refers to "a permanent personal quality where an individual is perceived as likeable, attractive, fun to be with, and easy going" (Triandis, Marin, Lisansky, & Betancourt, 1984, p. 1363). *Respeto* involves showing respect for others based on age, gender, and authority (Villarreal et al., 2005). According to Cox and Monk (1993), "Age connotes status, and younger people are expected to behave deferentially toward their elders" (p. 93). Additionally, respeto is of importance to the caregivers of older adults (Gallagher-Thompson et al., 2003). Children who do not adhere to this behavior are viewed as violating significant cultural norms (Cox & Monk, 1993). Thus, not caring for a loved one opposes the cultural value of respeto. For many Hispanics, breaking cultural traditions during a loved one's time of need is not an option. To uphold respeto, family members and friends may end up caring for a loved one with AD out of cultural obligation, even when it severely inhibits their own quality of life. *Fatalism* is the belief that one has no control over one's destiny (Unger et al., 2002; Villarreal et al., 2005).

Espiritismo is closely related to fatalism. As stated by Gallagher-Thompson et al. (2003),

> Espiritismo refers to a focus on religiosity as an important determinant of well-being, as well as a belief system in which the world is inhabited by both good and evil spiritual beings who can affect humans' health, happiness, and peace of mind. (p. 429)

Espiritismo may be problematic insomuch as this cultural belief potentially impacts the care a loved one receives. By viewing health conditions as having a spiritual cause, Hispanics may view AD as coming from God instead of from physiological origins (Gallagher-Thompson et al., 2003). That is, if some spiritual being meant for the person to become ill, then the caregiver may believe that she or he deserves the pain as a result.

Finally, *machismo* is reflected in the male qualities of "masculinity, male dominance, sexual prowess, physical strength, and honor" (Unger et al., 2002, p. 260). Traditional sex roles in the Hispanic culture tend to be clearly defined: an authoritarian husband and submissive wife (Cox & Monk, 1993; Vasquez, 2003). Large numbers of Hispanic women typically place the needs of their families before their own; caring for the home and the children (Cox & Monk, 1993). In most Hispanic households where AD is present, the adult daughter is likely to be the primary caregiver for her parents or her parents-in-law (Aranda et al., 2003). Because of this cultural obligation, Hispanic women may end up caring for a loved one with AD even when it severely inhibits their own well-being. Therefore, it is paramount to analyze caregiving, particularly Hispanic caregivers, when assessing AD across cultures.

Relational Health Communication Competence and Caregiving

Lay caregivers do not work in an information vacuum. They regularly need to coordinate efforts with health-care providers, extended family members, and possibly social-service personnel and/or clergy members. Competent relational health communication skills are imperative to help caregivers elicit cooperation with these key health-care partners (Kreps, 1988). Communication competence is defined as "the perceived tendency to seek out meaningful interaction with others, render support, be relaxed, appreciate others' plight, and turn-take appropriately" (Query & Kreps, 1996, p. 339). Individuals who perceive themselves to be competent communicators should be more likely to play a fundamental role in assembling and maintaining support systems, as well as be better enabled to influence an array of important health outcomes, such as quality of life, for individuals with AD and their caregivers (Query & Kreps, 1996).

Relational Health Communication Competence Model

The Relational Health Communication Competence Model (RHCCM) describes a process by which communication competence influences health outcomes across challenging terrains (Kreps, 1988). The model depicts a wheel with spokes representing an individual's social-network members (e.g., health-care providers, family members, support-group members), and the hub of the wheel represents the individual health-care consumer. The preceding network shapes the quality of an individual's health-care journey. The terrain represents communication contexts with the presumption that more challenging terrains require higher levels of communication competence (Query & Kreps, 1996). That is, "high levels of communication competence positively influence health communication goals, such as increased interpersonal satisfaction, therapeutic communication outcomes, cooperation between providers and consumers, social support, and effective information exchange" (Kreps, 1988, p. 353). Individuals with insufficient levels of communication competence, however, will influence the wheel to stand still, or worse, roll backwards, failing to achieve key health communication goals (Kreps, 1988).

The RHCCM (Kreps, 1988) serves as the theoretical framework for this article. Previously this model has helped theoretically ground studies examining communication competence, social support and interaction, perceived stress, and cognitive depression (Query & Kreps, 1996). Additionally, the model has been used as a template for studying a variety of contexts and populations, including: Hispanic lay caregivers for loved ones with AD (Weathers, 2008); support groups (Query & Kreps, 1996; Query & Wright, 2003); retirement community residents (Query & James, 1989); nontraditional students (Query, Parry, & Flint, 1992); recovering alcoholics (Hurt, 1989); OB-GYN client interactions (Sanmiguel, 1992); battered ethnic women (O'Brien, 2006); and breast cancer survivors (Bible, 2006).

Social Support

According to Albrecht and Adelman (1987), "Social support is a process inextricably woven into communication behavior" (p. 14). This intertwined nature of communication and social support influences a variety of health outcomes among distinct populations (Query & Wright, 2003). For example, Rapp, Shumaker, Schmidt,

Naughton, and Anderson (1998) found that the role of social competence is helpful in creating social relationships associated with support and well-being among caregivers of individuals with dementia. AD, as well as many types of related dementia, can cause pervasive memory loss, confusion, and other symptoms (Alzheimer's Association, 2010), all of which present formidable communication challenges.

Social support may be viewed as "emotional, instrumental, and financial products delivered through the web of friends and acquaintances that surround an individual" (Query & James, 1989, p. 167). Those friends and acquaintances are an individual's "social support network" (Query & James, 1989, p. 167). Here, "social support network" is examined in terms of network size and social support satisfaction.

Social network size. Social network size has been associated with communication skill levels, or communication competence. Heller (1979) stated, "Competent persons, who are more immune to the adverse effects of stress, are also more likely to have well-developed social networks as a direct result of their more general social competence" (p. 361). Query and James (1989) assessed communication competence, social network size, and social support satisfaction among elderly support-group members. Results indicated that individuals with high communication competence maintained larger social networks than low communicatively competent individuals. These findings suggest that "elderly individuals who perceive themselves as competent communicators are better able to develop and mobilize social networks to meet their needs than elders who view themselves as less competent communicators" (Query & James, 1989, p. 177).

Query et al. (1992) further supported this finding, reporting that nontraditional students, who viewed themselves as high in communication competence, had larger social networks than nontraditional students who viewed themselves as less competent. Similarly, Weathers (2008) found that Hispanic lay caregivers who reported high communication competence levels also reported a greater number of social network relationships than Hispanic lay caregivers who reported low communication competence levels.

Taken together, these findings suggest that "individuals who perceive themselves as competent communicators are better able to build and activate social networks to deal with life transitions than are individuals who view themselves as less competent communicators" (Query & Kreps, 1996, p. 340). In addition to size, social network satisfaction is also vital for mediating health-related challenges.

Social network satisfaction. In addition to its role as a mediator of life events, social support satisfaction is thought to be related to communication skill levels (Query et al., 1992). Some evidence suggests, for example, that the absence or underdevelopment of these skills may trigger feelings of dissatisfaction, since less competent individuals will be less likely to elicit and provide social support during times of need (Query et al., 1992).

According to Query and James (1989), "the ability to send, receive, and interpret symbolic messages appears to be essential to meaningful interaction and the expression of social support" (p. 177). The extent to which Hispanic lay caregivers are supported and perceive themselves as being helped is due, in part, to their abilities to negotiate life obstacles through ongoing communication with significant others. Social support satisfaction is thus forged in part by communication competence level, since many Hispanic lay caregivers strive to maintain a network of care providers that can aid them across the many trials and tribulations of AD and related dementias.

Perceived Coping

In the case of lay caregivers in general, and AD caregivers in particular, depression is prevalent. Wright, Hickey, Buckwalter, Hendrix, and Kelechi (1999) reported that over time, depression increased significantly for AD caregivers, with 21% evidencing moderate to severe depression at baseline, and 50% eliciting depressive symptoms one year later. Additionally, Kiecolt-Glaser, Dura, Speicher, Trask, and Glaser (1991) examined caregivers of individuals with dementia over a thirteen-month period. Results indicated that caregivers were more likely to experience depressive symptoms than individuals who were not caregivers.

While the preceding studies demonstrated that depression is prevalent in caregivers of individuals with dementia—AD in particular—additional findings suggest that it may be more so in Hispanic caregivers for individuals with AD. For example, Covinsky et al. (2003) measured patient and caregiver characteristics associated with depression among primary caregivers for individuals with AD. Results indicated that 32% of caregivers reported six or more symptoms of depression and were classified as depressed. Moreover, results demonstrated important ethnic differences in rates of caregiver depression, with the lowest rates in caregivers of Black patients, and the highest rate in caregivers of Hispanic patients. A similar study conducted by Cox and Monk (1993) explored the ways in which cultural values and norms influenced the experiences of Hispanic caregivers for individuals with AD. Findings revealed a markedly higher rate of depression and personal role strain among Hispanic caregivers.

In light of the preceding burdens of caregiving, stress in health communication research is often associated with perceived general coping. According to Leatham and Duck (1990), "Coping is the mobilization of personal resources, both psychological and tangible, to deal with life stresses" (p. 2). They further noted, "If those resources are governed by the individual under strain, it is part of the coping process" (Leatham & Duck, 1990, p. 2). Social support, however, may also be established by gaining resources controlled by others (Leatham & Duck, 1990; Query & Wright, 2003). Caring for an individual with AD can be an exceptionally stressful situation, however, the effects of stress on lay caregivers' well-being can be minimized to some degree by using effective coping practices. The preceding array of research, focusing on general coping strategies, further highlights the role of communication competence by indicating that such coping strategies are shaped by communication competence levels.

Future Health Interventions

Based on previous research, it seems reasonable to conclude that perceived communication competence levels directly affect perceived social support and perceived coping among Hispanic caregivers for loved ones with AD (Weathers, 2008). As such, these findings have important implications for Kreps's (1988) RHCCM and future health communication intervention programs. Potential health interventions for Hispanic lay caregivers for family members with AD are discussed below.

Increasing Caregiver Communication Competence: Education and Training

Given that communication competence is an important set of skills among caregivers, future interventions targeting Hispanic lay caregivers for loved ones with AD should consider attempts to increase communication competence skills through education and training, keeping in mind that other variables (such as social support and stress) may have mediating effects on the outcomes. Kreps (2006) verified the benefits of training programs to increase cultural competence and promote better health outcomes among minority populations. Specifically, we argue that caregiver education and training programs may help Hispanic lay caregivers improve the quality of their communication with friends, family members, and medical providers, which in turn may positively affect their health and well-being. Additionally, communication training programs may also increase Hispanic lay caregiver competence by teaching caregivers to take an assertive, active role in the care and treatment of their loved one.

CARE Plus, a training program implemented by researchers at Boston University, seeks to increase lay caregiver competence by teaching them specific skills for managing the behavioral symptoms of AD and for maintaining their own emotional well-being. Funded by the Alzheimer's Association International Research Grant Program, the CARE Plus intervention has shown a decline in problem behaviors and AD caregiver stress. The 90-minute, weekly educational sessions provide caregivers with information regarding AD symptoms, tips for improving communication and interactions with loved ones who have AD, and strategies for reducing caregiver stress. As participants in CARE Plus, caregivers learn valuable information about AD, specific techniques to help manage behavioral problems associated with AD, and often feel more emotionally supported because of their participation in the five-week intervention. Creating intervention programs with outcome objectives similar to those of CARE Plus, but that specifically target Hispanic lay caregivers, like other intervention campaigns that tailor messages to specific populations (Hornik & Ramirez, 2006), may prove to be the most effective way to increase caregiver communication competence, to reduce caregiver stress, and to effect positive health outcomes for individuals with AD.

Increasing Caregiver Social Support and Decreasing Stress

Caregiving Families

While conventional caregiving services and programs focus mainly on the needs of the primary caregiver, it is often the case that other members of the family are (to some extent) involved in providing care. In their analysis, Chiu et al. (2009) found that family members who provide care, along with the primary caregiver, also want information about AD, and need support services. This finding led to intervention programs advocating for a whole caregiving family, instead of advocating for only one, primary caregiver. In order to make such an intervention successful, consideration needs to be given to the behavioral mindset of the lay caregivers and their families. Hornik and Ramirez (2006) emphasized the need for interventions to be mindful that people are in different places, and may not be ready to change their opinions or behaviors at the same time. Cultural factors must be considered if intervention programs are to transform the ways in which Hispanic lay caregivers and their families care for loved ones with AD. Perhaps tailoring intervention messages so that they emphasize the benefits of working as a collective unit—the notion of a whole caregiving family—can become a reality. As a result, caregiving families may experience greater social support satisfaction and decreased levels of caregiving stress. Such an intervention warrants immediate exploration.

Narratives. Storytelling is a very powerful communication tool. As such, future health communication interventions should consider highlighting the importance of Hispanic lay caregiver narratives related to the negative as well as the positive aspects of their caregiving experiences. A large body of research confirms the potential value of narratives across challenging health-care contexts (Query & Wright, 2003). Specifically, narratives illustrate instances of supportive, mixed, and nonsupportive interactions, and help uncover how communication competence levels, among other things, shape the nature of lived experiences. Further, people are naturally storytellers. Thus, intervention programs that emphasize the emotional benefits of self-disclosure have the potential to influence Hispanic lay caregivers to communicate with their social networks more frequently. In turn, sharing regularly may reduce caregivers' stress by granting them permission to not carry the caregiving load alone.

Additionally, government and community-based interventions should consider developing narrative-based intervention programs. One way to implement such a program may be to utilize the different media (e.g., television, radio, new media) that are seen as credible in the Hispanic community, in order to share the health benefits from caregiver narratives within individuals' social networks (Hornik & Ramirez, 2006). Such intervention programs may provide Hispanic lay caregivers and caregiving families with the necessary skills to function effectively across life crises, and simultaneously provide them with the emotional support necessary for caring for a loved one with AD (Query & Wright, 2003).

Internet and online support groups. The rapid growth of the Internet has facilitated communication between people with common health concerns, and has allowed for the development of online support groups, which have attracted the attention of researchers interested in social support (Wright, 2000). A recent example has been advanced by Query, Jibaja-Weiss, Volk, Siriko, and Yamasaki (2005), in which efforts are underway to develop a mediated support tool, ultimately available via CD/DVD and through the Internet, targeting ethnically diverse caregivers for individuals with AD. Future interventions, however, should consider employing "enhanced" Internet caregiver support systems that send e-mail or text-message reminders to caregivers on a regular basis. Utilizing popular online social networking sites such as Facebook may also prove to be a useful channel for disseminating health information to Hispanic lay caregivers for loved ones with AD. As with other media technologies, the Internet allows caregivers to access online interventions on their own time, after juggling caregiving and other life responsibilities.

Targeting multilingual needs. It is vital that future health communication interventions address Hispanic caregivers' multilingual needs. A large number of Hispanics struggle with language barriers in the United States (Vasquez, 2003). As such, they are also likely to have difficulties communicating about health issues in English. With the number of new and emerging technologies readily available, future interventions should minimize language barriers by collecting and delivering health information to Hispanic caregivers in their native language. For example, Chiu et al. (2009) implemented an electronic design feature called "telling more" that assisted Chinese caregivers for family members with AD in better composing e-mails. The new, interactive design provided the users with a list of descriptor situations in both Chinese and English. Once selected, the descriptors were then converted into a draft e-mail, ready for editing or sending to their health-care provider. The design enabled users to create e-mail messages more effectively and efficiently.

Tailoring message content to reflect linguistic and educational norms by providing health messages in both English and Spanish may help interventions better target specific cultural groups (i.e., Hispanic lay caregivers), and enhance health outcomes (Hornik & Ramirez, 2006). Future intervention campaigns that utilize the Internet or online support groups should strive to execute message delivery in an accessible manner for the cultural group they hope to impact.

Conclusion

The RHCCM (Kreps, 1988) serves as a vehicle for "interpersonal interaction between consumers and providers of health care ... by guiding strategic health behavior, treatment, decision-making, and influencing psycho-social adaptation" (Kreps, 2001, p. 598). Given the many confirmations of the model's utility, it should continue to be a springboard for developing community-based, tailored interventions to help Hispanic lay caregivers cope with the trials and tribulations of caring for loved ones with AD. For health communication scholars, ideally, this paper will provide another weapon in their arsenal as they combat caregiver stressors. There is no question that the AD battlefield will continue to be fraught with many potential dangers. Drawing from Kreps's model, however, health communication scholars can greatly assist the "front-line warriors" who are lay caregivers (Query & Flint, 1996).

Bibliography

Albrecht, T. L., & Adelman, M. B. (1987). Rethinking the relationship between communication and social support: An introduction. In T. L. Albrecht & M. Adelman (Eds.), *Communicating social support* (pp. 13–16). Newbury Park, CA: Sage.

Alzheimer's Association. (2009). Alzheimer's facts and figures. Retrieved from http://www.alz.org/national/documents/report_alzfactsfigures2009.pdf.

Alzheimer's Association. (2010). Alzheimer's facts and figures. Retrieved from http://www.alz.org/alzheimers_disease_facts_figures.asp.

American Psychiatric Association. (2011). Latino mental health. Retrieved from http://www.healthyminds.org/more-info-for/hispanicslatinos.aspx.

Aranda, M. P., Villa, V. M., Trejo, L., Ramirez, R., & Ranney, M. (2003). El Portal Latino Alzheimer's Project: Model program for Latino caregivers of Alzheimer's disease-affected people. *Social Work, 48*(2), 259–271.

Bible, S. E. (2006). Assessing the role of communication competence and social support between cancer patients transitioning to survivor roles and their oncologists. Unpublished master's thesis, University of Houston, Houston, TX.

Chiu, T., Marziali, E., Colantonio, A., Carswell, A., Gruneir, M., Tang, M., & Eysenbach, G. (2009). Internet-based caregiver support for Chinese Canadians taking care of a family member with Alzheimer disease and related dementia. *Canadian Journal on Aging, 28*(4), 323–336.

Covinsky, K. E., Newcomer, R., Fox, P., Wood, J., Sands, L., Dane, K., & Yaffe, K. (2003). Patient and caregiver characteristics associated with depression in caregivers of patients with dementia. *Journal of General Internal Medicine, 18*(12), 1006–1014.

Cox, C., & Monk, A. (1993). Hispanic culture and family care of Alzheimer's patients. *Health & Social Work, 18*(2), 92–100.

DeNavas-Walt, C., Proctor, B. D., & Smith, J. C. (2010). Income, poverty, and health insurance coverage in the United States: 2009. Washington, DC: U.S. Government Printing Office.

Gallagher-Thompson, D., Haley, W., Guy, D., Rupert, M., Arguelles, T., Zeiss, L. M., Long, C., Tennstedt, S., & Ory, M. (2003). Tailoring psychological interventions for ethnically diverse dementia caregivers. *Clinical Psychology: Science and Practice, 10*(4), 423–438.

Heller, K. (1979). The effects of social support: Prevention and treatment implications. In A. P. Goldstein & F. H. Kanfer (Eds.), *Maximizing treatment gains: Transfer enhancement in psychotherapy* (pp. 353–382). New York, NY: Academic Press.

Hornik, R. C., & Ramirez, S. (2006). Racial/ethnic disparities and segmentation in communication campaigns. *American Behavioral Scientist, 49*(6), 868–884.

Hurt, H. T. (1989). The relationships among interpersonal skills, social support, and recovery from alcohol addiction. Paper presented at the annual meeting of the International Communication Association, San Francisco, CA.

Kiecolt-Glaser, J. K., Dura, J. R., Speicher, C. E., Trask, O. J., & Glaser, R. (1991). Spousal caregivers of dementia victims: Longitudinal changes in immunity and health. *Psychosomatic Medicine, 53*(4), 345–362.

Kreps, G. L. (1988). Relational communication in health care. *Southern Speech Communication Journal, 53*(4), 344–359.

Kreps, G. L. (2001). Consumer/provider communication research: A personal plea to address issues in ecological validity, relational development, message diversity and situational constraints. *Journal of Health Psychology, 6*(5), 597–601.

Kreps, G. L. (2006). Communication and racial inequities in health care. *American Behavioral Scientist, 49*(6), 760–774.

Kuhn, D. (2003). *Alzheimer's early stages: First steps for family, friends, and caregivers* (2nd ed.). Alameda, CA: Hunter House.

Leatham, G., & Duck, S. (1990). Conversations with friends and the dynamics of social support. In S. Duck, (Ed.) with R. *Silver, Personal relationships and social support* (pp. 1–29). London, UK: Sage.

Macias, E. P., & Morales, L. S. (2000). Crossing the border for health care. *Journal of Health Care for the Poor and Underserved, 12*(1), 77–87.

Marin, G., & Marin, B. V. (1991). *Research with Hispanic populations*. Newbury Park, CA: Sage.

O'Brien, C. (2006). Examining communication competence, social support, coping, and openness to disclose among abused women. Unpublished master's thesis, University of Houston, Houston, TX.

Query, J. L., Baker Stuart, J., & Golden, M. A. (2008). Caregivers coordinate cultures: Communication with Alzheimer's patients in bi-ethnic families. In K. B. Wright & S. Moore (Eds.), *Applied health communication: A sourcebook* (pp. 291–312). Cresskill, NJ: Hampton Press.

Query, J. L., & Flint, L. J. (1996). The caregiving relationship. In N. Vanzetti & S. Duck (Eds.), *A lifetime of relationships* (pp. 455–483). Pacific Grove, CA: Brooks/Cole.

Query, J. L., & James, A. C. (1989). The relationship between interpersonal communication competence and social support among elderly support groups in retirement communities. *Health Communication, 1*(3), 165–184.

Query, J. L., Jibaja-Weiss, M. L., Volk, R. J., Siriko, T. A., & Yamasaki, J. (2005). Towards developing a mediated support tool for ethnically diverse caregivers. Paper presented at the 13th Annual Alzheimer's Association Dementia Care Conference, Chicago, IL.

Query, J. L., & Kreps, G. L. (1996). Testing the relational model for health communication competence among caregivers for individuals with Alzheimer's disease. *Journal of Health Psychology, 1*(3), 335–351.

Query, J. L., Parry, D., & Flint, L. J. (1992). The relationship among social support, communication competence, and cognitive depression for non-traditional students. *Journal of Applied Communication Research, 20*(1), 78–94.

Query, J. L., & Wright, K. B. (2003). Assessing communication competence in an online study: Toward informing subsequent interventions among older caregivers with cancer, their lay caregivers, and peers. *Health Communication, 15*(2), 203–218.

Rapp, S. R., Shumaker, S., Schmidt, S., Naughton, M., & Anderson, R. (1998). Social resourcefulness: Its relationship to social support and well-being among caregivers of dementia victims. *Aging and Mental Health, 2*(1), 40–48.

Sanmiguel, L. M. (1992). The gynecologist-patient context: A model and research agenda. Paper presented at the annual meeting of the Speech Communication Association, Chicago, IL.

Sink, K. M., Covinsky, K. E., Newcomer, R., & Yaffe, K. (2004). Ethnic differences in the prevalence and pattern of dementia-related behaviors. *Journal of the American Geriatrics Society, 52*(8), 1277–1283.

Triandis, H. C., Marin, G., Lisansky, J., & Betancourt, H. (1984). Simpatía as a cultural script of Hispanics. *Journal of Personality and Social Psychology, 47*, 1363–1375.

Unger, J. B., Ritt-Olson, A., Teran, L., Huang, T., Hoffman, B. R., & Palmer, P. (2002). Cultural values and substance use in a multiethnic sample of California adolescents. *Addiction Research and Theory, 10*(3), 257–279.

U.S. Bureau of the Census. (1996). *Projections of the total and elderly populations, by age, race, and Hispanic origin: 1995 to 2050*. Washington, DC: U.S. Government Printing Office.

U.S. Bureau of the Census. (2006). Minority population tops 100 million. Retrieved from http://www.census.gov/PressRelease/www/releases/archives/population/010048.html.

Vasquez, M. J. T. (2003). Latinas. In M. Aguirre-Molina & C. W. Molina (Eds.), *Latina health in the United States: A public health reader* (pp. 345–370). San Francisco, CA: Jossey-Bass.

Villarreal, R., Blozis, S. A., & Widaman, K. F. (2005). Factorial invariance of a Pan-Hispanic familism scale. *Hispanic Journal of Behavioral Sciences, 27*(4), 409–425.

Weathers, M. R. (2008). Exploring communication competence, social support, perceived coping, and religious coping among Hispanic family members caring for loved ones with Alzheimer's disease. Unpublished master's thesis, University of Houston, Houston, TX.

Wright, K. B. (2000). Social support satisfaction, on-line communication apprehension, and perceived life stress within computer-mediated support groups. *Communication Research Reports, 17*(2), 139–147.

Wright, L. K., Hickey, J. V., Buckwalter, K. C., Hendrix, S. A., & Kelechi, T. (1999). Emotional and physical health of spouse caregivers of persons with Alzheimer's disease and stroke. *Journal of Advanced Nursing, 30*(3), 552–563.

5

Culture-Centered Constructions Among Health Providers in Rural Indiana: Setting Agendas for Interventions

Christina Jones and Mohan J. Dutta

Federal agencies are increasingly encouraging researchers to address cultural differences in population health. Group characteristics attributable to geographic location presumptuously fall under the heading of "culture" here. It seems appropriate, then, to expand this renewed interest to encompass rural health, especially considering the health disparities present between rural and urban areas across the various sectors of the globe (see Dutta-Bergman, 2004a, 2004b). However, few rural health researchers are asking the relevant cultural question: "Why does rural residence (culture, community, and environment) reinforce negative health behaviors?" (Hartley, 2004, p. 1676). Although the rural-urban divide may be theorized in a global context, examining the problem in the backdrop of national health policy allows for theorizing about the postcolonial politics of marginalization in health care within the very sites of capitalist hegemony here in the United States that carry out the global agendas of perpetuating health-care inequalities based on material (in)access to resource spaces in the mainstream.

A quarter of the U.S. population lives in rural areas, and rural communities commonly share characteristics such as high rates of elderly citizens (almost 20%) and high poverty rates (Ricketts, 2000). According to the National Rural Health Association (NRHA, 2010), on average, the per capita income in rural areas is almost $8,000 less than in urban areas, and nearly 24% of rural children live in poverty. Rural residents are less likely to have employer-provided health-care coverage or prescription drug coverage, and the rural poor are less likely to be covered by Medicaid benefits than their urban counterparts. Almost 45% of Medicare beneficiaries in rural communities are without drug coverage, compared to only 31% in urban areas (NRHA, 2010).

Rural residents are, on average, more often uninsured than urban residents (18.7% vs. 16.3%), and are more likely to report being in fair or poor health, having restricted activity, and having lower levels of access to a regular primary-care provider (Ricketts, 2000). Additionally, people who live in rural America also have been found to rely more heavily on the federal Food Stamp Program (NRHA, 2010). The Carsey Institute at the University of New Hampshire found that while 22% of Americans lived in rural areas in 2001, a full 31% of the nation's food stamp beneficiaries lived there. In 2001, 4.6 million rural residents received food stamp benefits. It is in the backdrop of this overall picture of inaccess that rural health-care providers continue to offer services for the underserved populations in the rural areas of the US. In this chapter, we present the work from a culture-centered, ethnographic project that sought to coconstruct the narratives of health-care delivery amidst the absence of structural resources in rural America (see Dutta & Kreps, Chapter 1, "Introduction," this volume; Dillon & Basu, Chapter 7, this volume). We frame this chapter in the context of the broad question: What are the experiences of health-care providers as they negotiate structural constraints in the rural US, more specifically, in the context of rural Indiana?

Rural Health Disparities in the US

The disparities experienced by rural populations play out both in terms of morbidity and mortality. Rural residents are at higher risk for a number of preventable injuries, with a 50% higher mortality rate for unintentional injuries than residents in large urban areas (Baker, Fitzgerald, & Moore, 1999), two-thirds of which are related to motor vehicle crashes (Baker et al., 1999; Maio, Green, Becker, Burney, & Compton, 1992). Although only one-third of all motor vehicle accidents occur in rural areas, two-thirds of the deaths attributed to these accidents occur on rural roads (Rural Healthy People, 2010). Rural residents are nearly twice as likely to die from unintentional injuries other than motor vehicle accidents as are urban residents. Such increased injury risk can stem from residential fires, suicide, unintentional firearm deaths, and occupational injuries (Baker et al., 1999). One reason for this increased rate of morbidity and mortality is that in rural areas, prolonged delays can occur between a crash, the call for EMS, and the arrival of an EMS provider. Many of these delays are related to increased travel distances in rural areas and personnel distribution across the response area. National average response times from motor vehicle accident to EMS arrival in rural areas was 18 minutes, or eight minutes greater than in urban areas (NRHA, 2010).

Rural residents also more often work in small businesses and in mining and agriculture, all recognized to have higher rates of traumatic injuries (Myers, 1990). Although farming-related injury mortality rates have decreased over the last decade, farming, mining, and construction remain the most dangerous occupations (Institute of Medicine Committee to Assess Training Needs for Occupational Safety and Health Personnel in the United States, 2000). Additionally, farm men, especially older farm men, were much less likely to use seat belts all of the time than other members of the rural community (Zwerling et al., 2001). Rural residents, especially women, suffer from high rates of depression (Coryell, Endicott, & Keller, 1992), and farmers have higher rates of depression and suicide, especially firearm-related suicide (Armstrong & Schulman, 1990). This may be attributable to the fact that 20% of nonmetropolitan counties lack mental health services versus 5% of metropolitan counties. In 1999, 87% of the 1,669 Mental Health Professional Shortage Areas in the United States were in nonmetropolitan counties and home to over 30 million people (Rural Healthy People, 2010).

Rural residents are at greater risk for the development of chronic diseases. Farmers are at increased risk for noise-induced hearing loss (Beckett et al., 2000) and degenerative joint disease (Thelin, 1990). Rates of selected cancers, birth defects, and Parkinson's disease associated with agricultural work and rural living have also been observed (Blair & Zahm, 1993). Cerebrovascular disease was reportedly 1.45 times higher in non-Metropolitan Statistical Areas (MSAs) than in MSAs, and hypertension was also higher in rural than urban areas (101.3 per 1,000 individuals in MSAs and 128.8 per 1,000 individuals in non-MSAs) (Rural Healthy People, 2010). Modifiable risk factors, such as obesity and smoking, are more common among rural residents, and are related to higher mortality rates and prevalence of chronic health conditions in rural areas (Eberhardt & Pamuk, 2004). Abuse of alcohol and use of smokeless tobacco is also a significant problem among rural youth. The rate of DUI arrests is significantly greater in nonurban counties. Of rural 12th graders 40% reported using alcohol while driving compared to 25% of their urban counterparts. Rural eighth graders are also twice as likely to smoke cigarettes (26.1% versus 12.7% in large metro areas) (Rural Healthy People, 2010).

Why are these disparities so prevalent? One answer may lie in conceptualizing rurality.

Statistically, miscalculation and misclassification of the rural-urban divide can lead to serious problems in the discretion of federal programs and funds. As Issermann (2005) detailed, the Office of Management and Budget (OMB) and the U.S. Census Bureau use the term, *rural*, in overlapping but contradictory ways. While the U.S. Census Bureau focuses its definition on the notion of separation of the rural from the urban per population size, the OMB focuses instead on the notion of the integration of a space with a nuclear, metropolitan area. A noticeable problem here is that referring to metropolitan counties as urban and all other counties as rural ignores the blending of urban and rural populations within a single county. The OMB points out specifically that "programs that base funding levels or eligibility on whether a county is included in a Metropolitan or Micropolitan Statistical Area may not accurately address issues or problems faced by local populations, organizations, institutions, or governmental units," (as cited in Issermann, 2005, p. 495). Additionally, the more we aggregate different types of rural areas, the less able we may be to pinpoint localized health-care and delivery problems at the state, regional, county, or town level.

Another contributor to the large disparity in health outcomes between those living in rural areas and urban areas may be that of health-care delivery. The traditional focus of rural health disparities centers upon the need for better access to care and equitable distribution of health personnel. The pressure of reform in the 1990s for health-care systems to transition from fee-for-service to managed-care systems blurred the lines between public and private providers, and rural health-care systems in different ways than urban systems, as there are fewer opportunities for cost savings due to the nature of the rural economy (Ricketts, 2000). This reform also led to government policy changes in delivery, including alternative hospitals under the guise of Medicare, and adjustments to Medicare payment systems that discriminate against rural providers. However, overall, Medicare payments to rural hospitals and physicians are dramatically less than those to their urban counterparts for equivalent services. This correlates closely with the fact that more than 470 rural hospitals have closed in the past 25 years (NRHA, 2010).

The skewed distribution of resources has led to long-standing problems in rural health care. When considering the distribution of physicians, rural America has 20% of the nation's population but less than 11% of its physicians (Ricketts, 2000). The Rural Healthy People 2010 report substantiates these findings still today (Rural Healthy People, 2010). When considering the number of specialists per 100,000 individuals, the rural rate of 40% pales in comparison to the 134% maintained in urban populations (Rural Healthy People, 2010). Rural health professionals and institutions have chosen recently to join into systems and alliances to cope with the turbulent environment of health-care policy and economics. These networks consist of multiple health-care providers who share common characteristics of rural health-care delivery: small population bases and limited resources. However, despite this trend, there is little strong, empirical evidence in support of the benefits of these networks and alliances for their members (Ricketts, 2000). The low population density in rural areas also creates special problems, since the critical mass of people is usually far less than required to economically or functionally support services or facilities (Martin, 1975). The NRHA (2010) reported that there are 2,157 Health Professional Shortage Areas (HPSA's) in rural and frontier areas of all states and U.S. territories compared to 910 in urban areas. Additionally, anywhere from 57% to 90% of first responders in rural areas are volunteers (Rural Healthy People, 2010). Also, as Ricketts (2000) pointed out, most rural areas are served by multicounty health departments. The absence of health departments altogether in some areas leaves residents without public-health services.

The hospital, a major health-care providing facility, is, in many rural areas, economically unstable, and causes further weakening of the health-care system. Compounding problems such as financial pressures to compete with larger hospitals, low occupancy rates, and a poor public perception threaten the existence of many hospitals. There are approximately 1,200 rural critical-access hospitals in the US. These smaller, rural-based hospitals often have fewer resources and less funding than larger urban hospitals. Researchers using indicators such as personnel, equipment, organizational systems, and quality improvement activities found that patients experiencing an acute myocardial infarction in rural hospitals in Kansas were less likely to receive standard care, and tended to have worse outcomes compared with their urban counterparts (Baldwin et al., 2004). Lutfiyya, Lipsky, Wisdom-Behounek, and Inpanbutr-Martinkus (2007) found that urban acute-care hospitals tended to perform better in more quality indicators than rural critical-access hospitals. Heart failure is more effectively managed in urban care settings than in rural critical-access hospitals. This study also revealed that urban acute-care hospitals were more likely to provide smoking-cessation counseling for heart-failure patients. Since rural residents are more likely to smoke, and counseling is a low-tech intervention, this seems to be an area that should be of particular interest for rural hospitals to address and improve.

Moreover, studies performed in diverse health settings, such as rural community health centers, consistently indicate that even in cases where a federal institution is bringing together a large number of individuals to work together within a common framework, results still indicate a need for interventions that adequately consider the barriers felt by rural health workers, such as staff solely devoted to the project and genuine commitment by the entire organization. For instance, Chin et al. (2004) detailed their findings from the Diabetes Collaborative quality improvement process, as part of the national Health Disparities Collaborative, where 19 Midwestern health centers formed quality-improvement teams that met regularly with the support of senior administrative leadership. Using a model entitled "Plan, Do, Study, Act," interventions were designed and implemented in small scales before being revised and implemented in broader populations. Additionally, regional cluster coordinators and staff used telephone conference calls, a computer listserv, feedback on required monthly progress reports, and regional meetings to assess the status of the intervention. Ultimately, over 30 different interventions were administered over the year (1998), including collaborations with community organizations and group cluster visits. However, many health centers still felt that they needed more resources to free staff time for the project, and some questioned the ability of such an intervention to progress beyond a segmented, "disease of the month" feel.

More current evaluation demonstrates that these barriers are still perceived. Physicians view patient education, financial issues, lack of services, and transportation to be significant impediments to providing necessary care (Siminerio, Piatt, & Zgibor, 2005).

Ultimately, answering the question of why and how rurality affects negative health behavior will require a clear understanding of rural culture and the inherent difficulties of public health delivery in rural areas. Disparities in access to care due to long transportation times, low population density, scarcity of resources, low funding for rural services, and difficulty in recruiting skilled health professionals are some of the challenges. These barriers suggest that public-health programs for rural populations deserve more defined consideration than simple application of urban systems to rural areas. We must devote continued effort to understanding the extent to which these problems within populations stem from structural factors, such as the lack of resources available to rural communities, and subsequent barriers faced by those working on rural population health issues, as well as how these problems function at the individual level, narrated through the individual experiences that negotiate the structural inequities.

When reviewing the literature in an attempt to investigate the barriers navigated by those providing public-health services to rural populations, it is easy for one to suggest that research is lacking in this area. Hart, Salsberg, Phillips, and Lishner (2002) brought this scarcity to light when they noted, "we know surprisingly little about these provider groups (public health professionals), including, for example, the needs of rural local health districts" (p. 226). When considering the barriers experienced by those interacting most closely with rural health disparities, the needs of physicians have been studied more thoroughly than any other kind of service provider (Dorsch, 2000). And, as noted previously, rural service providers will face a certain number of barriers merely functioning as part of the rural environment, including inadequate access to resources and information, and long distances inhibiting professional collaboration and idea- and information-sharing. Hart and colleagues (2002) suggested that intrastate, small-area, health workforce modeling and analysis methods are rudimentary, and need to be further refined to consider the possibilities for regional and even community-based professional collaboration. This kind of workforce analysis is important to research, policy, and the targeting of resources to the rural communities most in need of attracting and retaining health professionals of any kind. However, a significant research and policy challenge involves decisions about how to allocate resources for research and policy analysis across rural provider types, geography, and program type. In this chapter, we report the results from a larger ethnography on rural health disparities by coconstructing narratives with the public-health professionals and health providers working in rural settings.

Method

Participants

The key individuals interviewed as part of this study were selected for participation due to their status as health workers in a particularly rural region of the Midwest. For the purposes of health service and resource allocation, seven counties in this area are combined under the organizing framework of one rural health "district." The counties together comprise 2,894 square miles, and include two of the most rural and least-populated counties in the entire state. All seven counties have a higher percentage of residents that are over 65 than the state average, and the population is almost entirely Caucasian. Most of the residents in the seven-county area have a high school education, but much fewer have a college education (Shen, 2008).

Each county has its own county health department, except for two that share a health department. All seven counties either qualify by federal standards as a Medically Underserved Area or a MPA (Shortage of Physicians) (Shen, 2008). A central, more urbanized community rests in the very middle of the seven rural counties, and this area serves as a centralized location where residents from the outlying areas come to receive health care and medical attention for more acute health problems. There are three large hospitals in an eight-county area, and six satellite locations among the seven counties. There is one federally qualified, low-income health clinic for the region that serves residents from all seven counties. The dynamic nature of the flow of residents in this particularly rural area is evident; residents of the seven counties travel across county lines for their jobs, health care, and other activities.

In regard to health status, the heart health of the seven counties is significantly worse than the state and U.S. benchmarks (Shen, 2008). In a recent chronic health assessment conducted in 2008, the percentage of individuals with coronary heart disease in the seven counties was 12.3%, over twice the national rate of 5.2%. Men in particular, within the seven counties, have almost twice the rate of heart disease as women, more than 2.5 times the percentage of heart disease as men in the state, and more than 3 times the rate of men at the national level. In the two most rural counties within the region, one in five adults age 25 and older has diabetes. All seven counties have a higher percentage of overweight and obese individuals than the state average, with the two most rural counties having the largest percentage of obese individuals. Compared to the state and nation, the seven-county area has approximately the same health coverage as the nation and state, around 86%. The high cost of a doctor visit is usually a barrier to those without health insurance. However, over half of the individuals in the seven counties with health insurance indicated that cost was also a barrier, per responses on a recent county assessment (Shen, 2008). In one of the most rural counties, close to one in four residents is without health insurance. One explanation could relate to small employers who cannot afford to offer health insurance. There may also be fewer health providers overall in the area.

Participants for the present study were obtained through a combination of convenience and snowball sampling methods. Personal contacts were used to solicit participation from individuals working at the district's low-income health clinic and each county's primary health department. These contacts referred others to the project (e.g., regional diabetes center employees, county public-health nurses), who were contacted for participation. Seven interviews were performed with individuals working in conjunction with local health departments, including county nurses, diabetes center administrators, a nutritionist, and an area epidemiologist serving all counties of the health district.

Interview Schedule

The interview schedule included questions regarding health interventions in the state at large, access to materials and funding for health programs, perceived severity and susceptibility of rurally targeted health issues, motivators and support for possible interventions, and broad suggestions regarding where to start to improve the ways in which rural citizens are educated about their health. Informed consent was obtained, including consent to record the interviews for transcription. No incentives were used for participation (except perhaps the educational value of the project). Interviews were conducted at respective work locations of participants (e.g., health clinic or health department) or by phone to make participation more convenient for participants. Interviews were transcribed by one of the authors, and the transcriptions were double-checked by the second author.

Communicative Co-Constructions as Intervention

Structures of Inaccess

In the scholarship on the culture-centered approach to health communication, participant narratives in the marginalized sectors of the globe continually point toward the structural location of health-care inequalities, pointing to the need for addressing health inequalities at the structural level. The theme of the structural locus of health-care disparities continues to be carried out in the narratives of health-care professionals and workers working in rural settings and with rural populations, pointing toward the urgency of framing health interventions in the context of creation of structural capacities. The participants in our project eloquently share their stories of inaccess, in terms of the inability to provide quality care because of their location in a rural context. The geographic margins of the neoliberal economic structure are constituted in terms of the absence of fundamental health-care resources and opportunities. The practice of providing health is situated at the geographic margins because of the inaccess to resources. According to one of the providers in the project, the limited choice locally is situated in the backdrop of needing to travel in order to find access to quality care:

> The choice is either to take a local one or go to Indianapolis because there are good doctors down there, you just have to bite the bullet and go down. That's the biggest problem is that either if you want to stay local a lot of times you don't have a lot of choices, and you just have to, even some of the Indy doctors, some of them you have to wait 2 or 3 months.

The articulation here draws attention to the limited choices that are available locally, situated dialectically with a great deal of choices if one were to travel to Indianapolis, which is approximately two hours of driving. Therefore, securing health care is located amidst geographic inaccess, thus constituted amidst the narrative of needing to travel great distances to find access to health. Here is another narrative that points toward the structural inaccess:

> The doctors down in Methodist are good, but if you had a set of team doctors here, you could do it here and eliminate the travel expense and everything could be done here. But it's all being almost placed in major cities, like Indianapolis, Chicago, Minneapolis, versus when I grew up I didn't live in a big town but there was a family doctor and he would make the country rounds, if you need his help, he was there.

Once again, highlighted in this narrative is the notion of spatial inequality in the distribution of health resources. The location of health resources in large metropolitan centers creates spaces of inaccess in the rural contexts. The effects of inaccess are particularly evident during needs of emergency:

> Well, that farmer that I take up to Newton county ... he had a heart attack outside Kentland, where's the closest emergency room that he could get to there? If he went to, it was Illinois, it is north and west of Kentland up in Illinois, I can't think of the city, but it's always on the radio you hear it.... But anyway, if he was closer he could've got a better recovery. He was paralyzed on his right-hand side because he did not get to the hospital in time. And I think that's a big problem but, there's really nobody helping the rural community folks, really. How many doctors are in Kentland? You couldn't do it. Or even Fowler? Any of those towns. There's nothing.

The marginality of rural health is symbolized in the absence of health resources during emergencies such as heart attacks. The absence of emergency rooms is played out in the form of not being able to secure health services in time. Health interventions in this context are constituted as structural interventions that are directed at building health-care capacities in rural communities.

The theme of inaccess is carried out in the domain of access to preventive services, where health-care providers draw attention to the need for providing structural access to health resources. Referring to the health-care inaccess in rural communities, one of the providers points out:

> But, is that because they don't have access to healthy food, like grocery stores? Is it because they are traditionally agricultural and they still eat that way that they used to? They don't have parks and trails and places to go for exercise.

The absence of grocery stores in rural areas becomes a major structural constraint in the lives of rural participants in securing access to healthy food. Also, the absence of places to exercise is noted by the provider. Similarly, another participant notes:

> The other thing I noticed from the rural counties is that it is a very long way to drive, and these people think, "there's no way I'm driving all of the way in there for this," they don't get their eyes checked, they have sores on their feet, they have no one to check their blood sugar. So, those have been the little things so far that I have noticed from talking to them and taking care of them.

Also, the lack of seeking out preventive services such as getting screenings and getting their eyes checked is situated amidst the distance one has to travel to get access to these services.

The articulation of structural inaccess is situated amidst working through narratives of individual agency that emphasize individual choice, and point toward healthy decision making within the limited constraints.

> With the patients that we see here, it is the follow-up or lack of motivation to do it, to make the changes. They see some of the things that I recommend as hardships to them. We recommend more fruits and vegetables and they think that they have to go buy fresh fruits and vegetables and they think it costs more. I try to point out that you can buy a cheap bag of frozen vegetables for the same cost.

As the provider narrative here points toward the structural constraints and barriers, it does so amidst the articulation of individual agency, suggesting alternatives such as purchasing a cheap bag of frozen vegetables for the same cost. However, the delivery of prevention information among rural communities is fairly limited because of the limited resources and limited number of health staff in rural counties. Here is what one of the providers had to say:

> Individuals at health departments, none of them work full-time. And, mainly they have one focus and that's to give people shots. Because they are there more often for people to make an appointment, come in, get their shot. They aren't teaching anything or doing anything. But they

> aren't there enough. And, they don't hold classes. Not that that wouldn't be a bad place to do it but you would need to hire somebody to do it. That's not part of their job description at all.

In the context of prevention, the reference here is to the limited structural resources to support the delivery of prevention information or to engage in health education in rural communities. Given the emphasis of health workers in rural health departments on giving people shots, and given the nature of their part-time work, there are not enough structural resources to support preventive services. This point is further noted in the following articulation:

> The rural county health departments aren't set up for health interventions. I think it would be best in a doctor's office. The people who work at the health departments aren't skilled as public health educators or would be competent to do that.

Furthermore, drawing attention to the dearth of financial resources in order to carry out preventive services in rural areas, one of the participants notes, "and, I would say the funding. There's nothing available from the state." From a communicative standpoint, the emphasis is laid on nothing being available from the state. Similarly, another participant notes:

> We have a tight budget here, being a small county office. I'm sure that there's funds allotted for that, but we try not to use those unless we absolutely need to. We'd have to go through the commissions and council to get appropriations.

Here's another articulation that attends to the structural constraints of providing health in rural contexts, "If you would really get out there and start doing a whole lot of educational things, we would need more staff to be able to keep doing our home visits free of charge." It is amidst the articulations of the structural elements of health that the participants suggest creative ways of enacting their agency and structural solutions to the problems facing the rural margins of health care.

Structural Solutions at Rural Margins

Given the framing of rural health disparities as structurally situated, the health-care providers working in rural contexts draw attention to the need for structural solutions to the problems of rural inaccess. One such structural solution is the emphasis on incentivizing rural community practice among doctors by reducing the medical school costs of doctors who go to rural communities. This is what one of our participants noted:

> Well, the biggest thing is enticing doctors to go to rural communities by reducing their medical school costs. If they could go, I don't know all the rules in Texas, but if you are getting a 50% reduction and you sign up to go to Odessa, Texas for 5 years and after 5 years you owe nothing, I think that's a good program. Versus paying full price and going to Dallas or a major city to get enough income to pay your loans. But it has to be more at a state level than at a federal level because the problem with it at a federal level is that you've got 50 different states with 50 different possibilities, what works in Rhode Island is not going to work in North Dakota, and I think sometimes they mean well in Washington but they lose because they become too generalized. Same thing for, they always talk about mass transportation, good mass transportation is going to work from Boston to Washington, but tell that to the farmer in North Dakota, how is that going to help him? So, they get lost in things, that's why I think it needs to be pushed at the state level, because the state level has the universities and they can see that if they have a Baylor University Medical Center and a Texas Tech University medical center and they need doctors in these areas they can put a lot of emphasis on that by getting the good students in and then enticing them by doing these different things, but I just think that it has to be at a state level instead of at a national level.

In this instance, the creation of incentives at the state level becomes a way for promoting health-care practice in rural settings. Where one of the major problems facing doctors who practice in rural areas is financial, the focus, then, is on creating financial incentives for physicians by reducing medical school costs. The intervention in this context is redefined as an intervention at the structural level, creating greater financial incentives for physicians working rurally, as opposed to individual-level solutions.

Also, rural health-care providers often discussed working creatively through the limited structural resources in order to create health-care resources in rural areas. In one such instance, a provider shared with the project team the creative use of a bus to continually present the local community with information about diabetes:

> It could be the constant information in front of them, but gently. And, if this bus came and stopped and senior centers, and retirement villages, and churches, and made the rounds, on a regular basis, that it would bring attention to it so that those who have diabetes would come to it,

> have their continuing education, and get checked out. And, it might bring enough attention to it in the community that that would be good.

In the absence of mainstream media for creating access to health information, the conceptualization of a bus that would disseminate prevention information reiterates creativity amidst limited structural resources. Another healthcare provider thought of herself as a resource for information, noting, "my job is just to be a source of information to the public." It is in the midst of the limited structural resources that the rural health-care providers articulate their resourcefulness:

> Well, I'm by myself, so I call Maria[i] and go over to the regional diabetes center, and then I'll call the satellite clarion Arnett family centers. This is what I've been thinking about doing. There are 6 small towns. What are these places doing with diabetes? I don't know? So I'm trying to find out what they are doing with diabetes. Would that be a good place for the bus to go by and help them? Do people go there and get education? I don't know. But I have to do all of that myself. There's certainly nobody in those counties doing anything. I'm just calling around to figure out what they are doing.

Once again, articulated in this narrative are the creative ways in which rural health-care providers enact their agency in seeking to offer health-care resources for underserved communities at the rural margins of health care. In this instance, the barriers of inaccess are negotiated through the building of networks, in order to identify and mobilize health resources for the underserved rural communities. Noting the tenets of the culture-centered approach that foreground the local articulations of agency amidst structural constraints, the rural participants narrate the stories of seeking out resources in order to provide preventive support for rural communities.

Discussion

In this chapter, we engaged in an ethnographic project to listen to the voices of the rural health-care providers that work within and through health structures in order to create access to health resources in rural settings. The voices of rural health workers make salient the geospatial contexts of material marginalization, played out in the form of the location of the rural sectors at the material margins of health care. Our ethnographic account draws attention to the structural locus of health-care problems in the context of disparities, situating the communicative construction of disparities amidst material structures, and noting the absence of basic health capacities in rural contexts. As noted by Dutta (2008) in his work with the culture-centered approach in the backdrop of marginalization in the global South, the structural markers of marginalization are played out in the lack of access to fundamental health resources. The spatial marginalization of rural health is reflected in a variety of contexts, including the lack of staff, the inaccessibility of hospitals, the inaccessibility of emergency care, the lack of preventive resources, and the absence of health-enhancing opportunities for rural populations. The delivery of health care in rural contexts is situated amidst these structures of inaccess, and therefore, rural providers work through these very structures in order to create points of access.

The narratives in this project point toward the necessity for structural transformations in addressing rural health disparities. The health-care providers and public-health professionals continue to point toward the necessity for addressing health disparities as structural, suggesting that the emphasis should be placed on noting the structures and on seeking transformations in health-care policies so that greater health resources can be available in rural contexts. The identification of the problem as structural is noted in the articulation of the absence of doctors in rural areas, as well as in the long distances one has to travel to secure access to basic health-care services. It is against this backdrop that the participants discuss policy reforms such as state-based incentivizing of doctors through subsidies given in medical school for demonstrating a commitment to working in rural areas. Similarly, in pointing out how the preventive-service units in rural contexts are understaffed and overworked, the rural health-care providers and public-health professionals draw attention to the need for structural transformations in rural health care, ensuring the greater flow of resources.

Here, the emphasis on policy reform points toward the identification of greater resources in order to facilitate the flow of health providers to underserved rural areas of the US, and also points toward the redistribution of resources at the structural level. Worth noting here are the parallels between the narratives of inaccess in rural contexts in the US with similar narratives of inaccess in other rural contexts identified in culture-centered projects of rural health (Dutta-Bergman, 2004a, 2004c; Dutta & Basu, 2007). The locally centered stories of health disparities in this project suggest entry points for transformations in structures, based on the contexts of inaccess in rural spaces.

It is within these structural constraints that the participants discuss the creative ways in which they enact their agency in order to ensure that health care is available and accessible in rural areas. On the one hand, they emphasize the importance of changing the structures through the creation of resources that are specifically directed at the rural sectors; on the other hand, the participants draw attention to the strategies they use in their day-to-day activities in order to work with these limited resources. Given the limitations of the number of staff members available in rural areas, rural health-care providers and public-health practitioners serve as advocates for the rural communities, often utilizing their networks with state-level health structures to draw attention to the need for resources. Along similar lines, rural providers discuss the networks of solidarity they create in order to deliver health care amidst the structural absence of resources. The ethnographic analysis conducted in this chapter foregrounds the importance of attending to structurally situated health narratives as entry points for seeking out structural transformations in addressing health disparities. Similar to the chapters on the culture-centered approach in this book, the articulation presented in this chapter foregrounds the importance of continually rethinking the definition of communication intervention in the backdrop of the structural locus of health-care disparities.

Note

1. Referenced individual names have been masked to ensure the anonymity of study participants.

Bibliography

Armstrong, P. S., & Schulman, M. D. (1990). Financial strain and depression among farm operators: The role of perceived economic hardship and personal control. *Rural Sociology*, *55*(4), 475–493.

Baker, D. W., Fitzgerald, D., & Moore, C. L. (1999). Quality of care for Medicare patients hospitalized with heart failure in rural Georgia. *Southern Medical Journal*, *92*(8), 782–789.

Baldwin, L. M., MacLehose, R. F., Hart, L. G., Beaver, S. K., Every, N., & Chan, L. (2004). Quality of care for acute myocardial infarction in rural and urban US hospitals. *Journal of Rural Health, 20*(2), 99–108.

Beckett, W. S., Chamberlain, D., Hallman, E., May, J., Hwang, S., Gomez, M., Eberly, S.,...Stark, A. (2000). Hearing conservation for farmers: Source apportionment of occupational and environmental factors contributing to hearing loss. *Journal of Occupational and Environmental Medicine, 42*(8), 806–813.

Blair, A., & Zahm, S. H. (1993). Patterns of pesticide use among farmers: Implications for epidemiologic research. *Epidemiology, 4*(1), 55–62.

Chin, M., Cook, S., Drum, M., Jin, L., Guillen, M., Humikowski, C., Koppert, J., ... Schaefer, C. (2004). Improving diabetes care in midwest community health centers with the health disparities collaborative. *Diabetes Care, 27*(1), 2–8.

Coryell, W., Endicott, J., & Keller, M. B. (1992). Rapidly cycling affective disorder: Demographics, diagnosis, family history, and course. *Archives of General Psychiatry, 49*(2), 126–131.

Dorsch, J. L. (2000). Information needs of rural health professionals: A review of the literature. *Bulletin of the Medical Library Association, 88*(4), 346–354.

Dutta, M. J. (2006). Theoretical approaches to entertainment education campaigns: A subaltern critique. *Health Communication, 20*(3), 221–231.

Dutta, M. J. (2007). Communicating about culture and health: Theorizing culture-centered and cultural sensitivity approaches. *Communication Theory, 17*(3), 304–328.

Dutta, M. J. (2008). *Communicating health: A culture-centered approach.* Malden, MA: Polity Press.

Dutta, M. J., & Basnyat, I. (2006). The Radio Communication Project in Nepal: A critical analysis. *Health Education and Behavior, 35*(6), 459–460.

Dutta, M. J., & Basu, A. (2007). Health among men in rural Bengal: Exploring meanings through a culture-centered approach. *Qualitative Health Research, 17*(1), 38–48.

Dutta-Bergman, M. J. (2004a). Poverty, structural barriers and health: A Santali narrative of health communication. *Qualitative Health Research, 14*(8), 1107–1122.

Dutta-Bergman, M. J. (2004b). Primary sources of health information: Comparison in the domain of health attitudes, health cognitions, and health behaviors. *Health Communication, 16*(3), 273–288.

Dutta-Bergman, M. J. (2004c). The unheard voices of Santalis: Communicating about health from the margins of India. *Communication Theory, 14*(3), 237–263.

Eberhardt, M. S., & Pamuk, E. R. (2004). The importance of place of residence: Examining health in rural and nonrural areas. *American Journal of Public Health, 94*(10), 1682–1686.

Hart, L. G., Salsberg, E., Phillips, D. M., & Lishner, D. M. (2002). Rural health care providers in the United States. *Journal of Rural Health, 18*(5), 211–231.

Hartley, D. (2004). Rural health disparities, population health, and rural culture. *American Journal of Public Health, 94*(10), 1675–1678.

Institute of Medicine. (2000). Committee to Assess Training Needs for Occupational Safety and Health Personnel in the United States. Retrieved from http: //www.

iom.edu/Reports/2000/Safe-Work-in-the-21st-Century-Education-and-Training-Needs- for-the-Next-Decades-Occupational-Safety-and-Health-Personnel.aspx.

Issermann, A. M. (2005). In the national interest: Defining rural and urban correctly in public policy. *International Regional Science Review, 28*(4), 465–499.

Lutfiyya, M. N., Lipsky, M., Wisdom-Behounek, J., & Inpanbutr-Martinkus, M. (2007). Is rural residency a risk factor for overweight and obesity for US children? *Obesity, 15*(9), 2348–2356.

Maio, R. F., Green, P. E., Becker, M. P., Burney, R. G., & Compton, C. (1992). Rural motor vehicle crash mortality: The role of crash severity and medical resources. *Accident Analysis and Prevention, 24*(6), 631–642.

Martin, E. D. (1975). The federal initiative in rural health. *Public Health Reports, 90*(4), 291–297.

Myers, J. R. (1990). National surveillance of occupational fatalities in agriculture. *American Journal of Industrial Medicine, 18*(2), 163–168.

National Rural Health Association (NRHA). (2010). *Rural health in America.* Retrieved from www.ruralhealthweb.org.

Phillips, C. D., & McLeroy, K. R. (2004). Health in rural America: Remembering the importance of place. *American Journal of Public Health, 94*(10), 1661–1663.

Probst, J. C., Moore, C. G., Glover, S. H., & Samuels, M. E. (2004). Person and place: The compounding effects of race/ethnicity and rurality on health. *American Journal of Public Health, 94*(10), 1695–1703.

Ricketts, T. C. (2000). The changing nature of rural health care. *Annual Review of Public Health, 21*(1), 639–657.

Rivara, F. P., Grossman, D. C., & Cummings, P. (1997). Injury prevention: First of two parts. *New England Journal of Medicine, 337*(8), 543–548.

Rural Healthy People 2010. (2010). *Office of Rural Health Policy.* Retrieved from http://www.srph.tamushsc.edu/rhp2010/.

Shen, P. (2008). Assessment of chronic health indicators in north central Indiana. Retrieved from http://www.tippecanoe.in.gov/egov/docs/1251123766457.htm.

Siminerio, S. M., Piatt, G., & Zgibor, J. C. (2005). Implementing the chronic care model for improvements in diabetes care and education in a rural primary care practice. *The Diabetes Educator, 31*(2), 225–234.

Thelin, A. (1990). Hip joint arthrosis: An occupational disorder among farmers. *American Journal of Industrial Medicine, 18*(3), 339–343.

Wang, C. C. (1999). Photovoice: A participatory action research strategy applied to women's health. *Journal of Women's Health, 8*(2), 185–192.

Wang, C. C., & Burris, M. (1994). Empowerment through photo novella: Portraits of participation. *Health Education Quarterly, 21*(2), 171–186.

Zwerling, C., Merchant, J. A., Nordstrom, D. L., Stromquist, A. S., Burmeister, L. F., Reynolds, S. J., & Kelly, K. M. (2001). Risk factors for injury in rural Iowa: Round one of the Keokuk County rural health study. *American Journal of Preventive Medicine, 20*(3), 230–233.

6

Promoting Women Leadership as a Strategy for Reducing Health and Digital Disparities in Latino/a Immigrant Communities

Tamar Ginossar

This chapter describes the grassroots efforts of a Latina women community organization to reduce health and digital disparities in their community. La Comunidad Habla (Spanish for "the community speaks") has been working in the Southeast Heights of Albuquerque, New Mexico. This Southwestern border state is the fifth largest state in the union. It has rich history and vast, dramatic scenery. With a population of only two million people, it is one of the most rural and sparsely populated states. One of the unique characteristics of New Mexico is that it is a minority-majority state, in which Hispanics are the largest ethnic group. Hispanics comprise 45% of the state population, three times the national average. Non-Hispanic Whites are the second largest ethnic group with 41%, and American Indians approach 10% of the state population (U.S. Census Bureau, 2010). New Mexico Hispanics include descendants of Spanish colonists and recent immigrants from Latin America, predominantly from neighboring Mexico (Oetzel, DeVargas, Ginossar, & Sanchez, 2007). The state faces higher poverty rates than the national average and lower average income (U.S. Census Bureau, 2010). Consequently, its residents cope with myriad inequalities, including health disparities.

La Comunidad Habla (LCH) is based in Southeast Albuquerque, which is comprised of La Mesa Village and Trumbull neighborhoods. Local activists recently succeeded in passing a resolution to name the area "the International District," in recognition of the diversity of its residents. The area is considered the most ethnically diverse in the state. It is a portal for newly arrived immigrants, as housing is more affordable than in other areas, it is centrally located, and it is served by public transportation. Residents include recent immigrants from Mexico, Vietnam, Cuba, Afghanistan, and Africa, as well as more established Hispanics, Anglos, and African Americans. The community copes with different family and community problems. Many families that live in the area are young, highly mobile, and economically and socially stressed. More than a quarter of the families live below the federally determined poverty level. Poverty is particularly high for certain families. Over 40% of families with children under five years old, and nearly 60% of female-headed households with children live in poverty (U.S. Census Bureau, 2000a). The area has consistently ranked as the highest or second highest high-crime neighborhood in the city (U.S. Census Bureau, 2000b).

One of the area's biggest challenges is the high proportion of rental properties. About 80% of residents are renters, according to Louis Kolker, executive director of the Greater Albuquerque Housing Partnership (Kamerick, 2009). Local community activists explain that this high rate of rental properties increases residents' mobility, which in turn decreases their ability to create social networks and social capital in the community, and therefore increases the challenges to social change efforts (Simmons, 2002). The grassroots efforts to create safe, affordable housing in the area received national media attention in 2008, when the ABC television show, "Extreme Makeover: Home Edition," aired an episode that centered on the efforts of Gerald Martinez, a local pastor, to create such housing in the community. In addition to building his home, the show advanced the pastor's work by demolishing a block of condemned houses in the area, in preparation for building new homes.

The success in passing the Albuquerque city council resolution to increase development and rename the area "the International District" was one of many in fighting a stigma of "The War Zone" and reducing the high crime rate in the area. Like other successes in creating social change, it involved consistent, dedicated advocacy efforts and collaboration of different community organizations and community members. The community has a long history of social change initiatives, including health coalitions, community leadership boards, housing and economic development projects that involve community members, as well as much collaboration between socially and educationally oriented agencies (Kaufman,

2006, 2010). These community-based efforts aim at addressing social disparities, from access to health care to safe housing and availability of jobs (Wallerstein, 2006). These attempts address the interrelated disparities that have detrimental effects on individual and community health (Kreps, 2005).

The Digital Divide in a Low-Income Latino/a Community

With the dawn of the new millennium emerged an increased awareness of the potential impact of a new social disparity, that of the *digital divide*. Defined as the gap between those who possess access to information technologies (IT), mainly the Internet, and those who do not (Chang et al., 2004; van Dijk & Hacker, 2003), further understanding of its exact nature and its individual and social implications are still emerging. The digital divide is not a simple dichotomy between those who have physical access to the Internet and those who do not. It is also determined by additional factors, including the type of access that individuals and communities have (Hargittai & Walejko, 2008; van Dijk & Hacker, 2003). For example, Internet users with broadband access differ from those who are limited to dial-up connection. Also, those who have access to the Internet at a place other than their home have different types of access and therefore different uses of the Internet than those who have Internet access at home. The digital divide is also manifested in users' literacy levels and their Internet skills. Individuals' literacy levels determine their ability to understand content and benefit from online information in diverse contexts, including in health. Similarly, individuals' technical skills, such as their ability to create content, also shape their Internet use as well as their experiences and the gratification they are able to derive from it (Hargittai & Walejko, 2008).

Although the Internet has diffused rapidly nationally and worldwide, scholars point out that the digital divide in the US persists, especially as it affects low-income Latinos/as (Fox & Livingston, 2007; Tornatzky, Macias, & Jones, 2002). The fastest-growing minority group in the nation, Latinos are also the most digitally disadvantaged ethnic group. They continue to trail Whites and other minority groups in access to IT, including Internet use, computer ownership, and e-commerce activities. Like other social disparities, this technological disparity is complex, and is manifested in multiple ways. Latinos are severely underrepresented in the IT workforce. This underrepresentation is related to the education and training of Latino/a parents and their perceptions of the importance of technology (Tornatzky et al., 2002). The digital divide is also evident in the lack of culturally diverse content on the World Wide Web. There is a paucity of websites that attempt to meet the needs of low-income, low-literacy communities (Lazarus & Mora, 2000), and of websites in Spanish and other languages to meet the needs of limited English proficiency communities. Moreover, the existing websites that provide information in Spanish typically provide content that was translated from English, as opposed to developing content that will actually meet the targeted communities' needs (Tornatzky et al., 2002).

LCH group members' work to increase access to technology and health information is delineated in this chapter as well as in previous publications (Ginossar, 2011a; Ginossar & Nelson, 2010a, 2010b). The digital divide has severe implications for individuals' health. The digitally marginalized are the same vulnerable populations that suffer from health disparities (Dutta, Bodie, & Basu, 2007; Kreps, 2005; Kreps et al., 2007). In view of the importance of health information on the Internet to individuals' decision making and well-being, the digital divide is recognized as a public-health problem (Kreps, 2005). In the Albuquerque Southeast Heights community that is the focus of this chapter, the digital divide is evident in the everyday lives of people. Although estimates of Internet access in the community are not available, anecdotal evidence points to a significant gap between this community's need for Internet access and the availability of access (Dailey, 2010; Ginossar, 2011a). LCH group members indicate that family members and friends contact them regularly with requests to use their Internet access for necessary actions to keep their families safe, such as paying their utility bills online to avoid service termination. However, even La Comunidad Habla members, who train others in their community in Internet and computer use, do not have consistent access to broadband Internet connection. A recent report that the Federal Communication Commission (FCC) invited on broadband access in low-income communities examined LCH's work and perspectives, and noted the paradox of their limited Internet access in view of their role in the community (Dailey, Bryne, Powell, Karaganis, & Chung, 2010). This gap is evident in the long lines of community members waiting for their turn to use a computer at the small and crowded local public library. Clearly, the number of available computers does not meet the demand for them (Dailey, 2010).

Lack of Internet access can have significant and negative effects on individuals' daily life, as well as on their future. Access to the Internet and Internet-related skills allows users to participate in different social groups. Whereas in the past, social relationships and work typically centered on face-to-face interactions, they are changing rapidly to hybrids of offline and online organizations (Mason & Hacker, 2003). In the case of immigrants, the digitally excluded are not able to maintain their relationships with family members and friends in their countries of origin, which is likely influencing their emotional well-being in negative ways. Those with Internet access are able to communicate with their family members who are far away.

The digital divide is likely to have significant marginalization effects on immigrants' adjustment to their new country, including their access to employment. Many positions are not within the reach of the digitally marginalized, as they are advertised exclusively online. Many employers are now requiring that job seekers submit employment profiles through an online application process (van Dijk, 2003). Furthermore, the digital divide might be jeopardizing immigrants' ability to secure their lawful presence in this country. Although it is known that low-income immigrants are at risk of being digitally marginalized, it is the experience of the author that in 2008–2009, the U.S. Citizenship and Immigration Services required online scheduling for appointments, with no telephone or face-to-face alternatives for such scheduling for applicants' initiated appointments.

Current research on bridging the digital divide elucidates that although providing physical access to the Internet to the underserved is a necessary step in bridging this divide, it is only part of a solution. Even when provided free Internet access at home, Mexican-Americans were less likely than other low-income participants to seek health information online (Wagner, Bundorf, Singer, & Baker, 2005). Therefore, culturally appropriate interventions are needed to meet the information needs of these communities. Community-initiated efforts are likely to bring such culturally appropriate focus to interventions by increasing community motivation and engagement (Bodie & Dutta, 2008). However, most community-based interventions did not involve community members in the design, and instead largely considered them as target audiences (Kreps, 2005; Kreps et al., 2007). In contrast to more established, federally funded initiatives, LCH grassroots efforts that are described in this chapter utilized concepts from community media (Mody, 1991). The community media approach centers on community members as content creators, and on their empowerment through participation, as opposed to the mass-media model of passive audiences who consume content produced by large commercial corporations or government-appointed experts (Ginossar, 2011a; Ginossar & Nelson, 2010a, 2010b).

La Comunidad Habla: First Days and Working Model

La Comunidad Habla (LCH) is a community media group that was created in 2002 by Sara Nelson, a community activist who works and lives in Southeast Albuquerque. At the time, Sara worked as a case manager and later as a community outreach specialist for Young Children's Health Center (YCHC), a University of New Mexico community-based pediatric clinic in the Southeast Heights. The clinic was established by the University of New Mexico Health Science Center in order to provide medical care for this underserved community. It provides comprehensive, bilingual physical and mental health care, as well as community and social services. To secure services to this community, eligibility criteria for services consist of proof of residency in one of two specific zip codes.

Sara recognized that women in the community were interested in receiving training in technology, but were unable to access such training. To meet this need, she offered free, Spanish-language computer classes in the evenings. The classes took place at East Central Ministries, a Christian community development center, and later in two additional community centers. The large room in East Central Ministries' center, which is a former residential house, was converted into a computer lab. The first computers were donated to Sara by Los Alamos National Laboratories. Sara drove more than 130 miles to Los Alamos, and returned to Albuquerque with the donated computers in her car.

East Central Ministries also provided a room in a building across the street from the "computer lab" with designated a child-care area. The availability of child care was essential in reaching low-income Latina women. Following the success of these classes,

Sara created “La Comunidad Habla” (LCH). As a communication graduate student and an activist with community grassroots experience in the US and Latin America, she was interested in applying concepts of community empowerment and community media to the use of computers and the Internet.

Sara recruited seven women who successfully completed the computer class to participate in LCH. At the time of recruitment, these women, who were recent immigrants, were marginalized in multiple ways. As recent immigrants, their social networks in their countries of origin were severed, and they suffered from social isolation. In addition, most of them did not graduate from high school and had limited English proficiency. Consequently, they worked in low-income jobs that did not provide them with the opportunity to rise above the poverty level.

The women heard about the group in different ways. For example, Sonia had to volunteer in the community to be eligible for government assistance, and approached Sara for information about her options, whereas Veronica first heard about the computer class from Sara in the clinic, and later was recruited to LCH. To participate in the group, women had to formally apply to the position. This process included writing and submitting a resume, and filling up paper work. Although Sara knew the women and could have adopted a less formal procedure, she wanted them to have the knowledge of how to apply for a professional position and the positive experience of being accepted to one. This was the first time most of them had applied for such a position, and the experience provided them with important job-related skills, including writing a cover letter and a resume, which are essential in attaining employment. In addition, this process secured the women’s commitment to the project.

LCH developed a unique grassroots model for advancing health and technology in their community. First, although neither Sara nor the other women were initially trained in technology, they relied almost exclusively on their own work, in content creation as well as technical expertise. Sara, who does not have official training or education in technology, learned Web development, and in turn trained LCH members in doing so. They received limited technical assistance from experts in Web development and maintenance. A second strategy that LCH utilized was “training on the job.” The group received different contracts, and the women learned the skills while they were working on the projects. Additionally, their approach centered on providing community members with training that integrated technology use, advocacy, and navigating the health-care system. They developed this model and fully implemented it in their work on the online "Salud Manual" (see www.mycommunitynm.org).

The Online Salud Manual

The first major LCH project was the creation and maintenance of a community health website, the online "Salud Manual." The manual originally centered on an online version of a print resource guide on health-care access in New Mexico's Bernalillo County. To serve the low-literacy community, in addition to the website being bilingual, it was also written at low-literacy level, included simple graphics, and presented an entertainment-education approach in the form of illustrated stories. It was the fruit of collaboration with Southwest Creations, another local community organization. LCH was hired to create an online version of a health-care resource directory. This community-health website provided information for local providers and consumers about medical services. It supplemented the print manual that Southwest Creations first put together. The manual was created to meet the community's and providers' need for information about local health-care resources available for uninsured and underinsured community members. In addition to this information, LCH created education-entertainment content aimed at advocacy tools for community members. They generated their ideas following discussion between members on barriers to care that they experienced firsthand and from community members. They did not receive formal training in creating such materials, and simply applied their expertise and tacit knowledge generated from working and living in the community, rather than following a more structured, theoretically informed process. Sara Nelson, who has a graduate degree in communication, was generally knowledgeable about the education-entertainment (EE) approach, which centers on using entertainment to deliver health-promotion messages (Sood, Menard, & Witte, 2004). In Latin America, EE efforts often utilized "telenovelas," which are Latin American radio and television soap operas (Ginossar, 2010); and print "fotonovelas," or illustrated stories (Parlato, Parlato, & Cain, 1980; Valle, Yamada, & Matiella, 2006). These illustrated stories engage readers with different literacy levels by creating emotional identification with the characters and the narrative (Slater & Rouner, 2002).

Following this tradition of Latin fotonovelas, LCH leaders created educational, illustrated stories. The idea to create illustrated stories was based on the need the community expressed to have materials in the form of stories, as well as with limited text and more graphics. The goal of these stories was to educate community members on overcoming barriers to health care. Based on their personal experiences and experiences of family members and friends, they identified three important barriers that prevented community members from seeking health care. The barriers included lack of transportation, mistreatment in the system by personnel at the clinics, and intimidation of immigrants seeking health care for their children by requesting unnecessary documents as conditions for care. As mentioned previously, the group selected the barriers based on their direct and indirect experiences as community leaders and community members. They met in group discussions, and drew upon their personal experiences with being blocked from receiving health care. Based on these personal experiences, they wrote three story lines about community members who learn to overcome barriers. Although this process of identifying the barriers was not research-based, the group's observations are consistent with studies that examined barriers to care in low-income Latino/a communities (Flores, Abreu, Olivar, & Kastner, 1998).

LCH created the fotonovelas using a participatory process. First, they shared stories and identified barriers preventing community members from accessing health care. These barriers included lack of transportation, being asked for unnecessary documents, and being disrespected by health-care workers. Following the identification of these needs, they wrote stories that depicted a Latina woman learning how to overcome these barriers. LCH hired a local illustrator who provided them with black-and-white illustrations of the stories. They used "Photoshop" software to add color, and posted the stories to the website. These fotonovelas were also printed and distributed in the community until the funding for these prints ended. They can be viewed on www.mycommunity nm.org.

Community-Based Training

From 2004 to 2006, LCH members reached over 1,000 community members, and provided them with training using technology and health "Talleres," (workshops) in the community. Most of these workshops were two hours long. In the first hour of the talleres, LCH members led a discussion on how to navigate the health-care system, about health disparities, and about how to affect change. They used this discussion to introduce the print version of the directory, and instructed participants on how to use it to locate resources. In the second part of the talleres, they provided participants with basic technology skill training (Microsoft Office, and Internet basics), as well as with demonstrations of how to utilize the online directory on the group's website, and the entertainment-education tools (the fotonovelas). As part of the discussion of the health-care system, the talleres related to issues of access to health care and empowerment, by having participants engage in role-playing the stories in the fotonovelas. These role plays provided a way to elicit participants' own stories and experiences in seeking health care, and facilitated discussions concerning ways to overcome barriers. Community members identified with the characters in the fotonovelas, and the stories allowed for a nonthreatening way for participants to discuss potentially controversial topics. The stories that were written in a humoristic manner led to discussion on topics that are typically controversial or even "taboo." For example, some community members did not want to consider the option that disrespect from clinic personnel was related to racism. The group discussions following the fotonovelas and their enacting by community members allowed them to explore this option with other community members.

This strategy of sharing experiences and community-based discussions about the system and the barriers that community members face is consistent with the culture-centered approach (Dutta, 2008), as well as with community participation and the community-based participatory research approach. The culture-centered approach highlights the importance of culturally congruent messages in a multicultural world (Dutta, 2007) as a way to undo historical injustice and reduce disparities (Dutta, 2007, 2008; Dutta-Bergman, 2004a, 2004b, 2005). Consistent with these values, the community-participation approach promotes empowerment of marginalized groups through popular education and group dialogue (Israel, 2001; Wallerstein, 2006). This strategy originated in Latin America by Paulo Freire, and encourages members of marginalized communities to engage in critical thinking, and motivates them to take action in order to gain more control over their lives (Kane, 2001). In the case of LCH, the talleres elicited vibrant conversations between participants who shared their experiences. Sharing their individual experiences led community members to realize that they are not alone, and that their struggles are shared by others in their community. In turn, this understanding can lead to realization of the existence of systemic barriers and a beginning of contemplating ways to address these barriers (Minkler, 2005).

Personal Transformation of LCH Members

LCH's success in reaching community members and in increasing access to technology and health was impressive. An important aspect of this group's success was the transformation of LCH group members from motivated but marginalized women to true community leaders. As previously described, upon joining the group these women were marginalized on many levels. In addition to their low income, they suffered from having limited social connections, and limited English proficiency. To participate in the group, they had to overcome great barriers. A major challenge was availability of child care. Most LCH participants had young children, and finding child care was a condition for their ability to work. During their work on the online Salud Manual, paid child care had been arranged for them and participants in workshops and training, but at times caretakers were not available. At other times, some children refused to stay at the child-care facility, and went to the next room where their mothers worked, demanding their attention. In addition, some of the funding agencies of other projects, like the University of New Mexico, had policies that did not allow for funding child care, leaving the women stranded. The challenges to LCH's work generated by the issue of child care cannot be overstated. The women listed inconsistent child care as the greatest challenge they had to face in their work at LCH.

Some LCH women had to face an additional barrier in the form of resistance from their male partners, who objected to their participation in the group. Consistent with "machismo" cultural views, some male partners felt that participation in LCH distracted the women from focusing on their families, or from getting a job in more traditional places of work, such as restaurants. Although women were paid for their work at LCH, their partners did not necessarily consider their work at LCH to be a "real" job. Having LCH as a women-only organization, with many women-only classes was one of LCH's strategies for overcoming community patriarchy and mitigating potential objections to women's participation from male partners. In addition to these structural and cultural barriers to participation in LCH, LCH members went through personal and familial challenges that increased the difficulties they encounter. These included struggling with children with behavioral health problems, life-threatening events of family members in the US and Mexico, and other stressful life events. A few women left the group due to these challenges, or due to the relocation of their families, and other women were recruited.

Whereas the challenges of participation were great, so were the rewards. Working as community leaders in technology and health changed how they viewed themselves, as well as their role in the community. Their work in LCH allowed them to gain valuable skills. These skills included a high level of computer and Internet knowledge, such as Web development and the use of different software applications. It also provided them with experience in professional work, including group communication, addressing different costumers' needs, planning an agenda, strategic planning, and meeting goals. Moreover, LCH leaders gained diverse experience in public speaking. They developed these skills by leading community and group discussions and presenting in different settings, from community-based workshops to presentations at the university. The women indicated that these experiences changed how they thought of themselves. One LCH leader said, "it changed me in every way. I am not the same person." Another woman described how she worked as a janitor in a fast-food place when she took the computer class. Following her participation in LCH, she informed her supervisors that she had computer skills, and was promoted to be a cashier. Moreover, participation in LCH also facilitated the creation of social connections between members. A few of them studied together at the local community center, and received their General Educational Development (GED).

In addition to increasing their personal empowerment and professional skills, participation in LCH transformed the women into community leaders. Wherever they go in the community, people approach them. They recognize them from workshops and presentations, and often consult with them on different issues. In addition, family, friends, and acquaintances turn to them for advice on diverse topics, from questions about resources and navigating the health-care system, to requests for help and advice on technology-related issues. Their knowledge on resources available to community members is invaluable, as they acquired this knowledge from years of working in the community with different agencies and not-for-profit organizations, as well as from living in the community and communicating with a large number of community members. However, despite these successes, many of LCH's leaders continue to work in low-income jobs that do not utilize their unique skills in technology and community leadership.

Challenges to LCH's Work

Along with the personal and community success of LCH came pressing challenges to its model of operation. As mentioned elsewhere, between 2004 and 2006, LCH had benefited from secured funding from the Con Alma Health Foundation (for more about this organization, see Con Alma Health Foundation, 2010). This consistent funding allowed for much of the work described in this chapter. In addition, it secured a part-time position for all six women who participated in the group for the duration of the project. They were thus able to plan their personal and work lives. Once this funding ended, LCH began research and community-based work on much smaller, short-term, and often ad hoc contracts. These new projects further developed their scope of work and expertise, but also created increased financial instability, which contributed to the sense of stress in their life. The group was contracted in different capacities to participate in diverse projects, including community-based participatory research projects. However, these experiences also included much disappointment. Some grant proposals for projects that the group truly believed held potential value for the community, were not funded, and other projects were terminated due to cuts in public monies. An additional factor was Sara leaving her position with Young Children's Health Clinic to embark on a career as a consultant.

These changes and financial instability seemed to threaten the existence of LCH as a community group. However, the group proved to be remarkably resilient. Despite long periods of time with no regular meetings, they were able to maintain their collegial relationships and come together to work on projects when they received funding. These new projects with different organizations contributed to the continuous enrichment of their skills, as well as to their independence. Sara today is not involved in all of the projects, and some group members took leadership roles on new projects, in accordance with their interests and availability. Therefore, the group was successful in the process of capacity building of these community leaders. For example, two LCH members are teaching free community computer classes similar to those taught by Sara. These classes are now part of community initiatives by Southwest Creations, who use a similar model to LCH's by integrating health and technology in their curriculum. Remarkably, LCH was successful in creating a new financial model and in attracting community collaborators to its website. The website, now available at www.mycommunitynm.org, includes more social resources, and is funded by the community organizations listed in its directory.

When I began writing this chapter, two LCH women worked on one University of New Mexico-led, community-based participatory research project. During the writing process, I was notified that I was awarded two pilot grants from the University of New Mexico Clinical and Translational Science Center (CTSC) and from La Tierra Sagrada Society, which allowed me to continue working on community-based participatory research with this community. This research is based on needs identified in the community to receive information about childhood development and behavioral health.

The Promises of LCH Work for Reducing Digital and Health Disparities

Researchers have noted that in our new network society, information technology is an important resource. Those with the most access have the power to influence changes in social structures, and to benefit from these changes, whereas those who do not have access are increasingly marginalized (Mason & Hacker, 2003). LCH's work described in this chapter provides a model that has the potential to reverse this trend and increase active participation of groups that are typically excluded from the digital revolution. This model focuses on increasing Latinos/as' access to Internet technology, not only as users, but also as content creators. As a local community group operating within a context of tremendous disparities and challenges, the evidence for their success and for their effects on reducing disparities are not as easily measureable as in institutionally driven campaigns; or as in federally funded, academically driven research projects. In fact, lack of academic collaborations in the first years of its projects, and limited awareness and funding for consistent evaluation have been previously noted as limitations of LCH's work, as is the case for most community-led organizations (Ginossar, 2010, 2011a). However, these constraints should not undermine these women's achievements. These achievements are evident in a few ways. First, LCH members reached more community members than most large, federally funded outreach programs in this underserved community typically reach. Second, the group proved to be persevering, and is sustainable despite numerous changes and economic challenges. Moreover, with the increased diffusion of the Internet in the community, significantly more awareness is provided regarding the importance of access to information and the disparities associated with its absence. Finally, and perhaps most notably, LCH's success is evident in the personal transformation of its members, as described in this chapter. These stories of the individual growth of women that were previously isolated and lacked education and computing skills, to become community leaders who teach others how to use computers and the Internet demonstrate the potential of applying community media participation concepts to digital-inclusion initiatives in these communities. However, the current lack of consistent, stable funding for such social-justice efforts that will include securing physical access, technological support, and the ability to create content, limits the diffusion of LCH's unique and inspiring, grassroots efforts to other communities, and jeopardizes its own existence as a viable community group.

Acknowledgments

I would like to thank La Comunidad Habla members Sonia Medina, Azucena Molinar, Sara Nelson, and Veronica Salazar (in alphabetical order) for their help and collaboration.

Bibliography

Bodie, G. D., & Dutta, M. J. (2008). Understanding health literacy for strategic health marketing: eHealth literacy, health disparities, and the digital divide. *Health Marketing Quarterly*, *25*(1), 175–203.

Chang, B. L., Bakken, S., Brown, S. S., Houston, T. K., Kreps, G. L., Kukafka, R., Safran, C.,...Stavri, P. Z. (2004). Bridging the digital divide: Reaching vulnerable populations. *Journal of the American Medical Informatics Association, 11*(6), 448–457.

Con Alma Health Foundation. (2010). Con Alma Health Foundation. Retrieved from http://conalma.org/Dailey, D., Bryne, A., Powell, A., Karaganis, J., & Chung, J. (2010). *Broadband adoption in low-income communities.* New York, NY: Social Science Research Council. Retrieved February 11, 2013 from http://webarchive.ssrc.org/pdfs/Broadband_Adoption_v1.1.pdf.

Dutta, M. J. (2007). Communicating about culture and health: Theorizing culture-centered and cultural sensitivity approaches. *Communication Theory, 17*(3), 304–328.

Dutta, M. J. (2008). *Communicating health: A culture-centered approach.* Malden, MA: Polity Press.

Dutta, M. J., Bodie, G. D., & Basu, A. (2007). Health disparity and the racial divide among the nation's youth: Internet as a site for change? In A. Everett (Ed.), *Learning race and ethnicity: Youth and digital media* (pp. 175–198). The John D. and Catherine T. MacArthur Foundation Series on Digital Media and Learning. Cambridge, MA: The MIT Press.

Dutta-Bergman, M. J. (2004a). Poverty, structural barriers and health: A Santali narrative of health communication. *Qualitative Health Research, 14*(8), 1107–1122.

Dutta-Bergman, M. J. (2004b). The unheard voices of Santalis: Communicating about health from the margins of India. *Communication Theory, 14*(3), 237–263.

Dutta-Bergman, M. (2005). Theory and practice in health communication campaigns: A critical interrogation. *Health Communication, 18*(2), 103–122.

Flores, G., Abreu, M., Olivar, M. A., & Kastner, B. (1998). Access barriers to health care for Latino children. *Archives of Pediatrics and Adolescent Medicine, 152*(11), 1119–1125.

Fox, S., & Livingston, G. (2007). *Latinos online.* Washington, DC: PEW Hispanic Center and PEW Internet Project.

Ginossar, T. (2011a). Bridging the health and digital divide in a low income Hispanic community: Using community-based participatory research to advance communities' well-being. In M. Brann (Ed.), *Contemporary case studies in health communication: Theoretical and applied approaches.* Dubuque, IA: Kendall-Hunt.

Ginossar, T. (2011b). Media globalization and 'the secondary flow': Telenovelas in Israel. In D. I. Arredondo Ríos & M. Castañeda (Eds.), *Soap operas and telenovelas in the digital age: Global industries and new audiences.* New York, NY: Peter Lang.

Ginossar, T., & Nelson, S. A. (2010a). La Comunidad Habla: Using Internet community-based information interventions to increase empowerment and access to health care of low-income Latino/a immigrants. *Communication Education, 59*(3), 328–343.

Ginossar, T., & Nelson, S. A. (2010b). Reducing the health and digital divides: A model for using community-based participatory research approach to E-health interventions in low-income Hispanic communities. *Journal of Computer Mediated Communication, 15*(4), 530–551.

Hargittai, E., & Walejko, G. (2008). The participation divide: Content creation and sharing in the digital age. *Information, Communication and Society, 11*(2), 239–256.

Israel, B.A., Schulz, A.J., Parker, E.A., Becker, A.B., Allen, A.J., and Guzman, J.R. (2003). Critical issues in developing and following community-based participatory research principles. In M. Minkler & N. Wallerstein (Eds.), *Community-Based Participatory Research for Health* (pp. 56-73). San Francisco, CA: Jossey-Bass.

Kamerick, M. (2009, March 29). Neighborhood strives to replace 'war zone' stigma with international district. *New Mexico Business Weekly*. Retrieved from http://albuquerque.bizjournals.com/albuquerque/stories/2009/03/30/focus1.html.

Kane, L. (2001). *Popular education and social change in Latin America.* London, UK: Latin America Bureau. Kaufman, A., Derksen, D., Alfero, C., DeFelice, R., Sava, S., Tomedi, A., & Powell, W. (2006). The Health Commons and care of New Mexico's uninsured. *The Annals of Family Medicine, 4*(suppl 1), S22-S27.

Kaufman, A., Powell, W., Alfero, C., Pacheco, M., Silverblatt, H., Anastasoff, J., & Scott, A. (2010). Health extension in New Mexico: an academic health center and the social determinants of disease. *The Annals of Family Medicine, 8*(1), 73-81.

Kreps, G. L. (2005). Disseminating relevant health information to underserved audiences: Implications of the Digital Divide Pilot Projects. *Journal of Medical Library Assocation, 93*(4), S68–S73.

Kreps, G. L., Gustafson, D., Salovey, P., Perocchia, R. S., Wilbright, W., Bright, M. A., & Muha, C. (2007). The NCI Digital Divide Pilot Projects: Implications for cancer education. *Journal of Cancer Education, 22*(1), S56–S60.

Lazarus, W., & Mora, F. (2000). *Online content for low-income and underserved Americans: The digital divide's new frontier.* Santa Monica, CA: Children's Partnership.

Mason, S. M., & Hacker, K. L. (2003). Applying communication theory to digital divide research. *IT&SOCIETY, 1*(5), 40–55.

Minkler, M. (2005). Community-based research partnerships: Challenges and opportunities. *Journal of Urban Health, 82*(2), ii3–ii12.

Mody, B. (1991). *Designing messages for development communication: An audience participation-based approach.* New Delhi, India: Sage.

Oetzel, J., DeVargas, F., Ginossar, T., & Sanchez, C. (2007). Hispanic women's preferences for breast health information: Subjective cultural influences on source, message, and channel. *Health Communication, 21*(3), 223–233.

Parlato, R., Parlato, M. B., & Cain., B. J. (1980). *Fotonovelas and comic books: The use of popular graphic media in development.* Washington, DC: USAID.

Simmons, P. (2002). *Community partnership for health director.* Albuquerque, NM.

Slater, M. D., & Rouner, D. (2002). Entertainment-education and elaboration likelihood: Understanding the processing of narrative persuasion. *Communication Theory, 12*(2), 173–191.

Sood, S., Menard, T., & Witte, K. (2004). The theory behind entertainment-education. In A. Singhal, M. J. Cody, E. M. Rogers & M. Sabido (Eds.), *Entertainment-education and social change: History, research, and practice* (pp. 117–149). Mahwah, NJ: Erlbaum.

Tornatzky, L. G., Macias, E. E., & Jones, S. (2002). *Latinos and information technology:The promise and the challenge.* Claremont, CA: The Tomás Rivera Policy Institute.

U.S. Census Bureau. (2000). American factfinder. Retrieved September 10, 2004 from *http://factfinder.census.gov.*

U.S. Census Bureau. (2010). State & County QuickFacts. Retrieved from http://quickfacts.census.gov/qfd/states/35000.html.

Valle, R., Yamada, A., & Matiella, A. C. (2006). Fotonovelas: A health literacy tool for educating Latino older adults about dementia. *Clinical Gerontologist, 30*(1), 71–88.

van Dijk, J., (2003). A framework for digital divide research. *Electronic Journal of Communication/Revue de Communication Electronique, 12(1).* Retrieved September 30, 2012, from http://www.cios.org/getfile/vandijk_.

van Dijk, J., & Hacker, K. (2003). The digital divide as a complex and dynamic phenomenon. *The Information Society, 19*(4), 315–326.

Wallerstein, N. (2006). What is the evidence on effectiveness of empowerment to improve health? Copenhagen: World Health Organization (WHO), Regional Office for Europe, Health Evidence Network. Retrieved June 6, 2008 from http://www.popline.org/docs/1715/306857.html.

Wagner, T. H., Bundorf, M. K., Singer, S. J., & Baker, L. C. (2005). Free Internet access, the digital divide, and health information. *Medical Care, 43*(4), 415–420.

7

Preventing HIV/AIDS Through a Culture-Centered Health Campaign: The Sonagachi HIV/AIDS Intervention Program

Patrick J. Dillon and Ambar Basu

Pushed to the margins of society, poverty and social stigma characterize the lives of the nearly 2.9 million women who earn their living as commercial sex workers (CSWs) in India (Dandona et al., 2006). Similar to CSWs across the world, CSWs who reside in communities like Sonagachi in Kolkata (West Bengal), India are poor and live in often-unhealthy conditions. Denied access to civil society resources such as education, health, the police, and the law, CSWs experience harassment and exploitation at the hands of local law enforcement, brothel managers (madams), moneylenders, and the clients they serve. Their marginalized status is accentuated by the high incidence of venereal diseases and HIV/AIDS infections among them. The National AIDS Control Organization (NACO), India's premier government-sponsored AIDS management agency, estimates that 5.38% (i.e., 155,064) of female CSWs are living with HIV/AIDS (NACO, 2008). In some states in India, such as West Bengal, reports indicate that HIV infections among sex workers is rising, despite the fact that most cases go unreported (Nagelkerke et al., 2002).

Because HIV/AIDS poses a serious threat to the health and welfare of India and CSWs' occupation and social position make them vulnerable to HIV infection, health campaigns aimed at CSWs have been described as the "cornerstone" of efforts to curb the HIV/AIDS pandemic in India (Nagelkerke et al., 2002). The spread of HIV infection in India is generally tied to unprotected sexual activity, and the use of condoms in sex work encounters has traditionally been infrequent (Jha et al., 2001). To curb the potential spread of HIV through unprotected sex between CSWs and their clients, HIV health campaigns have consistently "targeted" this population. The majority of these campaigns have adopted a persuasive, message-based approach to combating HIV infection, usually by promoting greater condom compliance among CSWs through spreading awareness about the benefits of condom use, and increasingly, their availability (Basu & Dutta, 2008, 2009; Jana, Basu, Rotheram-Borus, & Newman, 2004; Nagelkerke et al., 2002). Such campaigns have typically been unsuccessful in increasing condom use or reducing rates of HIV infection among CSWs. These campaigns have also been criticized for framing CSWs' vulnerability to HIV as the result of lack of information or irrational individual choices, ignoring the cultural and structural contexts, such as poverty and stigma, which limit the ability of sex workers to negotiate the use of condoms with their clients (Basu & Dutta, 2008; Jana et al., 2004; Karim, Karim, Soldan, & Zondi, 1995). Furthermore, such top-down campaigns fail to extend opportunities for participation of CSWs in determining the agenda, structure, and communicative choices of the campaign (Airhihenbuwa, 1995; Basu & Dutta, 2008, 2009; Chattopadhyay & McKaig, 2004; Dutta, 2008).

Jana et al. (2004) and Basu & Dutta (2008) suggested a reconfiguration of CSW-oriented HIV/AIDS campaigns—from being top-down to being participatory—such that community members are able to articulate needs, mobilize resources, and take charge of the campaign. This approach highlights interactions between culture, structure, and agency within participants' living contexts, to document the health issues and potential solutions that are most relevant to community members (Dutta, 2008). They also advocate a shift in the focus of the campaigns from targeting risky individual behavior (such as engaging in sex work) in isolation, in favor of explicitly recognizing and addressing the cultural and structural factors that make CSWs vulnerable to HIV infection.

Stepping away from the traditional, top-down, message-transmission model of health campaigns requires new and innovative frameworks for the design and implementation of HIV/AIDS campaigns, particularly in marginalized spaces like CSW communities. Rooted in subaltern and postcolonial studies (Beverly, 2004; Guha & Spivak, 1988), the culture-centered approach to health communication provides theoretical, methodological, and pragmatic guidelines regarding the practice of participatory health communication in marginalized settings (Dutta, 2008). The culture-centered approach argues that health campaigns should resonate with a sense of a community's culture, while accounting for the structural realities in which health meanings and practices are embedded; in doing so, it foregrounds the agency of local communities as entry points for change. This approach highlights the agency of cultural participants, and emphasizes the importance of community participation in identifying health needs and promoting local solutions to health problems. Thus, the culture-centered approach offers an alternative to traditional models of health campaigns by creating entry points for listening to the histories and voices of those who have been traditionally marginalized, and by creating programs that meet the local needs of their communities.

Consistent with the tenets of the culture-centered approach to health communication, the Sonagachi HIV/AIDS Intervention Program (SHIP), a sex-worker-mediated HIV/AIDS prevention campaign in the Sonagachi red-light area of Kolkata, has been successful in fostering participation among local community members, and addressing structural inequalities as a means of reducing rates of HIV/AIDS infection (Basu & Dutta, 2008, 2009; Jana et al., 2004; Swendeman, Basu, Das, Jana, & Rotheram-Borus, 2009). This chapter will focus on the SHIP, and will examine how the culture-centered approach pans out in the context of the campaign. The first section of the chapter will provide a description of the principles of the culture-centered approach. The second section will present details about the SHIP and position it as an exemplary health campaign aimed at addressing health disparities. The next part of the chapter will analyze how procedures and strategies laid out in the culture-centered approach are in tune with the SHIP's successful planning and implementation. The chapter concludes with a discussion on how the culture-centered approach provides a viable framework for designing and implementing health campaigns that work to reduce health disparities in marginalized spaces.

The Culture-Centered Approach to Health Communication

Privileging the narratives that emerge through dialogue with members of marginalized communities, the culture-centered approach highlights the interaction between culture, structure, and agency in how health communication is theorized at the margins of society (Dutta, 2008). While structures constrain and limit the possibilities of health among underserved people, agency is enacted through communicative actions and processes that challenge, navigate, and attempt to change these structures. This line of thinking foregrounds the importance of understanding articulations of health by engaging local voices. The culture-centered approach offers opportunities to work towards eliminating health disparities and enacting social change by challenging dominant notions of social reality and cultural and behavioral norms. This section provides an overview of the major tenets of the culture-centered approach.

Culture

Culture can be defined as the communicative process of production and reconfiguration of webs of shared meanings, beliefs, practices, and rituals (Geertz, 1973) that influence attitudes, perceptions, and communication behaviors in a society (Airhihenbuwa, 1995). Culture involves a "continuous component that flows through the history of the culture, and a dynamic component that continually shifts as the culture responds to statewide, national, and global shifts in politics, economics, and communication flow" (Dutta, 2007, p. 311). The interaction between these continuous and dynamic elements of culture provides context for cultural meanings that are in flux, including meanings related to notions of health. The culture-centered approach stresses the need to theorize health communication in ways that are consistent with the cultural frameworks of local communities.

The culture-centered approach challenges approaches to health communication that reduce the role of culture to an emphasis on lifestyle choices, and treat it as a static variable to be measured by outside experts in the formulation of health campaigns. Such a disposition tends to frame cultures in marginalized spaces as static and rudimentary; cultures are then equated with beliefs, and cultural codes are treated as barriers to health behavior change. This process ignores the capacity of marginalized people, and positions them to be "educated" on ways to implement health practices that are deemed rational by the dominant culture.

The culture-centered approach also moves beyond culturally sensitive approaches to health campaigns that attempt to integrate cultural concepts as a means to improve the delivery of health-promotion efforts. Though culturally sensitive campaigns attempt to integrate notions of culture, cultural concepts and the agenda of the health campaign are still formulated by experts external to the local culture. A culturally sensitive health campaign may be able to accommodate cultural values, norms, and practices into its planning and implementation, but it maintains the view of the communities at the heart of the health campaign as its "target." Meaning-making takes place outside the target community, and the target community is only pulled in to contribute to the successful implementation of the experts' agenda.

Participation

Many contemporary health campaigns are described as "participatory"; however, these health campaigns are often incongruent with bottom-up forms of participation in marginalized communities (Dutta, 2008). In these applications, outside experts engage with community members of the target populations in order to identify potential barriers to the communities' acceptance of solutions to health problems identified from the outside. This process allows the outside experts to identify the most appropriate ways to diffuse their health messages in order to alter the individual beliefs, attitudes, and behaviors of the target population. Participation is characterized by using feedback from community members to design more effective health interventions.

In these cases, the process of structuring and delivering a health campaign remains located in Eurocentric ideals, as participatory health campaigns are delivered to marginalized spaces in the guise of technical assistance or technical cooperation programs (Airhihenbuwa, 2007a, 2007b). Community leaders, chosen by health experts from outside the community, are taught the skills to educate and diffuse the health knowledge created by the experts into the depths of the community. The portrayed, underlying aim is to set up a cooperative collaboration with the community through its so-called representative leaders to effect health-related changes among its members. What goes unsaid in the discourse on such health-promotion efforts is that the community leaders are not representatives of the community, its needs, goals, culture, and/or resources. Instead, they represent the goals and ideals of the experts who choose them to further expert-driven ideologies (Dutta, 2008).

The culture-centered approach seeks to transform this hegemony and status quo in health communication theory and praxis, particularly in the context of marginalized spaces. This approach questions the extant dominant formations of knowledge—the ideologies that frame these formations—intervenes politically in this process of knowledge production, and attempts to subvert it. By rupturing rationalist assumptions and biomedical solutions that undergird constructions of health and illness discourse in the dominant paradigm, the culture-centered approach offers an alternate paradigm to health communication. It locates health within constantly shifting frames of cultural production. Culture is theorized as the communicative process of meaning-making from within participant spaces, and the discursive canvas of health communication opens up to multiple methodologies of knowing what health is, what illness is, and how to negotiate illness and disease (Dutta, 2008). Culture is no longer viewed as a barrier to healthy habits, nor an enabler or even a resource to promote health practices relevant to the Western paradigm; culture as a participatorily constructed process is the source of knowledge and discursive practices on health. Health problems are identified by participants in the cultural space; participants engage with available structural resources and determine plans of action (Dutta-Bergman, 2004a, 2004b). The culture-centered approach centralizes the narratives that emerge from these participatory acts in marginalized spaces, and introduces them into discursive frameworks that have systematically chosen to erase them through one-way models built upon the assumed expertise of the health-communication scholar. These narratives act as resistive scripts, as they not only articulate subaltern agency, but also seek to subvert the dominant narratives on what constitutes health and illness (Basu & Dutta, 2008).

Participation of community members in the cultural production process is thus critical to the culture-centered approach. The participatory cultural process is tempered by, and in turn influences, structural frameworks as participants enact agency by engaging with the structures and thus shape cultural values, traditions, and norms. This notion of participation is opposed to that of participation through answering questions posed by so-called experts during formative research for a health campaign (Dutta, 2007, 2008). Here the modalities of participation, the strategies of communication, and the agenda of the participatory discourse are set by members of the culture, and not by the external expert. It also suggests how communication about culture in the realm of health materializes at the interstices of available structures and the agency of the participants.

Structure

Structures are forms of social organizing that provide or limit community members' access to resources that influence their health and well-being (Dutta, 2008). Structures that influence health include those at the micro-level, such as community medical services, community modes of transportation, channels of communication, and health-enhancing resources such as food, clean spaces, and spaces for exercising; meso-level resources, such as points of policy implementation, avenues of civil society organizations, and media platforms; and macro-level resources, such as national and international political actors and points of policy formulation, and national and global health organizations (Basu & Dutta, 2008). Structures at these multiple levels work in tandem to influence each other and the health outcomes of participants who create and reify such structures as well as engage with them to create discursive webs that frame their lives.

It is important to note that structures are embedded in material reality as well as in communicative practices, with one influencing the other. The emphasis in the culture-centered approach is to gain an understanding of the structures that potentially limit the possibilities of health for members of a community. The culture-centered approach to health communication seeks to foreground these structural factors as realities in the lives of marginalized people—realities that have a profound influence on their health—as a means of transforming local, structural realities in ways that will promote the health of community members.

Agency

Agency refers to the capacity of human beings to engage with structures that encompass their lives. This engagement allows individuals to make sense of the structures, to script texts of engagement with the structures, to live within these structures, and, at the same time, to create discursive openings to transform these structures (Dutta, 2008). The recognition and challenging of the structures that constrain populations' ability to pursue health is a primary step toward transforming structures in ways that will initiate meaningful change in communities. The culture-centered approach locates agency at its interaction with culture and structure, and notes that the human capability to devise communication strategies by making sense of the structures defining one's life helps to connect to the roots of a culture. It documents pathways that cultural members forge and follow to conduct their lives.

This agency then creates pathways for community members to use the resources at their disposal to change and transform structures that impede their ability to lead healthy lives. Thus, agency is enacted in its interaction with the structures at different levels, embodying a variety of communicative actions and processes that challenge, navigate, and attempt to change the structures that constrain the health of marginalized communities. Through this process, members of marginalized communities participate in social-change endeavors that transform structures in ways that increase their access to health-related resources, and hence help to reduce health disparities.

Context and Resistance

Context refers to the local surroundings within which cultural members enact agency. As a part of the day-to-day lives of individuals, contexts inform how members of communities construct health and illness within the larger, dynamic structures that surround them (Basu, 2010). Contexts embed individuals within their cultural texts, and offer opportunities to interact with these texts to enact change through communicative practices. Cultural context is located at the center of the culture-centered approach, emphasizing the meanings that emerge from within participant spaces. These localized vocalizations are at the core of social-change projects that acknowledge and derive expertise from local, marginalized knowledge contexts, ultimately challenging the discursive and material inequities emblematic of the dominant discourse on health communication that pays scant attention to them.

The culture-centered approach also articulates the ways in which marginalized groups resist structural constraints and challenge existing contextual frames. The emphasis is on resistance as a communicative act that challenges the dominant structures and seeks to transform them (Basu & Dutta, 2008). The centers of power that participate in the marginalization of community members are interrogated, and opportunities are sought for their transformation.

Resistance is constructed as a continually transformative, communicative process that emerges and evolves in relationship to structure (Dutta, 2008). Resistance, in this sense, is both an everyday practice enacted through day-to-day actions that challenge the dominant structures, and a practice of organizing and mobilizing for more overt forms of communication directed toward challenging structures. These acts of resistance serve the purpose of empowering marginalized communities, as individuals and groups express themselves and articulate their needs. In contrast to the top-down forms of empowerment that characterize mainstream health campaigns, the culture-centered approach views empowerment as a process where the community members are in touch with communication avenues that enable them to articulate their needs. Empowerment, then, helps to address the felt needs of marginalized communities, most of which are contextualized within spaces of health-related inaccess, inequities, and outcomes.

Voice, Dialogue, and Identity

The culture-centered approach suggests that knowledge and communicative practices on health are created in mutually reinforcing, dialogic spaces created by cultural participants and the researcher (Dutta, 2008). This represents a fundamental shift in the role of the researcher from an interventionist who plans and executes health campaigns to that of a listener and a coparticipant who engages in dialogue with community members.

The culture-centered perspective is founded on notions of solidarity. Instead of imposing health interventions on marginalized communities, the focus is to understand how health-communication processes are enacted in these spaces, to listen to the narratives that come out of spaces as members engage with their limited resources, to make meanings of health and lives, and to introduce the cocreated scripts into those mainstream discursive structures where they have been ignored. In other words, the coconstructive process of meaning-creation that the culture-centered approach to health communication emphasizes creates spaces for marginalized people to frame their own discourses on health rather than being told how to frame such discourses (Basu & Dutta, 2008). This creates openings for local voices to be heard—local, marginalized voices that often call for access to health structures that have been hard to come by, ultimately foregrounding a discourse that questions the dominant discourse that creates and sustains health disparities in the first place.

Thus, the culture-centered approach moves away from the dominant paradigm in health communication and offers a viable framework for addressing health disparities in marginalized spaces. In the following section, we examine how the SHIP embodies the tenets of the culture-centered approach while working to eliminate HIV/AIDS in the Sonagachi red-light district.

The Sonagachi HIV/AIDS Intervention Program

The Sonagachi CSW community in north Kolkata is more than 100 years old. Earlier called Sonagaji, the place derives its name from a Muslim religious preacher. It is the largest red-light area in Kolkata, with an estimated 9,000 CSWers (6,000 are full-timers who live in the area, and 3,000 are casuals who commute to the area to work). Similar to other CSW communities in India, Sonagachi had long been a testing ground for HIV/AIDS campaigns prior to the establishment of the SHIP. Most of these campaigns promoted a condom-use-based medical model for HIV/AIDS prevention. They operated under the assumption that spreading awareness about the high risk of HIV/AIDS infection in the community, the life-threatening dangers associated with such an infection, and how using condoms can help prevent such infections will lead to condom-use compliance among sex workers (Jana et al., 2004). These campaigns were largely unsuccessful.

After an epidemiological survey identified the Sonagachi community as one at particularly high risk of HIV and sexually transmitted infection (STI) transmission, the All-India Institute of Hygiene and Public Health initiated the SHIP in 1992 under the direction of Dr. Smarajit Jana. Since its inception, the SHIP has been lauded as an exemplar health campaign for its success in combating HIV/AIDS in Sonagachi. Despite being Kolkata's largest red-light area, the rate of HIV infection among CSWs in Sonagachi is around 11% (UNAIDS, 2002). This is in contrast to HIV seroprevalence rates of 50–90% among CSWs reported in cities like Mumbai, New Delhi, and Chennai (NACO, 2008). In addition to reduced rates of HIV infection, evidence suggests that condom use among CSWs in Sonagachi has also been rising, from 3% in 1992 to 90% in 1999. This success is largely attributed to the SHIP, which advocates a participatory, community-based approach based on solidarity among sex workers.

The success of the SHIP in increasing condom use and reducing HIV infection, as well as the sustainability of the project (it has operated for nearly 20 years), have brought much acclaim and financial backing. The SHIP has received funding from many prominent external agencies, including the Ford Foundation, the

Bill and Melinda Gates Foundation, and the World Health Organization (WHO). The WHO has even called the SHIP a “model” for STD/HIV intervention design, and HIV/AIDS campaigns based on the SHIP are being established in various areas of India and the rest of the world, including the Bill and Melinda Gates Foundation-funded Project Avahan HIV prevention campaign targeting high-risk populations in India (Swendemen et al., 2009).

The SHIP as an Exemplar Culture-Centered Health Campaign

Though the origins of the SHIP have been described as “largely unplanned” and “atheoretical” (Jana et al., 2004), in this section we describe how the success of the SHIP can be found to be aligned with the tenets of the culture-centered approach (Dutta, 2008). The work of the SHIP embodied the core criteria of the culture-centered approach in three primary ways: (a) it involved the local community in the identification of a health problem, and in the development of relevant health solutions to address the problem through active participation in dialogue; (b) it located the health communication effort in the local context, articulated through the voices of community members participating in the project; and (c) it addressed the structures surrounding the health issues through the mobilization of community members.

Promoting Dialogue and Community Participation

The culture-centered approach highlights the importance of the participation of community members in the articulation of health problems as a step toward achieving meaningful change. Thus, culture-centered health campaigns should involve participatory processes during the configuration of campaign agendas and processes. In contrast to previous HIV/AIDS campaigns in CSW communities, Jana et al. (2004) noted that the underlying principle of the SHIP was the belief that to address the health of CSWs in Sonagachi, the CSWs would need to take the lead role in the campaign. Though many local political leaders and experts expressed doubts that the CSWs were capable of fulfilling this leadership role, the planners of the SHIP remained committed to this campaign framework. This belief indicates that campaign planners acknowledged the ability of CSWs to resist and act within the context of their community, and to shape the steps of the health-promotion efforts.

Though it began with a conventional health-campaign agenda (focused on educating CSWs about health issues, promoting condom use, and free health-care services), the SHIP enacted this agenda with a focus on community participation. The SHIP began by hiring CSWs from the Sonagachi community to work as peer educators. In addition to engaging in health-promotion activities, including promoting condom use and utilization of the free health clinics, the peer educators were asked to serve as liaisons between members of the community and the SHIP. This included engaging in dialogue with other CSWs to discuss community problems. Members of the community were initially skeptical of the peer educators, but soon began to accept them as a means of making the leadership of the SHIP aware of community problems and ways of addressing them.

Moving forward, the SHIP attempted to foster other opportunities for community members to engage in dialogue with each other and the leaders of the SHIP. This included opportunities for community members to come together, brainstorm about community issues, and come up with solutions. This participation provided members of the community with a sense of control over the communicative processes that shaped decision-making and problem solving of the SHIP (Basu & Dutta, 2009). For instance, community members came together to make it known that one important step the SHIP should take was to figure out a way to provide financial security to sex workers. This would enable sex workers to worry less about providing for their families and have more power to negotiate condom use with their clients.

In addition to a sense of local control over the campaign's agenda, participation in the SHIP also provided the CSWs with a sense of solidarity with each other and the leaders of the campaign. As Niyoti (a member of the Sonagachi community) told Basu and Dutta (2009):

> "The problems are ours and we can solve them. You are not coming to tell us what to do. So here, we can sit together and figure out what works and what doesn't. And that makes me have faith in this [the SHIP]." (p. 96)

Participation in the SHIP forged collaborative ties among members of the community, as they were able to identify and address their own problems.

The SHIP has also fostered participation with the Sonagachi CSW community by gradually training members of the community to take on leadership roles in the organization. Social workers and public-health professionals initially filled management roles in the SHIP. In recent years, leadership roles, including the project director, have been filled by CSWs from the community, creating an avenue for localized vocalizations on health and living needs to be highlighted in the campaign. The structure of the SHIP also includes area-based committees of sex workers, who come together on a weekly basis, and who elect a Central Committee who have responsibility and authority to manage the work of the locally based committees. The SHIP encourages further community participation by offering leadership training and limiting the amount of time individuals can serve on a committee to no more than 2 years.

As evidenced by the examples above, community participation in the context of the SHIP extends beyond cultural participants responding to surveys or formative research that seeks their answers to questions set by experts from outside their cultural spaces. In the SHIP, the agenda that guides the community participatory process emanates organically from within the community, as opposed to being dictated by external agents. Through fostering culture-centered participation in Sonagachi, the SHIP was able to understand and address the needs of the community.

Locating Meanings of Health in the Local Context

The SHIP offers an example of a health campaign where cultural participants are at the core of the meaning-making process. The SHIP exemplifies the ability of CSWs in Sonagachi to engage with the context and structures that embody their daily lives, make sense of them, and organize communication platforms and strategies based on assessments of these contexts. This process materializes through a realization that context needs to be at the core of any sense-making exercise, and that community members should take charge of these sense-making processes. The SHIP focused on understanding health issues as unique to the community, informed by the structures, culture, and context of Sonagachi (Basu & Dutta, 2009).

In the case of the SHIP, this idea is manifest in an alternative discourse regarding HIV/AIDS prevention. In contrast to the dominant discourse of education about HIV and condom use, the local HIV/AIDS discourse involves the everyday living context of sex workers, which influences how they make sense of their lives, their health, and HIV/AIDS—combating marginalization and ostracism at the hands of police, goons, madams, pimps, and politicians, and contending with a lack of resources (e.g., money) as strategies to ensure better health. This discourse suggests the importance of communicating about structural barriers, rather than merely emphasizing individual-level behavior such as condom use.

The local HIV/AIDS discourse that emerged from the Sonagachi CSW community was one that located the individual behavior of condom use within the broader structures that constitute sex work. Because CSWs' livelihood is dependent on engaging in sex acts with clients, the CSWs' ability to negotiate condom use was inhibited by their dependence on clients. Jana et al. (2004) stated that prior to the establishment of the SHIP, CSWs in need of money were forced to borrow it from banks or moneylenders. The payment of these loans typically included charges of up to 50% interest. Under such circumstances, CSWs would often fall to the temptation to forego condom use in an attempt to make money for themselves and their families.

Furthermore, this local discourse also emphasized how the negotiation of condom use was constrained by community stakeholders (i.e., madams, pimps, police, men's social clubs) who exercised control over the CSWs (Evans & Lambert, 2008). For example, the CSWs emphasized that negotiating condom use with reluctant customers could be a time-consuming process. For CSWs, this was often time that they did not have available. Because CSWs often share rooms in brothels managed by madams, there was pressure to finish work rapidly so that they could take another customer, or to free the room for another CSW to use. Madams, suspicious that they might have engaged in multiple sex acts and were not paying their full dues, would chastise and punish CSWs who took too much time. Identifying structural issues like those described above allowed the SHIP to understand the contextual barriers that prevented CSWs from engaging in healthy behaviors.

Consistent with the culture-centered approach, the SHIP shifted its emphasis from educating sex workers about condom use to first attempting to understand sociostructural constraints that affect their health and HIV/AIDS behaviors. Communication within the community, through participation of members in the affairs of the campaign, to reach a consensus about health needs is central to the SHIP. In turn, as members of the community participate in the running of the SHIP, this process sets into motion a discursive practice that privileges localized group decision-making as a means of formulating and implementing local solutions to health issues.

Addressing Health Problems Through the Mobilization of Community Members

Once a community has been engaged in dialogue with health-campaign planners and a local understanding of health issues has developed, a culture-centered campaign should then focus on developing community-mobilizing strategies as a means of addressing the structural inequalities and inaccess that limit health (Dutta, 2008). This should ideally take place as strategies and tools are used to communicate with community stakeholders and policy-makers in order to bring about structural change.

As noted earlier, an important tenet of the culture-centered approach is agency. Agency is the capacity of individuals to participate in relationship with the structures that characterize their lives, and work within the constraints imposed by these structures as a means of securing resources and engaging in healthy behaviors (Dutta, 2008). Within the culture-centered framework, agency becomes meaningful as it is enacted in a particular cultural context where community-centered meanings are exchanged, constructed, and reconstructed. In the case of the SHIP, CSWs enacted agency through resisting structures that constrained their ability to engage in healthy behaviors by working together to provide resources, develop new community norms for engaging with clients, and support these norms through collective action.

This is best characterized by the establishment of a nongovernmental organization (NGO) called Durbar Mahila Samanwaya Committee (DMSC, see www.durbar.org/). This organization, which now runs the SHIP campaign, promoted CSWs' social and political rights through critical discussions of the stigma attached to sex work, and organizing them in their fight against structural and physical violence (Cornish, 2006). Many of the CSWs had the willingness to resist violence and exploitation prior to the establishment of the DMSC, but they could not do so because there were not many of them in the community who would come under one roof, trust each other, and resist the forces that marginalized them

(Basu & Dutta, 2008). The DMSC also worked to encourage political involvement of its members as a means of building a strong voting bloc, in order to leverage power over local politicians. Through the SHIP, the DMSC was able to provide a common platform for sex workers to launch and sustain their unified struggle against exploitation and violence.

In addition to the DMSC, a second example of the mobilization of the Sonagachi CSW community to solve local problems was the establishment of Usha, a cooperative-loan service. In contrast to banks and moneylenders, Usha granted small loans at low rates of interest and offered feasible payment arrangements. This provided the CSWs financial security, and meant that they could resist clients who insisted on not using condoms. Because they knew that they could go to Usha if they were in need of money, the CSWs could turn these customers away. Consistent with the culture-centered approach, the establishment of Usha further demonstrates how bringing community members together to participate in localized sense-making enables them to plan and implement strategies that address the contextual needs of the community.

Serving as a link between organizing and resistance, the SHIP also provided a platform for CSWs to address issues with other community stakeholders. Members of the SHIP staff began to serve as mediators in CSWs' disputes with landladies, customers, and neighbors. Members of the SHIP also began to engage in dialogue with stakeholders at various levels in the community. This included regular meetings with the most powerful groups in the red-light district (i.e., political leaders, pimps, men's social clubs). These meetings served to apprise these high-level stakeholders of the work of the SHIP, and seek their cooperation. The SHIP also engaged lower-level stakeholders, such as madams and babus (regular customers of sex workers), to garner their support and get them to participate in the SHIP's health efforts. Many were initially reluctant to support the SHIP for fear of losing their power in the community; however, the continued advocacy efforts of the SHIP and DMSC led many of these stakeholders to begin supporting the efforts of the SHIP.

Having fostered community participation in identifying important health issues and locating potential solutions within the sociocultural context of the community, the focus of the SHIP shifted to providing avenues for structural change. Consistent with the tenets of the culture-centered approach, the organizing of the SHIP campaign served as a point for organizing around health issues critical to the Sonagachi community, such as providing material resources and uniting to address exploitation. Through collective community action, the SHIP provided a means for the community to enact agency by actively resisting the structural inequities that constrained CSWs' ability to engage in healthful behaviors.

Conclusion

Commercial sex worker communities in India have long been targeted by health campaigns attempting to reduce the disparate level of sexually transmitted infections and HIV/AIDS cases found in these communities, which are mostly as a result of the marginalization they face and the structural and occupation-related disparities that impede their ability to stay healthy. Too often, these health campaigns have operated through a top-down, message-based approach where outside experts identify community health issues and solutions to these problems. Chattopadhyay and McKaig (2004) noted that such campaigns did not address the sociocultural barriers, such as lack of money, discrimination, and harassment by police and pimps, which sex workers say they face, that limit the acceptance of regular condom use. Furthermore, such campaigns have not achieved the desired results in terms of condom use or reduced HIV infections (Jana et al., 2004), and have been resisted by sex-worker communities (Basu & Dutta, 2008).

In contrast to these top-down, expert-driven health campaigns, the SHIP campaign has been successful in increasing condom use and reducing the incidence of HIV infection in the Sonagachi CSW community through the application of a participatory campaign that maps well with the framework of the culture-centered approach to health communication. More specifically, the SHIP has been successful in reducing health disparities through actively engaging in dialogue with community members in order to identify health problems important to the community and developing solutions to these problems together. Through this dialogic engagement, the SHIP was able to locate the health issues and potential solutions in the local sociocultural context by identifying structural inequities. The SHIP then worked with community members to resist the structures that limited their ability to be healthy, including setting up the DMSC and Usha. Through this engagement, the SHIP has been able to provide an avenue for community members to reduce their susceptibility to HIV/AIDS infection while also improving their overall quality of life by attending to the structures that constrained the health opportunities for sex workers.

The success of the SHIP challenges the dominant paradigm of health communication, which has often ignored the potential for agency in marginalized spaces (Basu & Dutta, 2008). It demonstrates that CSWs do continually engage with the structures that embody their contexts, and are able to organize themselves in participatory platforms to resist structures that create health inequities. Thus, the SHIP aligns with the culture-centered approach by foregrounding the ways in which members of marginalized spaces can participate in social-change endeavors.

Moving forward, the example of the SHIP also provides evidence that culture-centered health campaigns are uniquely positioned to address health disparities in marginalized spaces by foregrounding the participatory capacity of local communities. The culture-centered approach opens up possibilities for exploring the ways in which structures constrain, limit, and define the health experiences of marginalized communities by limiting their access to resources like health-care services and money. The culture-centered approach also recognizes that members of marginalized communities have the ability to understand the sociocultural contexts within which their health is enacted, and to work within these existing social structures to find avenues for solving health problems and work toward changing structural forces that constrain their lives. This recognition allows for the planning and implementation of health campaigns that meet the unique contextual needs of such communities.

Expertise in the culture-centered framework is located within cultural participants, instead of being attributed to the external experts who have limited knowledge of the local environment. Therefore, knowledge is articulated through dialogic engagement with community members, and through their involvement in identifying major structural issues facing the community. The work of the health campaign originates within the community and is driven by the community's understanding of what its major problems are, the ways in which these problems ought to be resolved, and the steps that ought to be taken (Dutta, 2008). Ultimately, the culture-centered approach offers a unique framework for health campaigns aiming to reduce health disparities in marginalized spaces by engaging in dialogue with community members to identify local health issues, and working with community members to develop solutions to meet their health needs.

A key characteristic of existing health campaigns and a prerequisite to substantial donor funding has been that the campaigns be "theoretically based" (Jana et al., 2004). The case of the SHIP demonstrates the efficacy of the culture-centered model of health campaigns as a viable theoretical vehicle to work with marginalized communities to engage in resisting structures that impede their ability to access resources and engage in health-promoting behaviors. Challenging the dominant discursive construction of marginalized communities as passive recipients of health education and aid, the example of the SHIP offers insights into the tangible ways in which health-campaign planners and practitioners can work with members of marginalized communities to resist dominant structures as they seek to reduce health disparities and, ultimately, to bring about social change.

Bibliography

Airhihenbuwa, C. O. (1995). *Health and culture: Beyond the Western paradigm*. Thousand Oaks, CA: Sage.

Airhihenbuwa, C. O. (2007a). *Healing our differences: The crisis of global health and the politics of identity*. Lanham, MD: Rowman & Littlefield.

Airhihenbuwa, C. O. (2007b). On being comfortable with being uncomfortable: Centering an Africanist vision in our gateway to global health. *Health Education and Behavior*, *34*(1), 31–42.

Basu, A. (2010). Communicating health as an impossibility: Sex work, HIV/AIDS, and the dance of hope and hopelessness. *Southern Communication Journal*, *75*(4), 413–432.

Basu, A., & Dutta, M. J. (2008). Participatory change in a campaign led by sex workers: Connecting resistance to action-oriented agency. *Qualitative Health Research*, *18*(1), 106–119.

Basu, A., & Dutta, M. J. (2009). Sex workers and HIV/AIDS: Analyzing participatory culture-centered health communication strategies. *Human Communication Research*, *35*(1), 86–114.

Basu, I., Jana, S., Rotheram-Borus, M. J., Swendeman, D., Lee, S. J., Newman, P., & Weiss, R. (2004). HIV prevention among sex workers in India. *Journal of Acquired Immune Deficiency Syndromes*, *36*(3), 845–852.

Beverly, J. (2004). *Subalternity and representation: Arguments in cultural theory*. Durham, NC: Duke University Press.

Chattopadhyay, A., & McKaig, R. G. (2004). Social development of commercial sex workers in India: An essential step in HIV/AIDS prevention. *AIDS Patient Care and STDs*, *18*(3), 159–168.

Cornish, F. (2006). Challenging the stigma of sex work in India: Material context and symbolic change. *Journal of Community & Applied Social Psychology, 16*(6), 462–471.

Dandona, R., Dandona, L., Kumar, G. A., Gutierrez, J. P., McPherson, S., Samuels, F., & Bertozzi, S. M. (2006). Demography and sex work characteristics of female sex workers in India. *BMC International Health and Human Rights, 6*(5). Retrieved from http://www.biomedcentral.com/1472-698X/6/5/.

Dutta, M. J. (2007). Communicating about culture and health: Theorizing culture-centered and cultural sensitivity approaches. *Communication Theory, 17*(3), 304–328.

Dutta, M. J. (2008). *Communicating health: A culture-centered approach.* Cambridge, UK: Polity Press.

Dutta-Bergman, M. J. (2004a). Poverty, structural barriers and health: A Santali narrative of health communication. *Qualitative Health Research, 14*(8), 1107–1122.

Dutta-Bergman, M. J. (2004b). The unheard voices of Santalis: Communicating about health from the margins of India. *Communication Theory, 14*(3), 237–263.

Evans, C., & Lambert, H. (2008). Implementing community interventions for HIV prevention: Insights from project ethnography. *Social Science & Medicine, 66*(2), 467–478.

Geertz, C. (1973). Thick description: Toward an interpretive theory of culture. In C. Geertz (Ed.), *The interpretation of cultures* (pp. 3–30). London, UK: Hutchinson.

Guha, R., & Spivak, G. (Eds.). (1988). *Selected subaltern studies.* Delhi, India: Oxford University Press.

Jana, S., Basu, I., Rotheram-Borus, M. J., & Newman, P. A. (2004). The Sonagachi project: A sustainable community intervention program. *AIDS Education & Prevention, 16*(5), 405–414.

Jha P, Nagelkerke, J. D., Ngugi, E. N., Prasada Rao, J. V., Willbond, B., Moses, S., & Plummer, F. A. (2001). Public Health: Reducing HIV transmission in developing countries. *Science, 292*(5515), 224–225.

Karim, Q. A., Karim, S. S. A., Soldan, K., & Zondi, M. (1995). Reducing the risk of HIV infection among South African sex workers: Socioeconomic and gender barriers. *American Journal of Public Health, 85*(11), 1521–1525.

Nag, M. (2006). *Sex workers of India: Diversity in practice of prostitution and ways of life.* New Delhi, India: Allied.

Nagelkerke, N. J., Jha, P., de Vlas, S. J., Korenromp, E. L., Moses, S., Blanchard, J. F., & Plummer, F. A. (2002). Modeling HIV/AIDS epidemics in Botswana and India: Impact of interventions to prevent transmission. *Bulletin of the World Health Organization, 80*(2), 89–96.

National AIDS Control Organization (NACO). (2008). HIV data. National AIDS Control Organization. Retrieved from http://www.nacoonline.org/Quick_Links/HIV_Data/.

Swendeman, D., Basu, I., Das, S., Jana, S., & Rotheram-Borus, M. J. (2009). Empowering sex workers in India to reduce vulnerability to HIV and sexually transmitted diseases. *Social Science and Medicine, 69*(8), 1157–1166.

UNAIDS. (2002). Epidemiological fact sheets on HIV/AIDS and sexually transmitted infections: India. 2002 fact sheet. Retrieved from http://www.unaids.org/hivaidsinfo/statistics/fact_sheets/pdfs/India_en.pdf.

8

Challenges and Dilemmas in Developing Health Communication Programs for Ethiopian Immigrants in Israel: The *Tene Briut* Health Promotion Organization

Nurit Guttman, Anat Jaffe, and Seffefe Aycheh

The sudden arrival of a relatively large, new "wave" of immigrants from Ethiopia to Israel in 1991 took many in Israel by surprise. In a dramatic airlift, within 36 hours, over fourteen thousand immigrants were airlifted from Addis Ababa to Israel. This event took place just a few days before the city was captured by rebel forces, and was presented to the Israeli public as a miraculous "rescue" mission. The Ethiopian immigrants were crammed into large aircraft whose seats were removed in order to hold a record number of passengers. The refugees landed in a country they viewed as the biblical "promised land" of "milk and honey," for which they had yearned over generations. Hence, the reality they encountered—which differed significantly from their expectations—amplified the usual challenges immigrants face. More so, upon arrival they found that their cultural heritage was not valued, and their darker skin color differentiated them physically, and even elicited prejudice and stigmatization (Ben-Eliezer, 2004). The Ethiopian immigrants came from a culture vastly different from contemporary Israeli society, and many continue to have limited economic means. These disparities contributed to the relatively high prevalence of health problems among this population, for which the health-care system was unprepared. These issues are the backdrop for the establishment of *Tene Briut*, an organization founded on the premises that cultural issues are at the heart of health promotion, and that the community itself should find ways to communicate to its members about health issues. This chapter begins with an overview of the situation of the Ethiopian immigrants in Israel, and describes *Tene Briut*'s activities, noting various challenges and dilemmas associated with the development of its health-communication strategies and materials.

These dilemmas might illustrate the kinds of challenges other organizations might face when they aim to promote the health of their community in a culturally appropriate way.

Context

Ethiopian Immigrants in Israel

Currently it is estimated that there are over 120,000 Ethiopian immigrants living in Israel. For several decades after Israel was founded in 1948, only a few Ethiopian Jews found ways to emigrate to Israel. In the mid-1970s a small number was transported secretly through the Sudanese border, because it has no diplomatic relations with Israel. The majority arrived in two mass waves of immigration. The first took place in the early 1980s, when many families left their villages and traveled by foot to Sudan, typically in harsh conditions that included attacks by bandits. Many did not survive the journey. The survivors were gathered in transition camps in Sudan and told to hide their Jewish identity. Many remained for years in the camps, and suffered from malnourishment and diseases. It is estimated that every family lost close relatives in this transition period.

The immigration process was renewed in the late 1980s when political changes in Ethiopia enabled enhanced diplomatic relations with Israel, and the community was threatened with economic changes and growing famine. Once again, families left their homes, this time traveling to Addis Ababa, where they were placed in transition camps, in which many suffered from malnourishment and diseases. As concerns grew for their welfare and political threats to the regime, the Israeli government decided, as described in the introduction, to embark in 1991 on a massive airlift operation. In the midst of military strife, within 35 hours, more than 14,000 immigrants were airlifted in crowded airplanes to Israel. One flight even experienced the birth of a child (Kaplan & Rosen, 1993). Thus, most Ethiopian immigrants experienced a prolonged and traumatic immigration process that involved harsh conditions, illness, and separation from and death of close relatives. During the transition period, some of the immigrants contracted infectious diseases such as tuberculosis and HIV. These circumstances contributed to stress and a sense of trauma in the Ethiopian immigrants as individuals and in the community as a whole, which remain with many today (Mashasha, 2008). Smaller numbers of immigrants continued to arrive from Ethiopia in recent years.

Ethiopian immigrants in Israel constitute an immigrant population with unique characteristics and needs. Although among the immigrants there were individuals with formal education and professional training, many came from a rural and traditional background with limited literacy skills. Upon arrival, most had to adjust to urban living conditions and to the Israeli educational, health-care, religious and welfare institutions. Most adults were unable to find employment, or found work mainly in maintenance or cleaning jobs. A large percentage (estimated to be nearly half) of immigrant families receive public assistance. A large majority of those over 65, even after two decades in Israel, cannot converse in Hebrew, and live in low-income areas with a high concentration of Ethiopian immigrants (Offer, 2004).

Compounding the hardships of immigration were the deep cultural gaps encountered by the immigrants upon their arrival. Most came from a vastly different social system, in terms of family relations, gender roles, and mutual support (King & Netzer, 2003; Varzwerger, 2003). Ethiopian immigrants in Israel also face the challenge of preserving their unique heritage and maintaining their sense of dignity (JOINT-Brookdale, 2001). The immigrants were shocked to learn that their Jewish religious practices were considered insufficient or inferior by established religious organizations and authorities. Their feeling of being stigmatized was manifested when a national scandal labeled, "The Blood Affair," erupted in 1996, when it was revealed that the Israeli blood bank systematically discarded blood taken from Ethiopian immigrants. Officials explained that this was done as a precaution, because infectious diseases such as HIV and hepatitis-B were appraised to be relatively prevalent among them (Seeman, 1999). This added to a general sense of discrimination among the Ethiopian community, which they attributed to their darker skin and to their culture and customs. Thus, it is very important to them to regain their pride and respect within their own community and among the larger Israeli public (Ben-Eliezer, 2004; Varzwerger, 2003).

Ethiopian Immigrants' Health and the Health-Care System

Researchers explain that migration-related stress affected practically every dimension of the Ethiopian immigrants' life, including family, work, community, housing, nutrition, religious practices, and bureaucracy (Reiff, Zakut, & Weingarten, 1999). Most immigrants had to adjust to a different type of health-care system, and to learn about and manage Western chronic diseases (e.g., asthma, diabetes mellitus, and high blood pressure) (Jaffe, Vardi, & Levit, 2001; Trostler, 1997). These illnesses were literally unknown to most in Ethiopia, and their management was not part of their cultural vocabulary (Guttman & Jaffe, 2002). Yet, they had become increasingly prevalent among the immigrants at a higher rate than in the general population, due to changes in living conditions, lifestyle, and nutrition.

Ethiopian immigrants often face various types of difficulties when encountering the health-care system, communicating with health-care providers in general, and in expressing medical symptoms, due to language limitations and differing cultural views regarding health and illness (Toledano et al., 2006). They also have a strong connection to a tradition of traditional healers (Myers-Joint-Brookdale Institute, 2001; Youngmann, Minuchin-Itzigsohn, & Barasch, 1999). These language and cultural barriers appear to have contributed to their relatively lower utilization of primary-care health services (Nirel, Rosen, & Ismail, 2000) and getting fewer diagnostic tests or referrals to medical specialists (Toledano et al., 2006). The issue of trust also emerges as a critical issue in the relationship between Ethiopian immigrants and health-care providers. Ethiopian immigrants express a feeling of being discriminated against, of not being taken seriously, of not being understood, of not getting adequate services, and even of getting "second rate" medications, and having a limited sense of efficacy in realizing health rights (Guttman et al., 2009; Toledano et al., 2006). Some still choose to return to Ethiopia to receive traditional medical care. Two state-supported liaison programs of health-care paraprofessionals are carried out in various community clinics to help meet doctor-patient communication challenges and provide HIV and tuberculosis treatment and prevention. However, many of the community clinics and medical centers continue to lack translation resources and cultural liaisons.

The *Tene Briut* Initiative

The *Tene Briut* initiative was established to address health promotion from the perspective of the Ethiopian immigrant community by drawing on the needs, culture, and strengths of its members, and by carrying out its activities within community settings. It began with an initiative to develop and implement culturally appropriate, outreach educational programs for the Ethiopian immigrants in community settings. Within less than a decade it developed into an outreach and advocacy organization that also created and implemented a medical translation and interpretation service. The organization underwent a transformation from a hospital-situated, community-outreach initiative headed by a non-Ethiopian immigrant physician to an independent organization with a board of directors that mostly consists of Ethiopian immigrants. It conducts several types of health-communication activities in community settings, mainly lectures and workshops; produces a monthly radio program on health issues in Amharic for the national radio station for immigrants; coordinates lectures and seminars for health-care providers on the Ethiopian immigrants' culture and health beliefs; and created and implemented a unique medical translation and interpretive service as part of its advocacy initiatives to promote equitable access to health care (Jaffe, Guttman, & Schuster, 2007; Schuster, 2007).

In a report of external evaluators, it was stated that *Tene Briut* "put the issue of the prevention and detection of chronic illness in the Ethiopian community on the map" of the Ethiopian community and of health-care decision makers, and also created a "box of tools" for this purpose. Another important contribution, according to the outside evaluators, is the creation of a group of indigenous professionals that became *Tene Briut*'s "Health Trustees," and who work within the community as role models. Their participation in the program also helped them to develop personally, and encouraged several to take prominent leadership positions outside the organization. This chapter describes the organization's strategic approach to health-communication activities, and identifies several dilemmas that might be inherent in the activities of organizations that aim to address health disparities among immigrants or minorities, in particular as they relate to issues of equity and advocacy.

The Beginning

The initial impetus for the establishment of the *Tene Briut* organization came from the growing frustration of a non-Ethiopian physician (Anat Jaffe, one of the authors of this chapter) who heads the Hillel Yaffe Medical Center's Endocrinology Unit. With a growing number of Ethiopian immigrants with problems associated with diabetes being treated at the medical center, she felt she was missing something important. After visiting homes of several of her patients, she realized that because of their unique culture and immigration experience it was essential that members of the community themselves should be involved in their community's health-care promotion. In March 1998 she gathered a group of Ethiopian immigrant, health-care-related professionals, who took it upon themselves to become Health Trustees by engaging in outreach health-communication activities in community settings. These practitioners were mostly hospital personnel (mostly male and female nurses) from diverse medical practice backgrounds and unfamiliar with health communication at the community level. However, they were committed to the idea of providing information in community settings, because they were already involved as unofficial liaisons, or were helping their own families in medical encounters. Some regarded themselves as lucky to have been integrated into Israeli society, and felt they had a mission to represent those less fortunate.

The Health Trustees

The group of nurses and other practitioners recruited to be Health Trustees received formal training in health communication, which provided them with new competencies in public speaking and community outreach. As evident in the graduation ceremony of the training class, which was attended by a government minister and local officials and dignitaries, their new role as Health Trustees enhanced their sense of pride and achievement. They became involved in material development, gave lectures in numerous community settings, appeared on a radio program, and were approached by members of the community as experts and role models. Most took upon themselves a leadership position in the organization itself, by becoming board members. Most also travelled as part of a group to the United States to present *Tene Briut*'s programs to professionals and community organizations. The project's evaluation report, conducted by outside professional evaluators, noted that "the Health Trustees, who are the main human resources of *Tene Briut*, had undergone an empowering experience and were enriched both personally and professionally." In fact, some of the Trustees became politically active in municipal politics and human rights advocacy organizations; one was elected to a municipal council and appointed to be in charge of women's rights and immigrant integration.

Mapping Needs

The initial step in developing its health-communication programs was to become familiar with the Ethiopian immigrants' beliefs and needs as they relate to health promotion. The main focus was on the prevention and treatment of diabetes, because *Tene Briut* belonged to the medical center's diabetes unit. The team met with nonmedical professionals and community workers within the Ethiopian community and with members of an Ethiopian immigrant community grassroots organization, to learn from them about their experiences and difficulties. These meetings helped the team members understand the difficulties members of the community have in grasping the implications of chronic and latent diseases (Kreps & Kunimoto, 1994). For example, they learned that community members held Israeli doctors in low esteem when they were unable to cure illnesses they described as "chronic," and the Ethiopian immigrants labeled them as "a disease of bad numbers." The grassroots organization's members also introduced the team to influential members of the community, who subsequently helped in evaluating the materials developed. In addition, a study was conducted with 179 Ethiopian immigrants regarding their beliefs, attitudes, and knowledge of health-promoting practices and of chronic illness (Guttman & Jaffe, 2002). The physiological part of the study consisted of physical examinations and blood tests, in order to document the relatively high incidence of chronic illness among the immigrants (Jaffe, Vardi, & Levit, 2001). The interviews were conducted in the homes of the Ethiopian immigrants by Ethiopian immigrant interviewers, who invited participants to visit the clinic and to have a physical exam and blood tests. This helped overcome the distrust they may have had with the health-care system. The staff administering the tests included Ethiopian immigrant practitioners who also gave lectures and answered questions while participants waited for the results. The research project became known in the community, and elicited trust in the program.

Health Communication Goals and Strategies

The overall goal of *Tene Briut* from its initial stages was to promote the health of members of the Ethiopian immigrant community by taking into consideration their culture and particular needs, including using culturally relevant metaphors and the importance of the cultural heritage. Specific goals at the outset were to:

1. Get members of the community to learn about chronic illnesses that were not familiar to them prior to immigration, especially diabetes.
2. Realize the risk in these illnesses.
3. Learn to adopt ways to prevent them.

Related goals were:

1. To foster a sense of self-efficacy at the individual and collective level.
2. To promote positive norms among community members to encourage them to adopt health-promoting practices.

As the organization evolved, it also aimed to achieve advocacy goals:

1. To put the health of Ethiopian immigrants and the disparities in appropriate health-care services on the national agenda.
2. To raise the issues of the need for linguistic services in particular.

For these purposes *Tene Briut* worked to develop and implement several communication approaches and strategies that included the production of print and audiovisual materials.

Partnerships with Community Organizations

One of *Tene Briut*'s main strategies to reach the Ethiopian immigrants in community settings was to incorporate lectures and workshops in ongoing programs in the community, such as sessions with women at well-baby clinics, senior citizens' social groups, etc. This approach was used both to strengthen the groups within the community and to more effectively promote the specific goals and objectives of the program by having them adapted to members of the particular groups within the immigrant population.

Using Culturally Appropriate Metaphors

The team was concerned about how to transmit new information and concepts, and after many deliberations it was decided to begin by using computerized presentation prototypes for lectures on health issues for the community members. This would enable them to show visual materials which could be adapted to particular audiences and changed over time. Thus, the team developed formal presentations to explain medical and physiological concepts using familiar situations from community life and metaphors, many from agriculture (for example, keeping channels open so that water can reach plants as a metaphor for the importance of glucose to reach the tissues and not only be in the blood). In order to revise the visual and text content of the presentation, questions and answers discussed during the presentations were documented. These included requests for clarification or the desire for more information than what was provided in the lecture. One initial revision was regarding basic physiology: In the initial lectures on nutrition, when the lecturer said that the human body was composed mainly of water, audience members became agitated. Despite their high esteem for the lecturer, the notion that their bodies were mostly water was difficult for them to accept. Consequently, the concepts were presented in a new way, and the explanation focused on the elements that constitute the body, and this was connected to the importance of various nutrients for growth and regular functioning of the body. This approach referred to their knowledge of body parts, which they knew about from their knowledge of farm animals. The *Tene Briut* team eventually developed a series of presentations on various topics, including nutrition, that referred both to the traditional Ethiopian diet as well as Israeli food, knowledge of the basic food groups, physical activity, oral hygiene, and diabetes detection, prevention, and treatment.

Appealing to the Story of and the Tribulations of the Immigration Experience

In response to a request from representatives of the community, the *Tene Briut* team produced a video about diabetes titled "Even a Meadow Can Be a Jungle" (drawing on an Ethiopian proverb). Although its purpose was to explain the nature of the illness and how to prevent it, nearly 30% of the 17-minute video was devoted to imagery of life in Ethiopia, the hardships in the journey to Israel, and the difficulties the Ethiopian immigrants have had in adapting to their new country. The video was formatted to be a story narrated by a traditional storyteller, and used photos, art, and music from Ethiopian villages. This concept was based on the suggestions of community representatives that having an Amharic-language video that incorporated familiar pictures, scenery, art, and music would awaken memories of their former homeland and prompt viewers to watch the video repeatedly. The narrator used many traditional proverbs to reinforce viewers' responsiveness to his recommendations.

In addition to referencing the past, the video connected the immigrant experience to current social and health issues. It mentioned the disappointment immigrants felt when their expectation of coming to reside in Jerusalem, which for them represented the "promised land," conflicted with the reality of being settled in crowded absorption centers or apartments in a country that was far from the golden ideal they imagined. Diabetes was presented as part of the challenges of immigration, and not as something bad that was brought from Ethiopia. It also depicted humorous scenes, such as the reaction of immigrants when they encounter the abundance of products, especially junk food and soft drinks in the huge supermarkets. The production of this video corresponds to the culture-centered approach of communicating about health problems from the perspective of the members of the cultural group, acknowledging institutional factors that marginalize them, and their sense of agency (Dutta, 2008). When the video was shown, it elicited strong, emotional reactions. For example, viewers in focus groups would put their hands on their hearts to express their strong emotions when asked about their reactions. It was the first Israeli video for most Ethiopian immigrants that showed the plight of their immigration to Israel. After the video was shown in the *Tene Briut* lecture activities, participants could purchase a copy for a nominal price, and many did.

Developing Resources for Health-Care Practitioners

Tene Briut embarked on producing resources on nutrition for health-care providers. This was based on the experiences of members of the community that the nutritional information they were given was not relevant for the foods they typically ate. This led to the realization that there was no detailed information about the

Ethiopian immigrants' diet or the nutritional values of their foods. Not only were Ethiopian immigrants not getting health information that was useful to them, they also felt that the information they received devalued their own food, which they considered healthy and an important part of their cultural tradition. This led the *Tene Briut* to conduct research on this topic, which included asking community members for their recipes and analyzing their nutritional content. The team learned that traditional foods continued to be a regular part of the immigrants' diet, but that substantial changes occurred in the quantity of these foods and the frequency with which they were consumed in Israel. A booklet called, "Not by the Ingera [Ethiopian type of bread] Alone: The Nutrition of Ethiopian Immigrants in Israel," was produced with the professional support of the Nutrition Department of the Ministry of Health and other organizations. One of its main goals was to introduce health-care practitioners to the immigrants' perspectives about food and nutrition, help them learn to respect the immigrants' traditions regarding their food, and to inform them about the nutritional value of these foods so that they would refrain from advising the immigrants to abandon eating their traditional foods and turn to a presumably healthier, recommended "Israeli" diet. The booklet included a description of cultural beliefs associated with foods, as well as relevant Ethiopian proverbs.

Monthly Radio Program

In 2003 *Tene Briut* utilized a grant it received from a private foundation to begin producing a monthly radio program on the Israeli national station for immigrants. The program was included in a national radio station's daily time slot allocated to programs for Ethiopian immigrants, which are broadcast mostly in Amharic. The radio program typically consists of interviews with health-care professionals for the main purpose of providing health information; it includes a question-and-answer session. It is particularly important to older populations and to new immigrants who do not know Hebrew. An evaluation study found that listeners, in particular older people, would change their schedules in order to be able to listen to the health radio program, and reported that they get important health information from it and that it has influenced their health practices. In 2009 it received an award from the association of journalists for the best radio program in its category.

Advocacy Activities

The research activities associated with *Tene Briut* provided data for advocacy activities, in particular to a collaborative initiative with four other advocacy organizations, including three civic- and health-rights organizations and an Arab health organization. A joint report and policy paper on the topic of health disparities calling for a national program to address health disparities titled, "Working Today to Narrow the Gaps of Tomorrow" was written and presented in June 2009 to policy-makers and the news media (Adva Center et al., 2010). A report on Ethiopian immigrants' non-government organizations noted that over the years these organizations tended to change their mode of operation according to the changes that took place within the community and its needs. It explained that the Ethiopian immigrant community moved from a society that wanted to preserve its differences to a society that is trying to address acculturation and integration issues. Further, that the immigrant community realized that acculturation was not sufficient for achieving equitable treatment in the Israeli society, and therefore the organizations moved to activities that aim to address the sources of social inequity (Balbachan, 2008). This move is reflected also in *Tene Briut*'s activities and orientation.

Challenges and Dilemmas

Communicating for the purpose of health promotion invariably raises various challenges and ethical dilemmas (Guttman, 2000). Several dilemmas were identified and discussed among the *Tene Briut* members in the development and implementation of the various materials and programs. For example, one dilemma was whether the inclusion of the story of the Ethiopian immigrants' journey to Israel was done in order to make the video on diabetes more attractive, or was its purpose to enhance the self-esteem and feeling of community pride among the immigrants? The answer given in the discussion that took place was "both." Examples of such dilemmas are summarized in Table 8.1. In this section we focus on three additional types of challenges and dilemmas that concern challenging traditional and newly acquired norms, diversity within the population, and adopting a culturally centered orientation or advocacy.

Dilemmas Regarding Challenging Traditional and Newly Acquired Norms

One of the major assumptions underlying *Tene Briut*'s community approach was that a supportive community can help change unhealthy practices, but that certain perceived norms can be detrimental to members' health. This meant that some of the traditional norms and values held by community members would need to be challenged. Specifically, this emerged regarding the issues of weight gain, physical activity, and gender roles.

Weight gain. In Ethiopia, gaining weight had been typically viewed as positive, and associated with wealth and success. Therefore, the immigrants did not view gaining weight as a health problem, and even viewed it favorably. To address this issue, a short satirical video was produced that featured a known comedian making fun of his own weight gain and the difficulties people encounter when they are overweight. The video could have been viewed as offensive, but usually was received as funny and amusing, and used to trigger a discussion on weight gain. It raised, however, a concern about stigmatizing people who have gained weight. Also, an ethical concern noted in health promotion is that one might "take away" from disadvantaged populations pleasures such as eating rich foods (MacAskill, Stead, MacKintosh, & Hastings, 2002). Therefore, the recommendations were not to deprive people from foods they enjoy, but to limit quantities and to differentiate among foods according to their nutritional values.

Walking. Another norm *Tene Briut* sought to challenge was community members' negative conceptions regarding physical activity for adults. Group discussions revealed that if people chose to walk to reach their destination this could be viewed as being "cheap" or as not having enough economic means to afford transportation. A newly acquired norm among some of the immigrants was that walking was something they left behind in Ethiopia. One way to address this challenge was to include images of positive role models walking, such as the professionals from *Tene Briut* team, as well as to present examples of native-born Israelis to emphasize that walking and other physical activity is the norm in Israel. The community lectures included tips on how to adapt physical activity to the daily routine of Ethiopian immigrants in Israel.

Gender role issues. The question of gender roles is a contentious issue among immigrants, as dramatic changes have taken place in family relations upon immigration. Men found that their traditional, dominant status had changed with the new economic and social circumstances in Israel. The issue of gender roles emerged in various aspects of program development, and perhaps most prominently in the development of print materials regarding foot care for older Ethiopian immigrants, in particular those affected by diabetes. It was decided to use the format of narrative in a series of pictures. The colors and illustration style were chosen to appeal to the older population. A tub for washing feet was drawn to remind viewers of the traditional custom of washing feet in Ethiopia, which is associated with honor and respect. The scenario was developed by the *Tene Briut* team members, who were given the task to ask community members about their reactions and preferences. At first it was suggested to use an older Ethiopian man as the character whose feet were being washed, but a dilemma emerged regarding who would be the character washing his feet. Some felt it would be important to show a woman washing the foot of a man, as customary in Ethiopia, because this was a culturally appropriate scene and would resonate well with the older immigrants. Others, mainly the women on the team, thought this should be an opportunity to challenge traditional gender roles, and they proposed that a man should wash the feet of a woman. As one of the women exclaimed: "It is time to change the image of Ethiopian women as subservient to men!" This proposal was rejected, because it was concluded that it would not resonate with the intended population and its message would be rejected. As a compromise, it was decided to have a young man help the older man wash his feet. This case illustrates trade-offs that need to be considered when a "culturally sensitive" approach is adopted regarding gender issues: Should traditional images that evoke memories and would be attractive to the intended population be used, or should this be used as an opportunity to challenge the norms of members' of traditional groups that differ from Western conceptions of gender equality?

Addressing Diversity Within the Immigrant Population

A challenge noted in the literature is to address the diversity within the immigrant group itself (Guttman, 2007; Kreuter, Lukwago, Bucholtz, Clark, & Sanders-Thompson, 2003). Within the Ethiopian immigrant communities there are individuals with limited literacy, and even those who speak different languages. The main approach used by *Tene Briut* to address this challenge is to "customize" its presentations and workshops to particular audiences (Kreuter et al., 2003), and to integrate its activities with those of organizations that work with specific population groups (e.g., new immigrants in absorption centers, young women in women's groups, or the elderly in programs for the elderly). This raises the challenge of not reaching those who are not involved in any of these activities.

Challenges and Dilemmas Associated With the Culturally Sensitive Approach

Many of the activities of *Tene Briut* could be characterized as adopting a cultural-sensitivity approach to convey and adapt health information in ways that resonate with the Ethiopian immigrants' experiences, values, norms, cultural views, and beliefs. This approach has been applauded primarily because it aims to respond to criticisms regarding how cultural and linguistic minorities have been treated in the health-care context (Resnicow, Braithwaite, Dilorio, & Glanz, 2002). *Tene Briut* was praised by its evaluators for creating an exceptional "toolkit" of culturally sensitive materials. However, as the organization evolves and adopts more advocacy goals, it faces the dilemma of whether, by providing culturally sensitive adaptations that do not require the system to make any significant changes, it might serve to absolve the establishment of making changes (Dutta-Bergman, 2005). In other words, do its activities mainly offer ways for the immigrants to cope with the existing circumstances and thus exempt the establishment from working to eliminate inherent disparities faced by the immigrants? The organization's goals and activities indicate its orientation aims to be "culturally centered": specifically, to elicit the views and voices of community members and frame needs from their perspective. This was exemplified early on in the community gatherings. In these meetings, members of the community could talk about health-care-related issues. These were probably the first and only community-level activities in which members of the community could voice their views about the way they felt they were treated in the health-care system in a public setting attended by health-care professionals. Thus, *Tene Briut*'s activities created a space for members of the marginalized Ethiopian immigrant community to express their views and concerns regarding health in a collective setting of dialogue (Dutta, 2008). For example, people said that they felt the medicines they get are inferior to those that Israelis get, that physicians do not want to touch them and refrain from doing a clinical exam, or do not respect their culture. These issues were raised and discussed during the community sessions. Thus, the meetings became a venue for raising and discussing community members' concerns, whereas originally they were intended to be health information dissemination activities. *Tene Briut*'s activities continued to enable community members to raise their concerns when its members became more active in advocating the need for specific translation services. They used the stories from this service to substantiate the argument for the need for legislation that would mandate medical translation services for linguistic minorities.

Additional Challenges and Potential Contradictions

Critics maintain that health-promotion initiatives that take place in diverse populations often reflect Eurocentric biases of individualism (Airhihenbuwa, 1995). By implication, another challenge for *Tene Briut* is whether its activities emphasize individualism and individual-level changes or advance community-level goals. It was proposed that rather than focus on a culturally sensitive approach, an emphasis should be on community rather than individual-level changes (Buchanan, 2000). According to this approach, activities would provide members with opportunities to express themselves and engage in critical dialogues that would enable members of marginalized groups to have their voices heard by themselves and by decision makers. This process is supposed to enable the definition of problems from group members' own perspectives, and help construct possible health-intervention solutions (Dutta, 2007). The goals and activities of *Tene Briut* aim to incorporate not only a "culturally sensitive" approach, but also a "culturally centered" approach, because it aims to present and to address the needs and goals of the community from its perspective, to enhance a collective sense of efficacy, and to influence policy-level changes. Further, its goals are to get the health-care providing organizations to provide services and programs to meet the particular needs of community members. The aim of achieving these goals creates a tension between the activities that help community members adapt to the current system and advocacy activities in favor of changing the system. This tension is illustrated in four additional challenges that concern community-level issues.

Preservation of and pride in the cultural heritage. Employing metaphors, themes, stories, images, and proverbs from the immigrants' cultural heritage was the central method of *Tene Briut*'s intervention approach. This could also serve to help preserve their cultural heritage by incorporating it into the materials and activities, and by presenting it in a respectable way. It could also help expose younger generations unfamiliar with their Ethiopian cultural heritage, or who even devalue it. *Tene Briut* also works to acquaint health-care providers and policy-makers with the Ethiopian immigrants' culture. Thus, a challenge for *Tene Briut* is how to go beyond using shared aspects of culture as a means to promote immigrants' health and to employ health-promotion activities as a means to create pride and help preserve their cultural heritage. A related challenge is how to address the issue of traditional medicine and traditional healers.

Limited public-media channels. Reaching linguistic minorities through radio and television, in particular those with low levels of literacy, is highly dependent on the availability of public-media outlets. The Ethiopian immigrant community is a relatively small minority, in a nation that consists of numerous minority groups. It is allocated only a few hours a week of programming in Amharic on public radio and television channels. These are the main media outlets used by older and new immigrants who are not proficient in Hebrew and do not have reading skills in general. Programs currently available in Amharic are considered to be a critical source for individuals to access information on health, welfare rights, and current events. The availability of the public-media outlets is limited to the Ethiopian immigrants, because of their restricted political power. *Tene Briut* managed to introduce a health program on the national radio channel, but only a monthly program. One of its challenges is to increase both the time allocated for the health program and the overall time for broadcasts for the Ethiopian immigrant population.

Concerns about stigma, paternalism, and discrimination. The Ethiopian community is highly concerned about the issue of stigma. "The Blood Affair" crystallized this concern, and Ethiopian immigrants often feel that they are, as a group, associated with stigmatized diseases, low levels of hygiene, crime, and other social inadequacies. For some, even discussions about the prevalence of relatively high rates of chronic illness constitute a threat to the community, because they believe it might further stigmatize it. In addition, members of the community feel they experience discrimination or that there is inequity in the services they are provided. Some express feelings of strong frustration or of lack of faith in the system, whereas others express feelings of apathy or fatalism (Guttman, Aycheh, Gesser-Edelsburg, Bloch, & Moran, 2012). Further, they feel that there is a paternalistic approach to their situation, and that decisions regarding them are made by others. This presents the challenge of how to discuss and address concerns about stigma and discrimination within and outside the Ethiopian immigrant community, and how to promote policies to address discrimination and inequity within and outside the health-care system.

Community-level changes. Because the Ethiopian immigrant minority is to a large extent economically disadvantaged and socially marginalized (Offer, 2004), one of the challenges is to advance community-level changes that increase the strength of the community as a whole. An Ethiopian proverb used in the programs of *Tene Briut* states the following: "When little spiders spin their webs together, the communal web is strong enough to catch even the biggest of lions." Thus, an important challenge is to advance changes associated with health promotion that would strengthen the community as a collective entity, and not only the practices of individuals. The goal is to strengthen the community to pressure the system to provide it with equitable health-care services, and also to create a change within the community itself.

Summary and Conclusions

The development of the *Tene Briut* organization is hailed within and outside the health-care system as an inspiring initiative that emerged to meet unique health-promotion needs of the Ethiopian immigrant population that were not addressed by the health-care system. It evolved from a medical-center-sponsored project, led by a non-Ethiopian physician, to an independent organization led by an Ethiopian immigrant staff. In this chapter we described several major health-communication activities, challenges, and dilemmas associated with these activities. These challenges might help illustrate similar challenges faced by other organizations that aim to promote the health of their community in a culturally appropriate way and to address social inequities of the system. An important question is whether these dilemmas might be inherent in the activities of a nongovernmental organization that aims to address health disparities of immigrants or minorities. In particular, do their activities provide "culturally appropriate" means for community members to adjust to the current system, and might this serve to reinforce inequities in institutional arrangements by enabling the system to remain as it is?

In the case of *Tene Briut*, most of its activities focused on communication to enhance behavior change within the Ethiopian immigrant Israeli community. Advocacy activities increasingly became an important part—mainly providing data on health-care disparities to influence policy-makers, and creating an interpretation service that was unavailable. The use of cultural themes and images was incorporated with a dual purpose: as a means to influence community members to adopt recommended practices, and to address the challenge of engendering a sense of pride among the immigrants and members of the younger generations.

Within the Ethiopian immigrant community, the organization faces challenges of how to promote cultural heritage and respect for tradition while challenging traditions that are considered inequitable, such as traditional gender roles or beliefs that contradict modern medicine. Within the health-care system it faces the challenge of providing resources and training to health-care professionals to get them to appreciate and understand the cultural heritage of the Ethiopian immigrants, and address prejudice and stereotyping within the health-care system. This also poses a challenge of how to institutionalize this type of training within the health-care system. Finally, there is the challenge of how to engage in advocacy while providing services that the system should have been providing. In the case of *Tene Briut*, it formed a collaboration with civic- and health-rights organizations to produce a policy document and advocate government responsibility for the provision of equitable services and allocations of resources to address disparities.

Table 8.1. Examples of Health Challenges, Strategic Methods, and Ethical Issues in Developing and Implementing Health Communication Programs for the Ethiopian Immigrant Community

Health Challenge	**Strategic Method**	**Ethical Issues or Dilemmas**	**Approaches to Address the Ethical Issues***
Reach members of diverse groups within the immigrant community	Form partnerships with other organizations and in the health-care system. Customize presentations for each group.	How to reach people who typically do not participate in organizational activities?	Conduct community-wide lectures unrelated to organizations. Conduct health fairs in community settings.
Draw interest	Emotional appeal of the "story of the immigration" in video. Traditional music and cultural artifacts.	Does the program use cultural symbols and appeals to emotions as a means to attract attention, or also to strengthen the cultural heritage?	Emphasize the importance of the culture. Discuss how various traditions contribute to the strength of the community and how they can be maintained.
Introduce new health-related practices	Appeal to traditional customs and norms. Emphasize personal responsibility to adopt health-promoting practices. Members of the organization as role models in the community.	Gender-role issues might arise, related to the traditional gender-roles (e.g., decision making, cooking, cleaning, child care, tending to others). How to integrate changes that will not compromise cultural traditions? How to avoid emphasizing norms of individualism and individual-level solutions?	Dilemmas are raised and discussed. Try to minimize and neutralize gender-role issues. Show younger people in more egalitarian activities. Discuss issues of solidarity within the community.

Health Challenge	**Strategic Method**	**Ethical Issues or Dilemmas**	**Approaches to Address the Ethical Issues***
Explain the notion of risk conditions and chronic illness	Use of agricultural metaphors. Clarify and expand answers for the questions on why chronic illnesses had become prevalent among the Ethiopian immigrants.	How not to demoralize people about the new illnesses? One concern was that people might be overwhelmed by the new illnesses. Concerns that the new illnesses will contribute to a sense of stigmatization.	Connect the travails with journey to Israel, and not problems inherent to the community. Health Trustees as role models of success and achievement.
Introduce new norms that challenge traditional norms	Use of humor and satire. Health Trustees as role models. Illustrate how improved biomedical measurements and a feeling of improved health are a result of adopting health promoting practices.	Would satire serve to insult or stigmatize people who have gained weight? What alternatives can be provided to the enjoyment of more abundance of foods in Israel?	Engage in discussions with the participants. Find ways to enhance cultural pride and alternative activities, in particular for the older adults. Show ways to enjoy traditional foods in healthier ways.
Health-care providers' lack of knowledge of the immigrants' beliefs, practices, and behaviors, their apparent lack of empathy, and language gap	Create written manuals and resources that include explanations and examples. Workshops and seminar for health-care providers. Point out difficulties of the medical staff in managing the care of Ethiopians. Establish an interpreting service. Integrate usage of the immigrants' language in written forms.	Do the organization's programs serve as a substitute for activities government agencies and the health-care organizations should be providing? How to acknowledge the difficulties that health-care providers themselves have in managing the care of Ethiopians?	Develop an interpretation services model intended to be adopted by the health-care system, and pressure the system to adopt this kind of service as part of the immigrants' health rights. Engage in advocacy activities to pressure health-care organizations to provide training for health-care practitioners. Emphasize services that are not covered by the general health system.

Health Challenge	**Strategic Method**	**Ethical Issues or Dilemmas**	**Approaches to Address the Ethical Issues***
Introduce and promote specific health- promotion practices	Develop and distribute materials and conduct seminars and lectures that are based on the experiences, culture, and questions raised by community members. Implement activities such as cooking classes that include women, men, and older children. Advise individuals how to prepare for and act in their interactions with health-care providers, how to present their problems, and how to assert their rights (enhance their confidence).	Is the program mainly addressing health issues that have been identified as problems from the point of view of the system?	Involve community members in decisions regarding the health-care issues that need to be addressed, from their own perspective, through various community programs and research. Work with other advocacy organizations to promote the concerns of the immigrants.

* These are partial or current ways of addressing the challenges and dilemmas.

Bibliography

Adva Center, Physicians for Human Rights-Israel, Association of Civil Rights in Israel, Galilee Society, The Arab National Society for Health Research & Services, & Tene-Briut for the Promotion of the Health of Ethiopian Israelis. (2010). Working today to narrow the gaps of tomorrow. Tel Aviv, Israel: Adva Center. Retrieved from http://www.acri.org.il/pdf/health-gaps-en.pdf.

Airhihenbuwa, C. (1995). *Health and culture: Beyond the Western paradigm.* Thousand Oaks, CA: Sage.

Balbachan, J. (2008). The Ethiopian community in the third sector: Trends in registration and changes in activities. Research report. Beer Sheva, Israel: Israeli Center for Third Sector Research, Ben-Gurion University.

Ben-Eliezer, U. (2004). Becoming a Black Jew: Cultural racism and anti-racism in contemporary Israel. *Social Identities, 10*(2), 245–266.

Buchanan, D. R. (2000). *An ethic for health promotion: Rethinking the sources of human well-being.* Oxford University Press.

Cohen, M. P., Stern, E., Rusecki, & Zeidler, Y. (1988). High prevalence of diabetes in young adult Ethiopian immigrants to Israel. *Diabetes, 37*(6), 824–828.

Drori, R., & Jaffe, A. (2007). Incidences and age at disease onset of Type 1 Diabetes Mellitus among Israeli Ethiopians are correlated with the duration of exposure to a new environment. Hadera, Israel: Israel Endocrine Society.

Dutta, M. J. (2007). Communicating about culture and health: Theorizing culture-centered and cultural sensitivity approaches. *Communication Theory, 17*(3), 304–328.

Dutta, M. J. (2008). *Communicating health: A culture-centered approach*. London, UK: Polity Press.

Dutta-Bergman, M. (2005). Theory and practice in health communication campaigns: A critical interrogation. *Health Communication, 18*(2), 103–122.

Guttman, N. (2000). *Public health communication interventions: Values and ethical dilemmas*. Thousand Oaks, CA: Sage.

Guttman, N. (2007). Challenges and dilemmas in social marketing of health issues in culturally and socially diverse populations: The case of Israel. In L. Epstein(Ed.), *Culturally appropriate health care by culturally competent health professionals*,(pp. 52–70). Tel Hashomer, Israel: Israel National Institute for Health Policy and Health Services Research.

Guttman, N., Aycheh, S., Gesser-Edelsburg, A., Bloch, L. R. and Avital, M. A Health rights information from the perspective of Ethiopian immigrants: Issues, barriers and policy recommendations. In B. Rosen, S., Shortell, and A. Israeli (Eds.). *Improving Health and Health Care: Who is Responsible? Who is Accountable*? (pp. 104-203). The Israel National Institute of Health Policy.

Guttman, N., Aycheh, S., Gesser-Edelsburg, A., Bloch, L. R., & Moran, A. (2012). Health rights information from the perspective of Ethiopian immigrants: Issues, barriers and policy recommendations. In B. Rosen, A. Israeli, & S. Shortell (Eds.), *Accountability and responsibility in health care: Issues in addressing an emerging global challenge* Hackensack, NJ: World Scientific.

Guttman, N., Aycheh, S., Gesser-Edelsburg, A., Bloch, L. R. and Avital, M. A Health rights information from the perspective of Ethiopian immigrants: Issues, barriers and policy recommendations. In B. Rosen, S., Shortell, and A. Israeli (Eds.). *Improving Health and Health Care: Who is Responsible? Who is Accountable*? (pp. 104-203). The Israel National Institute of Health Policy.

Guttman, N., & Jaffe, A. (2002, November). 'We didn't have it in Ethiopia!' Preliminary findings on attitudes and beliefs of Ethiopian immigrants to Israel regarding diabetes. Israel Society of Diabetes Mellitus 2001 (ABS-oral). Paper presented at the annual conference of the American Public Health Association, Philadelphia, PA.

Guttman, N., & Salmon, C. T. (2004). Guilt, fear, stigma and knowledge gaps: Ethical issues in public health communication interventions. *Bioethics, 18*(6), 531–552.

Jaffe, A., Guttman, N., & Schuster, M. (2007). The evolution of the *Tene Briut* model: Developing an intervention program for the Ethiopian immigrant population in Israel and its challenges and implications. In L. Epstein (Ed)., *Culturally appropriate health care by culturally competent health profession-*

als (pp. 121–142). Tel Hashomer, Israel: Israel National Institute for Health Policy and Health Services Research.

Jaffe, A., Vardi, H., & Levit, B. (2001). Diabetes in the Ethiopian Jewish community of Hadera: Prevalence, atherosclerotic risk factors. Israel Society of Diabetes Mellitus 2001 (ABS-oral). EASD 37th annual meeting, p403 2001(ABS-post). Third Jerusalem International Conference on Health Policy 2001(ABS), Glasgow, Scotland, UK

JOINT-Brookdale. (2001). Integration of Ethiopian immigrants in Israeli society: Challenges, policy, program and direction. Executive summary. Unpublished report [Hebrew]. Joint-Brookdale Institute, Israel.

Kaplan, S. and Rosen, H. (1993). Ethiopian immigrants to Israel: Between preservation of culture and invention of tradition: Jewish Journal of Sociology, 35(1), 35–48.

King, Y., & Netzer, N. (2003). *Selected data from census of Ethiopian immigrants in eight towns* [Hebrew]. Jerusalem: Center for Immigrant Absorption, Joint Institute, Brookdale.

Kreps, G. L., & Kunimoto, E. N. (1994). Effective communication in multicultural health care settings. Thousand Oaks, CA: Sage.

Kreuter M.W.; Lukwago S.N.; Bucholtz D.C.; Clark E.M.; Sanders-Thompson V. (2003). Achieving cultural appropriateness in health promotion programs: Targeted and tailored approaches. *Health Education and Behavior, 30*(2), 133–146.

Kreuter, M. W., Lukwago, S. N., Bucholtz, D. C., Clark, E. M., & Sanders-Thompson, V. (2003). Achieving cultural appropriateness in health promotion programs: Targeted and tailored approaches. *Health Education and Behavior, 30*(2), 133–146.

Kreuter, M. W., Strecher, V. J., & Glassman, B. (1999). One size does not fit all: The case for tailoring print materials. *Annals of Behavioral Medicine, 21*(4), 276–283.

MacAskill, S., Stead, M., MacKintosh, A. M., & Hastings, G. (2002). 'You cannae just take cigarettes away from somebody and no' gie them something back': Can social marketing help solve the problem of low-income smoking? *Social Marketing Quarterly, 8*(1), 19–34.

Macklin, R. (1999). *Against relativism: Cultural diversity and the search for ethical universals in medicine.* New York, NY: Oxford University Press.

Mashasha, M. (2008). Open wound. *Kav Haofek, 28*, 16–18.

McLeroy, K. R., Clark, N. M., Simons-Morton, B. G., Forster, J., Connell, C. M., Altman, D., & Zimmerman, M. A. (1995). Creating capacity: Establishing a health education research agenda for special populations. *Health Education Quarterly, 22*(3), 390–405.

Myers-Joint-Brookdale Institute. (2001). Integrating Ethiopian immigrants in Israeli society: Challenges, policy, plans and directions [Hebrew]. Jerusalem: Author.

Nirel, N., Rosen, B., & Ismail, S. (2000). *'Refuah Shlemah': An intervention program for Ethiopian immigrants in primary care clinics: Results of an evaluation study.* Research report: RR-357-00. Jerusalem: JDC-Brookdale Institute.

Nudelman, A. (1994). Health services to immigrant and refugee populations: Patient and provider cross-cultural perspectives. *Collegium Antropologicum, 18*(2), 189–194.

Offer, S. (2004). The socio-economic integration of the Ethiopian community in Israel. *International Migration, 42*(3), 29–55.

Reiff, M., Zakut, H., & Weingarten, M. A. (1999). Illness and treatment perceptions of Ethiopian immigrants and their doctors in Israel. *American Journal of Public Health, 89*(12), 1814–1818.

Resnicow, K., Braithwaite, R. L., Dilorio, C., & Glanz, K. (2002). Applying theory to culturally diverse and unique populations. In K. Glanz, B. K., Rimer, & F. M. Lewis (Eds.), *Health behavior and health education: Theory, research, and practice* (3rd ed., pp. 485–509). San Francisco, CA: Jossey-Bass.

Rosenberg, R., Vinker, S., Zakut, H., Kizner, F., Nakar, S., & Kitai, E. (1999). An unusually high prevalence of asthma in Ethiopian immigrants to Israel. *Family Medicine, 31*(4), 276–279.

Rubinstein, A., Goldbourt, U., Shilbaya, A., Levtov, O., Cohen, G., & Villa, Y. (1993). Blood pressure and body mass index in Ethiopian immigrants: Comparison of operations Solomon and Moses. *Israel Journal of Medical Sciences, 29*(6–7), 360–363.

Schuster, M. (2007). Linguistic and cultural mediation in public institutions [Hebrew]. *Hed ha ulpan hadash, 91,* 1–8.Seeman, D. (1999). "One people, one blood": Public health, political violence, and HIV in an Ethiopian-Israeli setting. Culture, Medicine and Psychiatry, 23, 159–195.

Seeman, D. (1999). "One people, one blood": Public health, political violence, and HIV in an Ethiopian-Israeli setting. Culture, Medicine and Psychiatary, 23, 159–195.

Shabtai, M. (2001). To live with a threatened identity: Life experiences with a different skin color of Ethiopian immigrant youngsters [Hebrew]. *Megamot, 97.*

Swirski, S., & Swirski, B. (2002). *Ethiopian Jews in Israel* [Hebrew]. Report 11. Tel-Aviv, Israel: Adva Center.

Toledano, Y., Givon, S., Kahn, E., Ayecheh, S., Guttman, N., & Jaffe, A. (2006). The use of healthcare services by Ethiopian immigrants and the quality of service they get compared to diabetes patients who are not Ethiopian immigrants [Hebrew]. In Gur & Bin-Nun (Eds.), *A decade to the Israel national health insurance law*. Israel Health Policy Institute.

Trostler, N. (1997). Health risks of immigration: The Yemenite and Ethiopian cases in Israel. *Biomedical Pharmacology, 51*(8), 352–359.

Varzwerger, R. (2003). The Ethiopian immigrant community: Current status, gaps, and claims about discrimination [Hebrew]. Document prepared for the Parliament Member, Michael Eitan. Israel Knesset, Jerusalem: Center for Research and Information.

Weingrod, A. (1995). Patterns of adaptation of Ethiopian Jews within Israeli society. In S. Kaplan, T. Parfitt, & E. Trevisan-Semi (Eds), *Between Africa and Zion* (pp. 252–257). Jerusalem: Ben-Zvi Institute.

Youngmann, R., Minuchin-Itzigsohn, S., & Barasch, M. (1999). Manifestations of emotional distress among Ethiopian immigrants in Israel: Patient and clinician perspectives. *Transcultural Psychiatry, 36*(1), 45–63.

9

1-2-3 PAP: A CAMPAIGN TO PREVENT CERVICAL CANCER IN EASTERN KENTUCKY

Elisia L. Cohen, Robin C. Vanderpool, Rick Crosby, Seth M. Noar, Wallace Bates, Tom Collins, Katharine J. Head, Margaret McGladrey, and Baretta Casey

The University of Kentucky Rural Cancer Prevention Center (RCPC) leverages the strength of a community advisory board, coupled with a planned collaboration of community members, public-health professionals, and researchers, to engage in targeted initiatives to reduce the burden of cancer confronting residents of the Kentucky River Area Development District (KRADD). To reduce rates of cervical cancer in this region, one core RCPC initiative includes a novel, multifaceted media campaign and community-based intervention designed to increase human papillomavirus (HPV) vaccination and Pap-testing among young adult women. As this chapter will describe in significant detail, the RCPC "Respect yourself, protect yourself" vaccination and Pap-test media campaign, along with a "1-2-3 Pap" DVD-based clinical intervention, offer two communication strategies to reduce cervical cancer disparities by increasing vaccination, cervical cancer screening, and adherence to the recommended vaccination and Pap-test schedule.

Here, we detail our goals, target audience, and intervention designed from community-based formative research to reduce the burden of cervical cancer in the medically underserved population of Kentucky Appalachia. Second, we describe our formative research findings and associated intervention development efforts to explain the strategic communication choices we made to better target our message strategy for this population. In so doing, we describe our efforts in campaign and message design, testing, and intervention planning. Finally, we conclude by describing our initial findings supporting our intervention and community-based media plan, and discuss areas for future research and program evaluation.

Context and Formative Research Informing the Community-Based RCPC Strategy

Although vaccines that protect young women against high-risk types of human papillomavirus (HPV) are clearly important to public health, this innovation is far more important in populations of women who are unlikely to receive adequate screening for cervical dysplasia and cervical cancer via Pap testing (Markowitz et al., 2007). Despite broad-based efforts to promote and provide Pap testing, cervical cancer is still a public-health concern in Kentucky, particularly in the 54 counties that comprise the Appalachian region of the state (Hall, Rogers, Weir, Miller, & Uhler, 2000; Hopenhayn, King, Christian, Huang, & Christian, 2008; Huang et al., 2002). From 2003–2007, the incidence rate for cervical cancer in Appalachia Kentucky was 10.78 per 100,000 compared to 9.22 for the state; the cervical cancer mortality rate for 2002–2006 in Appalachia Kentucky was 2.91 per 100,000 compared to 2.69 for the state (Kentucky Cancer Registry, 2010). Data from the 2008 Behavioral Risk Factor Surveillance System (BRFSS) indicate that 18.3% of Kentucky women age 18 and older have not had a cervical cancer screening in the past three years (CDC, 2009). Notably, according to 2006 Kentucky BRFSS data, 22% of women in Appalachia Kentucky have not had a cervical cancer screening in the past three years, and 32% of women residing in the eight-county Kentucky River Area Development District—our intervention catchment area—have not participated in cervical cancer screening, the highest district percentage in the state (KDPH, 2006).

Consistent with the findings for Pap testing, we initially suspected that uptake of the HPV vaccine among women most underserved by the health-care system would be quite low in contrast to women enjoying the advantages of well-developed health-care-delivery systems in suburban and urban America (Brewer & Fazekas, 2007; Fazekas, Brewer, & Smith, 2008; Lyttle & Stadelman, 2006; Yabroff et al., 2005). Additionally, young adult women in rural, medically underserved communities are not the usual recipients of target, HPV-marketing campaigns undertaken by pharmaceutical companies (Hopenhayn et al., 2008), and we suspected that uptake may be far lower for young women residing in poverty, as is quite typical among women living in southeastern Kentucky (Schoenberg, Hopenhayn, Christian, Knight, & Rubio, 2005).

To better gauge the need for intervention to promote HPV-vaccine uptake in southeastern Kentucky, we conducted an uptake study of 500 young women (18 to 26 years of age) in our catchment area. From March 2008 through September 2009, a research assistant recruited female patients in any of five regional health clinics located in five rural counties of southeastern Kentucky. During that same time period, a second research assistant recruited women attending a local community college (with buildings located in four of the same five counties used for the clinic sample). Community college women were selected to offset what would have otherwise been a purely clinic-based sample of young women. Women were eligible if they were not pregnant, were 18–26 years old, and had not been vaccinated with Gardasil (the only HPV vaccine approved for use at the time). To avoid self-selection bias, the project was called the "Women's Health Study." Volunteers were told that "the purpose of this survey is to learn more about why women would or would not accept the HPV vaccine if it was made available to them." After providing consent, women completed a questionnaire, and they were then compensated with a $25 gift card for their time. The fact that Gardasil also would be provided at no cost was not advertised or disclosed until after women completed the questionnaire. Before leaving, the research assistant provided women with a voucher to receive three doses of free Gardasil (including a waiver for associated costs of service) at the clinic they were recruited from, or, in the case of college women, a centrally located clinic. These coupons were coded with an ID number that matched the ID number recorded on women's questionnaires. Redeemed coupons were used to create a set of medical records indicating Gardasil uptake. The number of women redeeming the voucher for each dose of vaccine served as the study outcome variables.

Uptake of dose 1 was approximately 25%. Uptake of dose 2 was less than 10% of the entire sample, and uptake of dose 3 was less than 5%. In contrast to a similar study in an urban area of Kentucky, we realized that rural uptake was only half of what we achieved in the urban area. More striking, rural follow-up rates (especially dose 3) were extremely poor, in contrast to uptake in the urban area (Vanderpool, Casey, & Crosby, 2011). These findings are counterintuitive to Hopenhayn, Christian, Christian, and Schoenberg's (2007) findings that 92% of women, age 18–29, residing in two Appalachian Kentucky counties would accept the HPV vaccine for themselves, as well as to the findings of Fazekas et al. (2008) that 66% of women in rural North Carolina were more likely to get the vaccine if it were free.

Follow-up qualitative telephone interviews with a small sample of participants (n = 16) who participated in this initial uptake study revealed that young adult women in the area lack information about HPV and the vaccine, specifically the link between HPV and cervical cancer, the number of shots required, and vaccine efficacy and safety. One participant stated, "It's [HPV vaccine] just like new cars that come out; they're always going to have a bunch of recalls. You know you ought to wait until after everything's been recalled and fixed and everything." Noted barriers to completion of the three-dose series included lack of reliable transportation ("My husband's always at work and that's the only vehicle we've got and I don't drive"), and an appointment reminder system. Interview participants endorsed the idea of taking the HPV vaccine out into the community to raise awareness and knowledge among the population. Consequently, we determined that the intervention effort should be twofold:

1. A community-wide, social-marketing program to promote vaccine initiation, and
2. An individual-level intervention to promote return for receipt of doses 2 and 3.

As an example of our efforts, we have engaged local business, including three different Walmarts in our catchment area, and the local community colleges to assist with vaccine promotion and uptake.

A Campaign to Increase HPV Vaccination

To address these needs, the RCPC has embarked on a multiyear sustained effort to increase HPV vaccination and improve cervical screening rates and adherence to medical protocol in the KRADD. The project timeline included formative research in year 1 (2009) to develop and test culturally sensitive media messages, and a social-marketing campaign and randomized control trial in years 2 and 3 (2010–2011) to determine the feasibility and efficacy of a social-marketing campaign and DVD-based intervention, in improving HPV vaccination and cervical cancer prevention outcomes. If effective, the strategies deployed in this eight-county district of southeastern Kentucky could be expanded to the Appalachian region and adapted for use in statewide cervical cancer prevention efforts. Here, we present health-communication strategies and supporting research addressing two aims. The first study aim is to test the efficacy of a diffusion of innovations program designed to promote HPV-vaccine uptake among women residing in the KRADD. The program will bring women to a point of care for receipt of dose 1. The second aim is to test the efficacy of a postvaccination (post-dose 1) counseling program designed to promote women's return for dose 2 and dose 3, as well as their future receipt of regularly scheduled Pap testing. Four hypotheses stem from these two aims:

Hypotheses from Aim 1

H1: Incidence rates of pre-invasive cervical cancer will be reduced between 2008 and 2014 in the KRADD as compared to the nearby, but not adjacent, Cumberland Valley Area Development District.

H2: Two population-level markers of protective behavior (rates of Pap testing and the number of HPV vaccinations given) will be significantly greater in the KRADD compared to the nearby, but not adjacent, Cumberland Valley Area Development District.

Hypotheses from Aim 2

H3: Women randomized to a postvaccination (post-dose 1 of the HPV vaccine) counseling protocol will be significantly more likely than those receiving standard-of-care vaccination to complete the vaccine series.

H4: Women randomized to a postvaccination (post-dose 1 of the HPV vaccine) counseling protocol will be significantly more likely than those receiving standard-of-care vaccination to return for the next scheduled Pap test.

Program Planning and Research to Enhance Effectiveness of the RCPC Strategy

Community-based participatory research—an orientation to research that Wallerstein and Duran (2006) described as incorporating community theories, participation, and practices into long-term, collaborative partnerships between academics and community members—is clearly well-suited to developing strategies designed to transcend local barriers to HPV-vaccine uptake. In this study, a highly committed and eager Community Advisory Board (CAB) was (and is still) engaged in the research. The programs developed for each aim have a "local flavor," but are also quite adaptable to other parts of rural Kentucky, rural Appalachia, and rural America in general.

The RCPC deployed a nonequivalent control group design (NCGD) to address the first research aim. The nonequivalent area development district selected for the comparison necessitated by hypotheses #1 and #2 is the Cumberland Valley Area Development District (CVADD). Given the two hypotheses, three outcome variables will be assessed at the population level (meaning that data will not be collected from human subjects) to evaluate promotion of HPV vaccination and Pap testing among women in the KRADD. Population-level data will be collected from existing surveillance systems and medical records over the ongoing (multiyear) effort. The social-marketing program consists of varied media (television, radio, social media, closed-circuit school-based, and billboard) messages and outreach (with efforts including outdoor hog roasts held monthly in rotating counties within the KRADD). Details of public-service announcement radio and television advertisements are available in Appendix A. Messages included information that the HPV vaccine would be provided to women residing in these counties at no cost at hog-roasting events and on a regular basis by traveling nurses with the Rural Cancer Prevention Center (headquartered in Hazard, KY).

The second aim includes a feasibility study and evaluation of a novel intervention we describe in some detail. In consultation with the CAB, we wanted the intervention to be scalable without losing fidelity, thus we opted for a DVD. The women recruited into this trial receive their first dose of the HPV vaccine as a part of the outreach efforts used in the social-marketing program. Thus, the randomized controlled trial (RCT) can be viewed as a smaller (substudy) of the NCGD design.

Only young women, ages 19 to 26, will enroll in the RCT. Recruitment and enrollment will occur at the outreach events and the Center for Excellence in Rural Health and include as many as 500 newly vaccinated young women. Any female 19 years of age or older, residing in the study counties and eligible to receive the vaccine (i.e., within the FDA age limit, not pregnant, and not reporting a hysterectomy), will be eligible for inclusion. Those less than 18 years of age or younger will be excluded simply because free vaccination is available to most of the population through the state-run Vaccines for Children program, and adherence is dependent on parental consent. Women who do not speak English (extremely unlikely in the KRADD) will be excluded from study participation.

After enrolling in the RCT and receiving the first dose of the HPV vaccine, women will be randomized into treatment or control (standard care) conditions. Standard care involved information given to women about the HPV vaccine's risk and benefits, the need to return in two and six months for doses 2 and 3, as well as hotline information to report any adverse effects. However, the treatment condition is a novel, 13-minute DVD designed to inform, motivate, and support HPV vaccination schedule adherence and routine Pap-testing behavior. All women also received follow-up reminder vaccination phone calls.

As detailed below, we recruited women both from regional colleges and community-based clinical settings to participate in formative research that iteratively produced core messages for use in the DVD intervention and broader media campaign (to address aim 1). Additionally, eight informal key informant interviews with rural clinic staff members were used to better understand the problem of nonadherence to vaccine scheduling from the clinician's perspective.

Cervical Cancer Prevention Message Development and Testing

During the 2009–2010 academic year, formative individual and group interviews and message testing was conducted to better understand the risk perceptions and attitudes toward HPV and the HPV vaccine among Kentucky women. These interviews included 83 in-depth interviews with Kentucky college students aged 18–26 who were HPV-vaccine early adopters and nonadopters. These interviews provided formative data for researchers' understanding of common attitudes and beliefs Kentucky women had about the vaccine. These interviews, along with CAB feedback, then helped guide the key questions asked in our population-specific research. Five small-group, focus-group discussions drew participants who had resided within our Appalachian Kentucky catchment areas within the past year; these young adult women were eligible if they reported having one dose of the HPV vaccine, were 18–26, and from clinical, community college, and university settings. We focused on young adult women having only one dose of the vaccine by specifically recruiting such individuals within our local clinic population in Hazard, Kentucky. Details of these research activities are described elsewhere (Cohen & Head, in press; Head & Cohen, 2012). We deployed a methodological strategy of "engaged elicitation" for focus-group discussions, whereby we both interviewed research participants for their HPV- and HPV-vaccine-related attitudes, beliefs, knowledge, and related behavioral intentions and behaviors, in addition to testing and modifying drafted DVD and core media messages based on participants' feedback.

These projects supported the researchers' decision that a short 10–15-minute DVD and related media (radio and television advertisements), in addition to social marketing (through social networks, local hog roasts, and information passed out at local events), should focus on an information-motivation-behavioral skills approach. This approach to message design has been widely used in HIV-prevention projects focusing on increasing condom-use behavior (Fisher & Fisher, 2002), and more recently, in improving adherence to health recommendations (see Ferrer, Morrow, Fisher, & Fisher, 2010). Central themes included informing and motivating young women, and helping them to consider the benefits of vaccination and pap tests (as part of a pelvic exam), return to receive doses 2 and 3 of the HPV vaccine on schedule, enhance self-efficacy, and overcome obstacles to vaccination schedule adherence. The slogan, "1-2-3 Pap," was coined by the lead author as an inclusive motto for the educational DVD, that both emphasizes the importance of the three-step vaccination sequence and the need for women to begin scheduling routine Pap testing if they have not already done so. These DVD-related messages also became integral to social-marketing and media efforts targeted to enhance HPV-vaccination and Pap-testing demand in the region.

General Media Message Recommendations (for TV, Radio, and DVD Materials)

In our formative research, women consistently described how cervical cancer seemed to be a distal threat; however, the more looming threat to young women was the abstract threat to their immediate reproductive health caused by HPV. In fact, women reported a certain willingness to go to a health-care provider to "stay on top of" their reproductive health, even if they felt a certain stigma in asking their providers about how to protect themselves against STDs, or in receiving a prescription for birth control (indeed, women described driving to the next county to receive reproductive health care in order to maintain their privacy). An important message for TV, radio advertisements, and the DVDs that resonated with women in the community was that

> One of the scariest things about HPV is that you don't normally have any visible, physical symptoms, so there's no way to know if you have it. That's why it's important for every woman to take steps to protect herself by getting vaccinated with the HPV vaccine and getting regular Pap tests. (See Appendix A, "One of the scariest.")

Eastern Kentucky women also described how, when they saw advertisements (from Merck) about the HPV vaccine, they thought they were not the target audience. Indeed, they considered the women and girls featured in the ad to be different from women who looked and spoke like they did; women in HPV-vaccine marketing materials appeared younger, of higher income, and from a nonrural population. Therefore, we found a local, cancer-control program advocate who was an identifiable face from local television news who could explain to viewers the importance of HPV vaccination and cervical cancer screening to women like them. The news anchor agreed to introduce and conclude the DVD introduction, as well as record material to be used as teasers for our radio and television spots.

Young adult women also described how they did not know a lot about HPV or the HPV vaccine, and they were suspicious of advertising promoting the vaccine from drug companies. In fact, in one focus group, the women asked us pointedly, "do you work for the drug companies?" We clarified that we did not, but that we thought it was one of two important strategies to prevent cervical cancer (the second we mentioned was routine pap testing and pelvic examination). Importantly, when we told these women that the population-level surveillance data indicated that women in Eastern Kentucky were 40% more likely to get cervical cancer than women in the rest of the United States, and that Kentuckians were generally less accepting of the novel HPV vaccine than people in the rest of the United States, they thought that these were "important statistics" to address. Indeed, it was clear from the formative research that there was a need to use television and radio (public-service announcement) strategies to broadly inform people in the region of the Rural Cancer Prevention Center's efforts to reduce cervical cancer by offering free HPV vaccinations and low-cost Pap tests. (See Appendix A for general messages designed for this purpose.)

In our formative research, women also made the consistent and important observation that while their parents were in control of their decision to vaccinate before they were 18, now that they were 19 or older, receiving the vaccine was "no one's business but their own." Specifically, women reacted positively to messages that supported vaccination and pelvic exams as an affirming health activity that could protect their "health for the long-term." Thus, one important message was that getting the free HPV vaccine and low-cost pelvic exams were ways in which women could "respect themselves, protect themselves" (a campaign slogan that the researchers felt would be mindful of these formative findings) from cervical cancer. Women also confirmed researchers' suspicion that after a woman received the HPV vaccine, they may find their decision criticized by romantic partners or family members. To address this concern, the DVD prompts a young woman's response:

> Unfortunately, relationships change, partners come and go, and life happens. But your body will always be your own. In addition, it might be helpful to remind your partner that all 3 doses of the HPV vaccination, along with regular Pap tests, are the best ways to protect yourself from cervical cancer. If you are in a long-term, committed relationship, this vaccine will help you to stay healthy for the long-term. Your partner should be supportive of you making this important decision to take care of your body.

Without exception, the respondents expressed that they liked the strength of the message, and identified with the independent and self-reliant spirit common to mountain culture. As one participant opined: "I mean, it's my health, it's not his. I mean, I gotta look out for myself." Furthermore, the group discussion revealed the everyday language women from the Eastern Kentucky mountain region would use to discuss cervical cancer and pelvic examinations. Unmarried women from smaller, rural Appalachian communities expressed concern about talking openly about sexual and reproductive health, generally, and sexual activity more specifically. Indeed, women in the groups expressed concern about information that would be posted about other women in their community on discussion forums (like Topix.com or Facebook.com).

Although many expressed confidence in their health-care providers (and noted specific encounters with nursing staff at local providers' offices), it was clear that cervical cancer and HPV were not words that they were comfortable using in everyday conversation. Moreover, women wanted to understand in plain language the link between HPV and cervical cancer, and wondered why the vaccine was advertised to young girls. They associated pelvic exams and HPV with "invasive procedures" and, even if they had not had a pelvic exam, understood them as a scary, uncomfortable procedure using tools, performed by an unfamiliar doctor.

To address these concerns, when describing how HPV infection increases the chance that abnormal cells will develop in the lining of the cervix, the DVD explains and graphically demonstrates how these abnormal cells "look like spots on your cervix and, over time, multiply and develop into cancer." We involved a local nurse that women indicated they trusted in a regional medical facility (a local Nurse Practitioner (NP)) to explain to the viewer,

> A Pap test is the tool used by doctors and nurses to do this. To do a Pap test, your doctor or nurse will use a small Q-tip-like brush to take cells from your cervix. It's simple, fast, and the best way to find out if your cervix is healthy.

Our participants felt that it was important that women understand that the Pap test is a quick, simple procedure using a tool that works "like a Q-tip." Thus, the women in the focus groups helped us clarify the description of the Pap test for our media materials. They also helped us develop a message that emphasizes that medical professionals (nurses in addition to doctors, as participants noted) would maintain patients' privacy. In the DVD, the NP counsels the viewer:

> Doctors and nurses do Pap tests every day, and they will walk you through the entire process. Because nurses and doctors are so experienced doing Pap tests, they know how to maintain your privacy so that you won't be exposed. Your entire lower body is draped, and the environment is very professional.

Our formative research also indicated that women who had been vaccinated or who had not been vaccinated (but were open to the vaccine) identified the vaccine as a "new opportunity" for their generation. At the same time, the vaccine's novelty made some women feel that it had not been adequately tested, rushed to market, or may have unknown safety risks. Some women, who received the first dose, became concerned about safety effects (through interpersonal or media sources) even after they had made the initial decision to vaccinate. To address these concerns, we felt that it was necessary to give women basic information on the vaccine's testing over the past decade, and on its safety effectiveness. Given the powerful potential of narratives, we incorporated this information into one young woman's personal story about her grandmother:

> Getting the HPV vaccinations is an opportunity that our mothers and grandmothers never had to protect themselves from HPV and cervical cancer. My grandmother passed away in 1977 from cervical cancer, before the HPV vaccine was developed. I never even had the chance to meet her. If this shot had been available to her generation, I may have had a chance to know her. The HPV vaccine available to our generation is safe, effective, and has been tested by doctors for more than ten years. Ours is the first generation that has access to two ways to fight cervical cancer. Make sure that HPV stops with you by getting all three doses of the vaccine and regular Pap tests.

These preceding messages were incorporated into our final DVD, but also were developed for 30-second, short radio and television advertisements (see Appendix A).

DVD-Specific Message Recommendations

The actual "1-2-3 Pap" postinitial vaccination counseling DVD that was developed is 13 minutes in length, with messages developed and tested by and for Eastern Kentucky women. The DVD is organized into 10 broad message segments (see Table 9.1), in addition to a roughly one-minute opening and closing with cues to action delivered by the local TV news reporter.

Across datasets, it was clear from formative research that young adult women had difficulty with the timing of the vaccination series. Transportation barriers were significant in the rural population, and for young adult women moving outside of the area for employment or higher education, there was a felt concern about the timing of the vaccination. Women reported not knowing the benefits of the second and third doses, and hearing that they needed to "start over" even if they were only a few days late. There was a clear need for message development to reinforce the message that three doses allow women to receive the full benefit of vaccination:

> Making sure you get all 3 doses will allow you to get the full benefits. So it's important to know when to get the next dose, and if possible, schedule an appointment with your healthcare provider so that you don't forget. If you forget to get your second and third shots when you're supposed to, it's OK! You can still go back later to get your second and third doses. But, the best protection comes from following the recommended schedule.

Moreover, the fact that clinical data made clear that many women were not adhering to the vaccine schedule, and that pap testing often was delayed, made it important that the DVD emphasize both the importance of vaccination and pelvic examination combined with the pap test for full protection against cervical cancer. Women in our Hazard, KY focus groups also reported that a credible, local clinic nurse was instrumental in convincing them to get their first vaccine and routine pelvic exam. Thus, our DVD utilized the local advocate as a spokesperson who delivered our medical information in the DVD. The NP explained:

> Remember, the HPV vaccine is not a substitute for your yearly exam—it's only half the battle. That's why completing your HPV vaccine series—3 shots over 6 months—combined with annual Pap tests is the best way to protect your reproductive health and prevent cervical cancer. After you receive your second and third vaccine shots, if you haven't begun having your yearly pelvic exam and regular Pap test, you should! A yearly exam and Pap test is recommended for women younger than 30 years.

This rhetoric was combined with other cues to action spoken by other young adult women from Appalachian Kentucky encouraging women to "commit" to getting all three HPV vaccine shots, and addressed barriers to vaccination (i.e., explaining how the three-shot series is the only way to ensure full protection, issuing reminders for the dosing schedule, and giving advice about what to say to their partners). From these focus-group findings, we developed the "1-2-3 Pap: Easy Steps to Prevent Cervical Cancer" campaign slogan.

In July 2010, the RCPC launched its two-part program (its media and social-marketing campaign effort and intervention with a RCT) designed to raise awareness of and increase demand for HPV vaccination and routine cervical screening in Eastern Kentucky. The "Respect yourself, Protect yourself" educational campaign and "1-2-3 Pap" intervention effort are an example of how health-communication researchers can turn meaningful communication principles and research into practice in an important health area.

DVD Message Evaluation

To provide formative evaluation of the DVD message, in September 2010 data were examined from the first twenty-five participants (aged 10–26, M = 22) enrolled in the counseling arm of the randomized control trial. Preliminary data analysis was conducted to discern participants' perceived message effectiveness (their attitudes and beliefs about the message) and intention to complete the full series of the HPV vaccine (participants viewed the DVD following randomization after receiving the first dose of the vaccine). Perceived message effectiveness was measured with an additive ten-item scale (alpha = .946) where viewers were asked to rate the video on Likert-type items (where 1 = "strongly disagree" and 5 = "strongly agree"). Questions included items assessing whether the "video kept my attention" (M = 4.28, SD = .89); "the video was believable" (M = 4.62, SD = .65); "I liked the video" (M = 4.20, SD = 1.08); "I identified with the women in the video" (M = 4.20, SD = 1.08); "the video was credible" (M = 4.48, SD = .65); "the information in the video was relevant to my own life" (M = 4.08, SD = 1.0); "gave me information or ideas" (M = 4.44, SD = .77); "gave me skills I can use" (M = 4.4, SD = .91); "gave me information that fits my needs" (M = 4.40, SD = .87); and "was made for women like me" (M = 4.28, SD = .84). The perceived message effectiveness scale was negatively skewed, indicating a range of scores from 25 to 50, where 75% of participants had a score above or equal to a 42 (Mo = 50, M = 43, Mdn = 47). All women in the pretest answered "yes" to the question of whether they intended to receive all three doses of the three-dose HPV-vaccine regimen. However, when asked how likely they were to return for dose two on an 11-point scale, where 0 is not likely at all and 10 was extremely likely, 6 women identified a number less than a "perfect 10." However, the range of scores was narrow, with no women indicating a likelihood of less than "8." The positive range of scores between 8 and 10 was similar when women were asked how likely they were to return for dose three; only one person surveyed provided a likelihood of 7. Furthermore, when participants were directly asked whether they thought the DVD would persuade "other women like yourself to return to receive the second and third dose of the HPV vaccine," 92% agreed or strongly agreed that the DVD would do so.

From July 2010 until December 2011, 350 young women enrolled in the RCT (although not all women are due for their second and third vaccination doses). Preliminary evaluation data suggested that women enrolled in DVD intervention may have a significantly ($p = .02$) greater return rate for all three doses (63.5%) compared to the rate observed for those in the control condition (45.6%). The study will be completed in December, 2012, six months after the last enrolled woman is scheduled to receive her third HPV vaccine dose. Research is ongoing to assess the intervention's effect on more distal behavioral outcomes, including Pap testing. Clearly, this preliminary data shows promise for the "1-2-3 Pap" informational and motivational, educational DVD to prevent cervical cancer among Appalachian Kentucky women.

Discussion

We offer the aims of this research as exemplary of how community-based communication strategies may be appropriate to achieving behavioral objectives that can be measured at the population level. That is, over the longitudinal period of our study, we can analyze the KRADD (intervention target case) as compared to nearby CVADD (control) regions to examine population-level markers of preinvasive cervical cancer (from registry data), and protective behaviors (rates of Pap screening and HPV vaccination) as evidence of campaign impact.

The project uses a community-based and culturally sensitive approach. That is, by using community-based participatory research, this work sought to develop intervention messages using an approach "responsive to the values and beliefs of the [local] culture" (Dutta & Basu, 2011, p. 324). In this case, researchers engaged community members to elicit and develop messages appropriate to the local culture, and deployed targeted messages to affect population-level health outcomes. Clearly, the RCPC is an exemplar of a funded prevention research center using targeted communication strategies to reduce the burden of cervical cancer in this medically underserved population. Within the next year, we will have conclusive data from our RCT to examine the effectiveness of the intervention (DVD). Although detailed theoretical grounding and testing of each message component of the media and "1-2-3 Pap" campaign was beyond the scope of this chapter, we offer conclusions from our formative research findings in Appalachia, Kentucky as a model of how such research may guide targeted, appropriate, scalable health intervention development efforts for underserved Appalachian regions and other rural regions of the United States.

Table 9. 1. 1-2-3 Pap Segment Outline

Segment number	Theme	Script exemplar
Opening	1-2-3 Pap is a special project designed to reduce cancer disparities among young adult women from Eastern Kentucky.	Local TV anchor and KY Cancer Control Specialist indicates "Women in Eastern KY are 40% more likely to die from cervical cancer than elsewhere in the United States."
1	HPV is a common disease.	Nurse explains that "many women who get cervical cancer when they are older, likely got HPV in their teens or 20s."
2	The HPV vaccine is effective.	Graphics and Nurse explaining: The HPV vaccine can protect women against the types of HPV that cause 70% of cervical cancers.
3	Granddaughter/grandmother narrative about cervical cancer.	A young woman from E. KY who lost her grandmother (she never knew her) to cervical cancer, explains that the vaccine is a novel opportunity for her generation.
4	The benefits of vaccination and Pap testing overwhelm the short-term consequences of each.	Nurse describes how the pain of a vaccine or the slight discomfort from a Pap test can prevent a lot of pain later, and even save women's lives.
5	1-2-3 Pap; reminder of vaccine and Pap schedule.	Nurse narrates over graphics and pictures of a young adult receiving a vaccine that 3 shots in addition to regular Pap testing provides fullest protection against cervical cancer.
6	Overcoming stigma associated with HPV vaccination.	Young adult women from E. KY explain how to talk to others who may question the need for a vaccine; women can remind their partners that all three doses of the vaccine, combined with regular Pap tests, will help women take care of their bodies and reproductive health for the long term.
7	1-2-3 Pap; remember to schedule your Pap test.	Graphics showing schedule and Nurse suggests that after women get their three doses of the HPV vaccine, "if you have not scheduled a pelvic exam and Pap tests that you should!"
8	A Pap test is private and routinely conducted by medical professionals.	The segment explains a Pap test, and shows the tools used by medical professionals in the process.

9	1-2-3 Pap; overcoming obstacles to vaccination.	Young women from E. KY describe the hurdles that women may face in getting all three doses of the vaccine. These segments describe how women may keep their decision (to vaccinate/get a Pap test) private, how they can talk openly with their loved ones about how HPV can be spread through skin-to-skin contact; how the safe and effective vaccine can protect against this common virus; and how the benefits of the HPV vaccine outweigh the pain.
10	1-2-3 Pap; scheduling efficacy.	Young women from E. KY remind the viewer that they should think through how, where, and when they can follow up by scheduling their next vaccination today.
Closing	RCPC information and cues to action.	Local TV anchor and KY Cancer Control Specialist thanks participants, and reminds them to schedule their next HPV vaccines and Pap tests by calling the 1-888 (toll-free) number listed on the screen.

Appendix A

RCPC Radio and Television Spots

"OUR GENERATION": 30
Music under:
(Young Woman from Eastern KY): "Getting the HPV vaccinations is an opportunity that our mothers and grandmothers never had to protect themselves from HPV and cervical cancer. My grandmother passed away in 1977 from cervical cancer, before the HPV vaccine was developed. If this shot had been available to her generation, I may have had a chance to know her. The HPV vaccine available to our generation is safe, effective, and has been tested by doctors for more than ten years."
TAG (YOUNG WOMEN FROM EASTERN KY) Call 1-866-686-7272 now.

"ONE OF THE SCARIEST": 30
Music under:
(Nurse): "One of the scariest things about HPV is that you don't normally have any visible, physical symptoms, so there's no way to know if you have it. That's why it's important for every woman to take steps to protect herself by getting vaccinated with the HPV vaccine and getting regular Pap tests. You never know whether the odds will be on your side."
TAG (YOUNG WOMEN FROM EASTERN KY) To vaccinate, call 1-866-686-7272 now.

"98%": 30
Music under:
Nurse: "More than 98% of cervical cancers are linked to an HPV infection. Many women who get cervical cancer when they are older likely got HPV in their teens or 20s. That's why it's important for every woman to take steps to protect herself by getting vaccinated with the HPV vaccine and getting regular Pap tests."
TAG (Young woman from Eastern KY): "Respect yourself, protect yourself. For information on how to receive a free HPV vaccine, call 1-866-686-7272 today."

"PARTNER": 30
Music under:
(Young woman from Eastern KY): "Many women who get cervical cancer when they are older likely got HPV in their teens or 20s. That's why you should get the HPV vaccine, and tell your partner if you are in a committed relationship that this vaccine will help you to stay healthy for the long-term. Your partner should be supportive of you making this important decision to take care of your body. Respect yourself, protect yourself. To find out how to receive a free vaccine, call 1-866-686-7272 today."

"PAP TESTS": 30
Music under:
Nurse: "It's important for every woman to take steps to protect herself by getting vaccinated with the HPV vaccine and getting regular Pap tests. Some women experience slight discomfort when getting the Pap test, especially when getting their first Pap test."
TAG (Young woman from Eastern KY): "But not protecting yourself against cervical cancer is much worse, and can lead to surgery, lifelong infertility, or even death. Respect yourself, protect yourself. Call 1-866-686-7272 today."

"RCPC": 30
Music under:
Chastity Gayheart: "Did you know that women in Eastern Kentucky are 40% more likely to get cervical cancer than women in other parts of the United States? To help reduce this statistic, we are doing a special project aimed at reducing cervical cancer in Eastern Kentucky."
TAG (Young woman from Eastern KY): "The Rural Cancer Prevention Center can help women in this listening area receive free HPV vaccinations and low-cost screening for cervical cancer. Help us make KY cervical-cancer free. For more information, call the Rural Cancer Prevention Center at 1-866-686-7272."

"RESPECT YOURSELF, PROTECT YOURSELF": 30
Music under:
(Young woman from Eastern KY): "Cervical cancer is a serious disease. This cancer is caused by a virus called HPV. People have different opinions about the HPV vaccine and Pap tests. For me, these are two very important ways I plan to protect and respect my body. How about you? Respect yourself, protect yourself. For free HPV vaccinations and low-cost cervical cancer screening, call the Rural Cancer Prevention Center at 1-866-686-7272 today."

Acknowledgments

The study described in this chapter was supported by Cooperative Agreement Number 1U48DP001932-01 from the Centers for Disease Control and Prevention. The findings and conclusions in this article are those of the author(s) and do not necessarily represent the official position of the Centers for Disease Control and Prevention.

Bibliography

Brewer, N. T., & Fazekas, K. I. (2007). Predictors of HPV vaccine acceptability: A theory-informed, systematic review. *Preventive Medicine, 45*(2–3), 107–114.

CDC. (2009). 2008 Behavioral risk factor surveillance system. Retrieved from www.cdc.gov/brfss.

Cohen, E. L., & Head, K. J. (in press). Identifying knowledge-attitude-practice gaps to enhance HPV vaccine diffusion. *Journal of Health Communication.*

Dutta, M., & Basu, A. (2011). Culture, communication and health: A guiding framework. (pp. 320–344). In T. Thompson, R. Parrott, & J. F. Nussbaum (Eds.), *Routledge handbook of health communication* (2nd ed.). New York: Routledge

Fazekas, K. I., Brewer, N. T., & Smith, J. S. (2008). HPV vaccine acceptability in a rural southern area. *Journal of Women's Health, 17*(4), 539–548.

Ferrer, R. A., Morrow, K. M., Fisher, W. A., & Fisher, J. D. (2010). Toward an information-motivation-behavioral skills model of microbicide adherence in clinical trials. *AIDS Care, 22*(8), 997–1005.

Fisher, J. D., & Fisher, W. A. (2002). The information-motivation-behavioral skills model. In R. J. DiClemente, R. A. Crosby, & M. C. Kegler (Eds.), *Emerging theories in health promotion practice and research: Strategies for improving public health* (1st ed., pp. 40–70). San Francisco, CA: Jossey-Bass.

Hall, H. I., Rogers, J. D., Weir, H. K., Miller, D. S., & Uhler, R. J. (2000). Breast and cervical carcinoma mortality among women in the Appalachian region of the U.S., 1976–1996. *Cancer, 89*(7), 1593–1602.

Head, K. J., & Cohen, E. L. (2012). Young women's perspectives on cervical cancer prevention in Appalachian Kentucky. *Qualitative Health Research, 22*(4), 476–487.

Hopenhayn, C. R., Christian, A., Christian, W. J., & Schoenberg, N. E. (2007). Human papillomavirus vaccine: Knowledge and attitudes in two Appalachian Kentucky counties. *Cancer Causes and Control, 18*(6), 627–634.

Hopenhayn, C., King, J. B., Christian, A., Huang, B., & Christian, W. J. (2008). Variability of cervical cancer rates across 5 Appalachian states, 1998–2003. *Cancer, 113*(10), 2974–2980.

Huang, B., Wyatt, S., Tucker, T., Bottorff, D., Lengerich, E., & Hall, H. (2002). Cancer death rates—Appalachia, 1994–1998. *Morbidity and Mortality Weekly Report, 51*(24), 527–529.

KDPH. (2006). Kentucky behavioral risk factor surveillance system 2006 annual report. Retrieved from http://chfs.ky.gov/dph/info/dpqi/hp/brfss.htm (no longer accessible).

Kentucky Cancer Registry. (2010). Cancer incidence and mortality rates in Kentucky. Retrieved from http://www.kcr.uky.edu/.

Lyttle, N. L., & Stadelman, K. (2006). Assessing awareness and knowledge of breast and cervical cancer among Appalachian women. *Preventing Chronic Disease, 3*(4), A125.

Markowitz, L., Dunne, E., Saraiya, M., Lawson, H., Chesson, H., & Unger, E. (2007). *Quadrivalent human papillomavirus vaccine: Recommendations of the ACIP*. Atlanta, GA: Centers for Disease Control and Prevention.

Schoenberg, N. E., Hopenhayn, C., Christian, A., Knight, E. A., & Rubio, A. (2005). An in-depth and updated perspective on determinants of cervical cancer screening among central Appalachian women. *Women Health, 42*(2), 89–105.

Vanderpool, R., Casey, B., & Crosby, R. (2011). HPV-related risk perceptions and HPV vaccine uptake among a sample of young rural women. *Journal of Community Health, 36*(6), 903–909.

Wallerstein, N. B., & Duran, B. (2006). Using community-based participatory research to address health disparities. *Health Promotion Practice, 7*(3), 312–323.

Yabroff, K. R., Lawrence, W. F., King, J. C., Mangan, P., Washington, K. S., Yi, B., Kerner, J. F.,...Mandelblatt, J. S. (2005). Geographic disparities in cervical cancer mortality: What are the roles of risk factor prevalence, screening, and use of recommended treatment? *Journal of Rural Health, 21*(2), 149–157.

10

Vulnerable Old Women of Northern Ghana at Risk of Being Ostracized for Witchcraft: How Wise Village Leadership Brings Them Back From the Brink

Torill Bull, Maurice B. Mittelmark, and Peter Sanbian Gmatieyindu

This book explores social and cultural factors that contribute to health inequalities, and describes how communication can foster better health outcomes for at-risk groups. This chapter's contribution is its focus on leadership behavior and processes, as it presents a case study from Ghana, showing how wise leadership practices, verbal and especially nonverbal, can rescue vulnerable people from the threat of isolation, ostracism, and banning.

For most people, the word, *witch*, brings connotations of evil forces, of the Medieval Age, of times long past. Not so for people in contemporary northern Ghana, where witchcraft is still a force to be reckoned with. Threat connected to witchcraft is maybe most potently felt by the most vulnerable, least protected women in local communities, especially elderly widows. The threat does not lie in being on the receiving end of spells, but in being accused of witchcraft, being mistreated, isolated, and ostracized, and being banned from their villages. Several thousand women in northern Ghana live unwillingly in the confinement of witches' camps, resulting from witchcraft accusations.

This chapter does not discuss the reality of wizardry, but presents a case from the Bole District in the Northern Region of Ghana, where an intriguing leadership intervention protects women at risk from the devastating fate of being declared a witch. The chapter opens with an overview of the case's context. Next, the leadership intervention and some of its communication aspects are described and discussed. The concluding part of the chapter discusses the role of "phronetic leadership," in successful community-based health interventions.

The Case Context

Aishetu lives in a village of about 800 inhabitants in the Bole district in the Northern Region of Ghana. Even if Ghana is doing well compared to other countries in sub-Saharan Africa, most of the benefits from the economic development have favored the south of the country. In the north where Aishetu lives, the few roads are rough and far between. Even the most basic health services are hard to access. Children die from preventable fevers. If complications arise when a woman is in labor, the only transport to qualified health care might be to carry her on the back of a bicycle over a distance of several miles (Bull & Mittelmark, 2010).

People in this sparsely populated part of Ghana live in small villages, where livelihoods come mostly from the juggling of subsistence farming and small-scale trading of goods and services. The houses in Aishetu's village are made of mud, most of them with thatched roofs. The most fortunate villagers have brick houses with roofs of corrugated steel sheets. Such houses are not so easily damaged during heavy rains and wind. Not that rain is a common complaint here—the rainy season is too short; the land is too dry. Making a decent crop takes hard work, a good portion of luck, and the benevolence of natural forces. Still, the north of Ghana is not a region without resources. There is gold in the ground, but the gold industry does not benefit the poverty-stricken local population much. There are also agricultural resources in the region, like the Shea tree which yields nuts for the valuable Shea butter production, which is an important source of income for many local women. However, the Shea tree is becoming scarcer, as even highly productive trees are felled for firewood and the production of charcoal.

The Bole district is inhabited by quite a number of different ethnic groups living together in villages and towns. Mostly, it is a peaceful district, with only occasional chieftaincy and land disputes to disturb the tranquillity. Chieftaincy is an important institution in the area. Though the area is governed by the District Assembly and the local government structures and institutions, chieftaincy plays a crucial role in the maintenance of peace and the promotion of development in the locality.

The tribes of the area have diverse and unique cultures, on display during marriage ceremonies, naming ceremonies, and funeral rites, and in forms of drumming, singing, and dancing. Christians,

Muslims, and Traditionalists live together in the same villages, and their similarities seem more salient than their differences; there is generally a sense of unity.

Ancestral beliefs and beliefs in spirits have long traditions in the area, and witch doctors have the power of communicating between villagers and the spirit world. The sacrifice of a goat may be necessary to pacify the gods, even if food is scarce, children are very hungry, and the goat might have relieved the household's hunger. There are many taboos for the Traditionalists, and many of the taboos seem to benefit men. A woman cannot eat the important protein source of egg, for instance, while men can. The culture is patriarchal, with decision power resting with men, who may have several wives.

Aishetu lives with her daughter and son-in-law. From where she sits outside the house she sees red dust covering the ground between the village houses, traced by a fine feathery pattern from brooms. Women in the communities take pride in having a well-swept yard. Guinea fowl peck at invisible specks on the ground and small goats satisfy their hunger from a thorny bush at the outskirts of the village. Two large trees create a beautiful green contrast against the blue sky, and an equally beautiful spot of shade on the red ground beneath them. Aishetu watches the children playing. She is tired. Aishetu is no longer young. The midday sun is hot.

Aishetu grew up in this village. She never went to school—there was no school, and if there had been one, Aishetu could not have attended. Her parents could not have managed without her help. From a very early age she assisted her mother around the home and in the fields. Life for women in Aishetu's village seems to be an endless row of demanding days. Starting early in the morning, water and firewood must be fetched, often demanding long walks. Children must be taken care of, the morning meal prepared for husband and children, and the house and yard swept. Next comes the walk to the little farm plot for a day of work under the scorching sun. Babies must be carried on the back, toddlers led by the hand, and far too often, a new little one is carried in the womb, all while the land must be cultivated with short hoes. The cultures of most of the tribes put heavy burdens on women (Bull, Duah-Owusu, & Andvik, 2010). Almost every household chore is designated for women. It is common to see a man and his wife returning from the farm with the woman carrying firewood or foodstuff, and carrying a baby at the back, while the husband walks behind her with just a cutlass or a hoe. That is his privilege, and the carrying of things, especially of children, is a female responsibility. Back home, the woman has to prepare dinner, take care of the children, and fetch water for the evening baths. The man has also worked hard during the day, but in the evening he may lie down or go visiting friends, only to come home when dinner is ready. His wife will even have prepared his evening bath for him. Both men and women work hard, but men get more rest. Even the nights bring little rest to women, with sexual obligations to their husbands, babies to feed, and frequent disturbances from children with coughs or diarrhea. In Aishetu's village, women go on bearing children from puberty to menopause. It is not uncommon to lose close to half of your children. To her grief, Aishetu has lost four of the seven children she gave birth to before her husband died.

Aishetu is a widow now, past 60 years of age. Her eyes are getting dimmer; her back is no longer straight. She has started to limp; there is a pain in her joints. She can no longer manage the hard work in the fields, so she depends on her daughter and her son-in-law for food and shelter. They struggle—there is never a surplus. The other wives of her son-in-law are annoyed by her presence, by the burden she represents on the scant resources of the household. And the story of Aishetu unfolds as an example of the destiny of thousands of vulnerable women in northern Ghana: as the child of one co-wife falls ill and dies, Aishetu is accused of bringing the death about by witchcraft. The evidence is undisputable. First, the mother of the child saw Aishetu in a dream the night before the child died. Second, when Aishetu is brought before the village chief, the elders kill a chicken to see which way it falls to the ground. It falls on its beak, confirming beyond any doubt that Aishetu is a witch. There is no way to appeal. Aishetu receives severe maltreatment in the village. She is driven away to live in a witches' camp.

Aishetu's story is an amalgamation of many stories. The case of witches' persecutions in contemporary northern Ghana is well documented by nongovernmental organizations (NGOs) such as World Vision (Boakye, 2009), the media (Abagali, 2010), and in scientific literature (e.g., Adinkrah, 2004; Badoe, 2005). The reports of who are accused as witches are fairly uniform. These are women who, in a strictly patriarchal culture, are not sufficiently under the protection of men. In some cases, they may be particularly strong women; "free" women with economic success, even if living without a husband. Locally, such a woman is sometimes called an "NGO," referring to the fact that she is not under governance of authority (in this case, of a man). What else could be the source of her success than witchcraft? If she had been living with her husband, his support would have been the obvious explanation for her success. Alone, there are no obvious reasons for her success.

However, it is far more common that the witches are the weakest, the most vulnerable in the community, women like Aishetu. Most of them are widows in late midlife, past menopause. They have few resources, no education, and little social support. The normal changes in an aging person sadly give rise to accusations of sorcery: wrinkles, sagging skin, weaker posture, eyes growing more opaque. Many elderly widows may have returned to the village and household of their fathers to live in the extended family of their brothers with wives. In these households they become the lowest in the status hierarchy, and are often viewed as nothing more than an economic burden. They are at the mercy of the extended family, and far too often, there is no mercy. Some have speculated that driving such a burdensome person away might have economic reasons. When an unhappy event triggers emotion, these elderly women become scapegoats, being accused of causing the event by witchcraft. In 1997, after a meningitis epidemic which took more than 500 lives in northern Ghana, three elderly women were lynched after having been accused of causing the outbreak (Badoe, 2005). At least 13 women have been killed in northern Ghana between 1995 and 2001 on the basis of accusations of witchcraft (Adinkrah, 2004). Elderly women have been killed by husbands, sons, daughters, other community members, and mobs. Faced with such a possibility, the only option is to flee the village in search of refuge. Many women find such refuge in the northern Ghana witches' camps, of which there are at least six: Kukuo, Nyani, Kpatinga, Gnani, Naboli, and Gambaga. Reports as to how many women live in these camps vary from one to eight thousand. According to some, alleged witches flee to the camps to find protection from violence, giving the camps the role of sanctuaries. Others maintain that witches are expelled to the camps and kept in confinement. Stories from women in the camps confirm both versions, depending on the case (Badoe, 2005). Some witches are said to maintain ties with family, others are said to be totally ostracized. Some try to return to their villages after the introduction of a new, more benevolent village chief, hoping for a more accepting attitude in the community.

Reports from life in a witches' camp vary from stories of serious mistreatment of women to a life which is basically a struggle for livelihood and survival. Exorcism is performed on their entrance to the camp, and in some camps women have to keep taking various brews to keep the evil forces at bay. Young girls from the witches' family are at times sentenced to accompany the women in their imprisonment, giving them practical support. For these girls, growing up in a witches' camp interferes with their education, and inflicts a serious social stigma, thus reproducing vulnerability across generations of women. The social stigma and mental trauma of being declared a witch is profound. Your integrity, your social status, your social ties—all are taken from you. This is not only the experience of a very few unfortunates; most people in the region know of someone being accused of being a witch.

The Leadership Intervention

David Mensah grew up in one of the poorest villages of the Bole district, destined and brought up to become a fetish priest, following his uncle. By a series of events, however, David Mensah converted to the Christian faith and received opportunities for education. Together with local friends he developed a desire to fight poverty and deprivation in his native region, and the forerunner for what is now called the Northern Empowerment Association (NEA) was established (GRID-NEA, 2010). David Mensah happened to get an opportunity to travel to Canada and complete a PhD. In Canada he married a Canadian woman, Brenda, and together they returned to northern Ghana to continue the work of NEA (Mensah, 2003). Dr. Mensah is Development Chief in one of the local villages, and he has also been appointed as Acting Paramount Chief for a traditional area, comprising several villages. NEA is thus something of a rarity, as a locally originated NGO led by local people with a thorough knowledge of the culture and context. Dr. Mensah is a well-respected man in the local communities, which contributes significantly to the success of his development interventions.[1]

NEA has a multipronged approach to development in the Bole District. Activities include making improvements to water and sanitation, education, nutrition, income generation, peace building, health care, aquaculture, and agriculture. As part of a wider agricultural program, women in need are enrolled in women's cooperatives and provided with initial resources and training to succeed as small-scale groundnut farmers. This is one of many ways in which NEA is seen as a concrete resource to the communities, another significant contributing factor to the success of NEA's interventions. So, which interventions, and in which ways are they successful?

At the heart of all NEA activities is a concern for the most vulnerable of the vulnerable. Elderly widows have been identified as such a group. Dr. Mensah, knowing that the risk of witchcraft accusations is the beginning of a path that threatens a woman's physical and mental well-being, started attending to women who seemed to be at risk for being accused of practicing witchcraft. Noticing a woman at risk, he would appoint her as an agricultural consultant in her village. People who wanted access to various resources from NEA were told to contact this woman. Initially, this caused wild protests—after all, she might be a witch and they would not have anything to do with her! The dry reply from Dr. Mensah would be: "I'm sorry, that is the only way you can take part in our program—she is our representative in your village." In the next phase of his intervention, Dr. Mensah took these women at risk of accusation, one at a time, as passengers in his truck when he went on errands around the villages. Riding in the modern truck, on the seat next to the chief, was something that was really noticed. Slowly, the status of the elderly women in the villages changed. From being at risk of social exclusion and even lynching, the elderly women were given special seats of status when the communities gathered, as a recognition of the key positions they held in the villages.

We now consider some of the communication aspects of the leadership acts by local chief and NEA director, Dr. Mensah. Our approach here is, and must be, somewhat superficial. There are a number of scholarly lenses available to communications scholars, all of which could lead to quite different and contrasting analyses (Fairhurst, 2008). In fact, we do not claim to view the subject through any particular communications lens. However, as health-promotion researchers, we are influenced by some of the principle ideas in the public-health communications literature.

That literature points to several important elements in the health-communication process (Finnegan & Viswanath, 2004): there is a message, which is described by its content; there is a sender of the message; there is (are) a channel(s) for communicating the message; there must be one or more recipients of the message. The communication process also includes the effect of the message, or, to quote Lasswell (1948), "who says what in which channel to whom with what effect?"

In the leadership intervention described here, the sender of the message is Dr. Mensah, the respected local chief and NGO director. The recipient is the local community, from the most powerful to the most vulnerable. The channel of sending the message is behavioral, a few simple acts of Dr. Mensah's: He appoints poor women as agricultural consultants, and he invites them as passengers in his car when visiting local villages. What, then, is the content of the message? And what is the effect?

We offer a fairly basic interpretation of the content of the message, which could, of course, be explored in great depth. By his message, Dr. Mensah states: "I, the chief, respect these women, whom you may suspect are witches." Another part of the message is: "These women are protected by a strong male." Through this message Dr. Mensah addresses cultural aspects of relevance—the influence of chieftaincy, and the need for women in patriarchal cultures to be protected by men. Both of these message parts lead to a shift of status of the elderly widows in a way which is highly sensitive to the particular cultural context. In the present context, there is an apparent increase in status by association with a powerful man. However, the status of the women is also increased in a more specific and empowering way: they are given key roles in the villages as regulators of access to resources from NEA. This challenges the power structures in the communities, and forces an active position-taking on the part of the villagers: they cannot passively regard the women as having high or low status; they actually need to relate to them to access valuable resources.

Moving to the issue of communication effects, there are at least two outcomes with health significance. First, the elderly widows are not condemned as witches, they are not ostracized, and they are not expelled to witches' camps. Second, over time the regard for these elderly women changes in the villages to the degree that seats of prominence are provided for them during community gatherings.

Why does this simple communication intervention work? This brings us to a discussion of processes through which cultural norms can change within a community. Ensminger and Knight (1997) discussed this, exploring processes in an East African rural society. They concluded that "bargaining" is one of the vital processes through which cultural change takes place:

> Asymmetries in resource ownership affect the willingness of rational self-interested actors to accept the bargaining demands of other actors. Unlike other sources of bargaining power, asymmetry of resource ownership is a factor common to a wide range of bargainers in a society. This is a desirable feature of any theory of norm emergence; only factors that are widely shared will produce the systematic resolution of interactions necessary for the emergence of social norms. Social actors suffer significant costs for the failure to coordinate on social outcomes, but those costs need not be suffered uniformly. Actors who have either fewer alternatives or less beneficial ones than others will be more inclined to respect the demands of those others. In this way, the existence of resource asymmetries in a society can significantly influence the choice of a social norm. (p. 5)

Dr. Mensah's status in the locality was of prime importance in this story. Dr. Mensah was not an outsider with a message about the need to change local customs. He was an insider with credibility. He was widely respected, both as a maker of peace between tribes and as the first man from the area to obtain a PhD. He had shown through the years that he did not use the opportunities life had given him for selfish indulgence, but rather dedicated himself to his local area's development.

He also was a man with resources which could be shifted. The local villages could gain from being included in the various development projects of NEA, so there was a motivation to bargain (Ensminger & Knight, 1997). Third, there was a power asymmetry in favor of Dr. Mensah, increasing the likelihood that norms would change in the direction he exemplified.

One could ask the question of whether this power asymmetry means that the intervention was enforced on a community in a top-down, culturally insensitive manner. That is, of course, a position one could take. Alternatively, one could view this as a case with several levels of power involved. There is the bottom level of the powerless, vulnerable, elderly widows, subdued and threatened by the intermediate level of powerful individuals in the community. At the higher level there is the power of the chief, who "bows down" to share power with the least powerful in the community, thereby empowering them and enabling a stronger bottom-up process in the next round.

Phronetic Leadership in Community Interventions

This chapter presents a case study of community development focussed on how local leadership and communication can be a force for social change and social justice. Its documentation of leadership is unusual, in its location (Ghana), in its focus on local leadership, and in its attention to leadership *per se* rather than leader or leadership development. Leader development aims to create human capital by increasing a particular individual's personal power, knowledge, and trustworthiness, and her or his skills of self-awareness, self-regulation, and self-motivation (Day, 2001). Leadership development aims to develop social capital through improving the quality of interpersonal respect and trust, and social awareness and skill such as empathy, political awareness, team building, and conflict management (Day, 2001). The methods of leader and leadership development are many, emanating from management science and organizational psychology, and including systematic feedback on performance, coaching and mentoring, and formal programs. The study of leader and leadership development concentrates on the evaluation of the effects of these and similar interventions, and a very large literature on these subjects has been generated in the academic and popular genres.

The study of *leadership per se* focuses mostly on the stories of giants of their times—presidents and generals, captains of industry, discoverers, inventors, and conquerors—famous and infamous (Kellerman & Webster, 2001). There is strongly emerging interest in understanding the practices of everyday leadership that foster opportunities of change. Yet another, but much less developed, approach to the study of leadership focuses on the *phronesis* of leadership practice *in situ*. The leadership study in this chapter may be considered as a study of "phronetic leadership," therefore this discussion now turns to the concept of phronesis in general, then to the more specific concept of the phronesis of leadership. Phronesis is a concept introduced originally by Aristotle, further explored by Arendt, Habermas, and Dunne among modern philosophers, and re-illuminated more recently by Flyvbjerg (2001) and Halverson (2004).

Phronesis is one of Aristotle's three intellectual virtues, the others being *episteme* (scientific knowledge) and *techne* (craft/art, the application of scientific knowledge, e.g., technology). Episteme and techne are known by the contemporary terms, *science* and *technology*, but phronesis has no modern term. According to Flyvbjerg (2001), phronesis is deliberation about values with reference to *praxis*, a kind of applied ethics. The result of phronesis is ethical and effective action that is appropriate, given the simultaneous consideration of (a) general "rules" of conduct, and (b) the particular circumstances of the moment. Phronesis is exhibited in the form of particular instances of action, or wise practice, which comes from experience in combination with consideration, judgement, and choice; "phronesis requires the interaction between the general and the concrete" (Flyvbjerg, 2001, p. 56).

Phronetic leadership is characterised by wise leadership practices. Working in the arena of school leadership research, Halverson (2004) has captured the essence of phronetic leadership admirably:

> Policy makers, program designers, professional organizations and educational researchers each contribute different aspects of the knowledge necessary for local school leaders to improve student learning. Policy makers, for example, create expectations, guidelines, and resources to direct local education efforts. Program designers cull together packages of techniques and procedures designed to further educational goals. Researchers often consider what happens as a result of the intervention, providing valuable feedback on what did and did not go right. Each kind of knowledge is critical to creating conditions to improve learning for students. However, there is another crucial kind of "practical" knowledge necessary for school leaders to pull these kinds of knowledge together. The knowledge of how to apply general principles, generic tools, or wide-scale evaluation information to the idiosyncrasies of particular contexts constitutes a separate, and often under explored realm of leadership knowledge. (pp. 90–91)

That "practical" knowledge is phronesis, about which Halverson goes on to write:

> While the idiosyncrasies of each school may be safely ignored or bracketed away from a policy, program, or evaluation perspective, establishing

> the conditions for improving student learning amidst these idiosyncrasies is the primary work of local school leaders. Successful school leaders rely on their sense of the local particulars to determine, for example, which teachers have prior experience teaching special education, which librarians are related to board members, which children lack supportive families, and which policies can be safely ignored. As a result of experience and reflection, successful leaders use their sense of the details to "see" the problems of their schools as solvable within local constraints, and are able to develop successful action plans to address problems. Over time, the patterns of how successful leaders recognize and solve problems come to constitute the professional expertise of these school leaders. (p. 91)

Following from the above, both with regard to communication and many other acts, the tribal leadership by Dr. Mensah, documented in this chapter, is characteristic of phronetic leadership. Since good leadership is a fundamental prerequisite to community development (Chaskin, 2001), we wish to learn from exemplars, as described in this chapter, to inform leadership-development programs. Many such programs are implemented in communities in the Global South as part of community capacity-building projects, often funded and managed by public and private donors such as the United Nations Development Programme (UNDP, 2006). An analysis undertaken by UNDP of what works and what does not work in leadership development includes the following conclusion: programs should not focus on individual leaders, but rather should focus on building a culture of leadership and collective capacities (UNDP, 2006, p. 21). However, the study of individual leaders may be quite important for providing exemplars that turn leadership generalities into concrete illustrations.

In the study of school leadership, Halverson and others have developed a methodology for constructing leadership narratives based on the analysis of artifacts such as policies, programs, and procedures that school leaders produce (Halverson, 2004). In tribal settings in the Global South, this methodology is not feasible. However, as this chapter illustrates, the qualitative methods of observation and key-informant interview can produce compelling accounts of phronetic leadership at the local level. Not surprisingly, a good deal of phronetic leadership, as we have observed it, is revealed in communicative behavior, verbal and nonverbal. Who is selected to sit next to the chief as he drives his vehicle sends a powerful message to all, those that observe and those they talk to, about what they have seen. This is the picture that is worth a thousand words; arguably, not even the most brilliant speech by a chief about the importance of social inclusion could have the impact that his overtly inclusive behavior can have.

Note

1. The third author of this chapter spent several years coordinating NEA women's co-operatives, while the first author got acquainted with NEA during a research field visit, collecting data through observation, key informant interviews, and focus-group interviews.

Bibliography

Abagali, C. (2010, August 25). Child abuse in witches camps of Northern Ghana. Retrieved from http://www.ghananewsagency.org/details/Features/Child-Abuse-in-Witches-camps-of-Northern-Ghana/?ci=10&ai=19726.

Adinkrah, M. (2004). Witchcraft accusations and female homicide victimization in contemporary Ghana. *Violence Against Women, 10*(4), 325–356.

Badoe, Y. (2005). What makes a woman a witch? *Feminist Africa, 5,* 37–51.

Boakye, F. (2009). Ghana: Facing the challenges of life in 'witch camps.' World Vision International. Retrieved from http://beta.wvi.org/africa/article/ghana-facing-challenges-life-%E2%80%98witch-camps%E2%80%99.

Bull, T., Duah-Owusu, M., & Andvik, C. A. (2010). 'My happiest moment is when I have food in stock': Poor women in Northern Ghana talking about their happiness. *International Journal of Mental Health Promotion, 12*(2), 24–31.

Bull, T., & Mittelmark, M. (2010). Living conditions and determinants of social position amongst women of child-bearing age in very poor ruralities: Qualitative exploratory studies in India, Ghana and Haiti. IUHPE Research Report Series, Volume 5 Number 1. Paris, France: IUHPE. Retrieved from http://www.iuhpe.org/index.html?page=50&lang=en#books_rrs.

Chaskin, R. J. (2001). Building community capacity: A definitional framework and case studies from a comprehensive community initiative. *Urban Affairs Review, 36*(3), 291–323.

Day, D. V. (2001). Leadership development: A review in context. *The Leadership Quarterly, 11*(4), 581–613.

Ensminger, J., & Knight, J. (1997). Changing social norms: Common property, bridewealth, and clan exogamy. *Current Anthropology, 38*(1), 1–24.

Fairhurst, G. T. (2008). Discursive leadership: A communication alternative to leadership psychology. *Management Communication Quarterly, 21*(4), 510–521.

Finnegan, J. R. Jr., & Viswanath, K. (2004). Communication theory and health behaviour change. The media studies framework. In *Health behaviour and health education: Theory, research, and practice* (3rd ed., pp. 361–388). San Francisco, CA: Jossey-Bass.

Flyvbjerg, B. (2001). *Making social science matter: Why social inquiry fails and how it can succeed again.* Cambridge, UK: Cambridge University Press.

GRID-NEA. (2010). About. Retrieved from http://grid-nea.org/.

Halverson, R. (2004). Accessing, documenting and communicating practical wisdom: The phronesis of school leadership practice. *American Journal of Education, 111*(1), 90–121.
Kellerman, B., & Webster, S. W. (2001). The recent literature on public leadership: Reviewed and considered. *The Leadership Quarterly, 12*(4), 485–514.
Lasswell, H. (1948). The structure and function of communication in society. In *The communication of ideas* (pp. 215–228). New York, NY: Institute for Religious and Social Studies.
Mensah, D. (2003). *Kwabena: An African boy's journey of faith.* Belleville, Ontario, Canada: Essence.
UNDP. (2006). Leadership for human development (A UNDP capacity development resource). UNDP. Retrieved from http://lencd.com/data/docs/246-Leadership%20for%20human%20development.pdf.

11

A Case Study of the Community Liaison Pilot Program: A Culturally Oriented Participatory Pilot Health Education and Communication Program to Decrease Disparities in Minority HIV Vaccine Trial Participation

Robin Kelly, Al Hannans, and Gary L. Kreps

Ethnic minorities and marginalized, vulnerable populations in the United States have been disproportionately harmed by HIV since the AIDS epidemic began nearly thirty years ago. The Centers for Disease Control and Prevention (CDC) estimates that African Americans/Blacks and Latinos represent approximately 27% of the U.S. population, yet have higher rates of HIV/AIDS at all stages—from infection with HIV to death from complications related to AIDS—compared to other ethnic groups (CDC, 2010). African Americans/Blacks, who account for approximately 12% of the total U.S. population, have been particularly hard hit by HIV/AIDS. They represent nearly half of all new infections and 46% of people living with HIV; the AIDS case rate for Blacks per 100,000 people was 10 times that of Whites in 2007 (The Henry J. Kaiser Family Foundation, 2009).

Disparity Unto Death

According to existing research, minorities, in particular African Americans, suffer the greatest difference in life expectancy compared to non-Hispanic Whites. Studies have narrowed down the diseases to certain ones, such as cardiovascular diseases, hypertension, and HIV/AIDS (Frew et al., 2010). In fact, Wong et al. (2002) reported that when adjusted for age, sex, and level of education, Blacks fared worse than Whites for the majority of specific causes, with hypertension contributing the most, followed by HIV disease and diabetes. These diseases contribute more to the disparities in mortality rates than even homicide.

Prevention Responses

One of the best hopes for the prevention of HIV/AIDS and other diseases may be on the horizon. While various trials for microbicides and vaccine clinical trials have shown promise for HIV prevention, there is a disparity in the United States among those who participate in clinical trials to test the effectiveness of these innovations; there is very low minority representation in HIV clinical trials (Frew et al., 2010). Sullivan, McNaghten, Begley, Hutchinson, and Cargill (2007) found that one of the most commonly reported reasons for not participating in a clinical research study by minorities was related to a lack of information about the research. This health-education intervention was designed to respond to this lack of relevant information.

In reviewing additional barriers to participation, they are structural as well as personal. Individuals, particularly African Americans, face a variety of structurally situated barriers that prevent them from participating in clinical research. Funding for HIV/AIDS communication, research, trials, and prevention initiatives has been constrained by the 2007–2009 economic downturn in the US, resulting in less financial resources for HIV/AIDS prevention programs than in the past. The CDC, which funds 58%, or 337 million, of all prevention services directly or indirectly through state and local health departments, flat-funded or reduced its HIV prevention programming (NASTAD & The Kaiser Family Foundation, 2009). In addition, the federally funded AIDS Drug Assistance Program (ADAP), which provides approved AIDS medications for low-income people without insurance, served only one-third of those eligible in 2008 (NASTAD & The Kaiser Family Foundation, 2009). Given that minority groups, the impoverished, and marginalized populations continue to bear the brunt of this pandemic, the downturn in resources has created reason for concern among these groups that rates of mortality will increase and erase limited previous gains in HIV/AIDS prevention. Thus, improved HIV/AIDS information dissemination about the hope of vaccines for prevention seems necessary to bolster spirits and participation in trials by persons of color.

To combat this disparity, we have a case study of a pilot program focused on participatory engagement and community health promotion dissemination. Although there were challenges along the way, this case study reveals that it is possible to reach a large number of people through a simple process. Through this process, myths can be addressed, and the opportunity for increasing awareness and then support for HIV trials can be enhanced, which, hopefully, will lead to a decrease in health disparities in the participation rates of African Americans.

Description of the Innovative Approach

Through the use of a participatory approach—defined here as an approach where members of the community are engaged in decision-making and sharing, as well as ownership of the product, namely the message—the intent was to use social network theory to educate others about HIV vaccine research communication (Valente, 2010). This pilot project demonstrated that social-networking programs based on peer support, trust, and culturally sensitive messaging can be cost-effective, low-tech, and assessable by anyone of any age, making culturally appropriate message dissemination easier.

Voluminous research has shown that health messages shared by members of one's own social networks, especially among those with similar race and/or ethnicity, are generally well-received (Kreps, 2008; Kreps & Sparks, 2008). Our communication strategy demonstrated that such a methodology is not only based on the person, but also on the message tailored by and for each network. This is the case, particularly if the message builds upon key cultural attributes of the community and enhances receptivity and acceptance (Valente, 2010).

This study represents a descriptive report of a pilot program that trained seven community liaisons in five USA states to tailor an HIV-vaccine message for their networks using participatory engagement. The purpose of this initiative approach was to promote awareness of HIV-vaccine research to minority populations, to encourage participatory engagement concerning HIV/AIDS within minority communities, and to enlist community members in adapting and disseminating an HIV/AIDS message across their social networks. These community members in this pilot program disseminated relevant HIV/AIDS information to members of their social networks. These members were also invited to disseminate the health information to their social networks in an iterative, evolving process of information diffusion through social networks (Valente, 2010).

Health Education-Health Communication Case Study

The Community Liaison Pilot Program was a theoretically based, cost-effective model that stressed cultural sensitivity and participatory health communication. Based on the Diffusion of Innovation theory, strategic use of familiar communication sources and channels, as well as timing for dissemination were key components of this multilayer endeavor (Haider & Kreps, 2004; Rogers, 1995). The message of HIV/AIDS vaccine research awareness was the core of the program. The process through which this message reached targeted minority communities was a significant factor in this information-dissemination intervention. The program used liaisons to infuse the message of HIV-vaccine research through their current social networks, and instructed the liaisons in the innovative use of participatory engagement. This engagement allowed targeted members of each network to provide feedback about the messages, and then recommend the most appropriate channels of dissemination (interpersonal, print, etc.), given the culture of their network. The use of a printed message was a tool for liaisons in their discussions, which afforded the members of the network the opportunity for later reference with members of the target population. Timing, as theoretically defined, connoted the period of decision-making and acceptance of the message. Time thus referenced the recorded process through which an individual—including a liaison—transitioned from gaining knowledge about HIV/AIDS, to making a decision to adopt the information, and then actively disseminated the information to others (Crosby, Holtgrave, Bryant, & Frew, 2004). In this intervention, timing included identifying, recruiting, and supporting liaisons who would become early adopters of HIV/AIDS information, and engaging them in the processes of message tailoring and dissemination through participatory engagement.

Recruitment and Selection of Liaisons

The pilot program needed to recruit volunteer liaisons with diverse backgrounds, races and/or ethnicities, and geographic locations. The most salient requirement for participation was a commitment to the purpose of this pilot program, and a willingness to dedicate the necessary time to the program activities, such as HIV communication plan development, self-reporting, and program evaluation. Recruitment activities were conducted at major HIV/AIDS events sponsored by the National Minority AIDS Council (NMAC). This organization provided access to a national pool of activists, program planners, evaluators, health-care providers, academicians, and participants dedicated to HIV/AIDS prevention, care, and treatment. It is also the organization that sponsors the United States Conference on AIDS (USCA), the largest AIDS-related gathering in the US. This conference brings together over 3,000 people from various parts of the country who are interested in HIV/AIDS. The conference is a forum for building or reconnecting with networks, and for exchanging the latest information and cutting-edge tools to address the challenges of HIV/AIDS.

During the first month of the pilot, recruitment as well as education about the Initiative's Community Liaison Program (CLP) began. The program manager conducted a workshop at the 2008 USCA to inform attendees of the mission and goals of the National Institute of Allergy and Infectious Disease HIV Vaccine Research Education Initiative, and to recruit liaisons for CLP. Sign-up sheets were available at the workshop, and afterwards printed information about the CLP was distributed to participants at state and regional training events sponsored by NMAC in an effort also to recruit volunteers. At one point during the recruitment process, the program manager made a direct recruitment appeal to interested colleagues at a few Historically Black Colleges and Universities (HBCUs) who shared the information with others. Recruitment of liaisons took a little longer than four months, with the initial number of recruits swelling to 19, then dropping to 10 persons. Seven persons finally committed themselves to be part of the first cohort of liaisons.

Liaison Demographics

The seven volunteer liaisons who participated in the program were affiliated with various organizations, such as community- and faith-based organizations, clinics, AIDS-service organizations, and colleges or universities. The majority of the liaisons were African American females between 41 and 60 years of age who lived in primarily urban areas. There was one Latina and one Caucasian male liaison, both of whom made specific outreach to Latino communities. The liaisons were located in five states—Texas, Pennsylvania, Oklahoma, North Carolina, and Georgia. While all liaisons were not from states that had ongoing vaccine-research programs, nor were they representative of the entire nation, the funder, NIAID, was most interested in garnering support and debunking the myths among African Americans and Latinos through this volunteer pilot project. With more education and awareness, the hope was that more HIV-vaccine research trials could be conducted in other areas to accommodate growing support.

Program Implementation Timeline

During the first four months of liaison recruitment, a great deal of time and attention was devoted to refining the design of the training component for the liaisons and initiating the development of the training materials. Then, during the next four months, an advisory committee was recruited, composed of experts in health communication, program development, evaluation, and clinical research, with experience in fieldwork in communities of color. The liaisons worked individually with the program manager to craft plans for how they would integrate HIV-vaccine research education into their existing activities. They were also required to view two online, introductory tutorials about HIV/AIDS and HIV-vaccine research. In addition, research was undertaken to develop culturally sensitive and informative health messages about the benefits of HIV/AIDS-vaccine research and the importance of community support relevant to the targeted audiences.

The training program included a series of three audio conference calls with all the liaisons. The first conference call was an orientation session which included an in-depth review of the logic model developed for the training program. The session also provided the liaisons with an opportunity to introduce themselves to each other, and to share information about their individual networks and the strategies they planned to utilize to inform the members of their networks about HIV-vaccine research. The second conference call was devoted to reviewing the materials the liaisons would use to:

1. conduct focus groups to test the different variations of the message they would use to build community awareness and support for HIV-vaccine research, and
2. conduct a process-and-outcome evaluation of their individual health communication campaigns.

The focus-group sessions served dual purposes to:

1. engage members of their network in discussing the validity of the culturally sensitive message for their networks, and

2. discuss the most appropriate message channel for further dissemination into their social communities.

The use of focus groups also provided the liaisons with a snapshot of baseline information about the level of HIV knowledge and awareness that existed within their communities. The third and final conference call took place at the end of the pilot program. After reviewing the data collected from the program, the liaisons shared their views about the pilot program and what they had gained from the experience. (See timeline.)

An exhaustive literature review was conducted during the second four-month period of the project concerning the use of culturally sensitive materials with targeted audience members. This information was incorporated into the discussions of participatory engagement held with the liaisons about communicating within their networks. Previous research has shown that most successful interventions in HIV/AIDS prevention among African American communities were grounded in theory, provided skills training, and were culturally sensitive to the needs of the community, particularly African Americans (Beatty, Wheeler, & Gaiter, 2004; Jemmott, Jemmott, & Hutchinson, 2001). Prior research suggested that use of messages that stressed gains/benefits (gain-framed) or losses/costs (loss-framed) could influence the persuasiveness of the HIV/AIDS messages for minority audiences (Apanovitch, McCarthy, & Salovey, 2003). For example, a gain-framed message might be: If you use condoms, you increase your chance of staying healthy, or "if you use condoms you are lowering your risk of getting a sexually transmitted disease." A loss-framed message might be, "If you don't use condoms, you are at greater risk of sexually transmitted diseases." We tested the influences of message framing in this project.

Developing the Health Communication Message

During the training, each liaison received individual instruction about the fundamentals of planning and delivering their messages. Specific tips included:

- Information should be credible.
- Messages should be compatible with the community norms and vernacular.
- Messages should consider cultural factors that may be obstacles to participation, such as distrust of researchers and religious beliefs, as well as social stigma about HIV/AIDS, homosexuality, intravenous drug use, or sex workers.

- Incorporate aspects of the culture into the discussion.

Following the participatory engagement, the liaisons had seven weeks to deliver their messages to their communities. They were given a level of autonomy over the exact numbers of times that they should present their messages and the channels they should use to engage with their communities. This level of autonomy was central to the pilot program design, which was that the HIV/AIDS-vaccine message easily be integrated into existing service-delivery processes. Next, the liaisons were given two weeks to reflect on the effectiveness of their work in disseminating the message to their community, and report the results of the data they collected from their message-dissemination efforts. These data included the ages and gender of those who received the messages, as well as responses to the following questions: how many of those reached increased awareness of HIV-vaccine research, and how many were willing to share the message with others. (See timeline 11.1 below.)

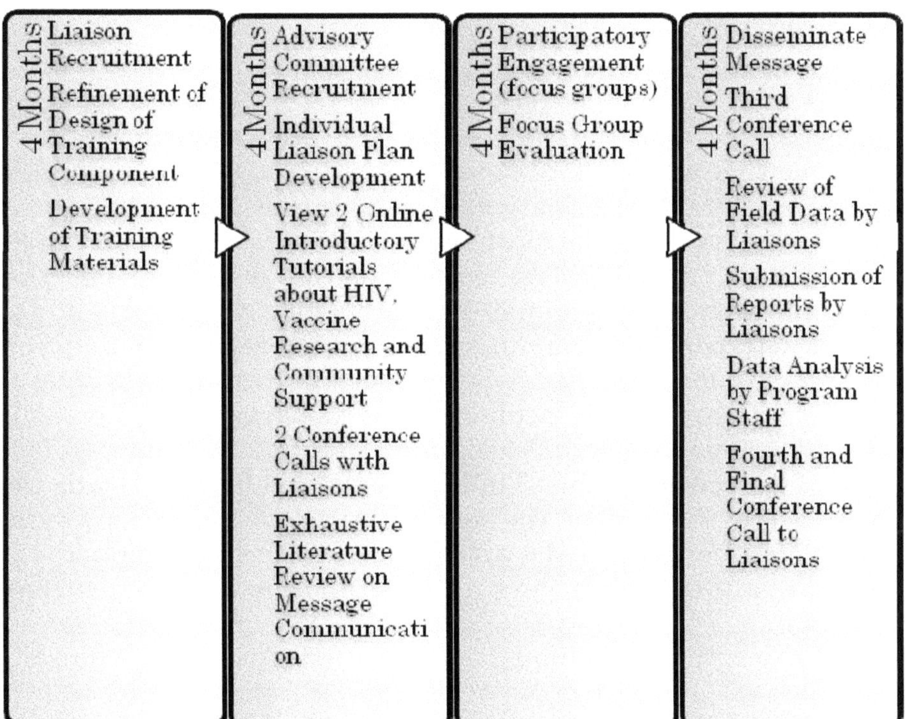

Timeline 11.1.

Program Evaluation

There was continuous process monitoring of the liaisons. Following the focus-group activity, all liaisons submitted reports on the participant responses to the following nine focus-group questions that the program team and liaisons developed prior to the focus groups.

1. What do you think is the main message of both variations?
2. Do you think the message is clear?
3. Do you think the message is believable?
4. Do you feel that the information presented was useful?
5. If someone that you know and trust in your community shared this information with you, how would you respond?
6. What are some things you like about the way the information is being presented?
7. What are some of the things you don't like about the way the information is presented?
8. Which variation (gain-framed or loss-framed) of the message do you think is most compelling?
9. What changes, if any, would you suggest?

These reports were analyzed by the program team using a technique that allowed for manual content analysis to note trends in the focus-group information. Results suggested that the loss-framed message would be most effective when used with our target audiences on printed pocket cards to be disseminated, and that gain-framed messages would be most effective when diffused through interpersonal communication. An interesting finding from the focus groups was that most participants were not willing to accept a message from someone, even if they trusted the person in other ways, but they would look at the merit of the message and decide if it were believable. Thus, not only should consideration be given to the training of the volunteers and their contacts and background, but also to the actual message to be communicated, determining if it is clear, and if useful information is transferred through it.

Following this community activity conducted by the liaisons, they were interviewed to determine what occurred and why in their individual focus group meetings. Most of the questions were open-ended and included items such as: Tell me a little about your focus-group techniques, how did you recruit the participants? Did you offer incentives? Where did you recruit the participants? There were also scaled questions such as: On a scale from 1–5, with 1 = needed improvement and 5 = the best—got great information, how well did the focus groups do? How were these groups recruited?

The recruitment methods varied by location and among different communities. Some invited participants directly; others made general announcements. Most provided food at the focus-group site, but offered no other incentives. There were consistent responses from the liaisons that their focus groups provided rich data, and that the liaisons appreciated the opportunity to engage in a participatory activity with their community. Specifically, most of the African American liaisons felt that they were assisting their community by sharing information about HIV/AIDS, and the one male Caucasian liaison felt that he was helping to give voice to the Latino community through the sharing of their thoughts about the message of HIV-vaccine clinical research.

Further outcome findings of the dissemination of the messages revealed an increase in self-reported HIV awareness among the high rates of HIV infection in their communities. Also, many of the network members became early adopters of the HIV-vaccine message and were willing to share that information with others. Another finding of interest was the rapid and extensive reach of the liaisons into their networks (see Table 11.1). Of the six community liaisons who submitted reports, five (83%) were African American. They were responsible for contacting 450 (70%) of the 644 individuals reached through the program, and recruited 58% of the 343 community residents willing to share the message. In addition, the one Caucasian liaison, who was able to use his fluency in Spanish and comfort with the Latino culture to relate to the population, recruited 14% of those willing to share information with others. He also was responsible for reaching out and providing HIV information on Latinos and vaccine research information to 25% of the Latinos contacted. This is very interesting, given the fact that there was not a Latino male among the liaisons, which was not intentional. Those who volunteered and showed commitment, or met the criterion of agreeing to devote time to the project were selected; there was not a volunteer who was a male Latino. Nonetheless, the contribution of this Caucasian liaison raises the question about the need for racial similarity by liaisons. It could be that shared values, culture, and culturally sensitive messages helped to produce the positive results in this project. The one Latina liaison recruited 29% of the Latinos who agreed to share the message and almost 60% of all the Latinos who were contacted in the program.

Table 11.1. Anticipated and Actual Numbers

	Anticipated Numbers to Be Reached	**Actual Numbers Reached**
Total Number of Liaisons	10	7*
Total Overall to Be Contacted	350	644
Total Number of African Americans Contacted	100	450
Total Number of Latinos Contacted	100	187
Total Number of Others Contacted		7
Total Number Who Reported an Increase in HIV Awareness	0–100	400
Total Number Who Reported That They Expressed a Willingness to Share the Info With at Least Three Other People	200	343

* Data available from only six liaisons.

Chart 11.1. Numbers by City, Contacts, Increased HIV Awareness, and Willingness to Share.

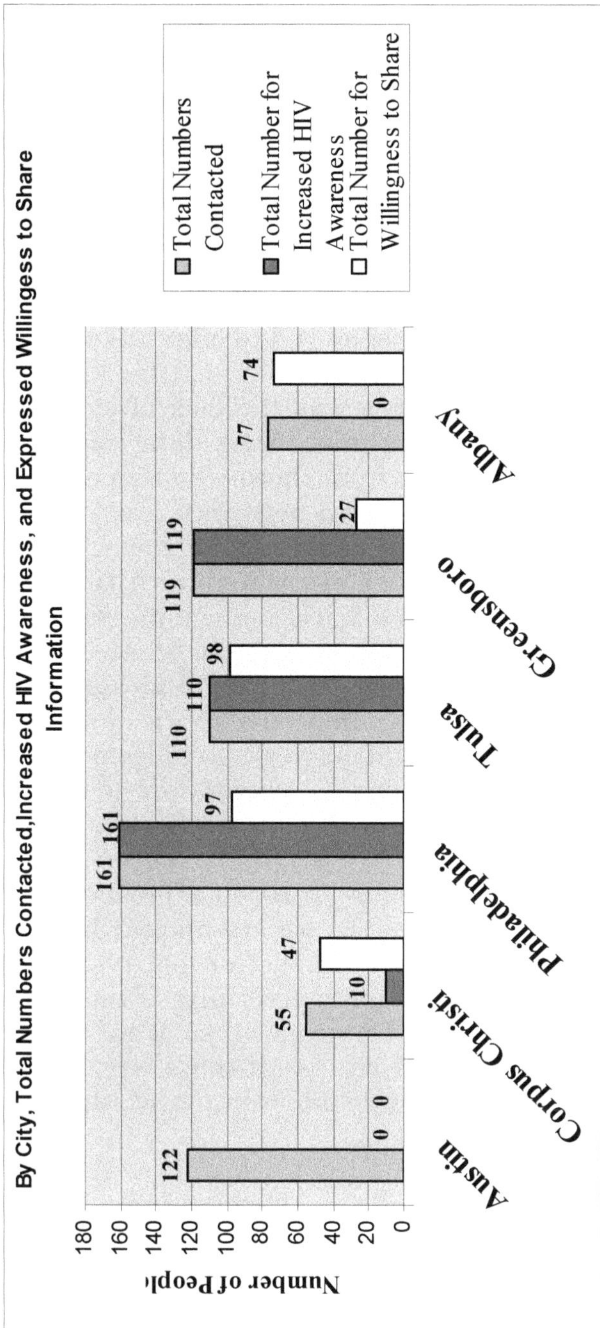

Limitations

We were only able to get data from six of the seven liaisons. As a pilot program, we were pleased overall with the volunteer response rate outcome. The initial NIAID funding was limited to message dissemination, therefore little money was available for the pilot program which included training liaisons, program reporting, and data analysis. In addition, NIAID only allocated a limited time for the information dissemination, so both funding and time allocation affected the chance to continue to recruit to achieve the anticipated number of liaisons and to offer any incentives or resources to the liaisons.

Another limitation was the lack of evaluation focus when initially developing the pilot. More data could be collected from the network, as well as from those who agreed to further disseminate the message to their own networks. The use of more standardized program evaluations would help to generalize findings, and to better understand issues associated with the network members' activities that lead to diffusion of the message. Also, more clarity in instruction and follow-up should be given about the use of the various communication channels. Additional data could have been collected about dissemination activities and barriers. For example, additional data could have been collected to clarify whether the use of printed messages—versus face-to-face, interpersonal communication—was a more or less effective communication channel for the liaisons.

While we were able to reach a broader audience through this dissemination technique, we are unclear how wide was the spread, or what was its direct impact on behavior change. There was no evaluation or follow-up to determine if those who agreed to further share the message did so, and if so, what was the outcome of their efforts? We also did not ask if those who did not know about vaccine trials prior to the intervention would now seek out more information on the trials.

Discussion

Overall, the outcome of this pilot program suggests that not just ethnicity but awareness of culture and the facts associated with the message were significant factors in the dissemination of HIV messages across social networks. The Latina liaison wrote that she visited the homes of the community members to engage with them in discussion, a more personal and culturally accepted form of engagement in that community, which allowed her to meet youth, as well as a number of family members who were not targeted but were at risk. Other liaisons met at sites commonly frequented by members of their communities, such as the public-housing meeting center. Anecdotally, one liaison reported that she made herself available to meet with community members, and wrote her personal mobile number and e-mail address on the handouts.

In keeping with prior research, the liaisons who closely resembled the target population were found to be effective in reaching members of their community and raising awareness about HIV/AIDS (Crosby et al., 2004; Frew, Archibald, Martinez, del Rio, & Mulligan, 2008; Frew, del Rio, Clifton, et al., 2008). However, the data also showed that there was a strong personal connection and high trust developed between the liaisons and their social-network members. The one Caucasian liaison was effective in disseminating health information to the Latino population due to his cultural and linguistic competencies, and several of the African American liaisons were also successful in disseminating health information to Latinos.

In terms of the demographics, many of the liaisons were not of the same age as those in their networks. The majority of the participants were between the ages of 30 and 50, while the liaisons were generally older, 40–60 years of age. However, age did not appear to be a barrier to dissemination of health information. This interpretation of the data suggests that the messages and the channels of dissemination can transcend generational differences in sharing relevant HIV/AIDS health messages. In addition, the speed at which the messages were disseminated within the communities suggested that established social-network relationships in the community played a significant role in effective communication.

Since African Americans and Latino Americans are most adversely affected by HIV/AIDS, this case study shows that community-based, participatory health education can be an important step toward increasing awareness about HIV/AIDS and the importance of HIV-vaccine research in these minority communities. In particular, the involvement of devoted, community-based messengers (liaisons) who conducted participatory meetings with community members and disseminated credible, culturally sensitive messages within social networks was found to be most effective. The outcome of this initiative builds upon and supports both the diffusion of innovation theory (Rogers, 1995) and social-network theory (Valente, 2010). The diffusion of innovation theory explains why interpersonal channels are effective at disseminating health information (Rogers, 1995; Haider & Kreps, 2004). Social network theory explains the power of personal relations in both sharing and spreading new heath information across established communities, especially with minority populations where there is often mistrust of outsiders (Valente, 2010). These theories underscore the importance of using established interpersonal relationships to promote acceptance and dissemination of relevant health messages. This study suggests that acceptance of HIV/AIDS messages were readily accepted when the source of the information was a member of the same social group, sharing similar cultural norms and values (Rogers, 1995).

The program showed that social networks can be effectively integrated and leveraged into HIV/AIDS education programs. The pilot showed that community volunteers (liaisons) would devote their time to learning participatory engagement, conducting focus-group meetings regarding messaging, developing work plans, implementing these plans, and sending in summary or evaluation reports of their efforts. A significant factor in the success of this intervention program was the interest and dedication of the liaisons. One liaison felt that her work with this pilot program was "her personal contribution to the end of AIDS in her community." Another liaison regarded the messages disseminated about vaccine trials as beneficial news and as a "beacon of hope." All the liaisons chose to participate in this program due to their desire to make a difference among African American and/or Latino populations to reduce the impact of HIV/AIDS in their communities. All who reported back integrated the messages into their existing practices and community work concerning HIV/AIDS. This project provides encouraging data suggesting that health-communication interventions employing culturally sensitive participation in crafting messages can successfully promote minority-community awareness and acceptance for HIV-vaccine dissemination within a relatively short period of time.

Acknowledgments

This work was supported by the National Institutes of Allergy and Infectious Disease [2145-5290-90].

Our thanks to the staff of the NIAID HIV Vaccine Research Education Initiative.

Bibliography

Apanovitch, A. M., McCarthy, D., & Salovey, P. (2003). Using message framing to motivate HIV testing among low-income, ethnic minority women. *Health Psychology, 22*(1), 60–67.

Beatty, L. A., Wheeler, D., & Gaiter, J. (2004). HIV prevention research for African Americans: Current and future directions. *Journal of Black Psychology, 30*(1), 40–58.

CDC. (2010). HIV among African Americans. Retrieved from http://www.cdc.gov/hiv/topics/aa.

Crosby, R. A., Holtgrave, D. R., Bryant, L., & Frew, P. M. (2004). Factors associated with acceptance of an AIDS vaccine: An exploratory study. *Preventive Medicine, 39*(4), 804–808.

Frew, P. M., Archibald, M., Martinez, N., del Rio, C., & Mulligan, M. (2008). Promoting HIV vaccine research in African American communities: Does the theory of reasoned action explain potential outcomes of involvement? *Challenge: The Journal of the Morehouse Research Institute, 13*(1), 61–97.

Frew, P. M., del Rio, C., Clifton, S., Archibald, M., Hormes, J., & Mulligan, M. J. (2008). Factors influencing HIV vaccine community engagement in the urban south. *Journal of Community Health, 33*(4), 259–269.

Frew, P. M., Hou, S. I., Davis, M., Chan, K., Horton, T., Shuster, J., Hixson, B., & Rio, C. (2010). The likelihood of participation in clinical trials can be measured: The clinical research involvement scales. *Journal of Clinical Epidemiology, 63*(10), 1110–1117.

Haider, M., & Kreps, G. L. (2004). Forty years of diffusion of innovations: Utility and value in public health. *Journal of Health Communication: International Perspectives, 9*(1), 3–11.

The Henry J. Kaiser Family Foundation. (2009). HIV/AIDS policy fact sheet: Black America and HIV/AIDS. Retrieved from www.kff.org/hiv.

Jemmott, L. S., Jemmott, J. B., & Hutchinson, M. K. (2001). HIV/AIDS. In R. Braithwaite & S. Taylor (Eds.), *Health issues in the Black community* (2nd ed., pp. 309–346). San Francisco, CA: Jossey-Bass.

Kreps, G. L. (2008). Strategic use of communication to market cancer prevention and control to vulnerable populations. *Health Marketing Quarterly, 25*(1–2), 204–216.

Kreps, G. L., & Sparks, L. (2008). Meeting the health literacy needs of immigrant populations. *Patient Education and Counseling, 71*(3), 328–332.

NASTAD, & The Kaiser Family Foundation. (2009, July). *The national HIV prevention inventory: The state of HIV prevention across the U.S.* Retrieved from http://health.utah.gov/cdc/hivsurveillance/hiv%20docs/National%20HIV%20Prevention%20Inventory%20FINAL.pdf.

Rogers, E. M. (1995). *Diffusion of innovations* (4th ed.). New York, NY: Free Press.
Sullivan, P. S, McNaghten, A. D., Begley, E., Hutchinson, A., & Cargill, V. (2007). Enrollment of racial/ethnic minorities and women with HIV in clinical research studies of HIV medicines. *Journal of the National Medical Association, 99*(3), 242–250.
Valente, T. W. (2010). *Social networks and health.* Oxford, UK: Oxford University Press.
Wong, M.D., Shapiro, M.F., Boscardin, W.J., & Ettner, S.L. (2002). Contribution of major diseases to disparities in mortality. *New England Journal of Medicine, 347*(20), 1585–1592.

12

Evaluation of the Influence of an Urban Community Park Revitalization on African American Youth Physical Activity

Gary L. Kreps, Melinda M. Villagran, Janey Trowbridge, Paula Baldwin, Yolanda Barbier, Amy Chang, Jung-mi Jun, Megan Tucker, Autumn Saxton-Ross, and Sieglinde Friedman

The Problem

The Centers for Disease Control and Prevention (CDC) established guidelines recommending that children and adolescents partake in one hour or more of physical activity each day (CDC, 2003, 2008). Yet, evidence shows that minority-group youth are falling below recommended physical activity levels, and new strategies are needed to promote physical activity for these youth (CDC, 2003). People living in low-income areas and in communities of color often have less access to recreation facilities, and face unique environmental challenges that may make it difficult for them to engage in the recommended amount of physical activity (Deutsche, 1986; Filner, 2006; Moore, Diez Roux, Evenson, McGinn, & Brines, 2008; Wilson, 1987). A growing body of research, based on ecological theory, suggests that building environmental support through the redesign of physical environments can encourage physical activity in both adults and children (Sallis, Bauman, & Pratt, 1998; Davison & Lawson, 2006; Duncan, Spence, & Mummery, 2005; Humpel, Marshall, Leslie, Bauman, & Owen, 2004; Saelens, Sallis, & Frank, 2003; Sallis & Glanz, 2006; Sallis et al., 2006; Transportation Research Board & Institute of Medicine, 2005). However, there has been limited research conducted, and results are mixed concerning the influences of public built environmental change on minority youth physical activity levels (Boarnet, Anderson, Day, McMillan, & Alfonzo, 2005; Evenson, Herring, & Huston, 2005; Merom, Bauman, Vita, & Close, 2003). Some research suggests that children's physical activity levels at school can be improved by increasing the accessibility of quality play spaces (Sallis et al., 2006; Stone, McKenzie, Welk, & Booth, 1998; Wechsler, Devereaux, Davis, & Collins, 2000). A recent study concluded that renovated play environments in urban areas result in higher utilization rates (Colabianchi, Kinsella, Coulton, & Moore, 2009). Our current study follows this program of research by examining whether redesigning and revitalizing an inner-city park in Washington, DC can help promote increased youth physical activity. This research is vital to designing, implementing, and sustaining public-health policies and environmental changes to urban community park spaces that could potentially increase the physical activity of minority youth in communities and enhance health outcomes.

The Value of Urban Park Revitalization

After years of decline and disuse of many urban communities, a confluence of local businesses, government agencies, and neighborhood groups have been organized to work together to help rebuild and revitalize recreational areas (Deutsche, 1986; Filner, 2006; Wilson, 1987). Collaborating institutions have been able to collectively reshape urban parks to help increase physical activity, security, and investment in revitalized neighborhoods (Ferré, 1987; Filner, 2006; Wilson, 1987). A significant benefit to urban park revitalization is the focus on community facilities, specifically public parks and recreation areas. When a park or playground has been renovated, children may be more likely to engage in physical activity there (Wilson, 1987). Park development is particularly important because commercial, physical activity-related facilities, such as health clubs or gyms, are less likely to be present in lower-income neighborhoods and in neighborhoods with higher proportions of racial minorities (Powell, Slater, Chaloupka, & Harper, 2006). The CDC (2008) suggested the use of park equipment to stimulate physical activity in children. Younger children usually strengthen their muscles when they engage in gymnastics, play on a jungle gym or climb trees (CDC, 2005). However, the ability of children and adolescents to partake in these activities may be limited because of their access to safe, well-maintained, and well-equipped recreational facilities (Wilson, 1987). Improving park facilities could be an important strategy to increase the physical activity levels of urban minority youth (Moore et al., 2008).

Increasing Physical Activity Levels

Research has suggested that with the implementation of park revitalization in urban areas, physical activity may increase among neighborhood children (Powell et al., 2006). Studies have also shown that there is an increasing prevalence of childhood obesity among preschoolers who live farther away from playgrounds and in unsafe neighborhoods (Burdette & Whitaker, 2004). Ecological models suggest that environmental barriers to physical activity may be an important factor that can be modified to facilitate more physical activity. Recent studies have shown that public parks are critical resources for physical activity in minority communities (Cohen et al., 2007). Residential proximity to the parks or recreation areas also influences the amount of physical activity that takes place at that location (Sallis et al., 1990). Therefore, the closer residents or families are to a neighborhood park facility, the more children will utilize the park to play (engage in physical activity).

Children and adolescents need three types of activities for health: aerobic activity, muscle strengthening, and bone strengthening (Filner, 2006). Childhood and adolescence are key times to form lifelong eating and physical activity habits (Brownson, Boehmer, & Luke, 2005). Furthermore, there are large racial and ethnic differences in the pervasiveness of overweight and obesity, suggesting that culturally sensitive approaches to promoting physical activity are needed (Brownson et al., 2005). Children in urban areas are at high risk of developing childhood obesity (Dietz, 1983). Both energy intake and energy expenditure are involved in the development of obesity, as shown by the effectiveness of behavior change in treating childhood obesity (Epstein, 1986; Saris, 1986). Addressing childhood obesity is complex, and requires a multipronged approach, including development and revitalization of recreational facilities. However, limited research has been conducted on the influences of urban park revitalization on increasing youth physical activity in urban areas, and most of the empirical literature was published over 20 years ago (Cohen et al., 2007; Deutsche, 1986; Ferré, 1987; Wilson, 1987). This current study builds upon this early literature and extends it to examining the influences of urban park revitalization on African American youth.

The Study

This study tested the relationship between the independent variable, environmental changes to park facilities, and the dependent variable, physical activity of African American youth, advancing the following major research hypothesis:

H1: Implementation of environmental changes to public park facilities in an urban area will lead to increased physical activity by African American youth.

The research team conducted observational research in two sites in Washington, DC: Marvin Gaye Park (MGP), a recently renovated park in a largely African American community, and a comparison park, Oxon Run Park (ORP), in a demographically similar urban community. A specific area within Oxon Run Park was selected for observation based on its comparable size and location to MGP to simulate baseline physical activity levels prior to the renovation of MGP.

Physical activity of African American youth from 6 to 14 years was measured at the two sites using direct field observation. The research team employed a predetermined coding procedure, based on the System for Observing Play and Recreation in Communities (SOPARC) (McKenzie & Cohen, 2006; McKenzie, Cohen, Sehgal, Williamson, & Golinelli, 2006). In 2008, the National Recreation and Park Association (NRPA), in association with other public and private funding sources, initiated the revitalization of MGP. MGP is located in an area which is 98% African American and where 25% of the families live below the poverty line (U.S. Census Bureau, 2000).

MGP, once referred to as "Needle Park," was a well-known drug haven littered with thousands of pounds of broken beer bottles, empty beer cans, trash, and used hypodermic needles (Ricard, 2009; Sullivan, 2006). Prior to the park revitalization, MGP was an open parcel of green space with no recreational equipment. In August 2009, community residents celebrated the opening of a new playground area within MGP, which includes a brand new play area with swing set and various modern playground activity stations including a cone-shaped activity center complete with a swing and climbing nets, two spin apparatuses, and an octagon-shaped piece of equipment with climbing nets. A special safety surface throughout the play area and park benches around the perimeter completed the installation. There also was landscaping work done to the park, as well as the introduction of an amphitheater that significantly changes the physical environment of MGP. The renovation represents a significant environmental change for the community served by the park, since the original condition of the park did not adequately support youth physical activity to any extent, and was not considered a safe location by the community residents.

The second site, ORP, was chosen as a comparison park because it simulates the approximate size and type of playground area at MGP prior to renovation, although it does not have all the debris that was part of the old MGP (Needle Park). The community where the park is located is demographically similar to the MGP neighborhood, where almost all the residents are African American, and the proportion of families living below the poverty line is 32% (U.S. Census Bureau, 2000). The comparison park's observation area has no equipment, as was the case with MGP pre-renovation. However, ORP is located in a safer area, which is to say that it had no community reputation as being unsafe, or did not demonstrate being inhabited by drug dealers or drug users. In many ways, ORP is a very conservative comparison area for this study, since it is much cleaner and safer than the old MGP area.

Observation Instrument

The study used an adapted version of the SOPARC observational guide, a validated tool for observing activity in parks, as a reference for a starting point for observing activity in the MGP park (McKenzie & Cohen, 2006; McKenzie et al., 2006). Strong validity and reliability measures have been established for the SOPARC tool, having been tested by observing 16,244 individuals in 165 park areas (McKenzie et al., 2006), as well as having been used in numerous other studies on observing physical activity and park use (Shores & West, 2009a, 2009b; Tester & Baker, 2009). SOPARC was designed to obtain observational data on the number of participants and their physical activity levels in community environments (McKenzie & Cohen, 2006; McKenzie et al., 2006). In addition, SOPARC provides contextual information on the setting in which the physical activity occurs.

SOPARC is based on momentary time sampling that involves observing groups and individuals at specified time intervals (McKenzie & Cohen, 2006; McKenzie et al., 2006). Like the SOPARC tool, this study's observation instrument includes two parts. The first coding sheet examines the overall conditions and activity in the park, and the second focuses on individual participants. The study's overall observation sheet asks for information on the location of the park, date and time of observations, weather-related and park conditions (including equipment), and total number of people in the park. The coding sheet includes space for recording gender, primary and secondary activities of the target audience, as well as their activity level.

The research team adapted the SOPARC tool to the objectives of this study and to the specific situation of the park and community in this study. The first and primary modification was the focus on park visitors between the ages of 6 and 14. Second, all the park participants were observed to be African American, so determination of race was not included in the instrument. Third, like the SOPARC tool, "sedentary" was operationalized as sitting, standing, and lying down. However, only one category, "active," was used for nonsedentary activity, measured by observing walking, running, playing structured or unstructured games, and using the equipment (spinning, swinging, and climbing). The limited nature of the activities that could be performed in the renovated MGP space did not warrant distinguishing between moderate and vigorous activity. In addition, a parallel section on the code sheet was included to measure activity in MGP outside the renovated playground area in an adjacent grassy space that was within observers' eyesight.

As for individual participants, the SOPARC tool is designed to observe the general level of activity (moderate or vigorous) of all participants in the park at 10-minute intervals. The coders observed all participants at different time intervals, and recorded activity of individual participants in-depth, one at a time, in the target age range (6 to 14 years), from the moment of their entry into the park until their departure. The research team noted the duration of each activity for each participant observed. In this way, the study could examine the use of the specific playground equipment and of the spaces without equipment, as well as the level of activity in those areas. The code sheet grid also enabled coders to note when sedentary activity was associated with a piece of equipment or even a structured game, such as sitting on the swing set or sitting on the sidelines watching children playing with a soccer ball.

Observation Procedures

Observation areas in both the control and intervention parks were visited, and the size, location, and boundaries of each space were determined. Detailed maps, including where the coders performing observations were to stand or sit, were developed. Coders at MGP observed from inside the playground area so that they could visually see all activities taking place there and in the grassy area outside the playground within eyesight. The coders at the MGP wore Washington Parks & People T-shirts to show that they were endorsed by a well-respected community organization. (Washington Parks & People is a nonprofit organization dedicated to restoring and enhancing public parks in the National Capital Region.) Researchers at ORP positioned themselves adjacent to the park area so that observation of the entire area was possible.

Six coders conducted the observations after participating in a two-part training program. The first part consisted of a classroom lecture and discussion session on observation techniques and use of procedures which are detailed in the instructions for completing the code sheets. The second part of the training included practice observations and completion of the code sheets at the intervention park. Overall activity was recorded five times during a two-hour session: once at the beginning, three times during the session, and once at the end. Observers started the individual participant coding process when a participant entered the park's observation area. Participant selection for coding was random, except when two or more children entered the park at the same time. In that case, the coders chose which child to observe and code. In order to reduce the possibility of bias in choosing which child to observe and code, none of the coders was from the community, nor did they know anyone from the community, other than representatives of the Washington Parks & People organization who served only in the capacity of community liaisons. Observers recorded the selected participant's total time in the park and the amount of time spent on each piece of equipment, as well as nonequipment activity both inside and outside the playground area as well as the level of activity. They also noted if the participants continued to be in the park after the two-hour recording session ended.

Following the training, the coders completed the code sheets independently for two practice intervals in MGP to achieve satisfactory reliability according to reliability formula suggested by Holsti (1969). The final, calculated inter-rater reliability score was .94 across all categories.

A total of 16 hours (eight two-hour coding sessions) were completed between October 3, 2009 and October 16, 2009. (See the coder schedule in Table 12.1.) Eight hours each were recorded at MGP and at ORP. Coding was curtailed or canceled for another eight hours of scheduled observation because of rain. Coding was later stopped after shooting incidents occurred within the MGP community. The observer schedule was modified so that park visitors and level of activity could be equally compared at the two parks at equivalent times of the day and week.

Table 12.1. Coder Observation Schedule

Observation Period	Park	Saturday	Sunday	Thursday
Oct. 3–Oct. 9, 2009	Marvin Gaye	11:00 a.m. – 1:00 p.m.	3:00 p.m.– 5:00 p.m.	3:00 p.m.– 5:00 p.m.
	Oxon Run	9:00 a.m.– 11:00 a.m.	9:00 a.m.– 11:00a.m.	3:00 p.m.– 5:00 p.m.
Oct. 10–Oct. 16, 2009	Marvin Gaye	11:00 a.m.– 1:00 p.m. (rain)	3:00 p.m.– 5:00 p.m.	3:00 p.m.– 5:00 p.m. (rain)
	Oxon Run	11:00 a.m.– 1:00 p.m. (rain)	3:00 p.m.– 5:00 p.m.	3:00 p.m.– 5:00 p.m. (rain)

Participants

Observers focused on African American children in both parks who were between the ages of 6 and 14, the target audience for the study. In some cases, the exact age of the participants was confirmed by the Washington Parks & People volunteers who were assigned to the observers. The coders recorded in detail activities for a total of 14 individual participants, nine males and five females. Approximately 50% of the participants were from 6 to 10 years old and 50% were from 11 to 14 years of age.

Data Analysis

As the sample size was small (N = 14), data were analyzed using summary statistics. The following outcomes were calculated:

1. The average number of people, regardless of age, at any one half-hour time interval at the two parks;
2. The most frequent activities engaged in by youth in the target age group by gender;
3. The proportion of youth observed in the snapshot intervals who were active versus sedentary;
4. The average amount of time spent in MGP by the individual participants in the sample;
5. The proportion of time individuals in the target age range engaged in active vs. sedentary participation; and
6. The proportion of time individual participants engaged in playground equipment vs. nonequipment activity.

The data analysis for outcomes is only computed on both parks on Outcome 1: the average number of people, regardless of age, at any one half-hour time interval at the two parks, as there were no activities occurring in ORP to be coded.

Results

Outcome 1: Average number of park visitors. The number of visitors to MGP playground recorded at 30-minute intervals varied from 0 to 48. Between 5:00 p.m. and 5:30 p.m., regardless of day of the week and amount of daylight (all observations were done during the day), the number of visitors dropped dramatically. The total participant count increased when a structured game, such as football, soccer, or baseball, was played in the grassy area adjacent to the MGP playground. The average number of visitors at any one time on the weekends (when there was no active precipitation) was 21. Only one observation was completed during the week, and the average number of visitors over four intervals was relatively low (4.5) with a range from 0 to 11, possibly because of conflicting after-school activities, such as sports or other extracurricular activities.

In contrast to the observations in MGP, the only activity among persons aged 6 to 14 observed in the comparison park, ORP, was two high school-aged children walking across the park to reach another destination for a six-minute period of time over the entire observation time period.

Outcome 2: Most frequently observed MGP activities. The research team recorded 148 observations of youth in the 6-to-14-year-old range who were participating in primary and secondary activities over the entire observation time period. However, the snapshot observations were taken at half-hour intervals, and individual participants could be double-counted if they continued to play in the park past 30-minute intervals. Approximately one-half were female (75) and one-half were male (73). The proportion of use of the playground equipment in relation to participation in nonequipment activities was 78% to 22%, or more than three and a half times as much. The results showed that 85% of the activities took place inside the playground; 15% of the activities were observed in the grassy area within eyesight of the observers.

The most frequent primary activity for females was use of the playground equipment, especially the swing set (53%), and to a much lesser extent, the cone activity (27%). Similarly, the playground equipment, especially the spin apparatus (35%) and the cone activity (30%), was used by the largest proportion of females engaged in secondary activities. As for primary activities for males, they participated in structured and unstructured games, such as baseball, football, and catch, inside and outside the playground, to a much greater extent than females (44% compared to 6%). However, for both primary and secondary activities, male park visitors also used the playground equipment. For example, 27% of those participating in primary activities used the swing set, and 29% used each of three pieces of equipment as a secondary activity.

Outcome 3: Proportion of active versus sedentary youth. The proportion of youth who were physically active while participating in the park in primary and secondary activities was 76.7%, while 23.3% were sedentary. Females were more active than males: 84% of the females were engaged actively in primary or secondary activities, while 70% of the males were engaged in physical activities.

Outcome 4: Average amount of time in park. In addition to snapshots of all visitors, data were collected on the individual participant sample of 14 participants, nine males and five females. Due to the small sample size, the results were not broken down by gender. The average amount of time spent in the park by all individuals in the sample was 25.13 minutes.

Outcome 5: Proportion of active versus sedentary individuals. The sample of observed individual participants spent an average of 77.3% of their time in the park engaged in physical activity, and 22.7% of the time in sedentary activities such as sitting, standing, or lying down in the park.

Outcome 6: Proportion of individual time on equipment versus nonequipment activities. Individual participants observed at MGP spent an average of 62.6% of their time in the park on the playground equipment and 37.3% on nonequipment-related activities. The swing set and the spin apparatus represented 48% of the overall park use and 66.7% of the park equipment use.

Even though the sample of individual participants (N = 14) was small, the results support the findings of the snapshot observations. For example, as noted above (Outcome 5), individual participants spent an average of 77.3% of their time in active physical participation, and 22.7% of the time in sedentary activities. These figures are virtually identical to the proportion of active versus sedentary youth participating in primary and secondary activities: 76.7%, to 23.3% (Outcome 3). However, the collective results in terms of playground equipment versus nonequipment activity, the collective measurement of 78% equipment versus 22% nonequipment activity (Outcome 2) does not match the individual result of 62.6% playground equipment versus 37.3% nonequipment (Outcome 6).

Discussion

The results of this study strongly supported the research hypothesis: Implementation of environmental changes to park facilities at MGP has resulted in increased youth physical activity. The data show significant differences between the renovated park and the comparison park, in terms of youth physical activity rates and intensity. In contrast to the comparison park, where very few visitors and minimal physical activity were observed, a sizable number of youth visited the renovated park, especially on weekends with good weather, an average of 21. (This number of visitors accounts for only the period of time when the park was being observed, not the total number of visitors per day). Most of the youth visitors to the park actively engaged in physical activity (76.7%), and used the park equipment (62.6%) for a fair amount of time, an average of 25 minutes.

According to the CDC's (2008) *International Consensus Conference on Physical Activity Guidelines for Adolescents*, it is recommended that "youth engage in one hour of aerobic physical activity per day, with three of those days including bone-strengthening activities such as jumping rope or running and three of those days including muscle strengthening activities such as gymnastics or pushups). Considering the fact that youth in this study spent more than 75% of their time (approximately 18 minutes per visit) participating in physical activity, the renovated park may help youth achieve the recommended level of physical activity. In addition, youth in the renovated park demonstrated a much higher level of physical activity in comparison to a previous study that investigated physical activity in 165 park areas using the SOPARC tool (McKenzie & Cohen, 2006). However, comparison of results is difficult because this study did not replicate all of the conditions of the McKenzie and Cohen (2006) research.

One interesting and unexpected result of this study is that females were more active than males: 84% of the females were engaged in either primary or secondary activities, while 70% of the males were physically active. This result contradicts previous studies showing that males tend to visit parks in greater numbers than females, and that males generally demonstrate a higher level of physical activity than females (Shores & West, 2009a).

However, this study has limitations in terms of research design and measures. First, there was no baseline comparison conducted at MGP prior to the renovation, and it may be that ORP does not exactly duplicate the conditions at the pre-renovated MGP. Second, the initial observation schedule of 40 hours was curtailed because of poor weather, and due to a gang-related, drive-by shooting in the neighborhood of the park. As a result, the sample size (N = 14) was smaller than planned, and was insufficient to conduct inferential statistics. A longer period of observation is also recommended for future research to examine the relationship between seasonal variations and days of the week on the one hand, and physical activity on the other. In this way, park designers and providers can develop and provide different activity programs depending on weather and other environmental conditions.

Third, the use of direct observation as the only research method has significant limitations. The park visitors were aware of the researchers' presence. Though the coders strived to avoid interaction with the park visitors, at times it was unavoidable. Furthermore, the activity in the park was very dynamic. It was initially challenging for coders to track the usage of the park facilities as well as the activity level of participants. Moreover, although some of the threats to internal validity due to researcher bias were controlled by limiting observation sessions to two hours, the coders were aware of the purposes of the research and could have been influenced by this knowledge (Frey, Botan, & Kreps, 2000). Thus, in future, research field observations should be supplemented with direct quantitative measures of physical activity and other unobtrusive measures of park utilization (Frey et al., 2000).

Fourth, an in-depth or qualitative analysis of community members' attitudes and behaviors could provide more insights into minority youth use of park facilities. The results of this study demonstrate that participants actively engage in physical activities, using the playground equipment during most of their time spent in the park. However, this research project did not address participants' motivations for using the park. Follow-up interviews and focus groups with youth, parents, and caregivers, as well as with community leaders could address additional questions, such as why youth in the target audience use the park and the equipment, and how they feel about the environmental changes to the park.

The fifth critical point to examine in future research is the collective or group activity of participants. Participant observation focused on conditions of the park and activities of individual participants. However, the research team noted that participants were more likely to act together with their friends and families. They continuously talked to each other, played as groups, and helped each other to use equipment (e.g., pushing younger siblings on the swings). These collective activities were not addressed in the observation code sheets. Finally, in future studies, the unit of analysis needs to be not only individuals but also groups, and more variables should be included to understand group behaviors.

As a part of a larger effort, it is recommended that this study be expanded to examine the continuing effects of this built environment over a longer period of time. For example, this particular study took place in the fall, and a more complete picture could be obtained by taking a yearlong look at the use of the park by the community children. A further development of this study could include the actual measurement of physical activity levels of the community children with the use of individual accelerometers that monitor and store the intensity of body movement.

Conclusion

Although this study's scope was somewhat limited by the effects of weather and neighborhood violence, the results obtained were significant. Comparing the physical activity level of the undeveloped park, ORP, to the activity level of the revitalized park, MGP, demonstrates the strong, positive effect on youth physical activity that built environments can have. As childhood obesity continues to be a challenge for the youth of today, studies like this and others that demonstrate positive effects of a built environment in an urban community revitalization project on its children's activity levels, continue to make contributions to the body of research surrounding this challenge. Healthcare professionals, landscape architects, park designers, program providers, and community leaders should be encouraged to consider parks as a catalyst for promoting a range of positive health behaviors among minority youth.

Improving the urban community with revitalized built environments appears to be a good step toward increasing youth physical activity levels. Additional research can provide a broader evidence base for directing youth physical activity interventions. Promising lines for research include longitudinal physical activity studies to determine the contribution of eating and physical activity behaviors to the development of obesity (Marcus et al., 2000; McKenzie et al., 1991). Interventions promoting physical activity for children (particularly obese children) should be carefully examined to determine their influences on self-efficacy related to exercise (Trost, Kerr, Ward, & Pate, 2001). Based on the current research, we strongly commend efforts to revitalize urban parks to promote increases in youth physical activity.

Bibliography

Boarnet, M. G., Anderson, C. L., Day, K., McMillan, T., & Alfonzo, M. (2005). Evaluation of the California safe routes to school legislation: Urban form changes and children's active transportation to school. *American Journal of Preventive Medicine, 28*(2), 134–140.

Brownson, R. C., Boehmer, T. K., & Luke, D. A. (2005). Declining rates of physical activity in the United States: What are the contributors? *Annual Review of Public Health, 26*(1), 421–443.

Burdette, H. L., & Whitaker, R. C. (2004). Neighborhood playgrounds, fast food restaurants, and crime: Relationships to overweight in low-income preschool children. *Preventive Medicine, 38*(1), 57–63.

Centers for Disease Control and Prevention (CDC). (2003). Physical activity levels among children aged 9–13 years—United States, 2002. *Morbidity and Mortality Weekly Report, 52*(33), 785–788. Retrieved from http://www.cdc.gov/mmwr/preview/mmwrhtml/mm5751a5.htm.

Centers for Disease Control and Prevention (CDC). (2005). Physical activity for everyone. Guide to Community Preventive Services website. Retrieved from www.thecommunityguide.org/pa/

Centers for Disease Control and Prevention (CDC). (2008). Physical activity for everyone. Retrieved from http://www.cdc.gov/physicalactivity/everyone/guidelines/children.html.

Cohen, D., McKenzie, T., Sehgal, A., Williamson, S., Golinelli, D., & Lurie, N. (2007). Contribution of public parks to physical activity. *American Journal of Public Health, 97*(3), 509–514.

Colabianchi, N., Kinsella, A., Coulton, C., & Moore, S. (2009). Utilization and physical activity levels at renovated and unrenovated school playgrounds. *Preventive Medicine, 48*(2), 140–143.

Davison, K. K., & Lawson, C. T. (2006). Do attributes in the physical environment influence children's physical activity? A review of the literature. *International Journal of Behavioral Nutrition and Physical Activity, 3*(1), 19.

Deutsche, R. (1986). Krzysztof Wodiczko's 'homeless projection' and the site of urban 'revitalization.' *October, 38,* 63–98.

Dietz, W. H. (1983). Childhood obesity: Susceptibility, cause, and management. *The Journal of Pediatrics, 103*(5), 676–686.

Duncan, M. J., Spence, J. C., & Mummery, W. K. (2005). Perceived environment and physical activity: A meta-analysis of selected environmental characteristics. *International Journal of Behavioral Nutrition and Physical Activity, 2*(1), 11.

Epstein, L. H. (1986). Treatment of childhood obesity. In K. D. Brownell & J. P. Foreyt (Eds.), *Handbook of eating disorders* (pp. 159–179). New York, NY: Basic Books.

Evenson, K. R., Herring, A. H., & Huston, S. L. (2005). Evaluating change in physical activity with the building of a multi-use trail. *American Journal of Preventive Medicine, 28*(2), 177–185.

Ferré, M. I. (1987). Prevention and control of violence through community revitalization, individual dignity, and personal self-confidence. *Annals of the American Academy of Political and Social Science, 494*(1), 27–36.

Filner, M. (2006). The limits of participatory empowerment: Assessing the Minneapolis neighborhood revitalization program. *State & Local Government Review, 38*(2), 67–77.

Frey, L. R., Botan, C. H., & Kreps, G. L. (2000). *Investigating communication: An introduction to research methods.* Englewood Cliffs, NJ: Allyn & Bacon.

Humpel, N., Marshall, A. L., Leslie, E., Bauman, A., & Owen, N. (2004). Changes in neighborhood walking are related to changes in perceptions of environmental attributes. *Annals of Behavioral Medicine, 27*(1), 60–67.

Marcus, B. H., Dubbert, P. M., Forsyth, L. H., McKenzie, T. L., Stone, E. J., Dunn, A. L., & Blair, S. N. (2000). Physical activity behavior change: Issues in adoption and maintenance. *Health Psychology, 19*(1), 32–41.

McKenzie, T. L., & Cohen, D. (2006). SOPARC. Description and procedures manual. Retrieved from http:// http://www.activelivingresearch.org/files/SOPARC_Protocols.pdf.

McKenzie, T. L., Cohen, D. A., Sehgal, A., Williamson, S., & Golinelli, D. (2006). System for observing play and recreation in communities (SOPARC): Reliability and feasibility measures. *Journal of Physical Activity and Health, 3*(1), S208–S222.

McKenzie, T. L., Sallis, J. F., Nader, P. R., Patterson, T. L., Elder, J. P., Berry, C. C., Rupp, J. W.,...Nelson, J. A. (1991). BEACHES: An observational system for assessing children's eating and physical activity behaviors and associated events. *Journal of Applied Behavior Analysis, 24*(1), 141–151.

Merom, D., Bauman, A., Vita, P., & Close, G. (2003). An environmental intervention to promote walking and cycling—The impact of a newly constructed Rail Trail in Western Sydney. *Preventive Medicine, 36*(2), 235–242.

Moore, L., Diez Roux, A., Evenson, K., McGinn, A., & Brines, S. (2008). Availability of recreational resources in minority and low socioeconomic status areas. *American Journal of Preventive Medicine, 34*(1), 16–22.

Powell, L., Slater, S., Chaloupka, F., & Harper, D. (2006). Availability of physical activity-related facilities and neighborhood demographic and sociodemographic characteristics: A national study. *American Journal of Public Health, 96*(9), 1676–1680.

Ricard, M. (2009, August 19). D.C. to celebrate improvements at Marvin Gaye Park. *The Washington Post.* Retrieved from http://www.washingtonpost.com.

Saelens, B. E., Sallis, J. F., & Frank, L. D. (2003). Environmental correlates of walking and cycling: Findings from the transportation, urban design, and planning literatures. *Annals of Behavioral Medicine, 25*(2), 80–91.

Sallis, J. F., Bauman, A., & Pratt, M. (1998). Environmental and policy interventions to promote physical activity. *American Journal of Preventive Medicine, 15*(4), 379–397.

Sallis, J. F., Cervero, R., Ascher, W., Henderson, K., Kraft, M., & Kerr, J. (2006). An ecological approach to creating active living communities. *Annual Review of Public Health, 27*(1), 297–322.

Sallis, J. F., & Glanz, K. (2006). The role of built environments in physical activity, eating, and obesity in childhood. *Future Child, 16*(1), 89–108.

Sallis, J. F., Hovell, M. F., Hofstetter, C. R., Elder, J. P., Hackley, M., Caspersen, C. J., & Powell, K. E. (1990). Distance between homes and exercise facilities related to frequency of exercise among San Diego residents. *Public Health Reports, 105*(2), 179–185.

Saris, W. H. M. (1986). Habitual physical activity in children: Methodology and findings in health and disease. *Medicine and Science in Sports and Exercise, 18*(3), 253–263.

Shores, K. A., & West, S. T. (2009a). The relationship between built park environments and physical activity in four park locations. *Journal of Public Health Management and Practice, 14*(3), e9–e16.

Shores, K. A., & West, S. T. (2009b). Rural and urban park visits and park-based physical activity. *Preventive Medicine, 50*(1), S13–S17.

Stone, E. J., McKenzie, T. L., Welk, G. J., & Booth, M. L. (1998). Effects of physical activity interventions in youth. Review and synthesis. *American Journal of Preventive Medicine, 15*(4), 298–315.

Sullivan, P. (2006, April 2). Getting its groove back. *The Washington Post*. Retrieved from http://www.washingtonpost.com.

Tester, J., & Baker, R. (2009). Making the playfields even: Evaluating the impact of an environmental intervention on park use and physical activity. *Preventive Medicine, 48*(4), 316–320.

Transportation Research Board & Institute of Medicine. (2005). *Does the built environment influence physical activity? Examining the evidence*. Washington, DC: National Academies Press.

Trost, S. G., Kerr, L. M., Ward, D. S., & Pate, R. R. (2001). Physical activity and determinants of physical activity in obese and non-obese children. *International Journal of Obesity and Related Metabolic Disorders, 25*(5), 822–829.

U.S. Census Bureau. (2000). American fact finder. Retrieved from http://factfinder.census.gov/home/saff/main.html?_lang=en.

Wechsler, H., Devereaux, R. S., Davis, M., & Collins, J. (2000). Using the school environment to promote physical activity and healthy eating. *Preventive Medicine, 31*(2), 121–137.

Wilson, D. (1987). Urban revitalization on the upper west side of Manhattan: An urban managerialist assessment. *Economic Geography, 63*(1), 35–47.

13

Reducing Health Disparities Using Health-Communication Interventions and Community Engagement

Muhiuddin Haider and Nicole I. Wanty

Health disparities are defined as the unequal access to health resources in a population that increases adverse health outcomes (Carter-Pokras & Baquet, 2002) during health emergencies, such as natural disasters and disease outbreaks. The keys to reducing health disparities are health-communication interventions which engage communities in

- health education,
- health communication and literacy, and
- training programs.

Numerous health-communication interventions have successfully engaged communities to reduce health disparities using health education (Fisher, Burnet, Huang, Chin, & Cagney, 2007), health communication and health literacy (Bennett et al., 2009), and training programs (Bolden, 2001; Kuhajda et al., 2006). For example, health-communication interventions have successfully engaged communities to control communicable diseases, such as small pox, polio, dengue fever, leprosy, and lymphatic filariasis.

We posit that health-communication interventions that successfully engage communities reduce health disparities among at-risk populations. Although community engagement in health-communication interventions is vital following a natural disaster because of health-related challenges, such as communicable diseases (WHO, 2006), this chapter focuses on community interventions that have successfully engaged communities to reduce health disparities among at-risk populations during epidemic, pandemic, and endemic cases of communicable diseases.

Importance to Public Health

This chapter focuses on health-communication interventions during epidemic, pandemic, and endemic cases of communicable diseases because of the enormous impact these cases have on entire communities and the world. In the 20th century, entire generations of Americans experienced mild to severe paralysis due to polio epidemics. The Global Polio Eradication Initiative has eradicated endemic polio everywhere except regions of four countries, including Nigeria, India, Pakistan, and Afghanistan (WHO, 2011). The World Health Organization (WHO) stated, "The eradication of polio is about equity in health and the moral imperative of reaching every child with an available health intervention" (WHO, 2011).

The Global Polio Eradication Initiative demonstrates that well-planned health interventions can reach even the most remote and disadvantaged populations in the world by using a multipronged strategy (WHO, 2011), highlighting the importance of using the most effective best practice(s) in a given community. Endemic polio, cases of reestablished transmission, and new outbreaks often occur in developing regions with high levels of health disparities because of the lack of monetary resources required to adequately vaccinate a population (WHO, 2011). Preventing an incapacitating disease, such as polio, reduces poverty and increases productivity and health in a population (WHO, 2011). A high level of community engagement is vital for a successful campaign; for example the WHO estimates that one out of every 250 people in a country has been involved in the polio eradication campaign.

Health Disparities

Health disparities are a result of poverty, low health literacy, lack of access to health information, disenfranchisement, unequal power, and prejudice (Carter-Pokras & Baquet, 2002; Kunitz, 2004). Health emergencies, such as epidemics and pandemics, threaten public health by placing sudden, intense demands on national and international health systems and resources. Developing countries have greater health disparities, and experience more adverse effects in a health emergency because they lack resources, emergency-management plans, and infrastructure (WHO, 2006). Health emergencies bring social systems to the point of collapse and paralyze public-health systems. The World Health Organization (WHO) verified more than 3,000 health-emergency events between January 2001 and November 2009. These events included emerging and re-emerging epidemic diseases and the release of biological agents. Emerging diseases are new or newly recognised pathogens (e.g., avian influenza, SARS, Ebola, Marburg). Reemerging diseases are previously controlled, well-characterized, outbreak-prone diseases (e.g., cholera, measles, shigellosis, yellow fever) that increase in prevalence. Release of biological agents can be accidental or deliberate (e.g., bovine spongiform encephalopathy and variant *Creutzfeldt-Jakob disease* (BSE/v CJD), smallpox, anthrax).

Public-health emergencies perpetuate and amplify health disparities. Health emergencies often expose existing weaknesses in public-health systems by:

- Draining resources, staff, and supplies away from other healt
- Stressing social and political systems, and
- Creating inappropriate and ineffective response behaviors (Brouwer, Akter, Brander, & Haque, 2007; Cordasco, Eisenman, Glik, Golden, & Asch, 2007).

Public-health systems are vulnerable from decades of social, environmental, and economic changes, including urbanization (e.g., crowding and slums), population displacement (e.g., refugee camps), global travel and trade, environmental exploitation and degradation (e.g., deforestation, water and air pollution), antibiotic resistance, emerging diseases, and poverty. Vulnerable public-health systems lead to a widening development gap between the developed and developing worlds, and increases health disparities. Given the breadth of health emergencies and the vulnerability of public-health systems, which disproportionately impact at-risk populations and increase health disparities, public-health organizations need a sustainable way to prepare at-risk populations.

Health-Communication Interventions

Health-communication interventions that successfully engage communities reduce health disparities among at-risk populations. Health-communication interventions have multiple components, including health education, communication, and health literacy, all of which are essential to reducing, eliminating, and preventing health disparities (Ensor & Cooper, 2004; Krauss, Godfrey, O'Day, & Freidin, 2006; Wu et al., 2008). The components of a strong health-communication intervention work synergistically. For example, health education employs various communication tools to increase health literacy in target populations. Health-education material may consist of environmental, physical, social, emotional, intellectual, and spiritual health (Donatelle, 2009, p. 4).

Health disparities are often perpetuated by false, misleading, and deceptive rumors or concepts associated with diseases. Health-communication interventions can address these rumors and false concepts to change behaviors, improve health, and reduce health disparities (Krauss et al., 2006; Wu et al., 2008). Health-communication interventions can lead to a reduction in poverty, provide better access to health information and health care, and reduce prejudice by dispelling false beliefs (Krauss et al., 2006; Wu et al., 2008). In return, a community whose participation is not hampered by rumors or false beliefs about a disease is more likely to participate in health-communication interventions and therefore reduce health disparities (Attree et al., 2011).

A successful health-communication intervention has multiple, critical components. First, one must determine the target audience and the purpose of the message to best achieve behavioral and social change. To do this, we can ask:

- Who will benefit from health education and communication?
- Why will this population benefit from health education and communication?
- What kinds of health education and communication are needed?
- What information does the population need to know?
- How can we effectively convey the information to the population?
- When is this population most receptive to health messages?

It is also vitally important to consider the resources that are available; what resources are needed; where the resources can be obtained and by what means; and how the resources will be used, maintained, and stored. While many times communities are not involved until the end stage of many interventions, we can solicit input during these stages using multiple methods, depending on the conventions of the community.

Once we have determined the basis for our health-communication campaign, and developed messages with the input of members of the community, it is important to pretest the messages using a social marking mix that utilizes the four "Ps":

1. Place
2. Product
3. Price
4. Promotion

Furthermore, during all stages of campaign development, we should focus on what the community can and is willing to contribute to disease control and prevention efforts. The objective of the health education and intervention is to contribute to the reduction of health disparities by engaging the community. This is an opportunity to develop a partnership with the community.

The Critical Role of Community

Community engagement during a health-communication intervention is critical to reducing health disparities (Zakus & Lysack, 1998). Local Ministries of Health and influential community leaders often have a great deal of knowledge about the health disparities in their community and what the community needs. Community members are also able to better relate with the target population, and to serve as trusted sources of information who are able to effectively and powerfully deliver health-related messages. For example, distrust of the authorities during Hurricane Katrina in New Orleans, Louisiana led to many residents ignoring the advice of public-health officials (Cordasco et al., 2007). In general, people are more likely to trust friends, family, and neighbors over outsiders, who may, in the minds of many, not have their best interests at heart. In the Dominican Republic, villagers had misgivings about dealing with outsiders and biomedical professionals, sometimes refusing to seek free care because the health officials asked questions about where they "lived and everything," and the villagers did not want to share personal information with a stranger (Quinlan, 2004).

In addition, involving community members as volunteer advocates for services and products can not only reduce costs of health initiatives, but also reach segments of the population that may not respond favorably to outsiders. Additionally, community members can also be employed as health advocates, which includes all of the above-mentioned benefits of volunteer advocates, and also provides a stable income for community members.

Community member engagement and involvement in a health-communication initiative also allows the community to take ownership of the health initiative. This empowers the community, and allows members to take control of their health; something that is not common is a population plagued by health disparities. Community engagement and ownership also proved to be the basis for continued, sustainable efforts to improve health following the departure of outside contributors, leading to a long-term reduction in health disparities (Figure 13.1).

Figure 13.1. Model of Sustainability in a Plan to Address Health Disparities with the Importance of Community Highlighted as the Central Module

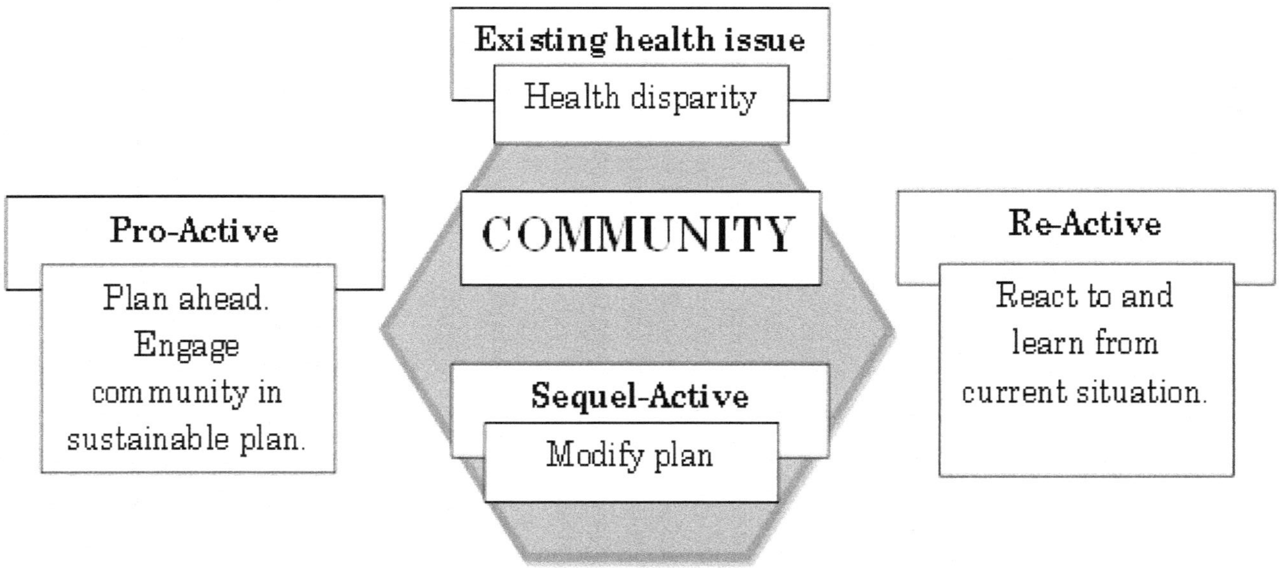

Sustainability—Evolving Plan to Address Health Disparities

It is in the best interest of all participants involved in a health-intervention campaign to involve the community in every step of the process, including:

- Formative research
- Conducting focus groups, interviews, and surveys with community members to provide essential information
- Message development and testing
- Implementation of intervention
- Community continuation

Ultimately, any outside campaign support (NGO, private-public partnership, etc.) will leave, and then it is up to the community to keep the program running with limited resources. Thus, if the local community is involved in the development and implementation of an intervention from the beginning, the intervention will be more likely to succeed after the community is left to continue it, because the intervention will be compatible with the local beliefs and resources. Not only can community involvement reduce health disparities, it increases economic development and improves social status.

Communication Methods to Engage Communities

There are two common communication strategies discussed in public-health communication:

1. Top-down communication
2. Bottom-up communication

Top-down communication is defined as communication efforts that come from authorities, such as Ministries of Health, and filter down to the community. Top-down communication requires increased political commitment and resource deployment to effectively increase response time and provide resources, but it also requires ethical considerations (Chakraborti, 2008), such as the communities' right to participate. Communities often participate in health-communication initiatives via bottom-up communication, defined as communication efforts that come from the community and work their way up through channels of authority to convey the needs of a community.

We suggest that there is a more effective communication strategy to engage communities and authorities during public-health initiatives. We call this method of health communication *transversal communication*. Transversal communication is defined as communication methods that cut across levels of authority to incorporate input from multiple stakeholders to synergistically accomplish the health initiative using culturally relevant and appropriate health interventions to increase effectiveness of the response, thus reducing health disparities. Stakeholders in transversal communication include the community, public-private partnerships, NGOs, the scientific community, the government and opposition parties, the media system, and the educational system (Figure 13.2).

Figure 13.2. Illustration of the Three Communication Strategies, With the Optimal Transversal Method to Engage Communities in Public-Health Interventions

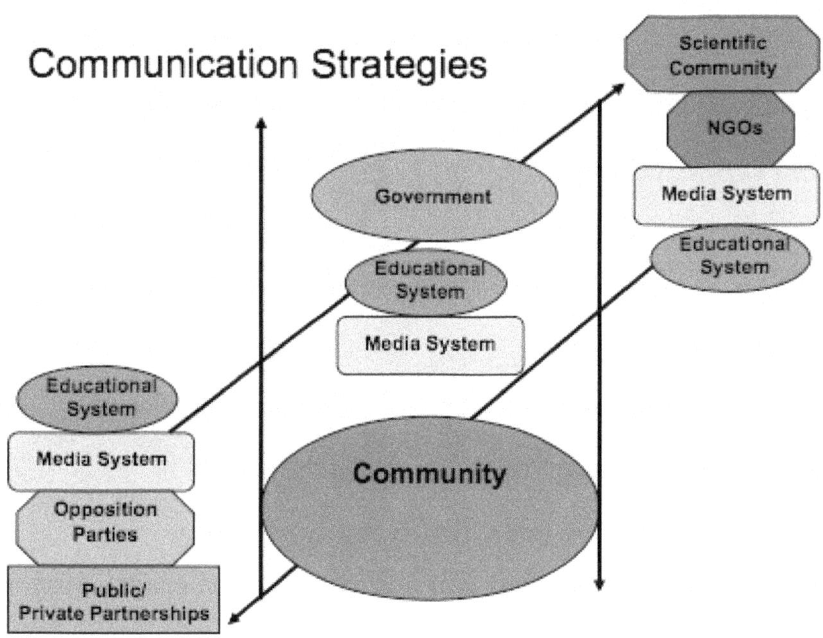

Affected communities are often on their own during the first stages of an epidemic or pandemic disease outbreak and then when the outbreak is over. Therefore, communities should be at

the center of any public-health communication initiative. Before and after outside assistance, a community must be prepared to respond, which is why community engagement during the development of public-health plans is important (Figure 13.3). For example, in Washington State, a collaborative model between state, local, and tribal health departments and academic institutions indicates that collaborative efforts of this magnitude may be applicable to emergency preparedness, general public health, and workforce development (Pearson, Thompson, Finkbonner, Williams, & D'Ambrosio, 2005).

Figure 13.3. Levels of Development of an Unprepared Community Versus a Prepared Community Before and After a Potential Disaster (e.g., Epidemic Disease Outbreak)

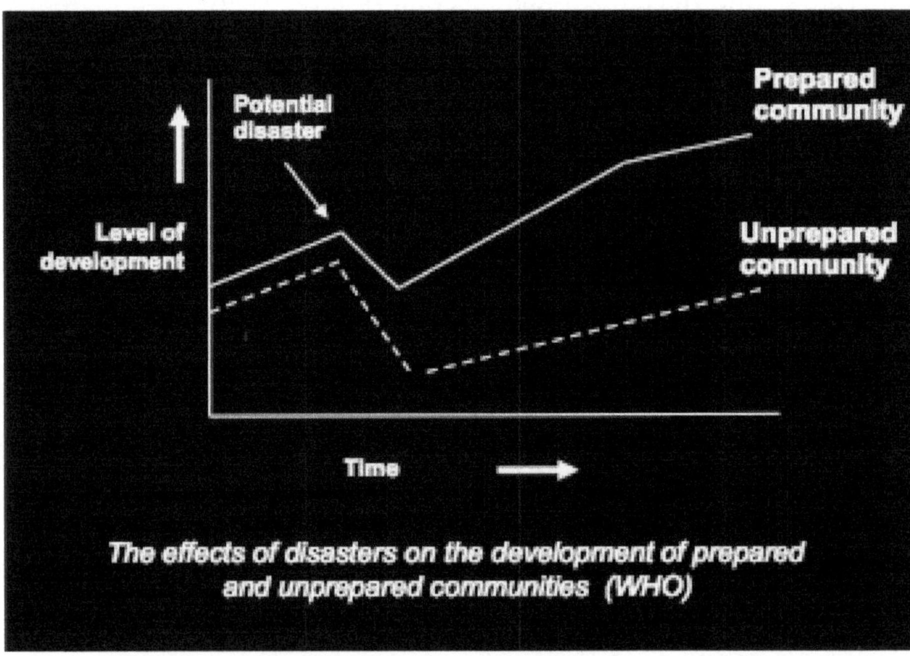

Filling the Gaps Between Knowing and Doing

There are many things that can slip through the cracks during intense and stressful times of epidemic, pandemic, and endemic outbreak response. As a result, there is a need to fill the gaps between knowledge and action. For instance, a community's understanding of diseases and the spread of disease is complex; yet, often messages for the community are very simplistic and lack depth. To develop an effective message, there is a need to understand how a community looks at disease before one can identify the changes that need to be made and messages can be targeted to affect behavior (Falade et al., 2006; Meyers, 1992; Netto, McCloughan, & Bhatnagar, 2007; Parsa, Kandiah, Abdul Rahman, & Zulkefli, 2006). Yet, often the community is not solicited for their experiences, opinions, and views. Although, it is recognized that the community should be active participants in the process, assumptions are often made on their behalf, as it is assumed that there is not any time to seek their counsel or that their beliefs are obvious. Eliminating these inconsistencies will increase response time that can greatly increase the number of lives saved during disease outbreaks and reduce health disparities.

Case Studies: Community Engagement in Action

Dengue Fever

Each year 50 to 100 million people are infected with the dengue virus, which causes dengue fever and the more severe dengue hemorrhagic fever with higher levels of illness and death (WHO, 2012a). Before 1970, only nine countries were affected by dengue epidemics, but now dengue is endemic in 100 countries, with ever increasing "explosive outbreaks" (WHO, 2012a). The disease is transmitted via a mosquito vector, primarily the female *Aedes aegypti* mosquito, which lives and breeds in urban habitats in man-made containers, such as tires and buckets (WHO, 2012a).

The *Aedes aegypti* mosquito feeds mainly during the early morning hours and before dusk, therefore measures such as bed nets do not reduce dengue transmission from the mosquito vector to humans. Strategies for controlling dengue focus on eliminating the mosquito vectors, such as insecticide space spraying, reducing vector breeding grounds, and reducing human-vector contact, because there are no specific medications or vaccines available to treat or prevent dengue (WHO, 2012a).

There are many challenges associated with dengue-control measures, which include promoting and sustaining the control measures, obtaining long-term commitment from individuals and institutions, and getting the required community participation in the application of integrated vector control measures. Community engagement and societal mobilization strategies are of vital importance for the prevention and control of dengue fever and dengue hemorrhagic fever because only widespread vector control can eliminate mosquito carriers and prevent dengue outbreaks.

The Malaysia National Dengue Control Program is a good example of a successful health-communication intervention that engaged a community to reduce dengue fever. First, the intervention included collaboration between the WHO and the State Health Department in the state of Johor (WHO, 2012b). In the state capital of Johor Bahru, 250,000 families are at risk for contracting dengue. In 2001, the WHO and the State Health Department implemented a communication for behavior impact (COMBI) vector control campaign (WHO, 2012b). The campaigned used a multi-pronged approach, including radio and newspaper ads, school-based mobilization, volunteer efforts to eliminate mosquito breeding grounds, educational material distribution at clinics, and traveling groups of vector-control promoters (WHO, 2012b).

The school-based mobilization used worksheets (Figure 13.4a & Figure 13.4b) that were distributed in schools to encourage children to carry out inspections with their parents. Three months after the campaign, approximately 90% of households were maintaining weekly inspections, and 95% of volunteers were continuing their work (Haider, 2010). This led to a 66% reduction in the number of expected dengue cases in Johor Bahru. Similar plans were carried out in Barbados, Bolivia, Brazil (four pilot sites), Colombia, Costa Rica (five sites), Dominican Republic, Ecuador, El Salvador, Honduras, and Paraguay (Haider, 2010).

Lymphatic Filariasis

About 1.3 billion people are at risk for lymphatic filariasis, with more than 120 million people infected. Lymphatic filariasis is found mostly in subtropical regions, which are among the world's poorest countries. To interrupt transmission of the worm that causes lymphatic filariasis, two drugs can be administered orally once a year (WHO, 2012c). These mass drug administration (MDA) campaigns need to reach a very high level of coverage, requiring at least 65% of an endemic population to participate (WHO, 2012d). This also requires that individuals and institutions commit to and sustain the actions for five consecutive years, which means persuading people who appear healthy to take medication year after year for the saturation level over the entire reproductive lifespan of the worm. When the last adult worm dies in the last infected person, the mass drug administration can be stopped, and lymphatic filariasis will be eliminated in that population. Of course, there is no way of knowing who is infected and who is not. It can take up to 10 years for the symptoms of the disease, including swollen limbs, to manifest. By the time symptoms do appear, it is too late to reverse the effect (WHO, 2012c).

Therefore, it is of vital importance for the entire community to be engaged in the public-health intervention to eliminate lymphatic filariasis. Various societal-mobilization strategies (Figure 13.4a & Figure 13.4b) that are based on the understanding of community beliefs are necessary to encourage community participation in the drug campaign, and to encourage individuals with lymphatic filariasis to seek disability management.

Mass drug administration (MDA) campaigns begin by mapping an area to determine populations most at-risk for infection. Once the target population is identified, multiple strategies are used to encourage participation in the campaign (Figure 13.6a), such as communication materials which encourage people to "take your LF treatment tablets from your Filaria Prevention Assistants on Filaria Day" (Haider, 2010); door-to-door medication distribution; and medication delivery through schools, workplaces, and other central points in a community (WHO, 2010). Numerous personnel, such as trained nonhealth personnel including community volunteers, community health workers, and social workers, successfully administered medications for the MDA campaign (WHO, 2010). The campaign reached over 50 million people (Haider, 2010). Examples of countries targeted include India: Tamil Nadu—target population of 28 million, with 74% coverage; Kenya—target population of 1.2 million, with 81% coverage; Sri Lanka—target population of 9.5 million, with 86% coverage; and Zanzibar—target population of 1 million, with 83% coverage (Haider, 2010).

Recommendations on Community Engagement to Reduce Health Disparities

Community engagement increases as the population becomes more informed about the benefit of their involvement during public-health emergencies (Attree et al., 2011; Dickert & Sugarman, 2005; Giddens, Fogg, & Carlson-Sabelli, 2010; Zakus & Lysack, 1998). Communities and individuals will engage and very often accept major disruptions to their lives if they feel like their knowledge and understanding of the situation is taken into consideration, they are offered new knowledge and understanding, they are involved in shaping the interventions, and they are allowed to trade off the benefits and burdens of engagement and behavioral adaptation. Increasing knowledge and awareness is very easy to do, but knowledge change does not necessarily translate into effective behavioral impact.

Communication should be integrated with initiatives that reduce obstacles—including misleading beliefs and conceptions—to practicing healthy behaviors. Changing policy—removing physical and economic barriers, rather than solely focusing the communication effort on individual attitude and knowledge—often enhances wider practice of recommended behaviors, such as reducing vector breeding grounds, taking preventive medications, and seeking disability management. Well-researched behavioral goals focus effort, and result in clear, consistent messages to support the development of sustainable behavioral practices. Strategic communication planning is a prerequisite to production of these materials.

Implementation requires not only community-wide involvement, but also commitment from local authorities. It may also necessitate some organizational restructuring to support multisectoral teams, since good possibilities exist for private-sector partnerships in this effort as well.

Summary

Community health interventions that engage the community are essential to reducing health disparities. Examples include the use of community engagement in the interventions utilized to prevent dengue and eliminate lymphatic filariasis. In any case, understanding the resources available, the barriers to providing and accessing care, and the perceptions of the community will help tailor the health interventions and the messages to change deleterious behaviors and encourage positive behaviors.

Figure 13.4a. Worksheet From the School-Mobilization Effort to Eliminate Mosquito Vector Breeding Grounds

BAHAGIAN B / *PART B*

SENARAI SEMAKAN PEMERIKSAAN RUMAH SETIAP AHAD SELAMA 30 MINIT
Checksheet for house inspection every Sunday for 30 minutes

Tandakan (✓) diruang 3 jika terdapat tempat yang dinyatakan di rumah/kawasan rumah anda dan di ruang 6 bagi tindakan yang telah diambil. Tandakan (✓) di ruang 4 jika tempat tersebut tiada dirumah/kawasan rumah anda.

Mark (✓) in column 3 if the listed places are found in your house/compound. Mark (✓) in column 6 for actions you have taken. Mark (✓) in column 4 if the listed places are not found in your house or in your compound.

1	2	3	4	5	6
TEMPAT DIPERIKSA / *PLACES INSPECTED*		**Ada** *Yes*	**Tiada** *No*	**TINDAKAN YANG DIAMBIL / *ACTIONS TAKEN***	(✓)
Talam Peti Sejuk / *Fridge Water Tray*	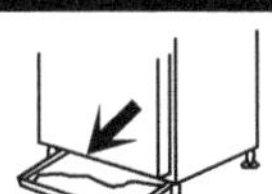			**Cuci setiap minggu /** *Clean every week*	
				Bocorkan / *Perforate the bottom*	
Pelapik Rak Pinggan / *Draining Tray*				**Buang air yang bertakung /** *Empty the stagnant water*	
				Jangan gunakan pelapik / *Do not use the tray*	
				Bocorkan / *Perforate the bottom*	
Perangkap Semut / *Ant Trap*	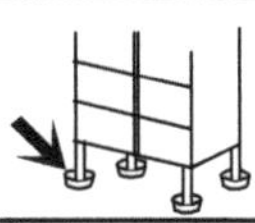			**Cuci setiap minggu /** *Clean every week*	
				Bubuh ubat pembunuh jentik-jentik / *Put larvicide*	
				Bocorkan / *Perforate the bottom*	
Kolah Air / *Water Storage*				**Cuci setiap minggu /** *Clean every week*	
				Bubuh ubat pembunuh jentik-jentik / *Put larvicide*	
				Pecahkan kolah / *Destroy the water storage tank*	
Pam Tandas / *Cistern*				**Kosongkan airnya /** *Drain away the water*	
				Tutup dengan rapat / *Close the lid tightly*	
				Bubuh ubat pembunuh jentik-jentik / *Put larvicide*	
Mangkuk tandas (rosak) / *Toilet bowl (non-functioning)*				**Bubuh ubat pembunuh jentik-jentik /** *Put larvicide*	
				Biarkan begitu sahaja / *Leave it as it is*	
Tong Drum / *Drum*	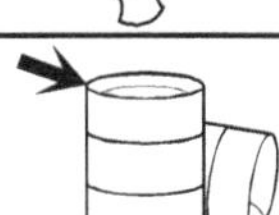			**Cuci setiap minggu /** *Clean every week*	
				Bubuh ubat pembunuh jentik-jentik / *Put larvicide*	
				Tutup dengan rapat / *Close the lid tightly*	
				Buang / *Throw away (Dispose)*	
Baldi / *Pail*	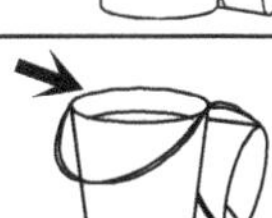			**Cuci setiap minggu /** *Clean every week*	
				Tutup dengan rapat / *Close the lid tightly*	
				Bocorkan / *Perforate the bottom*	

Figure 13.4b. Example of Education Materials Used in the School-Mobilization Campaign to Increase Home Inspections and Reduce Vector Breeding Grounds

1	2	3	4	5	6
TEMPAT DIPERIKSA / *PLACES INSPECTED*		**Ada** *Yes*	**Tiada** *No*	**TINDAKAN YANG DIAMBIL / *ACTIONS TAKEN***	(✓)
Tempayan / *Earthern Jar*				**Cuci setiap minggu /** *Clean every week*	
				Bubuh ubat pembunuh jentik-jentik / *Put larvicide*	
				Tutup dengan rapat / *Close the lid tightly*	
				Bocorkan / *Perforate the bottom*	
Jambangan Bunga / *Vase*				**Salin airnya setiap minggu /** *Change the water weekly*	
				Bubuh ubat pembunuh jentik-jentik / *Put larvicide*	
				Jangan bubuh air / *Do not put any water*	
				Buang / *Throw away (Dispose)*	
Pasu Bunga / *Flower Pot*				**Gemburkan tanah setiap minggu /** *Loosen the soil*	
				Buang / *Throw away (Dispose)*	
Alas Pasu Bunga / *Flower Pot Plate*				**Cuci setiap minggu /** *Clean every week*	
				Bubuh ubat pembunuh jentik-jentik / *Put larvicide*	
				Bocorkan /*Perforate the bottom*	
Pelbagai Jenis Tin / *Assorted Empty Cans*				**Kumpulkan dan lupus dengan sempurna /** *Collect and dispose properly*	
				Biarkan begitu sahaja / *Leave it as it is*	
Tayar / *Tyre*				**Letakkan ditempat yang terlindung /** *Place in the shed*	
				Bubuh ubat pembunuh jentik-jentik / *Put larvicide*	
				Susun dan tutup diatas / *Stack and cover the top*	
				Letak dimerata tempat / *Put anywhere*	
Takungan Hidroponik / *Hydroponic*	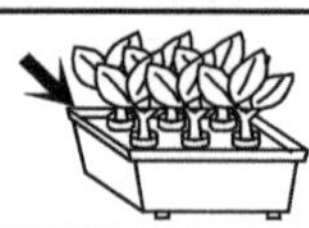			**Bubuh ubat pembunuh jentik-jentik /** *Put larvicide*	
				Biarkan begitu sahaja / *Leave it as it is*	
Kanvas, Plastik Semua Jenis / *Canvas And All Types Of Plastic*	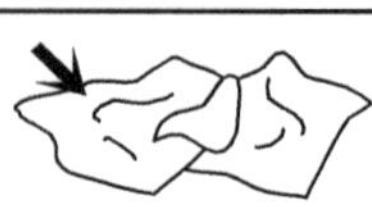			**Buang air yang bertakung /** *Empty the stagnant water*	
				Letak ditempat terlindung dari hujan / *Place in the shed*	
				Letak dimerata tempat / *Put anywhere*	
Bekas Minuman Binatang / *Receptacles For Animal Drinks*	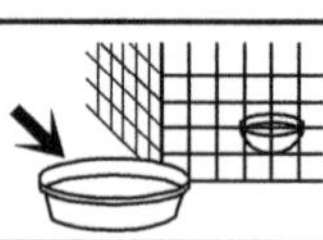			**Salin airnya setiap minggu /** *Change the water weekly*	
				Bocorkan / *Perforate the bottom*	
Tin Rendam Berus Cat / *Tin For Cleaning Of Paint Brushes*	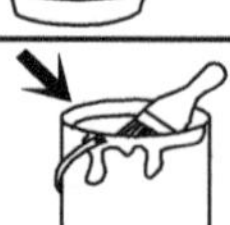			**Kosongkan airnya /** *Drain away the water*	
				Salin airnya setiap minggu / *Change the water weekly*	
				Bocorkan / *Perforate the bottom*	

Acknowledgments

The authors wish to thank Asiya Odugleh-Kolev and Audrey Babkirk for their contributions to this chapter.

Bibliography

Attree, P., French, B., Milton, B., Povall, S., Whitehead, M., & Popay, J. (2011). The experience of community engagement for individuals: A rapid review of evidence. *Health & Social Care in the Community, 19*(3), 250–260.

Bennett, I.M, Chen, J., Soroui, J.S., & White, S. (2009). The contribution of health literacy to disparities in self-rated health and preventive health behaviors in older adults. *Annals of Family Medicine, 7*(3), 204-211.

Bolden, A. J. (2001). Community-focused partnering. *Journal of Dental Research, 80*(11), 1960–1961.

Brouwer, R., Akter, S., Brander, L., & Haque, E. (2007). Socioeconomic vulnerability and adaptation to environmental risk: A case study of climate change and flooding in Bangladesh. *Risk Analysis, 27*(2), 313–326.

Carter-Pokras, O., & Baquet, C. (2002). What is a 'health disparity'? *Public Health Reports, 117*(5), 426–434.

Chakraborti, C. (2008). Pandemic management and developing world bioethics: Bird flu in West Bengal. *Developing World Bioethics, 9*(3), 161–166.

Cordasco, K. M., Eisenman, D. P., Glik, D. C., Golden, J. F., & Asch, S. M. (2007). 'They blew the levee': Distrust of authorities among Hurricane Katrina evacuees. *Journal of Health Care for the Poor and Underserved, 18*(2), 277–282.

Dickert, N., & Sugarman, J. (2005). Ethical goals of community consultation in research. *American Journal of Public Health, 95*(7), 1123–1127.

Donatelle, R. J. (2009). Promoting health behavior change. In R. J. Donatelle, *Health: The basics* (8th ed.) (pp. 2–28). San Francisco, CA: Pearson Education.

Ensor, T., & Cooper, S. (2004). *Overcoming barriers to health service access and influencing the demand side through purchasing.* Washington, DC: The World Bank.

Falade, C. O., Ogundiran, M. O., Bolaji, M. O., Ajayi, I. O., Akinboye, D. O., Oladepo, O., Adeniyi, J. D., & Oduola, A. M. (2006). The influence of cultural perception of causation, complications, and severity of childhood malaria on determinants of treatment and preventive pathways. *International Quarterly of Community Health Education, 24*(4), 347–363.

Fisher, T. L., Burnet, D. L., Huang, E. S., Chin, M. H., & Cagney, K. A. (2007). Cultural leverage—Interventions using culture to narrow racial disparities in health care. *Medical Care Research and Review, 64*(5), 243S–282S.

Giddens, J., Fogg, L., & Carlson-Sabelli, L. (2010). Learning and engagement with a virtual community by undergraduate nursing students. *Nursing Outlook, 58*(5), 261–267.

Haider, M. (2010, May). Community engagement in response to public health emergencies: A critical link. Paper presented at WHO Inter-Country Workshop on Health Education for the Eastern Mediterranean Region, Cairo, Egypt.

Krauss, B. J., Godfrey, C. C., O'Day, J., & Freidin, E. (2006). Hugging my uncle: The impact of a parent training on children's comfort interacting with persons living with HIV. *Journal of Pediatric Psychology, 31*(9), 891–904.

Kuhajda, M. C., Cornell, C. E., Brownstein, J. N., Littleton, M. A., Stalker, V. G., Bittner, V. A., Lewis, C. A., & Raczynski, J. M. (2006). Training community health workers to reduce health disparities in Alabama's Black belt—The Pine Apple Heart Disease and Stroke Project. *Family & Community Health, 29*(2), 89–102.

Kunitz, S. J. (2004). The evolution of disease and the devolution of health care for American Indians. In C. G. N. Mascie-Taylor, J. Peters, & S. T. McGarvey (Eds.), *The changing face of disease: Implications for society* (pp. 153–169). New York, NY: Taylor & Francis.

Meyers, C. (1992). Hmong children and their families—Consideration of cultural influences in assessment. *American Journal of Occupational Therapy, 46*(8), 737–744.

Netto, G., McCloughan, L., & Bhatnagar, A. (2007). Effective heart disease prevention: Lessons from a qualitative study of user perspectives in Bangladeshi, Indian and Pakistani communities. *Public Health, 121*(3), 177–186.

Parsa, P., Kandiah, M., Abdul Rahman, H., & Zulkefli, N. M. (2006). Barriers for breast cancer screening among Asian women: A mini literature review. *Asian Pacific Journal of Cancer Prevention, 7*(4), 509–514.

Pearson, D., Thompson, J., Finkbonner, J., Williams, C., & D'Ambrosio, L. (2005). Assessment of public health workforce bioterrorism and emergency preparedness readiness among tribes in Washington State: A collaborative approach among the Northwest Center for Public Health Practice, the Northwest Portland Area Indian Health Board, and the Washington State Department of Health. *Journal of Public Health Management and Practice, 11*(6), S113–S118.

Quinlan, M. B. (2004). *From the bush: The front line of health care in a Caribbean village.* Toronto, Canada: Wadsworth.

World Health Organization (WHO). (2006). Communicable diseases following natural disasters: Risk assessment and priority interventions. Retrieved from http://www.who.int/diseasecontrol_emergencies/en/.

World Health Organization (WHO). (2010). *World Health Organization Global Programme to Eliminate Lymphatic Filariasis: Progress report 2000–2009 and strategic plan 2010–2020.* Geneva, Switzerland: WHO Press.

World Health Organization (WHO). (2011, November). Poliomyelitis: Fact sheet. Retrieved from http://www.who.int.mediacentre/factsheets/fs114/en/.

World Health Organization (WHO). (2012a). Dengue and severe dengue: Fact sheet. Retrieved from http://www.who.int.mediacentre/factsheets/fs117/en/.

World Health Organization (WHO). (2012b). Dengue epidemiology: Malaysia. Retrieved from http://www.wpro.who/int/sites/mvp/epidemilogy/dengue/maa_profile.htm.

World Health Organization (WHO). (2012c). Lymphatic filariasis: Fact sheet. Retrieved from http://www.who.int/mediacentre/factsheets/fs102/en/.

World Health Organization (WHO). (2012d). Lymphatic filariasis programme. Retrieved from http://www.who.int/lymphatic_filariasis/disease/en/.

Wu, S., Li, L., Wu, Z. Y., Liang, L. J., Cao, H. J., Yan, Z. H., & Li, J. J. (2008). A brief HIV stigma reduction intervention for service providers in China. *AIDS Patient Care and STDs, 22*(6), 513–520.

Zakus, J. D. L., & Lysack, C. L. (1998). Revisiting community participation. *Health Policy and Planning, 13*(1), 1–12.

14

NUMANA FEEDS HAITI: A CASE STUDY OF A GRASSROOTS, COMMUNITY-BASED SOCIAL MOVEMENT TO "SAVE THE STARVING"

Deborah S. Ballard-Reisch, Rick McNary, Pamela O'Neal, Ashley Archiopoli, Bobby Rozzell, Melissa Granville, and Alyssa Ballard-Reisch

On December 29 and 30, 2009, almost 4,000 volunteers met in the small town of El Dorado, Kansas, and packaged more than 285,000 meals to feed school children in Haiti. That weekend was the formal launch of Numana, Inc., a start-up nonprofit whose mission is to "empower people to save the starving" (www.numanainc.com). What was unique about the Numana effort was that volunteers rolled up their sleeves, donned plastic aprons, gloves, and shower caps, and mixed, packaged, and prepared the food for shipment to Salvation Army schools in Haiti. At tables of 12–14 volunteers, rice, soy, freeze-dried vegetables, and a 21-vitamin/mineral tablet—a diet specially designed for the metabolism of people who are starving—were measured into six-serving bags, weighed, vacuum sealed, packed 36 to a box, and loaded on a truck. The truck would carry the food to Norfolk, Virginia, where it would be shipped by boat to Haiti. It was expected to arrive in six to eight weeks.

Several of my[1] graduate students from the Elliott School of Communication at Wichita State University, my son and daughter, my daughter's youth group, and I were among the volunteers who showed up to package food in one-hour shifts. Most of us stayed the whole day, and several of us came back the second day. What we experienced that weekend was the beginning of a social movement that in six months led over 120,000 people across the United States to package more than 20 million meals for Haiti, and the food packaging continues. We view Numana as a social movement, because it embodies "vision, identity, a sense of linkage and ongoing action" (Zoller, 2005, p. 345). While Zoller (2005) posited that a single organization is not a social movement in and of itself, Numana was the catalyst for a social movement that spread throughout the United States and resulted in large-scale efforts, grounded in their methodology, in the form of community activism to address the problem of global hunger.

Being at the start of this ground-up, community-based movement has allowed us to observe the unique dynamics involved in coordinated social action around one specific cause, feeding the starving. Numana met a need felt by many Americans, to do something tangible to make a difference for others. Health-communication strategies are valuable in analyzing the process through which Numana facilitated meeting that collective need. Specifically, research on grassroots social movements (Brown & Zavestoski, 2005; Griffin, 1952, 1969; Hoffman, 2003; Kolker, 2005), engaged communication scholarship (Barge, Simpson, & Shockley-Zalabak, 2008; Cheney, 2008; Putnam, 2009; Van de Ven, 2007), and community-based participatory research (Israel, Eng, Schulz, & Parker, 2005; Minkler & Wallerstein, 2002) inform this chapter.

Creating Context

As with any successful social movement, factors intervened that galvanized people to act, and propelled and strengthened the movement (Griffin, 1952, 1969). While social movement theory typically addresses marginalized or disenfranchised groups coalescing around an injustice they can no longer tolerate (Griffin 1952, 1969) and often grounded in conflict and protest (Della Porta & Diani, 1999), social movement theory also characterizes movements as composed of informal networks, where participants share beliefs and solidarity, and engage in collective action (Della Porta & Diani, 2006; Zoller, 2005). It is these characteristics that position Numana for study as a social movement. For Numana and the Feed Haiti initiative, one compelling factor that mobilized communities to action was the most violent earthquake in a century (U.S. Geological Survey, 2010a), an estimated 7.0 magnitude earthquake that hit Haiti 15 miles WSW of the capital, Port-au-Prince, on January 12, 2010. The earthquake occurred almost two weeks after that first food-packaging event and while those 285,000 meals were in transit to Haiti.

Hunger was endemic to Haiti prior to the earthquake. According to CIA World Facts (2010), "Haiti is the poorest country in the Western Hemisphere with 80% of the population living under the poverty line and 54% in abject poverty." Over two-thirds of the population has no formal employment, and most survive by subsistence farming. According to the Food and Agriculture Organization (FAO) (2009b), Haiti suffers from the fourth highest rate of hunger in the world. Because of its proximity to the US, the availability of partners who could assist in food distribution, the relative cost effectiveness of getting the food to Haiti, and the demonstrated need, Haiti was chosen by Numana as the first country to benefit from their food-packaging events. On January 12, 2010, the food from that first packaging event, targeted to feed children in schools, was on the road. The earthquake made the chronic problem of hunger in Haiti a crisis, and emergency relief was now required. Numana food became a critical contributor to the crisis food response in Haiti in the aftermath of the earthquake.

World Hunger

Eradicating poverty and world hunger is the number one UN Millenium Goal for 2015 (DPI, 2010). According to the UN's World Food Programme, one in seven people—over 925 million people worldwide—do not get enough food to be healthy, making hunger and malnutrition the number one risk to global health—greater than AIDS, malaria, and tuberculosis combined (DPI, 2010). The World Health Organization (WHO) identified 21 countries that "suffer from acute or chronic malnutrition" (WHO, 2009, p. 1). Of the world's hungry, 65% live in seven countries: India, China, the Democratic Republic of Congo, Bangladesh, Indonesia, Pakistan, and Ethiopia (FAO, 2008). Regionally, Asia and the Pacific are home to the largest number of hungry people; sub-Saharan Africa has the highest prevalence rate, with more than one in three people undernourished (FAO, 2010a, 2010b). In 2009 numbers, there were 265 million people hungry in sub-Saharan Africa, 642 million in Asia and the Pacific, 53 million in Latin America and the Caribbean, 42 million in the Near East and North Africa, and 15 million in developed countries (FAO, 2010b, p. 1).

While the majority of the hungry people worldwide live in developing countries, "hunger also occurs in the industrialized world" (FAO, 2010b, p. 1). According to the U.S. Department of Agriculture (DOA, 2009) one in six Americans, or 15% of the population, goes to bed hungry each night. This is the highest number since the DOA started keeping statistics in 1995.

Numana, Inc.

Numana, Inc. is a nonprofit, hunger-relief organization with the mission to "empower people to save the starving." Numana organizes large-scale, food-packaging events, at which volunteers prepare meals to be shipped to a country in need. Through meal-packaging events, Numana sends food to countries where intense food insecurity exists. These events are often funded by the Salvation Army or corporate sponsors, but at times the funds are raised by communities or groups hosting the events. At every international food-packaging event Numana hosts, bins are also available to collect food for local food programs. From January 2010 through June 2010, schools, churches, universities, cities, corporations, civic groups, youth organizations, and others across the United States hosted 44 food-packaging events with over 120,000 volunteers to package more than 20,000,000 meals for Haiti (www.numanainc.com).

Theoretical Orientations

Social Movements

A social movement can be defined as an organized and somewhat sustained process to advance a specific cause, interest, or pursuit (Encyclopaedia of Sociology, 1992), or, as Zoller (2005) noted, collective activism on a large scale to promote change over time. Della Porta & Diani (1999) noted that social movements involve movement members engaging in unconventional actions outside institutions. Numana's activities to empower people to save the starving, their community-grounded approach, and the scope of their impact on hunger in Haiti in the first six months following the earthquake qualify them for consideration as a social movement. Two other characteristics of social movements are particularly relevant to the Numana effort: culturally resonant collective action (DeGenaro, 2009) and the development of meaningful, culturally relevant themes (Benford & Snow, 2000; Kolker, 2005).

As noted by DeGenaro (2009), a critical component of culturally resonant collective action involves recognizing the human face of social movements. Polletta and Jasper (2001), Seidman (1994), and Kilgore (1999) advanced that those who engage in contemporary social movements are often motivated by ethics, values, and moral visions. This was certainly the case with Numana, as those initially involved in the movement came from a strong, common, rural Kansas, Christian faith community. They saw God's hand in their work. This movement, however, resonated not only with Christians, but ultimately, with secular and other religious groups as well. Mary Grigsby (2004) in discussing what she termed, "greater good movements," argued that participants in these initiatives desire to "define themselves as worthwhile and good people" (p. 53). Sandlin and Walther (2009) referred to this as "moral identity work" that leads to individuals seeing themselves as "legitimate social actor[s]" (p. 302) involved in what Hoffman (2003) called "change from below" or "grassroots social movements" (p. 75). Clearly, Numana had tapped into a need on the part of at least some Americans to do something tangible to help others. The community-based collaborative process of working with others to package and prepare the food for shipment reinforced the social action in which participants were engaged. These dynamics allowed volunteers to see themselves as legitimate social actors, members of a community engaging in collaborative action to make a difference in the lives of people in Haiti. This "change from below" process was empowering and reinforcing for participants. As some volunteers said, "I may not have money, but I can give my time."

Benford and Snow (2000) advanced that social movements evolve through the construction of symbols and meanings, what they called culturally resonant themes (p. 612), that are meaningful to a population and that mobilize collective action (Kolker, 2005, p. 138). Deborah Kilgore (1999) extended this analysis when she asserted that collective learning and a shared, collective identity emerge within social movements as a result of the "dynamic interaction and mutual development of individual meanings and shared meanings" (p. 200), leading to "cognitive, moral, and emotional connection with a broader community" (Polletta & Jasper, 2001, p. 285). This identity formation is a reciprocal process, as Finger (1989) pointed out, because personal identity "cannot be separated from a person's experienced life, nor from his or her social commitment"; thus, identity is at the same time individual and social (p. 21). Collective identity is crucial to the success of social movements (Polletta & Jasper, 2001).

By tapping into the collective positive experience of large-scale, food-packaging events, Numana encourages the formation of collective identity both structurally and informally. Structurally, Numana used two strategies to support this collective identity. First, the food-packaging events are open to anyone of any age and skill level. Thus, five-year-olds work alongside 97-year-olds and all ages in between in tasks that meet their capacities. Everyone can help feed Haiti. Second, Numana created a special cadre of volunteers called "green shirts" (later changed to "red shirts" for Salvation Army-sponsored events). "Green shirts" are repeat volunteers who have already experienced a food-packaging event and are then trained to lead groups, coordinate tables, assist with supply distribution during the event, setup, and tear-down, etc. This structure provides both critical support for managing food-packaging events and increases the commitment of individuals to future events by rewarding their participation and allowing them to take on some level of responsibility at future events, if they so desire.

In all cases, it is the volunteers who "own" the events. On an informal level, a term emerged to describe people who became committed to Numana and food-packaging events. They call themselves and refer to others as "Numaniacs." As in, "I'm a Numaniac," or "Oh, you're a Numaniac, too!" Thus, a culture of Numana, feed the starving, volunteers is created.

Engaged Scholarship

The first six months of the Numana project to feed Haiti is an example of engaged communication scholarship and community-based participatory research in action. Engaged scholarship "involves negotiation and collaboration between researchers and collaborators in a learning community" (Van de Ven, 2007, p. 7) which stresses reflexivity and colearning, and aims to develop "theoretical and practical knowledge" (Barge & Shockley-Zalabak, 2008, p. 264).

Putnam (2009) outlined four faces of engaged scholarship, two of which are particularly appropriate to this project—the activism and social justice face, and the collaborative learning face. Consistent with grassroots social movements as discussed above, the activism and social justice face emphasizes partnerships that promote community engagement, leading to community organizing, enabling decision makers, and getting access to information so that social problems, inequalities, and the interests of those who are neglected, unnoticed, or disenfranchised can be adequately addressed (Putnam, 2009). This type of collaboration is grounded in a critical and moral imperative to act to make a difference in the lives of those who have been economically, socially, politically, or culturally underrepresented or disadvantaged (Putnam, 2009).

The collaborative learning face emphasizes partnerships that acknowledge community expertise in understanding the socially constructed aspects of culture. "This face of engaged scholarship concentrates on activities related to building communities, civic participation, and working together to develop strategies for managing problems" (Putnam, 2009). In terms of process, collaborative learning involves cocreation of knowledge, attention to multiple voices, and a commitment to diverse perspectives (Putnam, 2009). Partners cocreate projects through colearning, conversation, and partnership in cyclical and iterative phases to address social problems (Barge, Simpson, & Shockley-Zalabak, 2008). Thus, collaborative research is grounded in partnerships, common understandings, solutions to community problems, and community-based participation in initiatives (Barge & Shockley-Zalabak, 2008).

On the U.S. side, launching a broad-based collaboration to feed the starving through food-packaging events hosted by diverse organizations and groups requires both a strong infrastructure to provide all materials necessary to package the food as well as transport it to and from events, a strong process plan to get the job done, and the ability to be flexible to the unique dynamics of groups and venues hosting events.

On the Haiti side, this type of effort requires knowledgeable, effective collaborators on the ground, and the ability to be flexible to the need for refinements in process and product that facilitate the distribution needs of those on the ground in Haiti.

These needs led to an iterative learning curve in which the main components of the Numana process were retained, but tweaked to meet the characteristics of venues and groups, as well as the needs of the Salvation Army and their partners on the ground in Haiti.

While Numana developed a strong core of local staff and volunteers, Rick McNary's goal from the outset was that organizers for each event feel ownership for their event and design, and facilitate the process consistent with local capacity. Except for initial events in El Dorado and at the Kansas Coliseum in response to emergency requests for more food from the Salvation Army, events were planned, organized, advertised, and facilitated by those who either contacted Numana to host events or were approached by Numana to do so. Events ranged from smaller and more focused in scale—like those held at family reunions, schools, and churches for their individual communities—or larger scale and open to the municipality or region, like WSU Feeds Haiti and the City of Wichita's Feed Haiti event which both pulled volunteers from throughout the region. The groups who hosted the events relied more or less on formal Numana support to help them coordinate their events. Typically, an event had a local coordinating committee or event management team who handled logistics around location, time, number of volunteers anticipated, number of packaging stations, desired number of meals to be packaged, fundraising to purchase food and supplies, etc.

As noted above, at least within the state of Kansas, a group of committed volunteers offered to travel to various locations and provide their services as "green shirts," assisting with everything from table coordination, to making sure packaging stations were adequately supplied with materials, to trash management. Event coordinators might choose to openly recruit "green shirts," or limit their events to their specific groups or communities. Thus, a culture around events, as well as a culture within each event, emerged to facilitate the food-packaging process and the ongoing commitment to "feed Haiti."

The food-packaging process did not change much from one event to the next, but the way the events were organized changed as appropriate from venue to venue. Local character was reflected in various processes across events. For example, at the Kansas Coliseum, people stood in line for up to two hours before being led to a table to package food. At WSU Feeds Haiti, the weather was bad, and volunteers were seated in an auditorium with music and occasional live entertainment as they waited to package. Some events allowed people to sign up for a particular packaging time. Others had a first-come-first-served approach. Computers were used at the Coliseum to sign in volunteers. Clipboards were used at WSU Feeds Haiti. At Andover Central High School, high school students worked alongside grade-school students. At WSU Feeds Haiti, the women's volleyball team, the men's golf team, the University President and his wife, the Mayor of Wichita, the full student bodies from two rural Kansas elementary schools, and

WuShock, the WSU mascot, packaged food alongside other volunteers.

In Haiti, ease of food distribution and the capacity of Haitians to carry the food took precedence. The 36-packet boxes were too heavy and unwieldy for people to carry, and they did not allow for the distribution of water. So, a modification was made in which 18 bags were packaged in the box, and the second 18 in a sturdy plastic bag inside the box so that volunteers could grab the bag, along with containers of water, and hand the bag and water to one family, and the box and water to a second family. Ultimately this method gave way to smaller cardboard boxes within the larger boxes that could be removed and distributed individually with water.

A more interesting dilemma arose when Haitians who received the food were suspicious of the colorful freeze-dried vegetables. Their diet consisted mainly of rice and beans, and many had never seen other vegetables, so they did not know what these colorful pieces were. There were even rumors that the vegetables contained mind-controlling drugs. Many people picked the vegetables out and threw them away. When this was discovered by Salvation Army personnel engaged in food distribution, they immediately pulled together community collaborators to identify the issue and appropriate remedies. As a result of input from the community, Numana replaced the colorful dried vegetables with dried beans, a more familiar food staple in Haiti. For more discussion on how these and similar cultural issues were negotiated, see Ballard-Reisch (2011).

Community-Based Participatory Research

Consistent with engaged scholarship, a community-based participatory research approach (CBPR) emphasizes the active participation of community members and scholars as contributing partners in a research process designed to address issues of mutual interest (Israel et al., 2005). All partners play critical roles, as community members contribute their expertise and insights about the community and relevant stakeholders while scholars provide disciplinary expertise and research skills (Wallerstein & Duran, 2006). With its emphasis on finding "local truth" rather than universal truth (Schwab & Syme, 1997), CBPR focuses on interventions that are appropriate to community dynamics and grounded in community infrastructure, capacity (Israel et.al., 2005), interests, and needs, as opposed to strategies that seek to insert interventions into communities (Minkler & Wallerstein, 2002). Grounded in the work of Israel, Eng, Schulz, and Parker (2005, pp. 7–10), seven steps in the community-based research process can be identified that are consistent with insights into the Numana Feeds Haiti initiative.

Engaged scholarship/community partnerships.

1. Create an environment of collaboration in which all participants work together to fill gaps in knowledge.
2. Maintain a focus on colearning in which everyone is empowered to inform everyone else throughout the research process.
3. Recognize that participatory research benefits both the academy and communities.
4. Community-based research is characterized by partners' commitment to sustain the relationship and carry out participatory initiatives.

Community-based process.

1. Recognize and build on existing infrastructure and community talent.
2. Utilize a "cyclical and iterative" process (Israel et al., 2005, p. 8) so that all collaborators inform the research process at each step and critical modifications can be made in a timely manner.
3. Disseminate "results to all partners and involve [*sic*] them in the wider dissemination of results" (Israel et al., 2005, p. 9).

Lessons Learned

Lesson #1: Engaged, Community-Based Projects Are Contagious

Numana efforts grew in a compounded manner. People did not just volunteer at one food-packaging event, they came back again and again to help. In addition, one event led to other events.

On Superbowl weekend, February 5–7, 2010, for example, with the support of the WSU administration and the Elliott School of Communication (ESC), I coordinated the WSU Feeds Haiti event with four WSU-ESC graduate students,[2] my daughter[2] and a friend of hers, and a community volunteer. All but two members of the team had taken part in at least one of the prior Numana food-packaging events. Three hundred and twenty-one "green shirts" also signed up to help, and joined 500 students, faculty, and staff from Wichita State University and 2,500 community volunteers in packaging over 641,000 meals at WSU's Metropolitan Complex.

This event led directly to two other packaging events, one sponsored by the Mayor and the City of Wichita, Kansas, March 20 and 21, 2010, which resulted in over 500,000 meals packaged. The other event occurred on May 6, 2010 at Andover Central High School in Andover, Kansas, where my daughter and her friend were students. Students from elementary, middle, and high schools packaged approximately 218,000 meals that day.

People come back to package again and again. Those involved in packaging events move on to plan events in their own communities.

Lesson #2: Building on Existing Infrastructure and Collaborative Relationships Increases the Effectiveness of Response in a Crisis. Building Those Relationships After a Crisis Is Much More Complex and Time Consuming

What made Numana different from other "feed Haiti" initiatives was their relationship with the Salvation Army. For over 60 years, the Salvation Army had been working in Haiti; at the time of the earthquake, they supported 49 schools throughout the country. Their existing infrastructure and relationships with communities, schools, and organizations, as well as the trust they had built over years of collaboration with the Haitian people and government, made it possible for them to distribute Numana food through existing, collaborative networks quickly and directly to people in need.

By midnight on January 15, three days after the earthquake, the food from the first, pre-earthquake packaging event in El Dorado had already been airlifted to Haiti by the U.S. Army's 82nd Airborne Division, and dropped by helicopter to Salvation Army sites. As the first food was delivered, an additional 700,000 meals were being packaged at a second event in El Dorado, Kansas. Later, the distribution system to Salvation Army posts, with the support of the U.S. Army, allowed Numana to know within a specific date range when shipments of food would be delivered and distributed. For example, during the February 5–7, 2010 WSU

Feeds Haiti event, packers were told that the food they packaged for the first of two trucks would be in the hands of Haitian families by Wednesday or Thursday of that week, February 9–10, 2010, while the second truck of food would arrive and be distributed later that same weekend.

Lesson #3: Numana's Strategy of Getting People Tangibly Involved in Doing Something at a Community Level to Make a Difference Is a Powerful Motivator of Individual and Collective Action

In her research on the politics of AIDS, Susan Chambré (2006) identified the "master frame" of New York AIDS policy as the assertion that "spending money would save lives" (p. 4). She concluded that this is "a central feature of American political culture, the fact that politicians and the public are indeed highly committed to spending money to save lives" (p. 4). The dynamic that Numana tapped into expands on this notion. As our country and citizens face difficult economic times, giving money is no longer as easy as it once was, and people are looking for other ways to make a tangible difference, to save lives, to feel good about the contributions they make to others. This is evidenced in the increase in walks, runs, and triathlons to raise money for good causes. It is evident in a different way in the social movement that has become Numana. Doing something to help others as part of a grassroots, community-based initiative leads to the development and strengthening of collective identity, and reinforces the moral imperative to do good works. Tangible indicators of the success of those efforts reinforces them. It was a powerful message to volunteers at the WSU Feeds Haiti event, for example, that hungry people in Haiti would be enjoying the food they were packaging that day when the volunteers were enjoying their own lunches on Wednesday or Thursday of that same week. Packaging the food with one's own hands is a more engaged process than writing a check. Knowing when people on the other side will be eating that food is an even more rewarding reinforcer.

Lesson #4: Flexibility Is Critical to Successful Grassroots, Community-Based Social Movements

Whether in Haiti, rural Kansas, or Chicago, a critical lesson from the Numana Feeds Haiti initiative is the necessity of flexibility.

The best plans are not workable in some contexts. Even the most well-thought-out strategies can have unintended consequences. Each venue has its own unique capacity and cultural dynamics, so the ability to retain critical elements of process while adapting to contextual imperatives is crucial to effective, coordinated action. This was evidenced both in the modifications in process made to accommodate U.S. partners, and in the revisions in packaging and food content made to meet the needs of both distributors and recipients of food in Haiti.

Conclusion

It remains to be seen how long the social movement that is Numana will last. Social movement theorists argue that movements have life cycles (Griffin, 1952, 1969). Currently, efforts are still going strong; since its first food-packaging event in December, 2009, Numana has sponsored 109 food-packaging events in 58 cities across 12 states. Almost 150,000 volunteers have packaged nearly 25 million meals to meet emergency food needs for people in Haiti, Kenya, Columbia, Sri Lanka, Mexico, Liberia, Mozambique, and the Philippines. In October 2011, over 4,000 students from 16 Kansas colleges and universities packaged just short of 600,000 meals for Somalis in Kenyan refugee camps. In March, 2012, 21 Kansas colleges and universities sent representatives to the second Kansas Hunger Dialogue where they affirmed a commitment to host the 8th Annual Universities Fighting World Hunger Summit in 2013. In March, 2012, Numana launched a community-wide community and communal gardens project in El Dorado, Kansas, on land donated by the city and cleared and planted by community volunteers with the aim of processing and distributing food locally to create the first "hunger free" community.

Ultimately, it remains unclear if the flexibility that marked the first six months of Numana will be sustainable, and the extent to which Americans will continue to be interested in feeding the starving. It is unclear whether cultural and political forces might impinge on Numana's ability to continue its work, or what trade-offs it may ultimately need to make to continue to pursue its goals, a circumstance that threatens many social movements (Rogers, 2009). It is even unclear whether or not Numana will ever be able to return to their initial vision of feeding children in schools, or whether emerging crises will continue to create more urgent food needs.

What is clear is that Numana has tapped a need in many Americans to engage in collective action to make a difference. What is clear is that involvement in these food-packaging events and other emerging efforts reaffirms the best in volunteers, and leads many to want more. What is clear is that over 20 million meals were delivered to Haitians after the earthquake from the hands of American volunteers with the support of the Salvation Army World Office and the U.S. Army 82nd Airborne Division in six months. What is clear is that regardless of the future, this is a significant social collaboration that has had a significant impact on many in the US and Haiti.

Notes

1. First person pronouns throughout refer to the first author of this chapter.
2. The graduate students mentioned here and my daughter are coauthors of this chapter.

Bibliography

Ballard-Reisch, D. S. (2011). Feminist reflections on food aid: The case of Numana in Haiti. *Women & Language, 34(1), 87–93.*

Barge, J. K., & Shockley-Zalabak, P. (2008). Engaged scholarship and the creation of useful organizational knowledge. *Journal of Applied Communication Research, 36*(3), 251–265.

Barge, J. K., Simpson, J. L., & Shockley-Zalabak, P. (2008). Introduction: Toward purposeful and practical models of engaged scholarship. *Journal of Applied Communication Research, 36*(3), 243–250.

Benford, R. D., & Snow, D. A. (2000). Framing processes and social movements: An overview and assessment. *Annual Review of Sociology, 26*(1), 611–639.

Brown, P., & Zavestoski, S. (2005). Social movements in health: An introduction. In P. Brown & S. Zavestoski (Eds.), *Social movements in health* (pp. 1–16). Malden, MA: Blackwell.

Chambré, S. (2006). *Fighting for our lives: New York's AIDS community and the politics of disease.* Piscataway, NJ: Rutgers University Press.

Cheney, G. (2008). Encountering the ethics of engaged scholarship. *Journal of Applied Communications Research, 36*(3), 281–288.

CIA World Facts. (2010). Central America and Caribbean: Haiti. https://www.cia.gov/library/publications/the-world-factbook/geos/ha.html.

DeGenaro, W. (2009). Politics, class and social movement people: Continuing the conversation. In S. M. Stevens & P. Malesh (Eds.), *Active voices: Composing a rhetoric of social movements.* Albany: State University of New York Press.

Della Porta, D., & Diani, M. (1999). *Social movements: An introduction.* Malden, MA: Blackwell.

Della Porta, D., & Diani, M. (2006). *Social movements: An introduction* (2nd ed.). Malden, MA: Blackwell.

DPI. (2010, September 20–22). We can end poverty: 2015 millennium development goals—Eradicate extreme poverty and hunger fact sheet. United Nations Summit, New York high-level plenary meeting of the General Assembly. Retrieved from http://www.un.org/millenniumgoals/pdf/MDG_FS_1_EN.pdf.

Encyclopaedia of Sociology. (1992). *Social movements. Volume 4.* New York, NY: Macmillan.

Finger, M. (1989). New social movements and their implications for adult education. *Adult Education Quarterly, 40*(1), 15–22.

Food and Agriculture Organization of the United Nations. (FAO). (2008). The state of food insecurity in the world. High food prices and food security—Threats and opportunities. Retrieved from ftp://ftp.fao.org/docrep/fao/011/i0291e/i0291e00.pdf.

Food and Agriculture Organization of the United Nations. (FAO). (2009a). More people than ever victims of hunger [News release]. Retrieved from http://www.fao.org/fileadmin/user_upload/newsroom/docs/Press%20release%20june-en.pdf.

Food and Agriculture Organization of the United Nations. (FAO). (2009b). The state of food insecurity in the world economic crises—Impacts and lessons learned. Retrieved from ftp://ftp.fao.org/docrep/fao/012/i0876e/i0876e.pdf.

Food and Agriculture Organization of the United Nations. (FAO). (2010a). Hunger: Basic definitions. Retrieved from http://www.fao.org/hunger/basic-definitions/en/ (no longer accessible).

Food and Agriculture Organization of the United Nations. (FAO). (2010b). Hunger: Frequently asked questions. Retrieved from http://www.fao.org/hunger/faqs-on-hunger/en/ (no longer accessible).

Goffman, E. (1986). *Frame analysis: An essay on the organization of experience.* Boston, MA: Northeastern University Press.

Griffin, L. M. (1952). The rhetoric of historical movements. *Quarterly Journal of Speech, 38*(2), 184–188.

Griffin, L. M. (1969). A dramatistic theory of the rhetoric of movements. In W. Rueckert (Ed.), *Critical responses to Kenneth Burke* (pp. 456–477). Minneapolis: University of Minnesota Press.

Grigsby, M. (2004). *Buying time and getting by: The voluntary simplicity movement.* Albany: State University of New York Press.

Hoffman, B. (2003). Health care reform and social movements in the United States. *American Journal of Public Health, 93*(1), 75–85.

Israel, B. A., Eng, E., Schulz, A. J., & Parker, E. A. (2005). Introduction to methods in community-based participatory research for health. In B. A. Israel, E. Eng, A. J. Schulz, & E. A. Parker (Eds.), *Methods of community-based participatory research for health* (pp. 3–29). San Francisco, CA: Jossey-Bass.

Kilgore, D. A. (1999). Understanding learning in social movements: A theory of collective learning. *International Journal of Lifelong Education, 18*(3), 191–202.

Kolker, E. S. (2005). Framing as a cultural resource in health social movements: Funding activism and the breast cancer movement in the US, 1990–1996. In P. Brown & S. Zavestoski (Eds.), *Social movements in health* (pp. 137–160), Malden, MA: Blackwell.

Minkler, M., & Wallerstein, N. B. (2002). Improving health through community organization and community building. In K. Glanz, B. K. Rimer, & F. M.

Lewis (Eds.), *Health behavior and health education* (3rd ed.) (pp. 26–50). San Francisco, CA: Jossey-Bass.

Nord, M., Andrews, M., & Carlson, S. (2009). Household food security in the United States, 2008. United States Department of Agriculture Economic Research Report Number 83. Retrieved from http://www.ers.usda.gov/publications/err-economic-research-report/err83.aspx.

Polletta, F., & Jasper, J. M. (2001). Collective identity and social movements. *Annual Review of Sociology, 27*, 283–305.

Putnam, L. (2009). *The multiple faces of engaged scholarship.* Paper presented at the Seventh Aspen Conference on Engaged Communication Scholarship, Aspen, CO.

Rogers, K. (2009). When opportunities become trade-offs for social movement organizations? Assessing media impact in the global human rights movement. *Canadian Journal of Sociology, 34*(4), 1087–1114.

Sandlin, J. A., & Walther, C. S. (2009). Complicated simplicity: Moral identity formation and social movement learning in the voluntary simplicity movement. *Adult Education Quarterly, 59*(4), 298–317.

Schwab, M., & Syme, S. L. (1997). On the other hand: On paradigms, community participation and the future of public health. *American Journal of Public Health, 87*(12), 2049–2051.

Seidman, S. (1994). *Contested knowledge: Social theory in the postmodern era.* Cambridge, MA: Blackwell.

United States Department of Agriculture. (2009). Access to affordable and nutritious food: Measuring and understanding food deserts and their consequences. Retrieved June 24, 2012 from http://www.ers.usda.gov/Publications/AP/AP036/AP036.pdf.

USGS. (2010a). Magnitude 7.0—Haiti region. Retrieved from http://earthquake.usgs.gov/earthquakes/recenteqsww/Quakes/us2010rja6.php.

USGS. (2010b). USGS CoreCast magnitude 7.0 earthquake strikes Haiti. Retrieved from http://www.usgs.gov/corecast/details.asp?ep=117.

Van de Ven, A. H. (2007). *Engaged scholarship: A guide for organizational and social research.* Oxford, UK: Oxford University Press.

Wallerstein, N., & Duran, B. (2006). Using community-based participatory research to address health disparities. *Health Promotion Practice, 7*(3), 312–323.

WHO (2009). Health Action in Crises. Highlights No 220 – 4 to 10 August 2008. Retrieved from: http://www.who.int/hac/donorinfo/highlights/highlights_220_4_10Aug08.pdf.

Williams, R. H. (1995). Constructing the public good: Social movements and cultural resources. *Social Problems, 42*(1), 124–144.

World Food Programme (WFP). (2010). World hunger. Retrieved from http://www.wfp.org/hunger.

Zoller, H. M. (2005). Health activism: Communication theory and action for social change. *Communication Theory, 15*(4), 341–364.

15

The Ozioma News Service: Targeting Cancer Communication to African American Communities

Elisia L. Cohen, Charlene A. Caburnay, Jon Stemmle, María Len-Ríos, Tim Poor, Barbara Powe, Susie Rath, Erin Robinson, Katharine Semenkovich, Glen T. Cameron, Douglas A. Luke, and Matthew W. Kreuter

In the United States, cancer is the second leading cause of death behind heart disease, and more than 1.5 million people were expected to be diagnosed in 2010. The National Cancer Institute (NCI) estimated that more than 11.4 million people in 2006 could be classified as cancer survivors (American Cancer Society, 2010b). Health disparities affect cancer rates in various populations, with African American men and women facing a disproportionate burden of cancer mortality across cancer sites (American Cancer Society, 2010a). Recent research shows a number of modifiable risk factors (including smoking, high blood pressure, elevated blood glucose, and body fat composition) explain a significant proportion of disparities in mortality from cancer, and some of the life-expectancy disparities in the US (Danaei et al., 2010). Thus, it is important that individuals understand that beyond psychosocial and health-care determinants of health disparities, a number of risk factors leading to disparities in cancer incidence and mortality include modifiable behaviors (Danaei et al., 2010, p. 13).

Among African Americans, whose health-care engagement involves confronting the legacy of discrimination in the health-care system, the communication gap between medical evidence, knowledge, and behavior is pronounced. Research demonstrates how African Americans are more likely to report distrust of health-care providers (Terrell & Terrell, 1981), more likely to express feelings of futility and fatalism when confronting cancer (Cohen, 2009;

Powe, 1995a, 1995b; Powe & Weinrich, 1999), and less likely to communicate openly about cancer (Cohen, 2009; Len-Ríos, Park, Cameron, Luke, & Kreuter, 2008). These communication gaps may make it less likely that African Americans will receive early cancer diagnosis, as research suggests that beyond socioeconomic status, modifiable sociocultural beliefs and attitudes are linked to cancer stage at diagnosis among African Americans (Lannin et al., 1998; Lannin, Mathews, Mitchell, & Swanson, 2002). Poor cancer communication and knowledge gaps also make it less likely that African Americans will receive evidence-based cancer treatments (Cooper & Koroukian, 2004; Lathan, Okechukwu, Drake, & Bennett, 2010; Masi & Gehlert, 2009), and have the outcomes typical of cancer survivors from the general population (Mayberry, Mili, & Ofili, 2002; Smedley, Stith, & Nelson, 2003).

Against this backdrop of cancer's disparate burden on the African American community, targeted communication strategies may be used to increase demand for cancer information services that may meet community needs, enhance cancer knowledge, correct cancer myths, and assuage cancer fears. Cancer information that is accurate, trustworthy, understandable, and effectively disseminated has great potential to help reduce cancer risk, incidence, morbidity, and mortality, as well as to improve quality of life (Kreps, 2003).

To improve cancer communication and to potentially reduce cancer communication disparities regarding cancer knowledge, attitude, and behavior outcomes, the St. Louis-based Center for Excellence in Cancer Communication Research (CECCR) developed the Ozioma News Service. Black newspapers are an important initial target audience for the CECCR's initial scope of work, given the magnitude of cancer disparities between African Americans and other U.S. groups, and the unique features of Black newspaper cancer storytelling in our formative research (Cohen et al., 2008; Caburnay et al., 2008).

Black Newspapers Are a Credible Source for Cancer and Health Education

There are a number of reasons why African Americans prefer Black newspapers to other media and interpersonal channels of communication for health and cancer information (Len-Ríos, Cohen, & Caburnay, 2010). First, given sociocultural reasons for distrust of medical professionals (Terrell & Terrell, 1981) and reticence to communicate openly about cancer with family and loved ones (Cohen, 2009; Powe, 1995a, 1995b), media are an important resource for African Americans to stay on top of their health. Our research demonstrates that Black newspapers play a unique role in the communication environment for African Americans, allowing African Americans to identify with and understand health problems afflicting their community (Cohen, 2011; Len-Ríos et al., 2010). Second, historically, Black newspapers gave voice to community concerns from a perspective that was different from the general-audience newspapers regarded as a "mouthpiece" of the White power structure during segregation (Bayton & Bell, 1951). This tradition continues today, as surveys of Black newspaper editors and reporters indicate a consistent belief in their mission to meet the distinctive needs of the African American community (Lacy, Stephens, & Soffin, 1991). Readers see this distinct style and focus of news coverage. In the US, there are more than 200 Black weekly newspapers, with a combined circulation of approximately 15 million (National Newspaper Publishers Association, 2013). While general-population newspapers have experienced a decline in readership and financial viability, the continued success of community newspapers such as the Black press can be attributed to several factors, including the ability to target selected audiences and focus on local news (Caburnay et al., 2008).

Indeed, our early research established that readers of African American newspapers trusted their newspapers, and relied on them for locally relevant, race-specific health and cancer information (Caburnay et al., 2008). Some scholarship reflects the view that African Americans evaluate the tone of the Black press more positively than the daily, general-audience alternative (Gibbons & Ulloth, 1982; Vercellotti & Brewer, 2006). Additionally, studies using unobtrusive measures of online news selection show that Blacks pay significantly more attention to the race of news story subjects than do Whites, and that they are more likely to seek out stories if the main character is Black (Knobloch-Westerwick, Appiah, & Alter, 2008). Our research supports this finding, indicating that individuals with greater Black self-identity (measured as the importance of respondents' racial identity to their self-concept) are more likely to seek out health and cancer information from a Black newspaper than from other media sources (Len-Ríos et al., 2010).

Moreover, individuals who have stronger media-dependency relationships with the Black newspaper for self-understanding are also more likely to turn to it for health and cancer news coverage (Len-Ríos et al., 2010).

Although African Americans turn to Black newspapers as a trustworthy source of cancer and health information, the stories within Black newspapers often do not report "good news." Moreover, in general, the "bad news" concerning cancer disparities is often racialized within news stories. Kim, Kumanyika, Shive, Igweatu, and Kim (2009) found that despite the decline from 1996 to 2005 in total number of articles focused on racial and ethnic health disparities in general U.S. newspapers, "three diseases (cardiovascular disease, HIV/AIDS, cancer) and 1 racial group [African Americans] dominate media discourse on racial/ethnic health disparities" (p. S227). Furthermore, among Black newspapers targeting an African American population disparately burdened by cancer mortality, our research (Politi et al., 2010) has shown that stories in Black newspapers are more likely to report health and cancer disparities than those in general-population newspapers ($p < .001$). Across study periods (January 2004–December 2006 in intervention and through December, 2007 for control communities), cancer stories in Black newspapers were more likely to report disparities by gender and race and/or ethnicity, with specific reference to Black/White cancer disparities ($p < .001$).

The focus on worsening disparities in cancer news stories, as well as the focus on cancer causes and incidence rather than prevention and treatment strategies is of particular concern, given that cancer news coverage may promote fatalistic beliefs about prevention because of information overload, reduced feelings of self-efficacy, or playing on self-fulfilling cognitive stereotypes (Gandy & Li, 2005; Goshorn & Gandy, 1995). That these disparities are empirically supported enhances journalists perceptions of their newsworthiness to minority audiences (Hinnant, Oh, Caburnay, & Kreuter, 2011), and may raise awareness among minority and underserved populations. How minority and medically underserved groups respond to such "disparity-cued" cancer communication, and whether that may prime a sense of fatalism or empower medically underserved populations deserves further attention. Neiderdeppe and Levy's (2007) general examination of fatalistic beliefs about cancer prevention indicate that stronger fatalistic beliefs were present in less-educated Americans, but generally weaker among both African Americans and Hispanics relative to Whites. However, controlled, experimental research among African American adults indicates that the way disparities are reported in media can influence public attitudes and intentions regarding cancer screening in a negative, unintended way (Nicholson et al., 2008).

In terms of actual cancer coverage, our formative findings also show that Black newspapers are more likely to report on sex-specific cancers (i.e., breast cancer and prostate cancer), and underreport lung and other cancers relative to both incidence and mortality (Cohen et al., 2008). Research also shows that stories are more likely to focus on diagnosis (i.e., celebrity), treatments, and causes, rather than on cancer prevention through modifiable risk factors and detection (Caburnay et al., 2008; Cohen et al., 2008). For African Americans already burdened by high rates of cancer mortality, the lack of easily accessible, localized, culturally specific, actionable cancer information is problematic. These findings support the initial assumption that undergirded our Ozioma-1 study: there is a need to improve cancer journalism to make cancer news stories culturally specific and locally relevant to African American news readers.

Given that media exposure may be an important factor in cancer awareness and knowledge, potential disparities exist between individuals who receive poorer quality cancer news and fewer recommendations to cancer-information resources in their cancer-information environment, compared with those who receive richer quality cancer news and referrals to personally relevant cancer-information resources. Stryker, Moriarty, and Jensen's (2008) research descriptively assessed how coverage in U.S. newspapers was similar to cancer-awareness trends in HINTS survey data. More recently, Slater, Hayes, Reineke, Long, and Bettinghaus (2009) "associated data on regional differences in media exposure to cancer prevention news stories with individual level data on prevention knowledge" (p. 516). Slater et al. (2009) examined a representative probability sample of U.S. newspapers to assess regional differences in cancer-prevention coverage, and knowledge of people in these regions. The study found that regional differences in U.S. cancer-prevention news coverage moderates the relationship between education and cancer-prevention knowledge. Moreover, attention to health news reduces knowledge differences associated with regional amounts of news coverage (Slater et al., 2009). This research was consistent with prior research indicating that attention to health news predicts greater knowledge of cancer causes (Viswanath et al., 2006); though some evidence also suggests that individuals with higher income and education levels are more likely to respond to media reports of celebrity cancer diagnosis and controversy with information-seeking strategies than are individuals having lower income and education levels (Niederdeppe, 2008).

A basic premise for the Ozioma-1 and Ozioma-2 interventions targeting African American communities is that the volume and quality of information on cancer prevention and treatment in the news can serve as both an enabling factor and a challenge, in terms of facilitating public empowerment and decision making. Our guiding framework is that information that is culturally specific, locally relevant, and delivered via a trusted media source can help to reduce cancer disparities. As one case study in an effective media-relations strategy to improve cancer communication serving African American communities, this chapter summarizes the efforts of the Ozioma News Service to translate evidence-based cancer knowledge in Black newspapers. Here, we describe two Ozioma News Service interventions. Each intervention is based on our iterative research establishing and augmenting how Black newspapers can serve as an important resource for health and cancer information for the African American communities that they serve. Then, we summarize the results of our Ozioma-1 study, and the evolution of the Ozioma News Service, to establish a partnership with the American Cancer Society (ACS). The result, Ozioma-2, is a novel, integrated effort to improve targeted cancer communication in local African American communities. Finally, the chapter concludes by summarizing preliminary, evaluative data on the Ozioma News Service as a model, media-based intervention that increases the availability of locally relevant, race-specific cancer information to African American communities, and describes areas for future research.

The Ozioma News Service

As detailed in Cohen et al. (2010), the Ozioma News Service is "a national service with three distinguishing features: (1) an exclusive focus on cancer news; (2) distribution to Black newspapers only; and (3) news releases that are customized for each community to which they are sent" (p. 569). Ozioma-1 refers to the operation of the Ozioma News Service from December 2005 through September 2007, where "we developed 444 Ozioma news releases—37 different releases each adapted for 12 different intervention communities—and delivered to local Black newspapers" (p. 570). Until Ozioma-1, no systematic attempts had been made to engage Black newspapers in covering cancer. In Ozioma-2, we expanded the number of communities in our study from 12 to 24, and thus increased the number of Black newspapers in our study from 24 to 45. While our current focus is on Black newspapers, here we describe the Ozioma-1 and Ozioma-2 concepts to demonstrate their adaptability for future research and dissemination efforts to target other racial and ethnic minority groups, special population subgroups, and other media (e.g., print, TV, radio, Internet). This approach is consistent with NCI's Strategic Plan for partnering with minority media to implement innovative and culturally appropriate approaches to disseminating research results to underserved populations (Strategy 8.5).

Ozioma-2 builds on the initial success of Ozioma-1 to test the effects of an enhanced Ozioma News Service that features highly localized information, quotes, and photographs supplied by lay journalists in collaboration with the American Cancer Society affiliate in each community. We are also developing an automated, Web-based version of Ozioma for journalists and media-relations professionals.

Ozioma-1: A Novel Effort to Improve Cancer News Reporting in Black Newspapers

In 2003, we began a novel effort to improve cancer news reporting. An interdisciplinary team of public health, journalism, public relations, and communication scientists sought to create locally relevant and race-specific news stories for 12 selected African American newspapers around the country. Twelve other African American papers served as the control, and did not receive Ozioma news stories.

The sampling frame for the study included large U.S. cities and standard metropolitan areas (SMAs) that had:

1. Cancer mortality rates for African American men or women that exceeded U.S. rates for African Americans; and
2. A local Black weekly newspaper.

The sampling procedure is described in detail elsewhere (Caburnay et al., 2008; Cohen et al., 2008). The 24 weekly newspapers in the final sample were located in 16 different states and most regions of the US, but predominantly in the Southeast and Midwest.

Study group randomization was stratified by three variables:

1. Large cities versus SMAs;
2. Newspaper circulation (<40,000 versus ≥40,000 for cities; <20,000 versus ≥20,000 for SMAs); and
3. Size of newspaper (≤20 versus >20 pages).

Newspaper circulation was obtained from the Gebbie Press. Size of newspaper was assessed for each individual newspaper by taking the average of the number of total pages (not including advertisement inserts) in one month of weekly newspaper issues.

To randomize into study group, we first identified 12 cities and 12 SMAs. Within the cities and the SMAs, we further classified them by newspaper circulation and newspaper size (resulting in 4 cells each for cities and SMAs). Each cell had either 2 or 4 cities or SMAs. In each of the 8 total cells, an equal number of newspapers were randomly assigned to either the intervention or control group, resulting in 12 newspapers (6 from cities and 6 from SMAs) in each group. Contamination in the sample was limited due to this stratification procedure for intervention communities (indeed, only one Florida newspaper in our control group later reran a version of an Ozioma story from a newspaper elsewhere in the state).

The Ozioma News Service is named after the Igbo/Nigerian term "Ozioma," or "good news." Rather than emphasizing that African Americans are faring worse compared to other groups, Ozioma releases emphasize progress and opportunity. This approach has been found to elicit positive emotional responses, increase interest in cancer screening, and counteract medical mistrust among African American adults (Nicholson et al., 2008).

Ozioma News Service stories were traditional press releases written using standard journalistic principles such as the inverted pyramid. This technique puts the most important information toward the top or beginning of the story, allowing editors to cut the story at any point for space limitations without compromising understandability.

All stories contained the following elements:

1. Localized cancer-related data—community, county, or state level, depending on the data available
2. At least one of six core messages:
   - Cancer is an important health issue for African Americans;
   - There are things you can do to prevent cancer;
   - Every African American should be screened for cancer according to recommended guidelines;
   - It is important to talk to family and friends about cancer prevention and screening;
   - Cancer is not a death sentence; and,
   - There are resources in your community to help in the fight against cancer.
3. Localized resources—typically in the form of an organization with a phone number, URL, and address pertaining to the topic covered in the story
4. Containing data specific to the local community or state (informational graphic) or for African Americans (using culturally matched photographs using race and location as proxies for culture)

Before the media intervention began, an introductory letter was mailed to each of the 12 Black newspapers in the intervention condition informing them about the Ozioma News Service, who we were, what we would be sending them, and when we would begin. Additionally, we followed up with each newspaper by telephone to learn about their preferred method of receiving the stories (almost all requested e-mail; one preferred fax), and who the best contact person(s) was (were) for the story.

A website was created for the news service to further add to the credibility and ease of accessing current and past stories, graphics, and photos. Each newspaper had its own user name and password.

The first cancer news release was distributed on December 11, 2005. Ozioma staff called the newspapers the day after to make sure they received the story. Papers were then called once per month for the first six months to confirm receipt of Ozioma news releases, to reinforce the Ozioma News Service brand as a trusted source for cancer information for African Americans, and to solicit health-related topics they perceived their readership would enjoy.

Ozioma staff held weekly team meetings to generate ideas for news releases. These ideas included responding to topics the newspaper editors reported their readers would be interested in, taking advantage of national breaking news stories, repackaging NCI press releases from the NCI Bulletin, and addressing monthly commemorative or seasonal themes (e.g., breast cancer awareness month, the Great American Smokeout, holiday overeating). This process resulted in an editorial calendar that addressed specific cancer types (e.g., breast, prostate), cancer risk factors (e.g., smoking, obesity, diet), and other cancer-related topics (e.g., celebrity cancer survivors, Black oncologists, family history). The research team further required that the final list of topics must have relevance for the local African American community, including the availability of community-, county-, or state-level health data related to the release topic.

In order to evaluate the effectiveness of the Ozioma News Service on cancer coverage, we content analyzed all of the published health- and cancer-related stories in the 12 Black newspapers receiving Ozioma releases, as well as the other 12 newspapers that served as controls. We coded these stories on journalistic criteria, and cancer-related stories were also coded on health-promotion criteria. Ozioma news releases were coded on both journalistic and health-promotion criteria (Caburnay et al., 2008; Caburnay et al., under review; Cohen et al., 2008; Cohen et al., 2010).

Preliminary summative results indicated that the number of cancer stories published in newspapers receiving Ozioma releases was 37 times more than the number of cancer stories published in control newspapers at 12 months postintervention. Stories published in newspapers receiving Ozioma releases were also more likely to be of higher story quality; they had more graphics, focused on the local community as a locale, contained action steps to improve one's own health, and included genetics or family history as a risk factor. Releases having a localized headline, localized data, other localized content, and a quote were significantly more likely to be adopted than releases that did not have these characteristics. In short, Ozioma increased cancer coverage, and local relevance was an important element of its success.

Ozioma-2: An Integrated, Localized Effort to Improve Cancer News Reporting and ACS Awareness

Building upon the success of Ozioma-1, we received funding for Ozioma-2 in September 2008. In developing the rationale for Ozioma-2, we turned to our previous results that found that

Ozioma releases with localized data, other localized content, and a quote were more likely to be adopted by newspapers than releases that did not have these characteristics. At the same time, we were seeking a way to test the Ozioma model within a more sustainable organizational structure. Ideally, this organization would have a national-level infrastructure with community-level impact. We envisioned how we might make Ozioma releases even more locally relevant with community members who could help provide this impact. The American Cancer Society, with its national reach and divisional and local affiliate infrastructure, was a great fit.

The addition of the ACS provided a national brand for cancer prevention, treatment, and survivorship, along with local affiliate offices to provide community-level acceptance with the newspapers and their readerships. The ACS consists of a National Home Office (Atlanta, GA) with 13 chartered Divisions and a local presence in nearly every community nationwide, with more than 3 million volunteers assisting in the work of the Society (American Cancer Society, 2010b). The Society set aggressive challenge goals focused on decreasing cancer incidence and mortality rates by 2015, and also increasing the quality of life among cancer survivors.

The Society is also committed to addressing the disproportionate burden of cancer among the medically underserved through advocacy, research, and programs. Within the National Home Office, the Office of Health Disparities (OHD) was formed in 2007 to lead the Society's strategic focus on reducing cancer disparities. The OHD seeks to: (a) Increase trust and credibility of the American Cancer society among the disadvantaged segment of the population; (b) Ensure effective disparities reducing practices through the provision of strategic guidance and resources; and (c) Enhance and strengthen our capacity to enable community-based outreach in diverse and low income communities (American Cancer Society, 2010a). Ozioma-2's partnership with three ACS Divisions and the OHD was designed to deliver community-relevant cancer messages to selected African American populations.

For Ozioma-2, the study drew a sample of Black newspapers from three ACS divisions. These three divisions were preselected by ACS as those that have the capacity and interest to engage in a community-based research partnership, and had access to large African American populations. In order to allow for geographic and divisional distribution and representation across the US, the sampling frame for Black newspapers was set by the ACS divisional structure. We recognized that the U.S. African American population and cancer incidence and mortality rates are distributed across different-sized communities; that Black newspapers' resources and infrastructures vary by the size, complexity, and competition of their media markets; and that local ACS cancer prevention and control resources and services may differ by ACS divisional structure. Therefore, to be included in the sampling frame, the newspaper had to have been published on a weekly basis, have a circulation $\geq$ 5,000 (to select newspapers with the largest reach), and serve a community with an African American population $\geq$ 50,000. All ACS divisions had cancer incidence or mortality rates for African Americans that are higher than the U.S. median. From each of the three divisions, we identified all-Black newspapers that met criteria for inclusion in the sampling frame, and noted the unique communities served by these newspapers (n = 36 communities). Each community within each division was then randomly assigned to one of the three study arms. If a community had more than one Black newspaper and is in an intervention group, all newspapers in that community receive the intervention. For control communities, the one with the largest circulation was selected. This method had two distinct advantages:

1. Each division has an equal number of papers assigned to each study arm, avoiding any divisional bias; and
2. Because only one newspaper is selected from each community, the chance of cross-contamination between arms was minimized.

For Ozioma-2, we are testing the feasibility and effectiveness of enhancing the local relevance of Ozioma news releases with authentic local voices and community-level information in an expanded set of newspapers and communities. These local voices, in the form of a local quote, photograph, and fact, are being supplied by lay journalists in each intervention community.

With the ACS's assistance, the Ozioma News Service worked to hire 12 "news specialists"—one for each intervention community in the study. Most news specialists were already volunteering for the ACS, and all were actively involved in the local African American community. In a one-day training held in November 2009, these 12 news specialists were instructed in basic journalism techniques, such as identifying, interviewing, and documenting sources. For each news release, news specialists were to obtain usable quotes and a photograph from two individuals in the community, plus a local fact or resource. These “enhanced localizing” features are embedded within the text of all Ozioma-2 enhanced localization arm press releases (see Figure 15.2). The news specialists were also given digital cameras and voice recorders, and instructed in using those devices to complete their tasks.

With the news specialists hired and trained, the three-tiered media intervention with 36 communities was implemented in the following manner:

- 12 communities received an enhanced news release with local quotes, photos, and resources provided by the news specialists in each community (see Figure 15.2)
- 12 communities received an evidence-based news release that was localized as in Ozioma-1 with local data and resources, but without a local quote or photo
- 12 communities acted as controls, receiving no news releases

Releases were prepared in much the same manner as Ozioma-1 with the same basic tenets in place. However, one change was that in the 24 intervention communities, all-Black newspapers in those communities were to receive the news releases and be included in the content analysis. Additionally, at the end of each enhanced news release, we included a statement that identified that community’s news specialist as a correspondent for the Ozioma News Service who has contributed to the story.

As in Ozioma-1, we contacted all newspapers to determine a contact person and how they wanted to receive the release. All 24 papers asked for the releases to be e-mailed. After these pieces were in place, we relaunched Ozioma (as Ozioma-2) in January 2010. Every two weeks, a new set of releases is sent to the newspapers and placed on the website. A total of 33 newspapers receive Ozioma news releases: 19 receive enhanced-localization releases, and 14 receive evidence-based releases.

As of September 24, 2010, we had sent out 19 news releases to each of the intervention newspapers, resulting in a total of 91 published stories. Among the newspapers receiving enhanced releases, we have identified 63 published stories in 15 out of 19 newspapers (79%), with a potential audience of 419,329 (as indicated by circulation). Among the newspapers receiving evidence-based releases, we have had 28 hits across 10 of 14 papers (71%), for a combined audience of 128,060. Enhanced releases lead to an average of 4.2 published stories per newspaper, while the average is 2.8 stories per paper for evidence-based releases. As in Ozioma-1, almost all of the 91 published stories in Ozioma-2 have been printed verbatim from the body of the original news release.

Conclusions

This chapter demonstrates the need for programmatic communication strategies designed to improve cancer communication for African Americans. Cleary, the case of cancer news reporting in Black newspapers is but one example of how newspaper storytelling practices can be improved in the public's interest. Our exemplar of the Ozioma News Service details one successful means of enhancing Black newspapers' storytelling practices. News media have a role to play in reducing cancer myths and cancer fears by providing actionable, culturally specific, and personally relevant information to individuals.

Currently, we are conducting a three-group, randomized trial to test the effects of enhancing the localization capabilities of the Ozioma News Service. In the next two years of study, we will have extensive preliminary data to help determine the degree of localization needed to produce changes in cancer coverage. Over time, we hope to develop best practices and evidence that show how community partners, such as the ACS, that have organizational structures that fit well with the purpose of cancer communication initiatives, can enhance their role in the community (and the community's awareness of the ACS, for example) by engaging in grassroots efforts to produce locally relevant, culturally appropriate cancer-information subsidies.

Clearly, beyond the cancer context, the Ozioma model of providing information subsidies speaks to the plausibility of other targeted health media relations strategies to improve communication about diseases disparately burdening underserved populations. In practice, local news affiliates (even beyond Black newspapers) seek out valuable, locally relevant, and trustworthy information subsidies to tell stories important to members of their community. National nonprofit organizations with local lead agencies, county health departments, and county extension offices often have the goal of raising awareness and educating communities from the local level. Given these congruent interests, the Ozioma case study demonstrates how health communication specialists can partner with such local agencies to enhance the reach and effectiveness of targeted communication strategies.

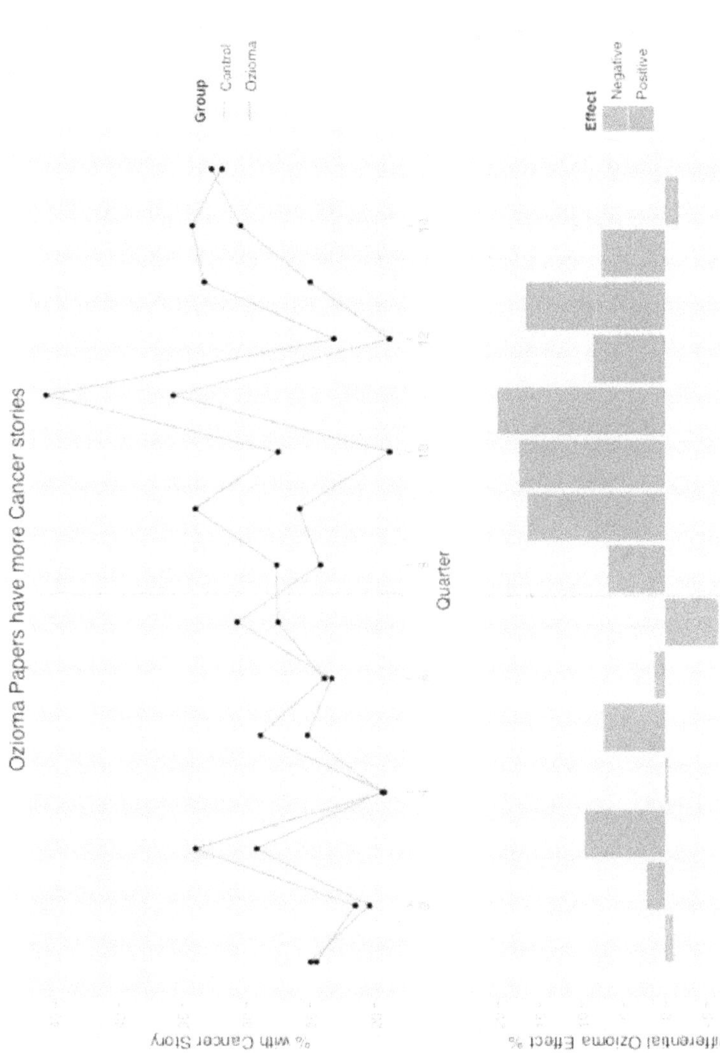

Figure 15.1. Percent of Black Newspaper Issues That Have at least One Cancer Article, by Time (Pre-/Postintervention) and Group (Ozioma vs. Control), and Net Effect Per Quarter

Figure 15.2. Sample Ozioma-2 Press Release With Enhance Localization Text (Highlighted) and Graphic

Ozioma
good news

For immediate release
Contact: Tim Poor
Phone: (314) 935-9398
ozioманews@gmail.com
Previous news releases for your community available at:
http://hcrl.wustl.edu/oziomanews/?cat=10

COLON CANCER SCREENINGS VITAL TO INCREASING AFRICAN-AMERICAN SURVIVAL RATE

A screening for colon cancer is a test most folks don't like to think about, but it's one of the most important medical tests for African-Americans to get. March is National Colorectal Cancer Awareness Month, so it's a good time to find out if you or members of your family are due for a checkup.

Colon cancer is the third leading cause of cancer death among African-American men and women in the United States. In Raleigh/Durham, an average of 50 African Americans died each year between 2002-2006, the most recent data available.

Talking to your doctor about colon cancer

It can be hard to bring up colon cancer with your doctor. Getting prepared can help.

before the visit

Write down any questions you have on a piece of paper, and bring it with you.

Find out your family history. Did your parents, aunts or uncles, or other relations ever have colon cancer, or other types of cancer?

during the visit

Ask any questions you have, such as:

"At what age and how often should I get tested for colon cancer?"

"Which screening test is the right one for me?"

Before you leave the doctor's office, make sure you understand everything. Don't be afraid to speak up:

"I don't understand what a colonoscopy is. Will you explain it to me a different way?"

"Can you write that down for me, or give me a brochure about it?"

Many of these deaths can be prevented, however, through colon cancer screenings, such as a colonoscopy. According to the American Cancer Society, when colon cancer is detected at an early, localized stage, the survival rate in African Americans is 84 percent.

Durham resident Marion Miles began screening for colon cancer at the age of 50. "There is a history of colorectal cancer in my family, so I started having screenings as soon as recommended and I changed my diet to include more fiber," Miles says.

In a colonoscopy, doctors can find and remove polyps – growths that can develop into cancer. The patient can be sedated during the procedure to minimize discomfort.

Acknowledgment

The project described was supported by Award Number P50CA095815 from the National Cancer Institute. The content is solely the responsibility of the authors, and does not necessarily represent the official views of the National Cancer Institute or the National Institutes of Health.

Bibliography

American Cancer Society. (2010a). Cancer facts and figures for African Americans, 2009–2010. Atlanta, GA: Author. Retrieved from http://www.cancer.org/acs/groups/content/@nho/documents/document/cffaa20092010pdf.pdf.

American Cancer Society. (2010b). Facts and figures 2010. Atlanta, GA: Author. Retrieved from http://www.cancer.org/acs/groups/content/@nho/documents/document/acspc-024113.pdf.

Bayton, J. A., & Bell, E. (1951). An exploratory study of the Negro press. *Journal of Negro Education, 20*(1), 8.

Caburnay, C. A., Kreuter, M. W., Cameron, G., Luke, D. A., Cohen, E. L., McDaniels, L., Wohlberg, M., & Atkins, P. (2008). Black newspapers as a tool for cancer education in African American communities. *Ethnicity & Disease. 18*(4), 488–495.

Caburnay, C., Luke, D., Cameron, G., Cohen, E., Fu, A., Choi, C.L., Stemmle, J., Paulen, M., McDaniels-Jackson, L., & Kreuter, M. (2012). Evaluating the Ozioma cancer news service: A community randomized trial in 24 U.S. cities. *Preventive Medicine, 54,* 425–430.

Cohen, E. L. (2009). Naming and claiming cancer among African American women: An application of Problematic Integration (PI) theory. *Journal of Applied Communication Research, 37*(4), 406–426.

Cohen, E. L. (2011). 'The good, the bad, and in-between news': African American women newsreaders' responses to St. Louis newspapers. *Speech and Theatre Association of Missouri Journal, 41,* 28–48.

Cohen, E. L., Caburnay, C. A., Len-Ríos, M. E., Poor, T. J., Cameron, G. T., Luke, D. A., Powe, B., ...Kreuter, M. W. (2010). Engaging ethnic media to expand the reach and effectiveness of communication strategies to reduce health disparities. *Health Communication, 25*(6), 569–571.

Cohen, E. L., Caburnay, C. A., Luke, D. A., Rodgers, S., Cameron, G. T., & Kreuter, M. W. (2008). Cancer coverage in general-audience and Black newspapers. *Health Communication. 23*(5), 427–435.

Cooper, G. S., & Koroukian, S. M. (2004). Racial disparities in the use of and indication for colorectal procedures in Medicare beneficiaries. *Cancer, 100*(2), 418–424.

Danaei, G., Rimm, E. B., Oza, S., Kulkarni, S.C., Murray, C. J. L., & Ezzati, M. (2010). The promise of prevention: The effects of four preventable risk factors on national life expectancy and life expectancy disparities by race and county in the United States. *PLoS Med, 7*(3), 1–13: e1000248.

Gandy, O.H. Jr., & Li, Z. (2005). Framing comparative risk: A preliminary analysis. *The Howard Journal of Communications, 16*(2), 71–86.

Gibbons, R. A., & Ulloth, D. R. (1982). The role of the Amsterdam News in New York City's media environment. *Journalism Quarterly, 59*(3), 451–455.

Goshorn, K., & Gandy, O. H., Jr. (1995). Race, risk and responsibility: Editorial constraint in the framing of inequality. *Journal of Communication, 45*(2), 133–151.

Hinnant, A., Oh, H. J., Caburnay, C. A., & Kreuter, M. W. (2011). What makes African-American health disparities newsworthy? An experiment among journalists about story framing. *Health Education Research, 26*(6), 937–947.

Kim, A. E., Kumanyika, S., Shive, D., Igweatu, U., & Kim, S-H. (2009). Coverage and framing of racial and ethnic health disparities in U.S. newspapers, 1996–2005. *American Journal of Public Health, 100*(1), S224–S231.

Knobloch-Westerwick, S., Appiah, O., & Alter, S. (2008). News selection patterns as a function of race: The discerning minority and the indiscriminating majority. *Media Psychology, 11*(3), 400–417.

Kreps, G. L. (2003). The impact of communication on cancer risk, incidence, morbidity, mortality, and quality of life. *Health Communication, 15*(2), 161–169 .

Lacy, S., Stephens, J. M., & Soffin, S. The future of the African-American press: A survey of African-American newspaper managers. *Newspaper Research Journal, 12*(3), 8–19.

Lannin, D. R., Mathews, H. F., Mitchell, J., Swanson, M. S., Swanson, F. H., & Edwards, M. S. (1998). Influence of socioeconomic and cultural factors on racial differences in late-stage presentation of breast cancer. *JAMA, 279*(22), 1801–1807.

Lannin, D. R., Mathews, H. F., Mitchell, J., & Swanson, M. S. (2002). Impacting cultural attitudes in African-American women to decrease breast cancer mortality. *American Journal of Surgery, 184*(5), 418–423.

Lathan, C. S., Okechukwu, C., Drake, B. F., & Bennett, G. G. (2010). Racial differences in the perception of lung cancer: 2005 Health Information National Trends Survey (HINTS*). Cancer, 116*(8), 1981–1986.

Len-Ríos, M. E., Cohen, E. L., & Caburnay, C. A. (2010). Readers use Black newspapers for health/cancer information. *Newspaper Research Journal, 31*(1), 20–35.

Len-Ríos, M. E., Park, S., Cameron, G. T., Luke, D. A., & Kreuter, M. (2008). Study asks if reporter's gender or audience predict paper's cancer coverage. *Newspaper Research Journal, 29*(2), 91–99.

Masi, C. M. & Gehlert, S. (2009). Perceptions of breast cancer treatment among African-American women and men: Implications for interventions. *Journal of General Internal Medicine, 24*(3), 408–414.

Mayberry, R. M., Mili, F., & Ofili, E. (2002). Racial and ethnic differences in access to medical care. In T. A. LaVeist (Ed.), *Race, ethnicity and health: A public health reader* (pp. 163–197). San Francisco, CA: John Wiley & Sons.

Moriarty, C. M., & Stryker, J. E. (2008). Prevention and screening efficacy messages in newspaper accounts of cancer. *Health Education Research, 23*(3), 487–498.

National Newspaper Publishers Association. (2013). About us. Available February 13, 2013 from http://www.nnpa.org/about-us/.

Nicholson, R. A., Kreuter, M. W., Lapka, C., Wellborn, R., Clark, E. M., Sanders-Thompson, V., Jacobsen, H. M., & Casey, C. (2008). Unintended effects of emphasizing disparities in cancer communication to African-Americans. *Cancer Epidemiology, Biomarkers, & Prevention, 17*(11), 2946–2953.

Niederdeppe, J. (2008). Beyond knowledge gaps: Examining socioeconomic differences in response to cancer news. *Human Communication Research, 34*(3), 423–447.

Niederdeppe, J., & Levy, A. (2007). Fatalistic beliefs about cancer prevention and three prevention behaviors. *Cancer Epidemiology, Biomarkers, & Prevention, 16*(5), 998–1003.

Politi, M., Caburnay, C., Cohen, E. L., Cameron, G. T., Cheng, M.-R., & Kreuter, M. (2010, April 26). Disparities communication in cancer stories in Black and general audience newspapers. Poster presented at the annual meeting of the National Cancer Institute Centers for Excellence in Cancer Communication Research, Philadelphia, PA.

Powe, B. D. (1995a). Cancer fatalism among elderly Caucasians and African Americans. *Oncology Nursing Forum, 22*(9), 1355–1359.

Powe, B. D. (1995b). Fatalism among elderly African Americans: Effects on colorectal cancer screening. *Cancer Nursing, 18(*5), 385–392.

Powe, B. D., & Weinrich, S. (1999). An intervention to decrease cancer fatalism among rural elders. *Oncology Nursing Forum, 26*(3), 583–588.

Slater, M. D., Hayes, A. F., Reineke, J. B., Long, M., & Bettinghaus, E. P. (2009). Newspaper coverage of cancer prevention: Multilevel evidence for knowledge-gap effects. *Journal of Communication, 59*(3), 514–533.

Smedley, B. D., Stith, A. Y., & Nelson, A. R. (Eds.). (2003). *Unequal treatment: Confronting racial and ethnic disparities in health.* Washington, DC: National Academies Press.

Stryker, J. E., Moriarty, C. M., & Jensen, J. D. (2008). Effects of newspaper coverage on public knowledge about modifiable cancer risks. *Health Communication, 23*(4), 380–390.

Terrell, F., & Terrell, S. L. (1981). An inventory to measure cultural mistrust among Blacks. *Western Journal of Black Studies, 5*(3), 180–185.

Vercellotti, T., & Brewer, P. R. (2006). 'To plead our own cause': Public opinion toward Black and mainstream news media among African Americans. *Journal of Black Studies, 37*(2), 231–250.

Viswanath, K., Breen, N., Meissner, H., Moser, R., Hesse, B., Steele, W. R., & Rakowski, W. (2006). Cancer knowledge and disparities in the information age. *Journal of Health Communication, 11*(1), 1–17.

16

Reducing Smoking Disparities for Hispanic Adolescents: Empowerment Through Media Literacy

Kathryn Greene and Smita C. Banerjee

Hispanics/Latinos in the United States bear a disproportionate burden of health and economic disparities on many levels. The Hispanic population, however, is not monolithic. As the proportion of Hispanics in the U.S. population increases, advertisers have seized opportunities for niche and/or target marketing for a wide range of products and services. One group and context where such advertising is of keen interest is in marketing cigarettes to younger Hispanics. In this chapter we report results from a media literacy project focused on empowering Hispanic adolescents to resist the effects of smoking advertisements.

Cigarette smoking continues to remain the leading cause of preventable morbidity and mortality in the United States among all ethnic groups (U.S. Department of Health and Human Services (USDHHS), 2004). The Substance Abuse and Mental Health Services Administration (SAMHSA, 2009) reported that more than 80% of smoking initiation occurs before 18 years of age. Recent results from the 2009 National Survey on Drug Use and Health (SAMHSA, 2010) indicated increasing levels of smoking initiation for adolescents. In 2009, 2.5 million people aged 12 or older had experimented with smoking in the past 12 months, significantly higher than previous estimates (2.1 million in 2004). Additionally, most new smokers in 2009 (58.8%) were younger than 18 years when they first smoked cigarettes. Clearly, some smoking initiation occurs before age 18. However, patterns of adolescent cigarette smoking differ substantially among racial and/or ethnic groups, and contribute to health disparities.

In terms of Hispanic adolescents' cigarette smoking rates, current smoking rates for Hispanic youths aged 12 to 17 years were 7.5%, greater than current smoking rates for Black (5.1%) and Asian (2.5%) youth, but lower than the rates for White youth (10.6%) in the same age range (SAMHSA, 2010). This finding suggests that besides targeting White youth, tobacco companies may have substantial interest in targeting Hispanic adolescents (Brown & Houseman, 2000).

Tobacco companies have targeted minority adolescents for some time. Targeted advertising and promotion of specific cigarette brands with names such as "Rio" and "Dorado" to the Hispanic American communities began in the 1990s, and reflect disproportionate targeting of this group by American tobacco companies (USDHHS, 1998). In 2000, the tobacco industry began to more aggressively market to Hispanic youth, as was evident from the launch of a $40 million advertising campaign targeting predominantly African American and Hispanic markets (Brown & Houseman, 2000). Thus, more effort was needed to focus antismoking efforts specifically at minority populations. But at what age should these minority antismoking efforts be targeted?

Prior research suggests that delaying the initiation of smoking may have significant public-health benefits (Rohde, Lewinsohn, Brown, Gau, & Kahler, 2003), and interventions aimed at interrupting smoking initiation may be significant, timely, and reduce the risk of smoking dependence (see also McNeill, 1991). The National Cancer Institute (NCI, 2010) reported that Hispanics have an earlier onset of cigarette smoking than Asians/Pacific Islanders and Blacks, while they have a higher but similar age of initiation as compared with Whites. These data emphasize the need for targeting minority youth in middle schools with preventive antismoking messages and/or interventions. We present an antismoking media literacy intervention developed and tested primarily with Hispanic middle school students (see Banerjee & Greene, 2006, 2007). Before describing the intervention in detail, we discuss Hispanic adolescent smoking correlates and the use of media literacy as the optimal tool for antismoking intervention efforts with these minority adolescents.

Why Do Hispanic Youth Smoke?

Smoking among Hispanic adolescents is associated with a range of factors. The National Cancer Institute's Tobacco Control Monograph #14 (Baezconde-Garbanati, 2001) reviewed factors related to culture, gender, acculturation, immigration status, socioeconomic status, and the historical and environmental factors that influence Hispanic/Latino adolescent smoking, and summarized,

> As Hispanic/Latino children grow and mature, traditional norms from their younger years that protect against smoking are challenged by the broader society. Societal engagement of Hispanic/Latino adolescents implies interactions with multiple ethnic and racial groups, various cultures, and individuals with varying social and cultural norms regarding tobacco use. The mixture of values and norms creates new expectations and challenges at a time when young people are truly beginning to define themselves and are searching for who they are as individuals. (Baezconde-Garbanati, 2001, p. 239)

The very nature of the challenges of normal adolescent development coupled with cultural factors create a particularly toxic combination for Hispanic initiation of smoking (and other risk) behavior. More recently, Lopez et al. (2010) proposed key factors that categorize Hispanic adolescents' reasons for smoking into two broad areas: intrapersonal factors (including personal attitudes and beliefs about smoking) and ecodevelopmental factors (including acculturation, family, school, and peers). Several aspects of family functioning, such as low parent-adolescent communication, low parental involvement, and low parental monitoring of peer activities are associated with smoking in Hispanic adolescents (Elder et al., 2000). Similarly, aspects of peer relationships, such as association with peers who smoke, has a positive relationship with smoking (e.g., Cowdery, Fitzhugh, & Wang, 1997), while friends' disapproval of smoking is associated with reduced likelihood of smoking among Hispanic adolescents (Gritz et al., 1998). Additionally, favorable attitudes and beliefs about smoking are associated with smoking in Hispanic adolescents (Elder et al., 2000).

However, these factors cannot be examined in isolation from the struggles of Hispanic adolescents to accommodate to some aspects of American culture. One culture-specific factor that has been consistently shown to have an effect on substance use among Hispanic adolescents is acculturation. Rogler, Cortes, and Malgady (1991) defined acculturation as a process by which members of one cultural group (e.g., immigrants from a particular country) adopt some beliefs and behaviors of another group (e.g., the host society). For Hispanics, acculturation means adopting beliefs and behaviors that conform to a "mainstream" American way of life (see Horigian, Lage, & Szapocznik, 2006). Acculturation affects adolescent smoking, with more-acculturated Hispanic adolescents engaging in greater substance-use behaviors than less-acculturated Hispanic adolescents (see, for example, Chen, Cruz, Schuster, Unger,

& Johnson, 2002). In particular, the values of individualism and rebelliousness portrayed in cigarette ads may be more receptively perceived by acculturated Hispanic adolescents. Additionally, Baezconde-Garbanati (2001) described how tobacco advertising and the entertainment industry present smokers as living a life of glamour and sexual prowess. These values portrayed in media appeal to a sense of manliness or "machismo" among Hispanic boys, and a sense of freedom and breaking away from traditional cultural and family norms for Hispanic girls. Such values and other pro-tobacco messages in media have been shown to predict smoking among Hispanic youth (Chen et al., 2002). All these findings suggest that tobacco prevention efforts need to be grounded in cultural and social realities of Hispanic adolescent life, including any cultural struggles and adaptation processes.

A Media Literacy Antismoking Intervention: Empowerment for Hispanic Adolescents

In this project, we used media literacy as a strategy to encourage Hispanic adolescents' critique of advertisements, to enable them to resist smoking advertisements and risky influence attempts more generally. Vargas (2006) pointed to the empowerment potential of media literacy for Hispanic adolescents who are not only in a transition from childhood to adulthood, but who are also transforming their cultural and national identities by using media among other cultural resources. Media is heavily used by Hispanic adolescents, and, in part, it helps them understand the societal and cultural nuances of the American life. A culturally oriented media literacy program not only has the potential to make adolescents critically aware of the techniques and motivations underlying media messages, but also to help them examine their own attitudes and beliefs shaped by media messages (see Vargas, 2006). However, before discussing the role of media literacy as an antismoking strategy, we address the question, "what is media literacy?"

Media literacy is an emerging avenue for smoking prevention (see Banerjee & Greene, 2006; Pinkleton, Austin, Cohen, Miller, & Fitzgerald, 2007; Primack, Gold, Land, & Fine, 2006) and other behaviors such as drinking (see Austin & Johnson, 1997; Chen, 2009; Greene et al., 2011) and body image (see Levine, Piran, & Stoddard, 1999; Watson & Vaughn, 2006). Media literacy traditionally incorporates skills of accessing, analyzing, evaluating, and communicating messages in a number of forms (Hobbs, 1998). Put simply, the purpose of media literacy education is to equip and empower young people to critically analyze media messages to become more aware of persuasive message strategies, structure, and claims. The website of the National Association for Media Literacy Education (NAMLE, 2010) summarizes the functions of media literacy in this way:

> Within North America, media literacy is seen to consist of a series of communication competencies, including the ability to access, analyze, evaluate, and communicate information in a variety of forms, including print and non-print messages. Media literacy empowers people to be both critical thinkers and creative producers of an increasingly wide range of messages using image, language, and sound. It is the skillful application of literacy skills to media and technology messages.

National organizations such as the American Academy of Pediatrics, the Centers for Disease Control and Prevention, and the Office of National Drug Control Policy have recommended media literacy specifically as an antismoking prevention strategy. The rationale is based on the engaging, novel approach with small-group interaction in combination with the potential to reinforce previously existing antismoking beliefs (see Pfau, 1995). These smoking prevention strategies rely on reinforcing existing negative beliefs regarding substance use such as smoking (e.g., cigarettes smell, smoking is dangerous). Voicing the same recommendation, Unger, Cruz, Schuster, Flora, and Johnson (2001) have suggested that media literacy programs should be developed to encourage adolescents' resistance to tobacco-marketing strategies.

Media literacy antismoking prevention strategies have typically been delivered in school settings, but clubs and organizations (e.g., YMCA or soccer clubs) are potential settings for future interventions. The term, media literacy intervention, refers to an experimental treatment that introduces specific concepts (in this case, specific concepts about smoking) to study participants (any age group, primarily youth) with the aim of increasing awareness of the meaning contained in media messages, promoting deeper understanding of the impact of media on society, and enhancing critical analytical skills. For example, the intervention could demonstrate:

1. How cigarette ads promote the idea that smoking is: cool, fun, depicts maturity, and asserts independence;
2. How smoking images attract youth; and

3. How smoking ads promote images of fun while underplaying harmful effects of smoking.

The general goal of these media literacy interventions is to provide people with the cognitive skills necessary to analyze and critique media, and eventually lead them to build upon these skills required to process media messages in a more active manner (Potter, 2004). Based on these acquired skills, adolescents should build defenses against the potential negative effects of media (Austin, Pinkleton, Hust, & Cohen, 2005), and be less susceptible to negative effects of advertising through these inoculation processes (see Banerjee & Greene, 2007; Greene, in press).

Media literacy antismoking interventions have demonstrated reduction in (or delayed increase in) smoking related attitudes, intention, and behavior for elementary and middle-school children (e.g., Austin et al., 2005; Gonzales, Glik, Davoudi, & Ang, 2004; Kupersmidt, Scull, & Austin, 2010; Pinkleton et al., 2007). These studies indicate that interventions can help young people understand the role of tobacco advertising in encouraging tobacco use among adolescents, and identify marketing strategies tobacco companies use in their advertisements (Beltramini & Bridge, 2001). Increasing attention to these types of outcomes may further transfer to the development of or change in attitudes that are maximally effective for smoking prevention efforts (see Banerjee & Greene, 2006, 2007). Unfortunately, these prior interventions are labor intensive for schools, and some require eight, 12, even 18 sessions or a semester to deliver. Additionally, this prior work has not focused on minority adolescents. We developed a brief intervention (75 minutes, or two class periods) targeting minority adolescents to focus on empowering them to analyze media and also to resist smoking influences.

The Intervention

The purpose of the media literacy antismoking intervention was to change smoking-related attitudes and intentions, and, using experiential learning, to forewarn participants against future smoking persuasion attempts and empower them to critique advertisers' messages. The intervention manipulated the strategies chosen for delivering the message: analysis of counter cigarette advertisements versus production of counter cigarette advertisements.

The analysis approach includes a more traditional view of media literacy in which students were involved in analyzing and critiquing cigarette advertisements and antismoking advertisements and billboards. The production approach involves students by first having them analyze and critique cigarette advertisements, and then involved them in creating their own antismoking posters. Prior studies on media literacy efficacy have utilized both strategies (i.e., analysis and production) in creation of media literacy curricula without identifying which component(s) contribute(s) to success. We should know which strategy (analysis or production) is most efficacious to focus intervention efforts (see Greene, in press). The National Cancer Institute (NCI) Monograph #19 (NCI, 2008) described the significance of the intervention under discussion here: "Aside from this study, there have been no empirical studies on the impact of this kind of media activism and related informal advertising on individual attitudes and behavior, tobacco industry activity, and media coverage" (p. 445).

Our study involved two approaches to antismoking intervention efforts. We designed our intervention with two assumptions:

1. Media literacy production was a more novel and creative way of teaching in a classroom situation, as compared to media literacy analysis, and
2. Media literacy production empowers students more than media literacy analysis, because in production students are encouraged to plan and create their own media and therefore may benefit from experiential learning.

With these assumptions, we hypothesized that the production approach would be more cognitively involving (as evidenced by greater attention and more positive workshop perceptions), and lead to stronger antismoking attitudes, norms, and intentions as compared to the analysis approach, and both the approaches would be superior to the control group. The design of the intervention is described below (see also Banerjee & Greene, 2006, 2007), followed by description of the workshop sessions.

Time 1 (Week 1), Pretest. All students were involved in a pretest that included questions on prior smoking behavior, attitudes, beliefs, and norms related to smoking, and future smoking intentions. Classrooms were then randomly assigned to analysis, production, or control groups.

Time 2 (Week 2), Analysis I. Students in both analysis and production groups participated in the same introductory media literacy analysis workshop that aimed at helping students understand the persuasion techniques employed by cigarette advertisers and the missing claims in smoking ads. Small groups were utilized for some activities in all workshops. The workshop began with an activity to identify smoking claims or persuasion techniques in cigarette ads found in youth-read magazines. Following the activity, we moved to explore other facts about smoking that are missing from cigarette ads. The workshop concluded with each group presenting their analysis of persuasion techniques in print cigarette ads to the entire class, followed by a class discussion recapitulating workshop themes. The workshop lasted for 40 minutes, including a brief questionnaire at the end with questions related to workshop involvement and evaluation. Students assigned to the control group were not involved in any intervention activities this week.

Time 3 (Week 3), Analysis II or Production. The workshops in week 3 were different for students in the analysis and production groups. The analysis workshop (Analysis II) aimed at helping students counter smoking ads by analyzing antismoking print ads. We started with reviewing smoking persuasion techniques discussed in the previous week, followed by introducing students to antismoking ads and discussing examples of how pro-smoking ads and antismoking ads differ. Students were then involved in a group activity to analyze antismoking print ads and billboards, followed by group presentations of advertisement analyses (each group received several ads and chose one to critque and present to the class). The workshop concluded with a class discussion recapitulating the analysis workshop.

The production workshop (Production) was aimed at helping students counter cigarette ads by planning and creating their own antismoking ads. We began by reviewing smoking persuasion techniques discussed in the previous session, followed by introducing students to antismoking ads, and demonstrating how pro-smoking ads and antismoking ads differ. The students participated in a group activity to create an antismoking poster. In this task, the students' goal was to plan and create a poster that represents the best way to share an antismoking message with someone their age or younger at their school (each group of students was given poster-size paper, pencils, and markers to complete the project). This workshop concluded with poster presentations, voting on the best poster, and a review of workshop themes.

Both the Analysis II and Production workshops lasted 40 minutes, and all students filled out a questionnaire, responding to questions related to perceived workshop involvement and evaluation. Students assigned to the control group were not involved in any intervention or evaluation activities this week.

Time 4 (Week 5), Posttest. Students in all conditions were involved in a posttest that included smoking behavior in the past 30 days, attitudes, beliefs, and norms related to smoking, and future smoking intentions (the two-week delay captured some potential delayed effects).

Study Participants

The participants in the media literacy antismoking intervention study were two hundred and sixty (N = 260)—male (n = 104) and female (n = 156)—students enrolled in sixth to eighth grades in two northeastern schools (in New Jersey and New York). The students ranged in age from 11 to 16 (M = 12.49, SD = 1.06). The sample was predominantly Hispanic (74%) and African American (13%) (other groups < 3% each).

Results

Data were analyzed for both between-participant and within-participant differences. Between-participant differences highlighted the difference between analysis, production, and control groups, and within-participant differences highlighted differences over time (the study occurred over five weeks; in order to establish intervention efficacy, results of pretest and posttest were compared for each of the groups).1 We describe the results in terms of overall significant effects (see Banerjee & Greene 2006, 2007 for detailed results).

Results for behavioral intention to smoke. The between-subject analyses demonstrate that behavioral intention to smoke was significantly higher for the control group compared with analysis and production workshops. The within-subject analyses reveal that the Production session was more successful than both Analysis I and Analysis II sessions in reducing intention to smoke. Overall, both analysis and production workshops were better than control alone, but the Production workshop resulted in least intention to smoke. This is a key finding, as we reviewed the prior negative outcomes of smoking previously. This finding indicates that the media literacy production workshop is a promising avenue to reduce health disparities by empowering Hispanic adolescents.

Results for attitude toward smoking. The between-subject analyses were not significant for attitude toward smoking for analysis, production, or the control group. The within-subject analyses revealed that only production workshop was successful in reducing attitude toward smoking from Time 1 to 4. Because smoking attitudes are key predictors in many theories of persuasion, this finding for Production superiority is also crucial, similar to the impact for the findings for intention to smoke.

Results for subjective norm. Overall, the between-subject analyses revealed no differences across groups for subjective norm for analysis, production, or the control group. The within-subject analyses reveal that all workshop types (including control) were successful in reducing favorable smoking norm from Time 1 to Time 4. That is, completing surveys or participating in either workshop was related to more positive (health-protective) perceptions of norms, or believing that others think that smoking is bad.

Results for attention. Participants reported paying more attention to the production workshop (at Time 3), keeping attention to workshop (at Time 2) constant. Because at Time 2 the Analysis I session was common to both the groups (analysis and production), the Production I session elicited more attention from participants than Analysis II session. Thus, the Production workshop was a more engaging activity, with a very "hands on" approach to involving these Hispanic adolescents (from a qualitative perspective, responses to the poster production exercise were enthusiastic; students were excited about and engaged in the planning activity). Paying more attention to the workshop is an indicator of more involvement, one way to establish the mechanism through which the production workshop functioned (see Greene, in press).

Results for overall workshop perceptions. Participants had more positive workshop perceptions (at Time 3) for Production workshop than Analysis workshop. At Time 2, the Analysis I session was common to both groups, thus the Production I session resulted in more positive workshop perceptions than the Analysis II session.

Discussion

We examined changes in behavioral intention to smoke, attitude toward smoking, subjective norm, and cognitive responses in order to understand the efficacy of two media literacy intervention approaches (analysis versus production) targeting primarily Hispanic adolescents. The findings from the study provided greater support in favor of the production approach in bringing about desired changes in intention to smoke and attitude toward smoking. Additionally, as expected, there was an overall support for production workshop, as it elicited more attention and more positive workshop perceptions than analysis workshop. These findings suggest that some cognitive processes (particularly attention and overall workshop perceptions) are important predictors of attitude and intention change, and the results justify the need to continue to unravel these mechanisms of change or specify the cognitive mechanisms involved in media literacy processing (see Greene, in press). If we are to continue to reduce health disparities in minority populations, we must first create effective interventions (such as the present example), and then ensure that we are certain which intervention features are maximally effective in producing change so that we can focus resources in further efforts. Understanding the process through which we reduce health disparities (i.e., how the intervention actually worked) will allow us to target other risk behaviors for these groups using similar strategies.

To address health disparities specifically, communication interventions must, at minimum, use examples that resonate with the target population, including but not limited to matching message sources (and clothing), language (including slang), and settings; and using targeted products and specific channels. For example, media literacy interventions targeting Hispanic adolelscents might use for analysis examples ads that are shown on

Spanish-language channels (e.g., Telemundo, Univision, or Television), or ads appearing in magazines such as *Urban Latino* and *Latino Magazine*. In some of these ads, Latinos may be shown smoking Hispanic brand cigarettes (e.g., Fortuna, Faros, or Deliciados). Beyond these message feature considerations, adapted communication interventions must meaningfully engage audience motivations for risk behavior such as smoking. For Hispanic adolescents, this deeper engagement may include challenging perceptions that most Caucasian adolescents smoke, yet also addressing the notion that "macho" and independent men smoke.

Rationale to Justify Findings

One of the findings of the study was greater effectiveness of the production session of intervention workshops in reducing smoking-related intention, as compared to the initial analysis and second analysis sessions. The production strategy also worked better in conferring resistance to smoking messages, and was more effective in changing attitude and reducing intention to smoke over time. The production session may have provided an opportunity for participants to engage in self-persuasion by creating their own antismoking messages, rather than simply analyzing antismoking messages designed by others. Because most of the students already had negative attitudes toward smoking (confirmed in pretest), both the workshops provided them with self-generated rationale confirming their attitude to be the "true" or correct attitude. The production group also publicly described and showed their posters to the class, which could also influence the process of change (an alternate explanation, the public commitment component).

Furthermore, we found that participants paid more attention to the production workshop (compared to the analysis workshop) and had more positive perceptions of the production workshop. Health-based classroom lessons are often taught by the instructor, and the medium of message delivery is usually through lectures, exercises, posters, and leaflets, among others (see Tingle, DeSimone, & Covington, 2003). The idea of creating their own medium (posters) for messages to peers or younger students may have resulted in active searching of other knowledge schemas (see Bordeaux & Lange, 1991) and greater attention paid to the production session. This strategy of creating their own poster can be applied to other novel activities such as writing a play, making a television or radio script, and designing brochures. The key, underlying feature in these potential interventions would be involving participants in the activities in novel ways, such as planning messages that require perspective taking and counterarguing (see Greene, in press), to provide skills to empower them to make effective decisions when exposed to advertisements.

Overall, greater support for production suggests that students like creating their own messages more than they enjoy analyzing messages designed by others (see Kubey, 2000). Creating their own media not only empowers adolescents, but also provides them with an opportunity to reflect on their learning (Tyner, 1998). According to Tyner (1998), the combination of production creates a spiral of success for the students whereby analysis informs production, which in turn informs future analysis. Adolescents are constantly exposed to a barrage of media messages, thus they need such media-literacy skills to successfully navigate both their media and interpersonal environments. Because the production workshop included both types of activities (analysis and production), it provided students with an expression of their learning, and the best opportunity to apply new knowledge to future media messages targeting them.

Implications

The results of these data reinforce the need for evidence-based media literacy interventions targeting minorities. First, this study is one of the few published works that have attempted to "measure" media literacy efficacy and processes of effects. By using attention and overall workshop perceptions as cognitive processes, this study demonstrates the need for unraveling the process of change in media literacy interventions. Second, this study points to a key factor in intervention and/or campaign design, namely, participant involvement in message creation. Engagement is a crucial feature to effectively connect with populations, especially populations that may be struggling to identify with some parts of U.S. culture. The present study demonstrated that the production workshop worked better than the analysis workshop in changing participants' attitude toward smoking and behavioral intention to smoke, we posit through involvement (attention measured here). Success of production over analysis media-literacy workshops suggests examining if participant involvement in message generation and/or design can lead to self-persuasion. This project is the first test comparing the two media literacy strategies to develop a brief, yet efficacious, media-literacy approach (see Greene et al., 2011 for a modified application to high school drinking). Third, self-persuasion is another aspect that has not been well researched in regard to health intervention and/or campaign efficacy. The present study provides support for one mode of encouraging self-persuasion (getting students involved in producing their own antismoking posters). The complex interrelation of health disparities and power highlights the significance of approaches to persuasion and message involvement that are not perceived by participants as directive or "telling" minority adolescents what to do. It is relatively easy to generate reactance in adolescents, especially in the context of health behaviors, due to extensive focus on long-term health consequences. Thus, we must continue to explore involving approaches to intervention activities for specific subgroups. Intervention and/or campaign efforts promoting self-persuasion clearly merit attention.

Future Research

There is still much work needed in the area of media literacy and its application to large, health-based campaigns and/or intervention for Hispanic adolescents. The present study shows that experiential learning in the form of producing antismoking posters is a successful way of involving Hispanic young adolescents and protecting them against future persuasive smoking messages. Future researchers should consider if media literacy can be used for self-persuasion, leading to an enduring attitude and/or norm change, and further, long-term change in behavior. Considering culture-specific attitudes and beliefs within the framework of media literacy will provide novel ways of addressing antismoking issues with minority populations, and potentially combat specific media targeting of this group. This could include, for example, comparing messages specifically targeting Hispanics with those targeting other groups. Furthermore, the exact mechanism of cognitive processes that explain change in attitude or intention due to participation in media literacy interventions needs exploration.

The sample for this project included predominantly Puerto Rican adolescents in the northeast United States who may not share the same cultural struggles as other Hispanic subgroups, for example, dealing with immigration or family legal-status issues. Thus, we would not presume that this present intervention success would generalize to other areas of the US (e.g., the southwest) or other Hispanic populations (e.g., Mexicans or Dominicans), because different Hispanic groups have different experiences. This notion could be tested, however, to address this question (and modifications made to the intervention for specific Hispanic subgroups). Using additional culture-specific issues (such as smoking ads exemplifying the image of machismo or acculturation) with male Hispanic adolescents may help make the interventions more target-appropriate for subgroups, and could explore differences in how male and female Hispanic adolescents resist pro-smoking messages. Additionally, the present study utilized English-language advertisements. All study participants read and spoke English fluently, and many were fully bilingual or English-dominant (even if their parents were Spanish-dominant). For some participants English was the primary language, yet others were Spanish-dominant. Future studies could utilize Spanish-language advertisements as well. The increasing availability of Spanish-language channels on cable television may make this an area to explore, although these channels may be less utilized specifically by Hispanic adolescents than by Hispanic adults; alternatively, acculturation might best explain language-channel useage and should be included in future research.

Our emerging research in this area of targeted interventions includes an NIH-funded (NIDA) project using a similar strategy with rural and urban adolescents' alcohol use (see Greene et al., 2011), focusing on different aspects of health disparities. This population is different from that described in the present study, but contributes to understanding ways we can better serve a range of minority populations. This new project includes measures of advertising skepticism and counterarguing, and seeks to unravel the mechanisms of cognitive change that lead to beneficial effects of media literacy interventions.

The Hispanic adolescents in the present study are the exact type of population in need of interventions to increase their resistance to advertising. Marketers increasingly target this group, as companies seek to diversify and expand their product marketing. Only by creating these types of targeted interventions can we expect to influence long-term health disparities in these communities. Through media literacy, minority adolescents can learn to analyze media messages more carefully, and become better aware of potential influences that can have a tremendous impact on many aspects of their future health. There is every reason to expect that learning to critique advertisements can generalize to other behaviors. For example, the current project utilized examples of cigarette advertisements, but we believe that learning specific advertising critiques, and producing and/or planning posters will allow these adolescents to also critique beer advertisements, fast-food marketing, and a host of other topics and contexts. Although this premise is untested in the present project, it could be a beneficial side effect. Not only did the current media-literacy project provide better outcomes for Hispanic adolescents' smoking, but it also provided tools to enable participants to become better engaged and productive societal members. By encouraging adolescents to be more skeptical of media messages and the intentions of advertisers, we are affecting health disparities in vulnerable populations. Continued research in this area would be beneficial on many fronts.

Notes

1. The procedures used for detecting within-participant and between-participant differences are summarized here. **Between-particpant analyses**: Step 1, to show that groups are equal on outcomes at baseline; Step 2, to show that groups are different on outcomes post-intervention; Step 3, to show that outcomes should change the same following the Media literacy introductory workshop; Step 4, to show that the Production and Analysis workshops generated different changes on outcomes. **Within-particpant analyses**: Step 1, to show that the control group is equal on outcomes at baseline and post intervention; Step 2, to show the cumulative effects of Introductory and Analysis sessions for analysis workshop; Step 3, to show the cumulative effects of Introductory and Production sessions for production workshop; Step 4, to show the difference in change in outcomes between Times 2 and 3 for analysis and production workshops.

Bibliography

Austin, E., & Johnson, K. (1997). Effects of general and alcohol-specific media literacy training on children's decision making about alcohol. *Journal of Health Communication, 2*(1), 17–42.

Austin, E. W., Pinkleton, B. E., Hust, S. J., & Cohen, M. (2005). Evaluation of an American Legacy Foundation/Washington State Department of Health media literacy pilot study. *Health Communication, 18*(1), 75–95.

Baezconde-Garbanati, L. (2001). Understanding tobacco-use research among Hispanic/Latino adolescents: A sociocultural perspective. In National Cancer Institute (Ed.), *Changing adolescent smoking prevalence: Where it is and why* (Smoking and Tobacco Control Monographs No. 14, pp. 227–241). Rockville, MD: National Cancer Institute.

Banerjee, S. C., & Greene, K. (2006). Analysis versus production: Adolescent cognitive and attitudinal responses to antismoking interventions. *Journal of Communication, 56*(4), 773–794.

Banerjee, S. C., & Greene, K. (2007). Anti-smoking initiative: Examining effects of inoculation based media literacy interventions on smoking-related attitude, norm, and behavioral intention. *Health Communication, 22*, 37–48.

Beltramini, R. F., & Bridge, P. D. (2001). Relationship between tobacco advertising and youth smoking: Assessing the effectiveness of a school-based antismoking intervention program. *The Journal of Consumer Affairs, 35*(2), 263–277.

Bordeaux, B. R., & Lange, G. (1991). Children's reported investment of mental effort when viewing television. *Communication Research, 18*(5), 617–635.

Brown, R., & Houseman, T. (2000, February). *The tobacco industry marketing strategies*. Paper presented at the Training of the Joint Ethnic Tobacco Education Networks of California, Sacramento, CA.

Chen, X., Cruz, T. B., Schuster, D. V., Unger, J. B., & Johnson, C. A. (2002). Receptivity to protobacco media and its impact on cigarette smoking among ethnic minority youth in California. *Journal of Health Communication, 7*(2), 95–111.

Chen, Y. (2009, May). *The role of media literacy in changing adolescents' responses to alcohol advertising*. Paper presented at the International Communication Association (ICA) Conference, Chicago, IL.

Cowdery, J. E., Fitzhugh, E. C., & Wang, M. Q. (1997). Sociobehavioral influences on smoking initiation of Hispanic adolescents. *Journal of Adolescent Health, 20*(1), 46–50.

Elder, J. P., Campbell, N. R., Litrownik, A. J., Ayala, G. X., Slymen, D. J., Parra-Medina, D., & Lovato, C. Y. (2000). Predictors of cigarette and alcohol susceptibility and use among Hispanic migrant adolescents. *Preventive Medicine, 31*(2), 115–123.

Gonzales, R., Glik, D., Davoudi, M., & Ang, A. (2004). Media literacy and public health: Integrating theory, research, and practice for tobacco control. *American Behavioral Scientist, 48*(2), 189–201.

Greene, K. (in press). The theory of active involvement: Processes underlying interventions that engage adolescents in message planning and production. *Health Communication*.

Greene, K., Elek, E., Magsamen-Conrad, K., Banerjee, S. C., Hecht, M. L., & Yanovitzky, I. (2011, October). *Developing a brief media literacy intervention targeting adolescent alcohol use: The impact of formative research*. Poster presented at the annual meeting of the American Public Health Association, Washington, DC.

Gritz, E. R., Prokhorov, A. V., Hudmon, K. S., Chamberlain, R. M., Taylor, W. C., DiClemente, C. C., Johnston, D. A.,...Amos, C. I. (1998). Cigarette smoking in a multiethnic population of youth: Methods and baseline findings. *Preventive Medicine, 27*(3), 365–384.

Hobbs, R. (1998). The seven great debates in the media literacy movement. *Journal of Communication, 4*(1), 16–32.

Horigian, V. E., Lage, O. G., & Szapocznik, J. (2006). Cultural differences in adolescent drug abuse. *Adolescent Medicine Clinics, 17*(2), 469–498.

Kubey, R. (2000). Media literacy: Required reading for the 21st century. *High School Magazine, 7*(8), 29–34.

Kupersmidt, J. B., Scull, T. M., & Austin, E. W. (2010). Media literacy education for elementary school substance use prevention. *Pediatrics, 126*(3), 525–531.

Levine, M. P., Piran, N., & Stoddard, C. (1999). Mission more probable: Media literacy, activism, and advocacy in the prevention of eating disorders. In N. Piran, M. P. Levine, & C. Steiner-Adair (Eds.), *Preventing eating disorders: A handbook of interventions and special challenges* (pp. 3–25). Philadelphia, PA: Brunner/Mazel.

Lopez, B., Huang, S., Wang, W., Prado, G., Brown, C. H., Zeng, G., Flavin, K., & Pantin, H. (2010). Intrapersonal and ecodevelopmental factors associated with smoking in Hispanic adolescents. *Journal of Child and Family Studies, 19*(4), 492–503.

McNeill, A. D. (1991). The development of dependence on smoking in children. *British Journal of Addiction, 86*(5), 589–592.

National Association for Media Literacy Education (NAMLE). (2010). *Media literacy defined.* Retrieved from http://namle.net/publications/media-literacy-definitions/.

National Cancer Institute (NCI). (2008). *The role of the media in promoting and reducing tobacco use.* Tobacco Control Monograph No. 19. NIH Pub. No. 07-6242. Bethesda, MD: U.S. Department of Health and Human Services, National Institutes of Health, National Cancer Institute.

National Cancer Institute (NCI). (2010). *Cancer trends progress report 2009-2010 update.* Retrieved from http://progressreport.cancer.gov/doc_detail.asp?pid=1&did=2007&chid=71&coid=703&mid#high.

Pfau, M. (1995). Designing messages for behavioral inoculation. In E. Maibach & R. L. Parrott (Eds.), *Designing health messages: Approaches from communication theory and public health practice* (pp. 99–113). Thousand Oaks, CA: Sage.

Pinkleton, B. E., Austin, E. W., Cohen, M., Miller, A., & Fitzgerald, E. (2007). A statewide evaluation of the effectiveness of media literacy training to prevent tobacco use among adolescents. *Health Communication, 21*(1), 23–34.

Potter, W. J. (2004). *Theory of media literacy: A cognitive approach.* Thousand Oaks, CA: Sage.

Primack, B. A., Gold, M. A., Land, S. R., & Fine, M. J. (2006). Association of cigarette smoking and media literacy about smoking among adolescents. *Journal of Adolescent Health, 39*(4), 465–472.

Rogler, L. H., Cortes, D. E., & Malgady, R. G. (1991). Acculturation and mental health status among Hispanics: Convergence and new directions for research. *Hispanic Journal of Behavioral Sciences, 46*(6), 585–597.

Rohde, P., Lewinsohn, P. M., Brown, R. A., Gau, J. M., & Kahler, C. W. (2003). Psychiatric disorders, familial factors and cigarette smoking: I. Associations with smoking initiation. *Nicotine & Tobacco Research, 5*(1), 85–98.

Substance Abuse and Mental Health Services Administration (SAMHSA). (2009). *Results from the 2008 national survey on drug use and health: National findings*. NSDUH Series H-36, HHS Publication No. SMA 09-4434. Rockville, MD: Office of Applied Studies.

Substance Abuse and Mental Health Services Administration (SAMHSA). (2010). *Results from the 2009 national survey on drug use and health: Volume I. Summary of national findings*. NSDUH Series H-38A, HHS Publication No. SMA 10-4586. Rockville, MD: Office of Applied Studies.

Tingle, L. R., DeSimone, M., & Covington, B. (2003). A meta-evaluation of 11 school-based smoking prevention programs. *Journal of School Health, 73*(2), 64–67.

Tyner, K. (1998). *Literacy in a digital world: Teaching and learning in the age of information*. Mahwah, NJ: Lawrence Erlbaum.

Unger, J. B., Cruz, T. B., Schuster, D., Flora, J. A., & Johnson, C. A. (2001). Measuring exposure to pro- and anti-tobacco marketing among adolescents: Intercorrelations among measures and associations with smoking status. *Journal of Health Communication, 6*(1), 11–29.

U.S. Department of Health and Human Services (USDHHS). (1998). *Tobacco use among U.S. racial/ethnic minority groups—African Americans, American Indians and Alaska Natives, Asian Americans and Pacific Islanders, and Hispanics: A report of the Surgeon General*. Atlanta, GA: U.S. Department of Health and Human Services, Centers for Disease Control and Prevention, National Center for Chronic Disease Prevention and Health Promotion, Office on Smoking and Health.

U.S. Department of Health and Human Services (USDHHS). (2004). *The health consequences of smoking: A report of the Surgeon General*. Atlanta, GA: U.S. Department of Health and Human Services, CDC. Retrieved from http://www.cdc.gov/tobacco/data_statistics/sgr/2004/index.htm.

Vargas, L. (2006). Transnational media literacy: Analytic reflections on a program with Latina teens. *Hispanic Journal of Behavioral Sciences, 28*(2), 267–285.

Watson, R., & Vaughn, L. M. (2006). Limiting the effects of the media on body image: Does the length of a media literacy intervention make a difference? *Eating Disorders: The Journal of Treatment and Prevention, 14*(5), 386–400.

17

PREVENTING SUBSTANCE USE AMONG MULTIETHNIC URBAN TEENS USING EDUCATION ENTERTAINMENT, SOCIAL MARKETING, AND PEER CONTENT

Sarah Diamond, Leslie Snyder, and Jean J. Schensul

Nearly half of all U.S. teenagers have tried an illicit drug prior to high-school graduation, with alcohol and marijuana being the most widely used drugs. Youth who use drugs risk different health consequences, depending on the type and combination of substances used, as well as the frequency and duration of use. Studies have shown that many youth who initiate use of illicit drugs in high school will stop using them, or use them less frequently, as they achieve the social-role changes attending adulthood in their mid to late twenties (Bachman et al., 2002). Yet, for a fraction of them, substance use escalates into problem use, or leads to other serious health complications, injury, or even death. Some research suggests that African Americans and Hispanics, despite generally similar rates of illicit drug use in adolescence compared with Whites, might be at greater risk for adverse health outcomes from substance use, particularly in later years as young adults.

This case study provides an overview of an intervention called "Xperience," which was designed to prevent alcohol and illicit drug use among multiethnic urban youth, ages 16–20. The intervention was developed and evaluated at the Institute for Community Research in collaboration with the Center for Health Communication and Marketing at the University of Connecticut through a three-year grant from the Centers for Disease Control and Prevention from 2005–2008. It was developed with significant input and assistance from African American and Hispanic (mainly Puerto Rican) community leaders and partners from Hartford, CT, including the directors and staff at Young Studios and MiCasa Family and Educational Center, Inc.

We begin with an overview of the research on racial and ethnic disparities in substance use and substance use-related health outcomes for African American and Hispanic youth in the US. Next, we provide some background regarding best practices in drug prevention and health communication that informed the design of the Xperience intervention. This is followed by an overview of the Xperience program design and curriculum, and a brief summary of some process and outcome evaluation results from the third year of our study. We conclude by considering implications of these findings for advancing public-health efforts to reduce substance use among low-income and minority youth living in urban areas.

Drug Use Among African American and Hispanic Youth

According to the 2010 Census, ethnic minorities constitute over one-third of the U.S. population, with Hispanics/Latinos (approximately 15.5 million) and African Americans (approximately 42 million) as the largest minority groups. National findings from the Monitoring the Future Survey (MTF) regarding patterns of substance use contradict media stereotypes that depict African American adolescents as heavier drug users than Whites. White high-school students are more likely than non-White students to have used alcohol and a broader range of illicit drugs and to use these drugs more frequently. African American students overall tend to report *lower* rates of use for illicit drugs compared with White and Hispanic students, including marijuana. Hispanic high-school students, on the other hand, tend to report *higher* rates of substance use when compared with African Americans, and, for some drugs, compared also with Whites. They tend to start using alcohol and marijuana at younger ages, and are at a slightly higher risk for using cocaine and heroin prior to high-school graduation when compared with Whites (Johnston, O'Malley, Bachman, & Schulenberg, 2010). This data, however, does not provide a complete picture of substance-use rates at the population level, particularly in urban areas where the school dropout rates are so high. For example, in the state of Connecticut, dropout rates were as high as 34%–42% among African Americans and Hispanics, respectively (Kantrowitz, 2010).

The National Survey on Drug Use and Health (NSDUH, formerly NHSDA), a representative household survey of the noninstitutionalized U.S. population, provides similar findings regarding substance-use patterns by ethnicity and race. A longitudinal look at NSDUH 1999–2000 data finds that Whites, African Americans, and Hispanic Americans have nearly equivalent rates for past-month illicit drug use (around 6.4%), once confidence intervals are taken into account. However, Hispanic Americans of Puerto Rican heritage, have statistically greater one-month prevalence of illegal drug use when compared with non-Hispanic Whites and Blacks (Cooper, 2003, p. 108). This survey included individuals who had dropped out of school, but neither the MTF nor the NSDUH contained data for individuals who were then detained or incarcerated. Individuals with a history of involvement in the criminal justice system are known to have higher rates of substance use than the general population, and are disproportionately Black and Hispanic.

It is well documented in scientific research that youth who begin using illicit substances before age 14 are most vulnerable to being diagnosed with a substance-use disorder in adulthood. Problem substance use can be predicted by other factors in childhood, including aggressive and shy behavior and school achievement. While genetic factors may play a role in propensities for addiction, research reveals that social and environmental factors can mediate the impact of family substance-use history on the development of problem substance use and other adverse health consequences.

According to the 2010 NSDUH, the rates for substance dependence among racial and/or ethnic groups were highest for American Indians or Alaska Natives (16.0%), and persons reporting two or more races (9.7%). They were also somewhat higher for Hispanics (9.7%), followed by Whites (8.9%), and lowest for Blacks at (8.2%). However, these rates, too, may be underestimates for the general population, due to those who are incarcerated or homeless not being included in the sample.

Structural Disadvantage and Drug Use Health Outcomes

A combination of economic, social, structural, environmental, and community-level factors contributes to varied health outcomes over the life course for substance users. Census Bureau data show that on average, racial and/or ethnic minorities tend to have lower incomes and less accumulated household wealth than non-Hispanic Whites. African Americans and Hispanics fare worse than Whites on most economic indicators, with disparities having increased during the most recent economic recession. Poverty rates among African Americans and Hispanics are more than double the percentage of White Americans and Asian Americans; 8.6% of Whites and 11.6% of Asian Americans are in poverty, while 24.7% of African Americans and 23.2% of Latinos are under the poverty line. Individuals living in poverty and minorities are less likely to access treatment services due to lack of or limited health-insurance coverage. Recent U.S. Census Bureau data show that 30.7% of Hispanics, and 18.9% of African Americans did not have health insurance, compared to 10.8% of Whites. Blacks and Hispanics also may experience poorer treatment outcomes due to gaps in culturally and linguistically appropriate health-care services (Brown, Ojeda, Wyn, & Levan, 2000).

African Americans and Hispanics are also more likely to live in racially segregated, densely populated, poverty-stricken, urban areas than Whites. Socioeconomic status, social networks, place of residence, and ethnicity and/or race each have been shown to independently predict health outcomes for substance users in studies throughout the US. Rates of problem substance use in the NSDUH survey were correlated with level of education (less education correlating with more problem use), current employment status (unemployed reporting more problem use), as well as with parole and probation status (those on parole or probation had nearly three times the rate of problem use, when adjusting for other factors). Rates for substance dependence or abuse among persons age 12 or older in 2010 were similar in large metropolitan counties (8.9 %) and small metropolitan counties (8.9%), and both higher than in nonmetropolitan counties (7.6%). While binge drinking is found to be more prevalent among individuals with greater economic means, poorer individuals who binge drink have a greater number of drinks on average (CDC, 2012).

Syndemics theory proposes that stigma, poverty, barriers to treatment and care, poor environmental conditions, lack of adequate housing, trauma, gun violence, etc. have a compounding, negative impact on health, which helps to explain racial and ethnic health disparities more generally (Singer, 2008). For example, Lillie-Blanton and colleagues found that Blacks were more likely than Whites to become crack-cocaine users. However, this difference no longer held when comparing Blacks and Whites living in the same types of neighborhoods (Lillie-Blanton, Anthony, & Schuster, 1993). Urban minority youth also generally have higher rates of illiteracy, greater risk of dropping out of school, greater exposure to the underground street drug economy, and higher rates of untreated posttraumatic stress disorder, all of which can increase the risk for problem substance use and other adverse health outcomes (Collins et al., 2010; Fagan, 1992; Storr, Chen, & Anthony, 2004). Although Black youth, in particular, have generally shown great resiliency in the face of these adverse conditions, some research suggests that more serious health problems related to substance use may not surface until later years of their lives, when they are confronted with the realities of chronic unemployment, unstable relationships, poor housing, and racial discrimination—suggesting a differential trajectory over the life course from Whites (CDC, 2012; Wallace, Bachman, O'Malley, & Johnston, 1995). This theory is supported by studies that show that alcohol use begins later among Blacks than among Whites, but once initiated, problem use is more likely to escalate during adulthood.

Research on substance use from the perspective of social determinants of health (SDH) also point to heightened risk for certain comorbid health problems among African American/Black and Hispanic substance users. A recent qualitative study of MDMA (ecstasy) users we conducted in Greater Hartford, CT, for example, found that those who were using ecstasy primarily as a coping mechanism to deal with undiagnosed or untreated trauma self-reported higher levels of problem use (Moonzwe, Schensul, & Kostick, 2011). Blacks and Hispanics experience disproportionate rates of HIV and mortality from homicide, which are associated with intravenous substance use and street-level drug dealing, respectively. Urban minority youth who use or deal drugs are also more likely to be arrested and adjudicated in court compared with suburban youth, in part due to racial profiling, policing concentrated in urban areas, and policies such as drug-free zones that disproportionately impact urban areas (Golembeski & Fullilove, 2005). Nationally, studies have found that African American youth are more than twice as likely to be detained for a drug offense than White youth (31% and 15%, respectively) (NCCD, 2007). Disproportionate incarceration is considered to be a critical intervening factor in heightened rates of HIV among Black and Hispanic substance users, along with other disparate health outcomes such as Hepatitis C.

Drug-Related Deaths

The U.S. Department of Health and Human Services Task Force on African American and Minority Health previously found that

African American mortality rates for chronic disease and cirrhosis were nearly twice as high as the rates among Whites (Flewelling, Rachal, & Marsden, 1992). Blacks have also been overrepresented in drug-related emergency room incidents reported in DAWN, particularly for cocaine-related and marijuana-related casualties. For amphetamines and for methamphetamine, Hispanics were overrepresented (Cooper, 2003, p. 108). However, DAWN data is collected mainly from a sampling of 24-hour ER facilities located in metropolitan areas in which the people utilizing the services are also disproportionately Black and Hispanic, thus it is problematic to generalize from this data to the general population. Also, DAWN reports do not provide standardized population estimates by race and ethnicity due to a large amount of missing data in these reporting categories.

It is important to note that other adverse outcomes due to substance use disproportionately impact Whites. Of critical and growing significance since 2002, is data showing that Whites are more likely to die of drug overdose when compared with African Americans and Hispanics. This is attributed in large part to their greater access to and use of prescription drugs (CDC, 2011). However, the actual number of overdose deaths for Hispanics may be underestimated by as much as 25%–35%, due to misidentification of Hispanics as White (Arias, Schauman, Eschbach, Sorlie, & Backlund, 2008), which could mean that overdose mortality rates are nearly equivalent across all three groups.

Overall, the highest rates of drug-related ER visits reported in DAWN are with individuals who fall within the 18–25 age range, pointing to the urgency of addressing substance use during these critical years of mid to late adolescence across all racial and ethnic groups (Cooper, 2003). Over half of the approximately 80,000 drinking-related deaths in the US each year also result from injuries that disproportionately involve young people (CDC, 2012).

Preventive Services

The majority of evidence-based, "universal," drug-prevention programs in the US have been developed and evaluated in suburban middle schools with majority White populations. Several of these interventions have also proven to be effective in urban middle schools with African American and Hispanic populations. Fewer evidence-based, primary drug-prevention interventions have been developed for high-school age youth despite the fact that this is when most people in the US begin to use illicit drugs more frequently, are initiated into using "harder" drugs such as ecstasy, cocaine, or dust (PCP and/or embalming fluid), and start to experience more drug-related injuries and deaths. Another need, particularly in the urban context, is for interventions that can be delivered in community-based settings with voluntary participation from youth, so as to potentially reach the large percentage of youth who drop out of school and who are at high risk of ending up in the criminal justice system. Last but not least, multilevel approaches to prevention may help us better address the combined individual, social, environmental, and structural conditions impacting substance trajectories and health outcomes for youth and young adults living in low-income, urban communities.

The Intervention Overview

Xperience Intervention Goals

Xperience was designed to address the need for effective primary drug-prevention interventions targeting multiethnic urban youth in their mid teens to early twenties. In designing the intervention we drew upon theory-based, behavior change approaches and scientific literature on best practices. Meta-analyses of evaluated programs have consistently shown that the most effective drug-prevention interventions utilize theory-based strategies, strengthen drug refusal and resistance skills, are interactive versus didactic, and foster antidrug peer social norms (Cuijpers, 2002). A multisite evaluation of 46 CSAP-funded programs for "at-risk," mostly minority youth identified five best practices, which included:

1. Behavioral skills emphasis,
2. Use of introspective learning,
3. Connection building focus,
4. Coherent program implementation practices, and
5. High service intensity (Sambrano, Springer, Sale, Kasim, & Hermann, 2005).

Research also shows multilevel interventions that include a community-wide focus (e.g., mass media, community task forces, health policy) to be more effective than classroom curricula alone (Cuijpers, 2002; Slater, 2006).

In general, youth exposed to some kind of substance-use prevention messaging in the media or in their environment are less likely to report past-month use of alcohol, cigarettes, or illicit drugs compared with unexposed youth (SAMHSA, 2010). Public Service Announcements (PSAs) aimed at drug prevention in the general public, however, vary considerably in their effectiveness, depending on factors such as how well they are designed and the degree of audience exposure they attain (Abroms & Maibach, 2008).

Many prevention specialists consider cultural tailoring to enhance effectiveness in both face-to-face and mass-media interventions targeting ethnic minorities. Cultural tailoring varies from surface, or superficial, to deep level (Resnicow, Soler, Braithwaite, Ahluwalia, & Butle, 2000), and also varies depending on how "culture" is defined. Some approaches emphasize historical or reconstructed cultural heritage (e.g., Afrocentrism), while others emphasize contemporary popular culture such as hip-hop.

Xperience Approach

Given the increased social and cultural pressures around initiation into illicit drug use during the adolescent years, we recognized that it was important to offer urban youth resources and fun-filled, social experiences that would both affirm and support their decision to avoid using alcohol and other drugs. Mass-mediated entertainment, and especially popular music, is recognized to play a large role in adolescent youths' lives, especially among African Americans and Hispanics who generally consume entertainment media at higher rates than Whites (Rideout, Foehr, & Roberts, 2010). It also is thought to influence youth attitudes, expectancies, and behaviors with respect to drugs. A growing body of work concerning media influence shows an association between popular music consumption and substance use (Diamond, 2006). Other research has shown a correlation between the frequency with which youth attend clubs, parties, and music concerts in general and their use of drugs (Schensul & Burkholder, 2005). Hence, the notion of an entertainment-education, drug-prevention program grounded in urban youth culture emerged.

Xperience was designed to counter pro-drug influences in urban youths' social and physical environment, while tapping into other pro-social dimensions of youth entertainment culture (e.g., socializing, expanding one's peer networks, exploring one's sexuality, relieving stress). A participatory approach, involving community leaders and youth from the target population, was taken to develop and implement our intervention. This was based on our belief that minority communities, which have historically experienced discrimination and exploitation in the name of science, should have greater representation in decision-making and leadership roles in scientific studies, and it was also a way to ensure that the intervention would be culturally congruent with urban youth culture and well-adapted to the community-based settings in which it was to be based.

We chose to use three social-marketing strategies:

1. Live entertainment shows,
2. CDs, and
3. Branded promotional items.

These were the main components of our drug-prevention program. Similar social-marketing techniques had been used very effectively to engage youth and strengthen outcomes in other health-promotion campaigns related to tobacco and obesity prevention (Evans, Price, & Blahut, 2005; Heitzler, Asbury, & Kusner, 2008). This approach is also supported by "narrative persuasion" theory, which claims that youth may be more receptive, exhibiting less counterarguing, toward prevention messages embedded in entertainment media with characters with whom they identify (Green & Brock, 2000; Moyer-Guse, 2008). Tailoring media messages to fit adolescents' sensation seeking has also been shown to enhance effectiveness (Donohew, Lorch, & Palmgreen, 1991).

We developed the Xperience Leadership and Craft Development Training Program curriculum to formally train youth to take on leadership roles in the intervention by creating and publicizing original works of art (e.g., music, poetry, dance, drawings, and graphic art) containing drug-prevention messages. The program guided youth through the process of producing a compilation CD and two live entertainment shows (a VIP and CD-release show) performed for their peers and families and the public at large. A street team of youth was trained to assist in marketing the shows. This approach was also intended to support youths' career aspirations and learning of new skills.

The intervention communicated its prevention messages through multiple communication channels and levels of influence, utilizing a combination of individual and small-group, peer-led activities; on-the-ground, community events; and social marketing and mass-mediated entertainment. Taking a strength- or asset-based approach, the Xperience program promoted alternative, drug-free activities in the city of Hartford and helped to raise the public profile of the sizeable percentage of urban minority youth who were nonusers or only infrequent users of alcohol and marijuana, and who supported nonuse as a lifestyle preference. This approach differed from the typical alcohol and drug-prevention programs in urban areas, which tended to be selectively focused on youth who were labeled as "high risk" or were indicated to be at highest risk for substance abuse (i.e., were frequent drug or poly-drug users).

The rationale for targeting and tailoring our intervention to this nonindicated risk group was as follows: For one, we knew from national data that the mid-to-late-teen years is the period when many abstinent youth begin initiation into using alcohol and marijuana, and others begin increasing their frequency of use and experimenting with new drugs such as ecstasy and dust. Those youth who began using alcohol and/or marijuana prior to age 14, and who continued to use on a regular basis, or who had used other hard drugs, were at highest risk for substance-use disorders *as adolescents*, and thus would likely require treatment services instead of a prevention program. Additionally, the program design aimed for the youth artists and street team to serve as drug-free role models for their peers and wider community. There was also concern on the part of our internal ethics review board that youth who were more heavily involved with drugs could have an iatrogenic effect on those abstinent or low-level-using youth in the intervention.

Xperience Outcome Goals

The intervention had four main outcome goals:

1. To reinforce the decision of the youth in the program who did not use drugs to continue to avoid the use of drugs altogether;
2. To encourage those who were currently "low-level" alcohol and marijuana users to reduce their use and resist initiation into

other drugs such as ecstasy, dust, and cocaine. Low-level use of alcohol was operationalized at five times per month, and for marijuana as less than seven times per month. This is actually considered a high level of use in other "universal" interventions targeting middle-school youth, but was considered "low" in our study, since it was on the lower end of the spectrum of use patterns for our targeted age group and population as determined by our prior research (Schensul, Institute for Community Research (ICR), & Hartford Animation Institute, 2007);

3. To support both groups of youth to become drug-free role models for their peers;
4. To expand opportunities for all youth in the city to participate in drug-free activities and affirm drug-free norms, and with the long-term goal of reducing adolescent drug use in the city as a whole.

Program Implementation: The Xperience Leadership and Craft Development Training Program

From 2007–2008, the Xperience program was implemented and evaluated at a Hispanic family service and education center located in south central Hartford, whose services had expanded to include other racial and ethnic groups in their catchment area, particularly a sizeable group of African refugees. The program was delivered to two different groups of youth (N = 13 and N = 17) during the fall and spring school semesters, respectively. The youth constituted a fairly representative group of young men and women from Hartford, including African Americans, African immigrants, West Indians, Puerto Ricans, and Whites. (The city of Hartford has a predominately African American (38.05%) and Hispanic (40.52%) population, with only 17.8% identifying as non-Hispanic White).

The Xperience training program took place during weekly, two-hour-long sessions for a period of sixteen weeks, followed by a post-show celebration/wrap-up session in week 17. In addition to referrals to the program from our host agency, youth were recruited from throughout the city via fliers, radio announcements, social networking sites, a program website, and direct outreach by program staff. Participants were predominately African American and Hispanic (Puerto Rican), but also included a smaller proportion of Caucasians, West Indians, and several recent immigrants from Sierra Leone.

The drug-free lifestyles concept for the program was communicated through information handed out about the program, through the program website, and through the use of the slogan, "For Those Who Choose Not to Use," which accompanied the Xperience logo. In order to be admitted to the program, youth performing artists were required to audition. During the audition, they had to express why they wanted to be part of the drug-free program and perform a 90-second routine demonstrating their artistic talent. The auditions were videotaped and reviewed by a five-person panel of judges (comprising program staff). Judging criteria included:

1. Talent and/or originality,
2. Overall presentation (creativity, appeal, performance), and
3. Content (supportive of drug-free lifestyles, age appropriate, has a positive message, message is clear and meaningful to others).

None of the youth who auditioned were excluded from admission to the program. If a candidate's artistic talent was judged to be unsatisfactory to be featured at the show, she or he was given the opportunity to join other program activities.

The program curriculum involved youth working together to promote drug-free lifestyles. The youth were divided into three teams: the performing artists, the visual artists, and the production ("street") team. Three facilitators, each with professional experience in one of the three areas and training in the intervention model, led the sessions.

The curriculum drew upon a set of theory-based, drug-prevention taglines, grounded in social learning theory, which had been pilot tested through focus groups with youth from the city. These taglines addressed factors known to predict drug use, specifically peer influence, drug outcome expectancies, and drug avoidance self-efficacy (aka refusal skills), and were as follows:

- Keep your reputation, avoid a drug situation (self-efficacy).
- Do it without drugs (self-efficacy).
- No Drinks, No Drama (expectancies).
- Help your friends choose not to use (social influence).
- They wouldn't dare (get you high), if they really cared (social influence).
- Do you really know what's in it? (expectancies).
- Don't waste my time/mind (expectancies).

The taglines served as guides for the youth as they developed their original works of art (for example, rap or R&B songs and spoken-word poetry, digital images and posters). They were also used on promotional items that were distributed to the program youth and in the community as an important component of the social-marketing effort for the program and the shows. For example, a metal key chain was engraved with the tagline, "No drinks, no drama"; and a water bottle had "Do you really know what's in it?" printed on the outside.

In preparation for the main performance event, and to increase the program's influence upon youths' peer networks, the youth in the training program were asked to invite three to four of their friends to attend a preview of the main event. These youth were treated as VIP guests of Xperience, and they each received a gift bag containing an Xperience CD from the prior year, an Xperience T-shirt, and three to four other branded, promotional items with the taglines.

Once youth completed the recordings of their works of art on a professionally mastered CD, the final Xperience CD-release show was marketed citywide. The live performance took place in a central location downtown, and was promoted via local media channels and activities of the street team. The show contained the following key components for delivery of the prevention messages: posters and banners, two show hosts (one youth and one adult) from the program, branded promotional giveaway items, live vocal and spoken-word performances, interpretive dance, and visuals projected on a screen during the show. A DJ played instrumental music intermittently during the show to add to the excitement, while avoiding any conflicting pro-drug messages in commercial music. An onstage dance competition and other strategies by the hosts were used to enhance audience participation.

Process Evaluation (Internal Evaluators)

Program staff documented and evaluated the Xperience implementation process. One to two program staff observed and took notes during each training session. The youth participants were also asked to complete comment cards following each session, and were interviewed using a video camera at different points during the program. The live shows were also documented using participant observation and videorecording. General information on the reach of the shows was gathered by tracking audience attendance. Process indicators of show acceptability included audience enthusiasm, as evidenced by clapping, participation in volunteer activities such as an onstage dance contest, and attendance throughout the event. Process data were analyzed to identify key events and reoccurring *themes* (both explicit and implicit) that were connected to the intervention theory and approach, implementation process, and outcomes (Braun & Clarke, 2006). In addition to desirable outcomes, we were also on alert for any possible iatrogenic effects of the program (i.e., any harmful impact, counterarguing, or other indirect signs of resistance to the messages, or adverse incidents associated with the shows).

Process Evaluation Lessons Learned

The intervention proved to be well-suited for the community-based, youth-serving agency, which hosted the program on a trial basis. The only major adjustment in implementing the program at this agency was that the upper age limit for youth admitted to the program had to be lowered from age 25 to age 20. In the two prior years, we had included artists up to the age of 25. However, our partner agency maintained separate programming for youth under and over age 18, so as to avoid exposing younger youth to risk behaviors of older teens, such as the social drinking of alcohol.

We observed evidence that the intervention group activities helped to strengthen youths' drug-free norms. Many of the activities were designed to elicit antidrug messages, or pro-drug-free lifestyle messages from the youth themselves, drawing upon their personal experiences and observations. For example, during week three, the youth artists were asked to present their storylines and/or concepts for their works of art, the mood, and the color they associated with their piece, and what equipment they needed for the show. Excerpted notes of their presentations are as follows (pseudonyms are used):

> My stage name is Kelly. My work of art is about a girl who has been around drugs most of her life. She wants to get away from the drugs—mood is hopeless at first and then hopeful.

> Shawna—My concept is minorities should stay away from the street. Tagline is "Do it without drugs." The mood is aggression, color is red; mike and mike stand.

> Tasha (presents herself with force)—Drugs are not going to take her anywhere and she wants to be known for something positive. Color silver White and light blue. Tagline is: "Do it without drugs." Need a mike.

The youth responded very enthusiastically to the works of art developed by their peers, clapping, complimenting, or providing words of encouragement to each other during the presentations.

Overall, youth did not show any overt signs of resisting the drug-prevention messaging while taking part in the program. They completed the group exercises with enthusiasm, and also agreed not to post any messages about alcohol or substance use on their Facebook or Myspace pages while in the program. Themes in their works of art tended to emphasize the negative influence drugs had on their family relations, romantic partners, and friendships, rather than the effect that drugs had on their physical health. When health consequences were mentioned, the reference was most often an extreme case of addiction, death from overdose, or violence committed toward others while using drugs. In conversations, youth also expressed concern over the consequences of drug addiction upon a person's physical appearance and attractiveness to others (e.g., weight loss, being unkempt, having a "dirty" appearance, etc.). We also observed informal interactions among the youth, which reinforced the program's harm reduction and/or abstinence aims. For example, one of the youth was overheard recommending a book to another youth in the program about a girl whose life "went down the drain because she went to Mexico and got hooked on crack." Another youth was overheard on her cell phone telling someone, "Black and Milds (a brand of cigars) are bad for you."

The biggest challenge we encountered was from several youth who were most enmeshed in the commercial entertainment industry. Several semiprofessional youth artists, who had initially expressed interest in joining the program, refused or were not allowed by their managing agents to sign our CD-recording contract, and thus were not enrolled in the program. One youth who did enroll had established his own business as a club and party promoter, and produced his own music. We discovered midway through the program that he was promoting an event on his Myspace page, which included an alcohol sponsor's advertisement. After pointing out that this advertisement was inconsistent with the program's mandate to act as role models for Xperience's drug-free messages, he agreed to remove the promotional advertisement and publicly apologized to the other youth in the program for having violated his agreement. He was visibly upset during his apology, and others in the room responded in an appropriately serious manner. The end result, we believe, was positive, in that it reinforced the group's emphasis on maintaining drug-free norms and provided support to this youth, who was at higher risk of substance use, to avoid it.

Both shows were extremely well attended by youth and supportive adults from the community. The first show, held in mid-January, 2008, reached site capacity of 200 audience members, and some people were turned away at the door. The second show, which was held in early June, 2008 also attracted a large turnout of over 185 audience members, despite the fact that it coincided with high-school graduation time. The audience members were very engaged throughout both shows, as evidenced by:

1. The loud applause and cheering in support of each artist,
2. Level of participation in an onstage dance contest,
3. Excitement over receiving promotional items that were periodically thrown into the crowd, and
4. Very few people exiting the show early.

Through participating in the Xperience program, youth vocal artists gained exposure for their prevention messages and artistic talent, as well as developed their civic leadership skills. In preparing for the VIP and CD-release shows, it was clear that several of the youth lacked confidence, either in their abilities to compose original lyrics to a song or words to a poem, and many were nervous to perform in public. Following the final show, most of the artists expressed pride in having overcome their insecurities. This achievement was recognized by their friends and family members, who heartily congratulated and praised them after the shows. The staff also reported positive feedback from audience members, saying how much they appreciated the youth's promotion of "positive" messages in their community. Several parents at each show asked staff if they could volunteer to help out with the next show or become involved in the program in other ways.

At the end of the three-year pilot study, the program had a mailing list of over 900 youth from the city who wanted to stay informed about future shows. Although the staff did not actively promote the Myspace page or Xperience website, 64 youth signed on as "friends" on Myspace and over 1,114 unique visitors had accessed our website (XperienceHartford.org) in less than a year's time. In addition, staff members and program youth distributed over 2,000 CDs and thousands of wristbands and other branded promotional items with drug-prevention messages throughout the city. During the study timeframe, staff commented that they had seen program and other youth in public venues wearing or carrying the Xperience promotional items, such as the bags, T-shirts, wristbands, etc. Youth in the program also came to the program afterschool with their Xperience bag, key chain, wristbands, etc. This indicated that the social marketing approach was successful in disseminating the drug-free messages, and in reinforcing youth identification with the Xperience brand and messaging.

Did the Program Work? Findings from the Summative Evaluation

A summative evaluation was conducted in 2007–2008 to test whether Xperience had an impact on youth alcohol and marijuana use. Because we were concerned about the short time period between the end of the intervention and the posttest, we also assessed the youths' behavioral intentions to use alcohol and marijuana in the next 30 days. *Behavioral intentions* focused on the likelihood of use in the next 30 days, and *actual behavior* assessed 30-day prevalence for alcohol, marijuana, and hard drugs.

An additional concern was whether the program was able to make a difference among youth at high risk of escalating their substance use due to environmental influences such as spending time with friends who use substances, or hanging out in clubs and bars. The *pro-substance social environment scale* used items on how often they went to clubs or bars, how often they attended house parties, and how many of their friends used marijuana or alcohol.

Based on communication research showing the importance of message saturation (also referred to as dosage or levels of exposure), we hypothesized that the youths' cumulative exposure to the drug-prevention messages through multiple channels (e.g., CD, branded promotional items, the shows) would mediate drug use, and social, cognitive, and behavioral outcomes. Because some youth in the comparison group were exposed to elements of Xperience, *Xperience exposure* was measured by an index of exposure to specific program elements: training (for members of the performance or production team), the website, the CDs, the wristband, the Xperience live shows, and the CD-release events.

Sampling method. The outcome evaluation used a pretest-posttest with comparison-group design. The sample consisted of youth who actively participated in the creation of the shows and CDs, their friends who attended the shows and received giveaways ("VIP" group), and the comparison group. The comparison group was recruited from youth attending other community-based afterschool programs, including several job-training programs, local YMCAs, youth church groups, etc. Youth received two coupons for free movie tickets as an incentive to participate in the evaluation. A unique identification code, based upon the birth date, three letters of the mother's first name, and area code, was used to track participants. The total sample of N = 116 ranged in age from 15 to 22, with a mean of 17.5. Almost three-quarters were female (73%), and most self-identified as African-American (70%), with a substantial number of Hispanics (26%), but few Whites (2%).

The evaluation was conducted in two waves, designed to assess the impact of the last two Xperience training cycles in fall and spring. Pretests were administered in person prior to the start of training (for youth artists and street team) and the performances (for their friends, the VIP group), and a month prior to the CD-release show (for the comparison group). Posttests occurred the week following the CD-release show, with an additional three-week window for completion of the posttest by those who missed the last session and for the comparison group.

Results. Baseline levels of alcohol and marijuana use were low, as expected, given the targeted sample (the averages were 1.9 days drinking per month, and 0.9 days smoking marijuana per month). Few teens at baseline (15%) had used other drugs—ecstasy, dust, or other substances—in the prior 30 days. There were no statistically significant differences by condition in changes in levels of alcohol, marijuana, or other drug use over time. However, there was a statistically significant difference in any other drug use, such that while 5% of the intervention group and 43% of the VIP group had used either ecstasy, dust, or some drug other than alcohol or marijuana in the past 30 days at baseline, no one in the intervention or VIP group used these drugs at post-test ($F(2) = 3.7$, $p = .03$).

The evaluation found that greater exposure to Xperience was associated with decreased intentions to use marijuana in the next 30 days ($\beta = -.25$, $p < .05$, $N = 84$). The pro-substance environment proved to be a moderator of the effectiveness of the program, in terms of marijuana and alcohol use. Youth in a pro-substance environment had greater baseline levels of substance use than youth not in a pro-substance environment. Youth in a pro-substance environment who were exposed to Xperience declined in their marijuana use intentions, marijuana use, and alcohol use. However, youth in a pro-substance environment who were *not* exposed to Xperience did not change their intentions to use marijuana over time, and slightly increased their 30-day use of marijuana and alcohol. Youth who were not in a pro-substance environment decreased their substance use over the course of the evaluation, whether or not they were exposed to Xperience.

Discussion

While national surveys have found similar rates of substance use among White, Black, and Hispanic youth, there is research evidence that points to racial and ethnic disparities in the use of certain drugs, access to preventive and treatment services, and a number of health outcomes related to different trajectories of substance use over the life course. Individuals living below the poverty line and in metropolitan areas experience disparities in substance use disorders and treatment access. Patterns of substance use and health outcomes of minority youth who use substances are influenced by a variety of individual, social, environmental, and structural conditions (e.g., poverty, incarceration, homelessness, unemployment, racism, and public versus private health insurance). Recognizing how these multiple factors intersect over the lifespan for African American and Hispanic youth living within urban contexts can help us to design more effective prevention interventions.

Xperience filled an important gap in drug-prevention programming for ethnically diverse urban adolescents. It built upon the positive elements of urban youth culture to promote drug-free community norms, and was successful in involving ethnically diverse youth in shaping prosocial, substance-free environments. The Xperience training curriculum and shows were culturally congruent; fun, engaging and appealing; and developmentally appropriate for adolescents from mid-teens to young adults. The program also provided occasions for professional skills building, social bonding, peer leadership, civic engagement, and creative self-expression oriented around shared, drug-free social norms.

The outcome evaluation showed that Xperience was effective in helping youth who participated in prodrug entertainment environments (attended house parties, had friends who used substances, and went to bars and clubs) to lower their intentions to use drugs. Xperience also appears to have changed the way these youth behaved in regard to lowering their use of drugs. Thus, the intervention may provide a measure of protection against environmental pressures to use drugs, something not claimed by other youth interventions.

Limitations of the study include the small sample size in the evaluation design, a common problem in evaluating many community-based, universal prevention programs. Although the sample was small, the effects were large enough to be statistically significant. It also proved difficult to obtain a comparison group that was equivalent in all ways except for their participation in Xperience. Finally, we were unable to conduct a long-term follow-up to determine if the effects of the intervention were sustained over time.

This peer-led, entertainment-education promises to be an effective environmental strategy to reduce risks of drug use associated with attending other risky, party-culture venues. The effect sizes are a marked improvement over more standard antidrug media campaigns, which average effect sizes of about $r = .02$, and widely replicated in-school interventions like DARE, which average about $r = .04$ (Snyder et al., 2004). The fact that the intervention intervened at multiple levels to address not only individual, but also peer-group, community, and environmental risks and protective factors may have contributed to its high level of effectiveness, and may lead to stronger longer-term outcomes. Our finding—that it is possible to increase resistance capacity and reduce substance use among high-school age youth who are exposed to high-risk party environments—is worth trying to replicate in future research. Due to the intervention design centering upon *peer-produced*, drug-prevention products (including CDs, branded promotional items, visual arts, and live-entertainment events), it can easily be adapted to other volunteer afterschool programs in diverse community settings, as well as schools and colleges.

Bibliography

Abroms, L., & Maibach, E. (2008). The effectiveness of mass communication to change public behavior. *Annual Review of Public Health, 29*(1), 219–234.

Arias, E., Schauman, W. S., Eschbach, K., Sorlie, P. D., & Backlund, E. (2008). The validity of race and Hispanic origin reporting on death certificates in the United States. *Vital Health Statistics, 2*(148), 1–23.

Bachman, J. G., O'Malley, P. M., Schulenberg, J. E., Johnston, L. D., Bryant, A. L., & Merline, A. C. (2002). *The decline of substance use in young adulthood: Changes in social activities, roles, and beliefs.* Mahwah, NJ: Lawrence Erlbaum.

Braun, V., & Clarke, V. (2006). Using thematic analysis in psychology. *Qualitative Research in Psychology, 3*(2), 77–101.

Brown, R. E., Ojeda, V. D., Wyn, R., & Levan, R. (2000). *Racial and ethnic disparities in access to health insurance and health care.* Los Angeles, CA: UCLA Center for Health Policy Research.

CDC. (2011). *CDC health disparities and inequalities report—United States, 2011.* Atlanta, GA: U.S. Department of Health and Human Services (DHHS), Centers for Disease Control and Prevention.

CDC. (2012, January 10). Binge drinking is bigger problem than previously thought. Retrieved from http://www.cdc.gov/media/releases/2012/p0110_binge_ drinking.html.

Collins, K., Connors, K., Davis, S., Donohue, A., Gardner, S., Goldblatt, E., Hayward, A.,...Thompson, E. (2010). Understanding the impact of trauma and urban poverty on family systems: Risks, resilience, and interventions. Retrieved from http://fittcenter.umaryland.edu/LinkClick.aspx?fileticket=jFOaDJRM1P8%3D&tabid=147.

Cooper, L. (2003). Drug use among racial/ethnic minorities (revised). Retrieved from http://archives.drugabuse.gov/pdf/minorities03.pdf.

Cuijpers, P. (2002). Effective ingredients of school-based drug prevention programs—A systematic review. *Addictive Behaviors, 27*(6), 1009–1023.

Diamond, S. (2006). What's the rap about ecstasy?: Popular music lyrics and drug trends among American youth. *Journal of Adolescent Research, 21*(3), 269–298.

Donohew, L., Lorch, E. P., & Palmgreen, P. (1991). Sensation seeking and targeting of televised antidrug PSAs. In L. Donohew, H. E. Sypher, & W. J. Bukoski (Eds.), *Persuasive communication and drug abuse prevention* (pp. 209–226). Hillsdale, N.J.: Lawrence Erlbaum.

Evans, W. D., Price, S., & Blahut, S. (2005). Evaluating the truth (R) brand. *Journal of Health Communication, 10*(2), 181–192.

Fagan, J. (1992). Drug selling and licit income in distressed neighborhoods: The economic lives of street-level drug users and dealers. In A. Harrell & G. E. Peterson (Eds.), *Drugs, crime, and social isolation* (99–146). Washington, DC: Urban Institute Press.

Flewelling, R. L., Rachal, J. V., & Marsden, M. E. (1992). *Socioeconomic and demographic correlates of drug and alcohol use: Findings from the 1988 and 1990 National Household Survey on Drug Abuse.* Rockville, MD: National Institute on Drug Abuse.

Golembeski, C. & Fullilove, R. (2005). Criminal (in)justice in the city and its associated health consequences. *American Journal of Public Health, 95*(10), 1700–1706.

Green, M. C., & Brock, T. C. (2000). The role of transportation in the persuasiveness of public narratives. *Journal of Personality and Social Psychology, 79*(5), 701–721.

Heitzler, C. D., Asbury, L. D., & Kusner, S. L. (2008). Bringing 'play' to life—The use of experiential marketing in the VERB (TM) campaign, *American Journal of Preventive Medicine, 34*(6), S188–S193.

Johnston, L. D., O'Malley, P. M., Bachman, J. G., & Schulenberg, J. E. (2010). Monitoring the Future national results on adolescent drug use: Overview of key findings, 2009. Bethesda, MD: National Institute on Drug Abuse (NIDA).

Kantrowitz, J. (2010, June 10). Graduation data reveal dropout crisis in Connecticut. Retrieved from http://blog.ctnews.com/kantrowitz/2010/06/10/graduation-data-reveal-dropout-crisis-in-connecticut/.

Lillie-Blanton, M., Anthony, J. C., & Schuster, C. R. (1993). Probing the meaning of racial/ethnic group comparisons in crack cocaine smoking. *Journal of the American Medical Association, 269*(8), 993–997.

Moonzwe, L. S., Schensul, J. J., & Kostick, K. M. (2011). The role of MDMA (ecstasy) in coping with negative life situations among urban young adults. *Journal of Psychoactive Drugs, 43*(3), 199–210.

Moyer-Guse, E. (2008). Toward a theory of entertainment persuasion: Explaining the persuasive effects of entertainment-education messages. *Communication Theory, 18*(3), 407–425.

National Council on Crime and Delinquency (NCCD). (2007). *And justice for some: Differential treatment of youth of color in the justice system.* Oakland, CA: Author.

Resnicow, K., Soler, R., Braithwaite, R. L., Ahluwalia, J. S., & Butle, J. (2000). Cultural sensitivity. *Journal of Community Psychology, 28*(3), 271–290.

Rideout, V. J., Foehr, U. G., & Roberts, D. F. (2010). *Generation M2: Media in the lives of 8- to 18-year-olds.* Menlo Park, CA: Henry J. Kaiser Family Foundation.

Sambrano, S., Springer; J. F., Sale, E., Kasim, R., Hermann, J. (2005). Understanding prevention effectiveness in real-world settings: The national cross-site evaluation of high risk youth programs. *The American Journal of Drug and Alcohol Abuse, 31*(3), 491–513.

SAMHSA. (2010). Results from the 2009 national survey on drug use and health: Volume I. Summary of national findings. Rockville, MD: Office of Applied Studies.

Schensul, J. J., & Burkholder, G. J. (2005). Vulnerability, social networks, sites, and selling as predictors of drug use among urban African American and Puerto Rican emerging adults. *Journal of Drug Issues, 35*(2), 379–408.

Schensul, J. J., Institute for Community Research (ICR), & Hartford Animation Institute. (2007). Rollin' and dustin': Pathways to urban life styles. Retrieved from http://www.incommunityresearch.org/panelsexp.htm.

Singer, M. (2008). *Drugging the poor: Legal and illegal drugs and social inequality.* Long Grove, IL: Waveland Press.

Slater, M. D. (2006). Combining in-school and community-based media efforts: Reducing marijuana and alcohol uptake among younger adolescents. *Health Education Research, 21*(1), 157–167.

Snyder, L. B., Hamilton, M. A., Mitchell, E. W., Kiwanuka-Tondo, J., Fleming-Milici, F., & Proctor, D. (2004). A meta-analysis of the effect of mediated health communication campaigns on behavior change in the United States.,*Journal of Health Communication, 9*(1), 71–96.

Storr, C., Chen, C., & Anthony, J. (2004). 'Unequal opportunity': Neighbourhood disadvantage and the chance to buy illegal drugs. *Journal of Epidemiology and Community Health, 58*(3), 231–237.

Wallace, J. M., Jr., Bachman, J. G., O'Malley, P. M., & Johnston, L. D. (1995). Racial/ethnic differences in adolescent drug use: Exploring possible explanations. In G. Botvin, S. Schinke, & M. Orlandi (Eds.), *Drug abuse prevention with multiethnic youth* (pp. 59–80). Thousand Oaks, CA: Sage.

18

THE IMPORTANCE OF FORMATIVE RESEARCH IN MASS-MEDIA CAMPAIGNS ADDRESSING HEALTH DISPARITIES: TWO KENYAN CASE STUDIES

Ann Neville Miller, Julie Gathoni Muraya, Ann Muthoni-Thuo, and Leonard Mjomba

As compared to their urban, middle-income counterparts, women and children living in rural areas and urban, informal settlements in Kenya face special challenges in accessing health information and services. Television and print health information is often confined to towns and cities, and shortages of health workers and supplies hamper most rural health-care systems (CBS, 2004). At the same time, rapid urbanization has put unprecedented strain on the existing resources in the cities, and has resulted in high rates of unemployment, poverty, and poor health outcomes, especially among women and children living in informal settlements (African Population and Health Research Center, 2002; Mutua-Kombo, 2001; Ngimwa, Ocholla, & Ojiambo, 1997). Only middle- and upper-class Kenyans who live in the largest urban centers have ready access to information about prevention and care, and can take advantage of a range of early detection technologies and treatment options at private hospitals.

Health communicators who aim to mitigate disparities in access to health information must begin by understanding key characteristics of the specific disadvantaged groups—cultural beliefs about health and illness, health-literacy characteristics, barriers to health care or healthy behaviors, and a host of other factors. If they do not develop this kind of knowledge, they risk perpetuating existing gaps between the rich and the poor (Dutta-Bergman, 2004a, 2004b, 2005), and ultimately effecting little positive change (Atkin & Freimuth, 2001; Noar, 2006; Snyder, 2007). Research that investigates these issues is referred to as formative research. Formative research also includes testing of specific campaign messages before they are disseminated to determine whether they are understood, liked, and accepted by target-audience members; and identifying appropriate media channels for the message. Methodologically, formative research can entail reviewing information about other communication efforts, conducting in-depth interviews and focus groups, doing household observations, constructing surveys, or running pilot studies. It occurs either before health communication is designed, or while it is being disseminated; either way the purpose is to help "form" the communication effort. When it is done well, formative research provides an opportunity for stakeholders from the community to participate in strategizing toward positive change in their own communities.

Unfortunately, thorough formative research is an ideal that some organizations may feel they have neither the time nor the money to indulge in. Building in thorough formative and summative (outcome) evaluation can take up to 20% of the budget for a health communication effort, and that may be a price that is difficult for small organizations in particular to afford. Organizations both large and small may also feel pressured to move quickly in order to demonstrate effectiveness to donors, or may believe that the needs they are planning to address are so urgent that they cannot justify waiting for the results of formative research before taking critical health messages to marginalized populations. In the remainder of this chapter we illustrate the importance of formative research by describing two very different health-related, mass-media efforts in Kenya. In both examples we conclude that inadequate formative research has led to less than optimal results in addressing health disparities, because audience responses to specific messages were not sufficiently investigated before campaigns were rolled out.

The first of these was a large, internationally funded, multimedia effort promoting abstinence among urban adolescents: the *Nimechill* campaign. The second is a small, locally funded print effort attempting to educate Kenyan women about breast cancer signs, symptoms, and early detection measures. Our description of these efforts is drawn from research reports about our interviews and focus-group discussions with members of both target audiences. In both of these studies we showed campaign messages to members of the targeted population for the health communication and solicited their responses (Muraya, Miller, & Mjomba, 2011; Muthoni-Thuo & Miller, 2010). Readers who are interested in more details about our methods and findings are referred to those articles.

In analyzing these efforts we want to make it clear that we are not singling out these organizations as either less thoughtful or less dedicated to quality communication and to their target populations than other, similar organizations. Two of the authors of this chapter have worked closely with the organizations in various capacities, and were impressed with the vision and dedication of the personnel there. In fact, it is because we suspect that their approach is fairly typical of well-intentioned individuals and organizations that we think it is worth describing our findings so as to point out the critical importance of formative research. In the sections below, we first give background on the two case studies and describe responses from target-audience members about their communication. At the end of the chapter, we draw conclusions about these efforts, and mention implications for formative research on similar projects.

The *Nimechill* Abstinence Campaign

The *Nimechill* abstinence campaign was the first large-scale, mass-media, youth abstinence campaign in Kenya. Developed by an international global health organization, the campaign targeted urban and peri-urban youth between the ages of 10 and 14, with mass-media components running from September 2004 to April 2005. The primary objective was to increase the percentage of youth who chose to abstain from sex, by changing three perceptions correlated with abstinence: social norms, self-efficacy, and behavioral intentions to remain abstinent. The twofold communication objectives of the media campaign were: (a) to create a norm where abstinence was viewed as a cool, smart, and responsible choice by using youth role models who go against the norm to have sex; and (b) to reduce norms and peer pressure regarding having sex.

The campaign featured different close-up visuals of urban youth aged 14 to 16 who were all flashing the "chill" symbol—index and middle fingers raised in the form of a "V." Texts on the visuals were written in first person, expressing the models' intention to abstain from sex, and ending with the phrase "*nimechill.*" The word *nimechill* is an idiomatic expression in *sheng*, the local slang used by most urban youth in Kenya, meaning, "I am abstaining." The logo for the campaign consisted of a cartoon drawing of a hand making the chill sign.

The *Nimechill* campaign used a multichannel approach, incorporating television and radio spots, print advertisements, posters, and billboards. Ads were developed by a professional advertising agency in Nairobi on the basis of formative research with Nairobi youth. In addition, articles about youth appeared regularly in national newspapers, and discussions on chilling were featured on radio stations. T-shirts with the *Chill* branding and messages such as "young, beautiful and chilling" or "handsome, intelligent and chilling" were handed out at youth events. A peer-education component involved the formation of *Chill* Clubs at 12,500 primary schools across Kenya. Although the mass-media component of the campaign ended in April 2005, a number of the *Chill* clubs were still in existence at the time we were writing this chapter.

Evaluation following the completion of the mass-media campaign revealed that 85% of youth recalled the *Nimechill* messages. Also, the proportion of Nairobi youth aged 10 to 14 who self-reported never having had sex increased during the nine months of the campaign, from 88% to 92%. This behavior change, however, showed no relationship to any level of exposure to the *Nimechill* campaign. Given that the messaging in the campaign was intense, albeit less than a year in duration, the lack of evidence for association seems surprising.

In presenting their evaluation, campaign developers were unable to pinpoint the reason that being exposed to the campaign showed no statistical relationship with behavior change. Several possibilities, however, come to mind. On the one hand, it could be that the campaign was one among many aspects of a trend toward later sexual initiation among Nairobi youth, or that the abstinence message it propounded ended up being spread through interpersonal communication. Either of these situations would make it difficult to tease out the effects of the campaign itself on youth behavior. Or, it could be that the campaign was not effective because objectives of the campaign were out of touch with participant cultural realities. Alternatively, objectives of the campaign might have been culturally appropriate, but the text and images in the campaign might have been constructed in such a way that audiences derived inconsistent or confusing contextual meanings.

Our research team (two Kenyan researchers and one American who had lived in Kenya for 13 years) held focus groups and interviews with Nairobi youth to explore these last two possibilities.

Groups were divided according to age (10- to 12 year-olds and 13- to 14 year-olds) and by socioeconomic status (youth from low-income areas and middle-income schools), and discussions were moderated by one of the Kenyan researchers. We showed the youth four of the five major images from the campaign in poster form, and asked them to respond to a series of questions about them. The same images had also been used on billboards, in magazines, and in other media outlets. We provide a description of these posters below.[1]

Content of the *Nimechill* Posters

Poster 1: Three youth in a matatu. The first poster depicted a close-up of three youth leaning out of the window of a *matatu* (minivans that serve as the most common form of public transportation in Kenya), flashing the two-fingered *chill* sign. The headline read: "Sex? No way, *tumechill*" (We are abstaining). Body text read: "We won't be taken for a ride," followed by the tagline, "*Ni poa ku chill*" (It's good to abstain). The logo in the corner depicted a hand with two fingers making a "V" sign designated to symbolize *chilling*.

Poster 2: Boy and basketball. The second poster showed a young boy with a cocky smile flashing the chill sign. The boy was holding a basketball, dressed in trendy clothes and a cap turned around backward, and obviously standing on a basketball court. The headline read: "Sex? *Siwezi, nimechill*" (Sex? I won't. I'm abstaining). Body text read: "*Bado siko ready. Ni poa ku chill*" (I'm not ready. It's good to abstain.)

Poster 3: Three girls. The third poster showed three teenage girls with pleasant smiles all flashing the *chill* sign. The girls were seated on a rustic piece of wooden furniture that could not be seen in its entirety. The headline read: "Sex? Not now, *tumechill.*" Body text read: "We know what the consequences are. *Ni poa ku chill.*"

Poster 4: Three boys. This poster depicted three schoolboys leaning over a railing and flashing the *chill* sign. All of the boys were wearing serious expressions. The headline stated: "Sex? *Zi, tumechill*" (No way, we are abstaining). Body text read: "We know better. *Ni poa ku chill.*"

Youth Response to the Posters

For a Western audience, the combination of catchy tag lines and text with attractive, youthful models in the *Nimechill* campaign might have made for a straightforward message. In fact, for two of the researchers on the project who were educated in the West, the initial impression of the campaign posters was appealing. Indeed, the youth we interviewed endorsed the objectives of the campaign, in terms of the realities of sexual behavior in their cultures. They were also able to articulate many of the associations that campaign creators presumably intended for the posters to spark in the minds of viewers. For example, several youth commented that the boy on the basketball court was probably supposed to convey the idea that involvement in sports could be an alternative to early initiation of sex. "He's spending his time in games that can keep him from bad behavior," one explained. Similarly, they suggested that the use of three boys in school uniform on Poster 4 implied that school was an environment in which teenage sex was practiced: "These things they start in school." "Some schools they share the same bathroom, girls and boys, and they get into relationship. They ask permission to go to the toilet at the same time and go to do bad things." They also recognized that abstaining enabled students to make the most of their educational opportunities. "If they engage in sexual activities when they're in school, they won't continue with their education," explained one youth.

However, participants gave this sort of overall positive comment only for Poster 4. The visuals on the other posters were problematic for them. Their objections to these three posters revolved around three aspects of the visuals: the *chill* symbol itself, the appearance of the models, and the choice of poster background.

The chill *symbol.* The *chill* logo, recall, consisted of the image of a hand with index and middle fingers held up in a "V" shape. Models in all of the posters also flashed the *chill* symbol as they looked into the camera. We were unable to determine the history behind the creation of this symbol for the campaign, but unfortunately, to our youthful interviewees it conveyed the exact opposite of abstinence. Most said it reminded them of open legs; they suggested abstaining would have been better expressed with fingers either together or crossed: "That's not the way to chill. You should chill with fingers closed and not open, facing straight up."

Appearance of the models. With the exception of the boys in school uniform in Poster 4, youth said the clothing, facial expressions, and hair of the models in the visuals were typical of sexually active youth, not youth who were abstaining. The models in the campaign were dressed stylishly—presumably in line with the campaign communication objective to create a norm where abstinence was viewed as cool. However, Nairobi is arguably in the midst of a major shift toward Western fashion. The models in the photos were at the leading edge of a trend that not every preteen or young teen had bought into. Furthermore, many high schools in Nairobi have strict dress codes even beyond the use of uniforms, such that certain hairstyles, jewelry, and clothing are forbidden. Thus, our interviewees objected to both female models in Poster 1. Many volunteered that one of them looked too old, and wondered why she was wearing a ring that looked like it might signify engagement or marriage. Her knowing, slightly rebellious expression struck them as "a bit naughty": "She's had sex so many times. There's a way it can show. The way she's smiling. It's a naughty smile." Others said her age was inappropriate for the audience: "She's 23 years old. She doesn't look young. She's wearing a ring. She's married or engaged." The second model on the same poster had a sweet expression, but sported dreadlocks, a style that is growing more popular but was not universally accepted in Nairobi. "This hair looks like rasta," one youth explained. "That hair is not good. It is confusing. It is dirty. Rasta people are dirty." "It's like they're just joking," summarized one youth. "When you say something, you should say it even with the way you dress. I don't agree with the picture."

Similar problems arose with the dress of the girls in Poster 3. Most of the interviewees thought the models' facial expressions were friendly, and conveyed the idea that you could be happy even when you did not follow the crowd. But they also said that all three models were wearing too much makeup ("They should rub off some of the makeup. Be yourself. Be natural. It doesn't help the message." "Makeup is for the big people and the married people, to attract their husbands."). They also opined that two of the girls were indecently dressed. For example, a participant explained, "The green girl could be a problem. [She] could draw attention there [pointing to the model's panties showing above her low-rise jeans]. It's like playing with fire. She attracts them and then when they come she says, 'I've chilled.'" About a second model another participant said, "The one in a red top is too *mtaa* [from the hood]. She looks like *Koinange* women [prostitutes] with the off shoulder."

Youths also objected to the appearance of the boy in Poster 2. The boy looked too young, they said, even as young as eight years old. And his clothing was too old for him: "How can he have so many rings and he's saying *bado siko* ready? Rings are for married people. Children should not be wearing bling." "His dressing style is weird. How can he play basketball dressed like that?" Another commented drily on the poster text (Sex? I won't. I'm abstaining. I'm not ready.), "In a few years he'll be ready. Maybe he has some girlfriends out there." Most also said the model's dressing—the sunglasses, the "bling," the angle of his cap—indicated that he was from *mtaa*, a slang word meaning low-income areas on the eastern side of Nairobi. "He's somebody from *mtaa*. He understands more about sexual activity. That is where they are most exposed more than in the *posh* [middle-income areas]," one participant stated. "I don't like his facial expression," another said, "He seems not so happy about abstaining. It is to show *hoodness*. It's the backstreet places."

With this poster especially we observed a noticeable difference in response between low-income and high-income interviewees. Low-income youth referred to the boy as a *homie*. They explained that *homies* hang out at video places and typically carry knives. "The ad isn't very convincing," one explained, "Just the way he's dressed and the way you know people like rappers, they don't abstain. When you look at him you think of them." In contrast, a few participants in the middle- and/or upper-income groups picked up on the communication objective of the campaign designers, saying they thought the model conveyed the message that, "I can dress cool, but still *chill*," and might resonate with people who dressed as the model did. "It is good that someone who wears like that gives such a message, because it can speak to people who look like that." This group said that you would not find a nice-looking basketball court in *mtaa*; the boy looked more like a *barbie* (person from an upper-income neighborhood). They decided that the second poster was targeting *barbies*, and the first poster was targeting people from the *mtaa*.

Choice of background. Once again, only the setting of Poster 4—a school environment—garnered general approval from the youth we interviewed. Some youth interpreted the fact that the models in

Poster 1 were in a *matatu* as incongruous with an abstinence message. *Matatus* are known for their loud music and sexually explicit videos playing inside. The *matatu* culture in Kenya is associated with promiscuity and drug abuse, and *matatu* touts often prey on, who may be offered free rides in exchange for sexual favors. "Why are they in a *matatu* schoolgirls?" one participant questioned, "*Matatu* may represent the bad people driving them to a bad way. It is somehow hard to abstain in a *matatu*." Others took a more positive view, and said that the models' being in a *matatu* showed that they were resisting temptation. A few conjectured as to why the boy in the poster appeared to be in a school uniform whereas the two girls were in regular clothing. They concluded that the girls had convinced the boy to skip school, an activity that would be in character with the *matatu* context. ("They are not *chilling*. Two ladies and one boy, it looks like they've forced him for the ride.") Some were confused because they could not see the interior of the *matatu* through the tinted windows: "The drawings [photo] of the matatu. They may look nice, but you don't know the meaning of it. What's on the other windows? Because you're saying *chill*, but you don't know what the drawings mean."

But the biggest problem with context was in Poster 3. Some of the youth in our focus groups could not even state the intended meaning of this poster: the promotion of abstinence. They understood the poster to be talking about secondary abstinence. This confusion arose in part because of the furniture on which the girls were sitting. Although the furniture is actually a style of sofa that is occasionally sold by roadside craftsmen in Nairobi, most participants thought it was a bed. The rough, handcrafted look of the sofa appeals more to expatriate than Kenyan tastes, and is rarely seen in Kenyan homes. Youth could not decipher what they were seeing, and ended up confused as to how three attractive girls could be sitting on a bed talking about *chilling*. They wondered if the girls were waiting for someone in a hotel. "It is confusing," one participant complained, "They are sitting on a bed. Why couldn't they sit somewhere else? Or stand?" Participants mused aloud whether the statement, "We know the consequences [of sex]," was an admission by the girls of past mistakes. "I think when they say that they know the consequences," one remarked, "maybe they engaged in sex with boys and then they were left."

The problems with the visuals seem surprising, considering that the strategy team at the public-health organization was mostly Kenyan. One part of the explanation may be that although the campaign was strategized by a public-health agency, creative work—as is typical—was contracted out to Kenyan advertising professionals (Muraya, 2009). The outlook of ad agency personnel is often more aligned with the international advertising industry than with vulnerable rural and urban poor who most need to understand health messages. More generally, adult professionals in Nairobi are urbanites who may tend to value straight talk, assertiveness, and mediated messaging. Low-income youth, even as they emulate Western dress and product preferences, may be more inclined toward an oral cultural approach which is characterized by interpersonal communication and indirect speech. Beyond that, it is easier for advertisers to focus health media campaigns on the urban, upscale market. Urban residents are savvy at interpreting witty social marketed HIV and AIDS health messages that urge audiences to avoid risky behavior. Middle-income urban youth, especially, live in a media-rich environment that makes Western values and fashions far more accessible to them than is the case with rural and low-income youth. A situation analysis conducted by the National Aids Control Council of Kenya (NACC, 2007) regarding the content of various HIV/AIDS interventions among youth emphasized this very point, concluding that not only were messages targeting Kenyan youth often confusing and contradictory, but also low-literate youth and those without access to broadcast media were seldom considered in campaign planning.

Breast Cancer Print Information in Kenya

The *Nimechill* campaign was a well-funded, visually sophisticated, multimedia campaign targeting Nairobi youth across all socioeconomic strata on the issue of sexual behavior. We now turn to consider a very different health communication effort, one that is mostly locally funded, employs a single media channel with unsophisticated graphics, and targets Kenyan women about breast cancer awareness and early detection. The primary researcher on this project was a Kenyan who had worked for years in health information repackaging for rural Kenyan women; the secondary researcher was an American academic who had lived in Kenya for 13 years. To set the stage for this case, we begin with information about women's breast cancer knowledge, attitudes, and early detection behavior in Kenya.

In Kenya, as in much of sub-Saharan Africa, women typically do not seek medical attention for breast cancer until their cancer is very advanced. They may resort to consulting with a doctor about their breasts only if they experience pain or notice a discharge from their nipples (Kenya Breast Health Programme, 2002, 2003). This late presentation for treatment leads to poorer prognosis, less successful treatment, and lower survival rates (World Health Organization, 2009). The situation is undoubtedly due in part to the fact that population-wide routine breast cancer screening is currently unavailable in Kenya; the number of mammography units in the national health-care system is grossly inadequate (Nairobi Cancer Registry, 2006). In fact, although the Kenyan Ministry of Health maintains a network of clinics and hospitals throughout the country that provide basic health services, there is only one operational cancer treatment center in the country, at the large government teaching hospital in Nairobi (Kaberia, 2009).

Clinical breast examination (CBE) by trained health personnel, and breast self-examination (BSE), in which women have knowledge of cancer symptoms and examine their own breasts monthly for changes, are less expensive alternatives for early detection of cancer if women seek medical advice about any abnormalities they find. But for them to be effective, women must have accurate cancer information. Unfortunately, breast cancer has received little emphasis in the big picture of health communication in Kenya. Women's health issues, such as family planning, HIV/AIDS, malaria prevention, nutrition, and child health have garnered much attention in the media, but noncommunicable or lifestyle diseases such as cancer, diabetes, and heart diseases have been notably absent from discussion. Only one local organization, the subject of our study, was actively involved in promoting public awareness of breast cancer at the time we investigated this issue. Their activities attain a high profile during the month of October, when the organization holds the "pink ribbon campaign." Ongoing, mass-media, educational efforts during the rest of the year are mostly limited to print communication materials about breast cancer, that is, brochures and posters in English that are distributed primarily in urban centers around the country.

Little is known about whether these vehicles are conveying to the target audience the messages that their creators intended (Muthoni-Thuo, 2008). Of particular concern is whether the messages in the materials—written in English—are culturally relevant, and whether the target audience has a sufficient level of health literacy to access, understand, and apply the information contained in them. With an estimated 2.8 million Kenyan women nonliterate, an unknown but undoubtedly larger number only marginally health literate, and many who are functionally literate in Swahili or other languages but not literate in English, the materials (again, written only in English) may be inappropriate for large segments of the population.

Perspective of Organizational Personnel

We interviewed the executive director and communication officer of the organization, so as to more fully understand their perceptions about the strengths and weaknesses of their print materials, and to get a better sense of the context in which the materials were used. Founded in 1999 under the auspices of Young Women's Christian Association (YWCA) and later registered as a local nongovernmental organization in 2003, the organization in question remains, so far as we are aware, the only organization in Kenya whose primary focus is breast cancer awareness. It is entirely locally staffed. Interviewees explained that activities of the organization take place mainly in the capital city of Nairobi. Representatives of the organization give breast cancer awareness talks in schools, churches, slums, women's groups, and workplaces. They also make occasional appearances in the Kenyan media. Events typically feature general information on breast cancer and BSE, including distribution of print materials on BSE. A group of survivors, called the "reach to recovery group," has been trained to do outreach to breast cancer patients and their families, and they often give testimonials at breast cancer-related gatherings.

The bulk of materials dissemination is conducted during the October screening activities, when materials are given out in clinics, schools, and women's groups. The brochure, posters, and breast health guide (see descriptions below) are aimed at informing the general public about breast cancer and BSE. The booklets are geared toward helping breast cancer survivors cope with life after diagnosis, and include information on treatment, diet, relationships, and exercise. Getting the materials out is challenging, especially in the rural areas, because doctors and nurses, who are the key volunteer personnel, are busy, and often do not manage to distribute materials provided to them. Materials have been printed in bulk, and the organization appeared to have a huge stock of them when one of us made a site visit.

According to the director, rather than subscribing to the prevailing notion that civil-society groups should only work with the poor, her organization chose to create awareness of breast cancer among people of all socioeconomic statuses. By her description, print information materials are not tailored to specific audience segments.

Posters and other print information were developed in two ways, neither of which involved formative research. First, some were adapted from materials that organization personnel collected from international breast cancer conferences in Europe and the United States. Second, some pieces were developed in-house through consultation with the organization's board of management, which is mainly composed of oncologists. The information officer detailed the progress on specific pieces:

> The breast health guide was also done in 2005; it was created and printed in the same year in a span of a few months. The posters were also done in 2005 and the leaflet series was done last year [2006] in a span of a week actually.... There are thousands of them and they will last us a very long time. We normally give them out in clinics, schools—just put them up.

When asked if target-audience members were involved in the development of materials, the communication officer replied that the organization had relied entirely on the input of the experts on their board, because audience members did not know much about breast cancer. By her admission, the organization also had not seen the need for evaluation of the effectiveness of the print materials after production: "No we have not done a research on our materials; it's something we can do now that you mention it. We can do a study to see if we are doing okay, if we are giving the impact that we needed." She did note that some positive feedback had come in regarding the layout of a BSE brochure and a breast health guide booklet. Women seem to like the easy-to-understand graphics explaining self-examination procedures, and the illustrations of mammography helped remove the myth that the screening involves "a machine that crushes people's breasts." Suggestions had also come in that more visuals should be included on some of the more text-heavy materials. The communication officer admitted to receiving regular requests to have print materials translated from English to Swahili and other Kenyan languages, but said the organization had not yet managed to do so. Organizational personnel listed the overall strengths of the materials as the fact that visuals were photographs of Kenyan breast cancer survivors and volunteers, rather than stock images. They also believed that the information was presented in a way that was understandable to most people, the posters were highly visual, and contact information would make it easy for readers to make further inquiries if they wanted to.

Given that the organization had not conducted formative research on the print pieces with members of the target audience, we set out to get feedback from the women they were trying to reach. We conducted eight focus groups with low- and middle-income, rural and urban Kenyan women. Rural participants were drawn from two locations that were selected because they are populated by two of the largest Kenyan ethnic groups (Kikuyu and Kamba), and because in the capacity of the first author as an information officer who repackages health information for rural residents, she already had a degree of entrée into both communities. Urban participants were drawn from two informal settlements and two middle-class neighborhoods. Finally, for each of the resulting four subpopulations, two focus groups differentiated on the basis of age were conducted, with members of younger groups ranging in age from 20 to 35 years, and of older groups ranging from 36 to 60. Discussants were identified with the assistance of community leaders. Each group had six to seven participants.

We asked women in our focus groups to respond to all of the print materials produced by the organization. First they discussed the pieces separately, then we asked them to comment on overall strengths and weaknesses of the materials. In this way, we were able to assess how close a match existed between audience perceptions of the materials and creator and/or disseminator expectations of the materials. We provide a description of the materials below.

Poster 1: Six faces. This poster was titled, "Anyone Can Get Breast Cancer: Get a Clinical Breast Exam Today." It depicted the faces of six people: one elderly woman, three young women, and two men. The main message on the poster recommended the age at which one should carry out BSE, clinical breast exams, and mammography.

Poster 2: Woman lying down. This poster depicted a young lady lying down with a thoughtful look on her face. The message declared, “She could be a wife…a friend…a mother…a sister or a daughter…but first she is a WOMAN! And she needs you as much as you need her.” The smaller print message on the poster indicated that breast cancer is curable if detected early.

Poster 3: Woman's face. The third poster depicted the face of an attractive young woman smiling back at the camera. It read, “Breast cancer: Early detection can save your life.”

Poster 4: Human eye. This poster was titled, “Breast cancer: Be on the lookout,” and featured a large image of a human eye. A subheading read, “The physical signs and symptoms of breast cancer,” and underneath five symptoms were listed.

Brochure. This was a three-panel, black and white brochure entitled, “Breast self-exam (BSE): A personal plan of action.” The cover presented faces of four young women. The inside of the brochure described the purpose and procedure of performing BSE in textual and pictorial format. It also contained brief information on signs and symptoms of breast cancer and guidelines on other early-detection strategies.

Booklet 1: Coping with diagnosis. This was a 12-page booklet with a photograph of flowers on the cover, entitled, “Breast cancer and you: Coping with a diagnosis.” The booklet gave guidelines on how to cope with physical, emotional, and social issues surrounding a diagnosis of breast cancer.

Booklet 2: Breast care. A second 12-page booklet also featured a photograph of flowers on the cover, and was entitled, “Your guide to breast health care.” The booklet contained information about breast cancer risk factors; BSE procedure; signs and symptoms of breast cancer; and guidelines for other early-detection strategies, including a pictorial of the mammography procedure. The booklet also incorporated testimonials from local breast cancer survivors.

Women's Response to the Print Materials

Our findings reveal four major areas in which the expectations of the organization about their print information materials on breast cancer differed from responses of their target audience: (a) the accessibility of the language used on the materials; (b) the value of the visuals on the materials, especially the posters; (c) the usefulness of contact information regarding the organization; and (d) the segment of Kenyan women for whom the materials are appropriate.

Accessibility of the Language Used

The responses of our participants indicated that language was perhaps a more serious issue than organizational personnel had assumed it to be. The majority of the women in our focus groups, particularly rural women and low-income urban women, found at least some of the materials confusing, and said they would prefer having them in Swahili rather than English. Some older rural women said that they would prefer the materials to be written in their vernacular language. Furthermore, although the communications officer cited the simplicity of the materials as one of their strengths, even middle-income urban women—the most highly educated among our participants—complained that some of the pieces were loaded with too much information and used difficult terms. Women pointed out that terms like "retraction of the nipple" and "lymphedema" were too complicated. Furthermore, they asserted, the font with which one booklet was written was illegible.

Value of the Visuals

The language difficulties participants had with the pieces were compounded by problems with some of the visuals. Although organization personnel were aware that text-heavy brochures were not getting positive feedback from their audience, they assumed that posters, which were primarily visual, were more effective. Our respondents gave a strong, positive affirmation to only one poster, the poster with photographs of six different people. "On this one, I first asked myself, do men and young people too get breast cancer?" One woman recounted. "This makes you more curious to know more. I didn't have this kind of information before. For men we know they are more affected by cancer of the throat."

Visuals on the other posters did not communicate anything about cancer to most participants. "The lady on this other poster you would think she is just sleeping. If you didn't know how to read you wouldn't get the message," one participant remarked about Poster 2. Although some older, urban women said that the use of young models was appealing and could give people hope that cancer was treatable, rural women, both young and old, thought posters with faces of beautiful women overpowered the message about breast cancer. Some even suggested that if a poster of the beautiful woman's face was displayed around local shopping centers, it would end up as wall hangings in people's homes: "If these posters are put up, they will be confused. Others like the young boys might tear it off and have them as pinups on their walls." The poster with the giant eye was especially confusing to most rural women, who thought that persons who could not read were likely to see it as "something to do with eye infections."

When asked what visuals they would like to see on the print pieces, women mentioned familiar-looking faces, or culturally sensitive visuals of breasts. Officials at the organization appeared to be aware of this type of desire, informing us that they used only photographs of cancer survivors and volunteers on their materials. None of our participants, however, picked up that the photographs were of real women who had survived cancer. The visuals that elicited positive responses from our discussants were the graphics that illustrated the steps of BSE, which was also one of the pieces that the communication officer mentioned having received the most positive feedback on. In this case, the visuals were clearly tied to the subject of the piece.

Contact Information

Although participants noticed the contact information for the organization on the posters, most said that the information was not helpful to them. They did indeed want guidance on where to get more information about breast cancer, but they needed a more local, accessible place to turn to. Although contact information for the organization in Nairobi was provided, the organization was geographically out of reach for low-income women, and none except a few middle-income urban women had ever heard of it. Once again, the organization seemed to have identified a legitimate target audience need, but their attempt at meeting that need was not very effective.

Appropriate Target Audience

Finally, the breast cancer organization under study has not, to date, differentiated its messages to any particular audience segment. This is understandable, given the extreme financial constraints under which the organization operates, and our findings highlight the desperate need for additional funding for breast cancer awareness in Kenya. Nevertheless, it is important to recognize that contrary to the stated intentions of organizational personnel that their current print materials should be relevant to a large range of women in Kenyan society, only middle-income, urban women had the literacy level and fluency in English to fully benefit from the materials. With a stock of print materials that they believe will last for years, the best move for the organization may be to focus use of those posters and brochures to reach middle-income women in the cities.

In addition, the print medium itself was also more appropriate for these same women than for other groups. Several urban, middle-income women mentioned that mass-media campaigns had encouraged them to find out more about breast cancer. These women were familiar with breast cancer awareness month. "I saw the campaign once and I was asking my friends, what is this on Lang'ata road?" recalled one woman. "Then I realized it was about breast cancer, and at that time [I realized] it was serious." Middle-income women suggested supplementing existing print materials with messages on vernacular FM stations. In contrast, rural women suggested that face-to-face communication strategies, such as meetings in churches or women's merry-go-round groups, were likely to be the most effective ways to disseminate information about breast cancer and breast cancer detection. (Merry-go-rounds are small, micro-loan societies that, as one participant explained, also serve social and normative functions: "During these merry-go-rounds we attend, we visit one another's house and see what others are doing keeping their homes well. Women teach one another through these experiences.")

The flip side of the good news regarding the perhaps unintentional segmenting of middle-class, urban women by the organization, is that findings from other research indicate that this is the group with the most accurate knowledge of breast cancer among Kenyan women already; they are in least need of breast cancer informational materials (Mutua-Kombo, 2001; Ngimwa, Ocholla, & Ojiambo,1997). In other words, for rural and urban low-income Kenyan women, the situation with respect to access to breast cancer information is actually worse than it appears, because what little breast cancer information is available within the Kenyan environment is not accessible to them. A young, rural participant observed, "Like here I can see on the walls, posters about HIV/AIDS, very many, but not on breast cancer." As a result, their breast health, like their general health, is likely to be poorer than that of their urban, middle-class counterparts, and their cancer mortality rate higher. One locally funded breast cancer organization, however, cannot realistically take up the challenge of creating and implementing a strategy based on the differing informational needs of these women absent additional funding from the government or local or international organizations.

Conclusion

It is important to stress that our interviews cannot be construed as indicating that the problems our interviewees identified caused the *Nimechill* campaign to be ineffective. As a post hoc exercise, our study cannot speak to causation, or even to the larger question of whether the campaign in fact contributed to a larger trend toward later sexual initiation. Nor can our focus-group discussions with rural and urban women, both rich and poor, be considered representative of women in the entire nation of Kenya. Furthermore, we only examined breast cancer print materials used throughout the year by the breast cancer organization. A full picture of current breast cancer communication in the country would have to assess the activities of breast cancer awareness month in October especially.

Even so, our discussions with Nairobi youth about the *Nimechill* abstinence campaign, and with Kenyan women about breast cancer information highlight the concern that when health communication proceeds in the absence of strong formative research its effectiveness may be limited. Although it entails additional initial costs, formative research is critical in identifying target-audience knowledge, attitudes, priorities, preferences, and abilities. In the case of the breast cancer print materials, lack of pretesting resulted in a mismatch between the expectations of the organization and the perceptions of many target-audience members, especially with respect to literacy and language issues. Our observation is that many small, community-based, and faith-based, health-related organizations find themselves in a similar position—they have neither the human nor financial resources to conduct thorough formative research, with the result that problems with their communicative strategies may continue unrecognized for years.

The *Nimechill* abstinence campaign was a much better funded and more extensive effort by an international health agency. Nevertheless, the lesson is similar: even locally targeted campaigns employing local health experts and creative talent can produce messages that are culturally misunderstood if formative research is not thorough. Ultimately, for the youth we interviewed, the first communication objective of the campaign—to create a norm where abstinence was viewed as a cool, smart, and responsible choice by using youth role models who go against the norm to have sex—would not likely have been achieved, because the youthful models were not credible.

Even as each of these health communication efforts targeted a wide swath of Kenyan society—the *Nimechill* campaign addressed urban and peri-urban youth in Nairobi, and the breast cancer informational materials targeted Kenyan women both rural and urban, poor and well-to-do—noticeable differences arose between audience segments in interpretation of some of the materials. With the breast cancer pieces, much of the difference boiled down to varying literacy levels in English. In the case of the *Nimechill* campaign, results may be partly due to the fact that although health-communication campaigns are strategized by public-health entities, creative work is often contracted out to urban-based, Western-educated, Kenyan advertising agency professionals. Low-income youth may not only experience different realities with respect to their lifestyle choices, but they may also be more inclined toward different sorts of communication than their more well-to-do peers. For formative research to accomplish its goals, campaign strategists would be well-advised to identify and work with creative personnel who are conversant with the processes by which target-audience members interpret messages. When this does not happen, rural and poor urban audiences are most disadvantaged, because there is often little shared meaning between them and the professionals crafting the messages.

This brings us to a general concern about health interventions in the developing world. For one of us who is involved on a daily basis with grassroots, rural health promotion efforts, it is evident that there is a mindset among some health communicators that nonliterate and semiliterate persons may not be capable of understanding their own health concerns, or even of identifying what health issues are most pressing for their communities. The natural outgrowth of this perspective is the assumption that there is no need to work with such people extensively to define their own health problems. We want to be clear that we are not imputing this sort of condescending attitude to the creators of the campaigns described in the case studies in this chapter. But the tendency is all too common, and it leads to shallow formative research that is nonparticipatory in fact though it frequently remains participatory in name. Not surprisingly, strategists like this do not pick up on the cultural undercurrents that may militate against acceptance of the mass-media message in the end.

In the case studies we have described in this chapter, the misinterpretation of posters by low-income women and youth suggests a serious problem for the resolution of health disparities. Around one-third of residents in Nairobi are believed to live below the poverty line, and many more just above it (Stifel & Christiaensen, 2005). Most rural residents—even middle-class people—have appreciably lower access to health information than do their urban counterparts (Mutua-Kombo, 2001). Our investigations support the concern expressed by the Kenyan government that many communication initiatives do not reach nonliterate segments of Kenyan youth (NACC, 2007). With vastly different experiences than middle-income youth, Kenyan low-income youth may need to be targeted as a separate audience segment in some health-communication strategies. The same is likely true of Kenyan women.

This leads us to our final, and perhaps most important, point: it is possible that confusion about the meanings in the posters from both campaigns might have been minimized if health communicators had fully engaged women and youth from various socioeconomic strata in campaign planning. Indeed, the situation analysis conducted by the National AIDS Control Council (NACC) of Kenya concluded that, in general, youth were not adequately consulted in the design and planning of communication programs that targeted them. In hierarchically organized societies such as Kenya, youth are often reticent to open up to adults, especially when it comes to sensitive topics such as sex. At the same time their health-related knowledge and attitudes differ markedly from those of the adults who are planning health messaging. Acknowledging both youth and women as experts in their own health concerns, and involving them in every phase of campaign development, is the best way to build shared meanings between them as target-audience members and health communicators. It may take longer, but ultimately, it is a more efficient use of funding than the alternative: the production of ambiguous visuals regarding critical health behaviors that are only appropriate to a slender slice of the audience.

Note

1. At the time this chapter was written, images of three of the posters were available at the following web addresses:
   http://www.psi.org/news/0606d.html (Three Youth in a *Matatu*),
   http://www.psi.org/resources/pubs/kenya-abstinence.pdf (Three Girls),
   http://www.psi.org/resources/pubs/kenya-abstinence.pdf (Three Boys).

Bibliography

African Population and Health Research Center. (2002). *Population and health dynamics in Nairobi's informal settlements: Report of the Nairobi cross-sectional slums survey (NCSS) 2000*. Nairobi, Kenya: Author.

Atkin, C. K., & Freimuth, V. S. (2001). Formative evaluation research in campaign design. In R. E. Rice & C. K. ATkin (Eds.), *Public communication campaigns* (pp. 125–145). Thousand Oaks, CA: Sage.

Central Bureau of Statistics (CBS) [Kenya], Ministry of Health (MOH) [Kenya], Kenya Medical Research Institute (KEMRI) [Kenya], Centers for Disease Control and Prevention (CDC) [Kenya], & ORC Macro. (2004). *Kenya demographic and health survey, 2003: Preliminary report*. Calverton, MD: CBS, MOH, and ORC Macro.

Dutta-Bergman, M. J. (2004a). Poverty, structural barriers and health: A Santali narrative of health communication. *Qualitative Health Research, 14*(8), 1107–1122.

Dutta-Bergman, M. J. (2004b). Primary sources of health information: Comparison in the domain of health attitudes, health cognitions, and health behaviors. *Health Communication, 16*(3), 273–288.

Dutta-Bergman, M. J. (2005). Theory and practice in health communication campaigns: A critical interrogation. *Health Communication, 18*(2), 103–122.

Kaberia, J. (2009). Kenya lacks cancer facilities. Retrieved October 13, 2009 from http://capitalfm.co.ke/news/Kenyanews/.

Kenya Breast Health Programme. (2002). *Women and breast cancer knowledge.* Unpublished report, The Kenya Breast Health Programme, Nairobi, Kenya.

Kenya Breast Health Programme. (2003). *Estimated incidences of breast cancer among women, age 35-60 in Kenya.* Unpublished report, The Kenya Breast Health Programme, Nairobi, Kenya.

Muraya, J. G. (2009). *Culture-sensitivity vs. culture-centeredness in an abstinence campaign in Nairobi, Kenya.* Unpublished master's thesis, Daystar University, Nairobi, Kenya.

Muraya, J. G., Miller, A. N., & Mjomba, L. (2011). An analysis of target audience member interpretation of messages in the Nimechill abstinence campaign in Nairobi, Kenya in the light of high- and low-context communication. *Health Communication, 26*, 516–524.

Muthoni-Thuo, A. (2008). *Breast cancer knowledge, attitudes and information needs of Kenyan women aged 20-60 years in Kikuyu and Matiboni locations.* Unpublished master's thesis, Daystar University, Nairobi, Kenya.

Muthoni-Thuo, A., & Miller, A. N. (2010). An assessment of print media breast cancer informational materials in Kenya. *African Communication Research, 2*, 331–349.

Mutua-Kombo, E. (2001). Women's groups and information provision in rural Kenya. *Library Review, 50*(4), 193–196.

Nairobi Cancer Registry. (2006). Cancer incidence report, Nairobi 2000–2002. Nairobi, Kenya: Author. Retrieved from http://www.healthresearchweb.org/files/CancerIncidenceReportKEMRI.pdf.

National Aids Control Council (NACC). (2007). *Youth HIV/AIDS Communication: A situation analysis.* Nairobi, Kenya: Author.

Ngimwa, P., Ocholla, D., & Ojiambo, J. (1997). Media accessibility and utilization by Kenyan rural women. *International Information and Library Review, 29*(1), 45–66.

Noar, S. M. (2006). A 10-year retrospective of research in health mass media campaigns: Where do we go from here? *Journal of Health Communication, 11*(1), 21–42.

Sansom, C., & Mutuma, G. (2002). Kenya faces cancer challenge. *The Lancet Oncology, 3*(8), 456–458.

Snyder, L. B. (2007). Health communication campaigns and their impact on behavior. *Journal of Nutrition Education and Behavior, 39*(2), S32–S40.

Stifel, D., & Christiaensen, L. (2005, March 21). Tracking poverty in Kenya: Methods and measures. Retrieved from www.worldbank.org/afr/padi/tracking_poverty_kenya.pdf.

World Health Organization. (2009). *Breast cancer screening.* Retrieved from http://www.who.int/cancer/detection/breastcancer/en/.

19

5 A DAY, THE RIO GRANDE WAY: INTERNET INTERVENTION AND HEALTH INEQUITIES IN THE UPPER RIO GRANDE VALLEY

David B. Buller and W. Gill Woodall

Disparities in health and health care are among the most intractable challenges in public health and medicine (The Agency for Healthcare Research and Quality, 2010; Committee on Understanding and Eliminating Racial and Ethnic Disparities in Health Care, 2002). Populations defined by race, ethnicity, socioeconomic status, and geography disproportionately experience chronic disease and have poor access to public-health interventions and medical care to prevent and treat them (Kagawa-Singer & Kassim-Lakha, 2003; Loerzel & Bushy, 2005). Gary Kreps has noted that disparities in health information coincided with disparities in health and health care (Kreps, 2005, 2006). He and others have called for health-communication researchers and practitioners to redouble efforts to close these gaps to reduce these inequalities (Thomas, Fine, & Ibrahim, 2004).

The past decade has witnessed a seismic change in the public-information environment, with the advent of the Internet. From its early days, the potential for the Internet to reach and help populations experiencing health and health-care disparities has been championed. Especially exciting is its ability to span barriers created by distance, disability, and stigma (Griffiths, Lindenmeyer, Powell, Lowe, & Thorogood, 2006); extend services from resource-rich to resource-poor regions; and reduce demands on limited health-care services (Miyasaka, Suzuki, Sakai, & Kondo, 1997). Also, websites with dynamic, multimedia interfaces can assist Americans with low health literacy, and tailor content to the circumstances of populations facing inequalities (Loerzel & Bushy, 2005). This led us to study an Internet intervention for improving dietary behavior among adults in a rural region of the American Southwest in 2000–2004. In this chapter, we review how we tried to address challenges to communicating with a population underserved both by the health-care system and by the telecommunications infrastructure.

Overview of the 5 a Day, the Rio Grande Way Project

The six-county Upper Rio Grande region in Colorado and New Mexico is rural. Its largest cities have fewer than 50,000 residents, and several counties have population densities under six per square mile. The population is comprised of three ethnic groups—non-Hispanic Whites, Hispanics, and Native Americans. Compared to the nation, residents are older, less affluent, and less educated (Buller et al., 2001). Some residents experience low English literacy. Many Native Americans and Hispanics are long-term residents. Several of these characteristics are associated with increased health and health-care disparities (Kagawa-Singer & Kassim-Lakha, 2003; Loerzel & Bushy, 2005).

The goals of the study were to (a) produce a web-based nutrition program promoting increased fruit and vegetable consumption, and (b) test its efficacy for increasing fruit and vegetable intake by adults. Eating fruits and vegetables is beneficial to both genders and at all ages for the prevention of diabetes, cardiovascular disease, and cancer (Boffetta et al., 2010; Carter, Gray, Troughton, Khunti, & Davies, 2010; Lock, Pomerleau, Causer, Altmann, & McKee, 2005; Pomerleau, Lock, & McKee, 2006; van't Veer, Jansen, Klerk, & Kok, 2000).

The project began in 2000, with formative research (Buller et al., 2001; Slater, Buller, Waters, Archibeque, & LeBlanc, 2003); production of the website, *5 a Day, the Rio Grande Way* (Zimmerman, Akerelrea, Buller, Hau, & LeBlanc, 2003); and methods for providing Internet access (Medina, Rivera, Rogers, Woodall, & Buller, 2006). The website was launched in 2002, and 755 adults were recruited to a randomized trial between January 2002 and January 2004 (Buller, Woodall, et al., 2008). The sample was predominately female (88%), contained three ethnic groups (65% Hispanic, 9% Native American, 36% White), and had a wide range of ages and education levels. After pretesting, participants were randomized to an intervention group (n = 380) with immediate access to the *5 a Day, the Rio Grande Way* website or a wait-list control group (n = 375). Posttesting occurred four months after randomization.

Addressing Communication and Health Inequities in Website Design

Initially, we explored the experiences of Valley residents with the Internet and diet. It was apparent that they experienced inequities in the Internet infrastructure. Dietary behaviors of the diverse population were surprisingly similar, although content of local interest and about local people was an essential design element.

Existing Internet Service and Experience

At the outset, Internet technology was just arriving in the Valley, and many residents had limited, if any, access. A collaborating organization, La Plaza Telecommunity Foundation, was constructing a transmission system, providing public access, and training residents to use the Internet in northern New Mexico. In southern Colorado, Internet service was provided by fledgling commercial providers. Another collaborating organization, Valley Wide Health System, operated federally funded clinics for low-income residents, and provided some public Internet access. Throughout the region, the telephone network provided the Internet through dial-up modems, often with speeds no better than 28.8KB. Broadband service was uncommon. Many computers had limited RAM and small, low-resolution monitors. A 1998 survey found substantial inequities in technology skills and experience (Buller et al., 2001), low Internet access (less than half had used the Internet; a third had never used it), and inequities in computer ownership and use (lowest among Hispanic, female, older, and less educated residents and outside towns).

Limitations to Internet service and computer hardware led to several production decisions. A chief goal was to author Web pages that loaded in 10 seconds or less over a 28.8KB connection. This required very small files, and made most streaming video or audio impossible. Instead, messages were carried by static graphics and text. The small monitors meant that Web pages could not have a large amount of text or numerous graphics and still be easily readable.

We considered two structures for the website—a series of sequential content modules (i.e., hierarchical structure), or freely navigational content areas (i.e., node structure where users could link to pages of their choosing). The node structure was selected to balance the necessity to engage users against the ability to deliver specific information in predetermined amounts. The low Internet bandwidth made sequential content modules slow to load, and risked creating unfavorable user experiences. Many commercial websites used a node structure, so it was generalizable. However, it was difficult to define a "use session," and impossible to ensure that all users would be exposed to a basic amount of nutrition content.

Formative Focus Groups With Adult Residents

The website needed to reflect the culture and experiences of the diverse Valley residents (Kreuter & McClure, 2004). To gain insight, we conducted focus groups with Hispanic, White, and Native American adults (Buller et al., 2001). Many participants, regardless of race, ethnicity, age, or socioeconomic status, reported that they and their families ate foods available in local stores, and expressed few concerns about their nutritional value. Residents cited availability and quality of foods at local stores; taste; personal preferences; cost; and lack of skills for planning, preparing, and storing fruits and vegetables as barriers to eating these foods.

We had expected to observe ethnic differences in foods eaten, but participants were reluctant to discuss traditional foods. Instead, they made general comments, such as that food is sacred and not to be wasted; a family celebrates richness and happiness with food; and food shows that a family cares for one another. We expected to hear about unique foods eaten during festivals and celebrations in the region, but again, residents were hesitant to discuss them. Instead, they described preparing typical holiday foods, such as casseroles, hamburgers, turkey, and salads. Several adults did say that people were "feasting" more often, and it was impolite to refuse food, suggesting many frequently overate. Still, most meals were eaten during routine days, not at celebrations, and adults wanted information on typical rather than traditional foods.

Another unexpected revelation was the broad popularity of home gardening. Gardens were a source of fruits and vegetables for many people, and adults were interested in learning how to improve gardening skills.

We investigated dietary information needs. Most wanted advice on vitamins and minerals; effect of food on the body; how to read food labels; what to eat and avoid to maintain health and prevent disease (e.g., cancer, cardiovascular disease, and diabetes); quality of "processed" foods; genetics and diet; emotion and diet; food preparation; weight loss; and fast, simple, and easy recipes. Many adults indicated that they needed information on how to improve children's diets when competing with less healthy fast foods. Adults received health and nutrition information from a variety of sources, e.g., family (especially mothers), friends, media (books, newspapers, magazines, television), traditional and nontraditional health professionals (physicians, nutritionists, herbalists, health-food stores), and formal education. Only a few adults reported using the Internet for dietary information in 2000.

Several findings from the focus groups guided production of the *5 a Day, the Rio Grande Way* website. There was a single regional diet based on foods available in local stores. Traditional foods were consumed infrequently. Instead, residents ate a diet comprised of simple, convenient recipes and a sizable proportion of "fast food." They wanted simple, fast, and easy to prepare foods. The diet of children was important. Gardening advice was a unique opportunity to promote fruits and vegetables.

Website Features and Content for the Multicultural Population

The *5 a Day, the Rio Grande Way* website was based on two theories of health behavior change—social cognitive theory (SCT) (Bandura, 1986) and diffusion of innovations theory (DIT) (Rogers, 2003)—and results of the formative research. Design goals were to be informative, educational, and persuasive by (a) provide necessary skills (SCT principle) and knowledge (DIT principle) on how to eat healthy; (b) convince users that fruits and vegetables were simple and compatible with current diets, easy to try, and advantageous (i.e., increased response efficacy in SCT and positive innovation attributes in DIT); (c) create beliefs that users were capable of changing their diet (i.e., increased self-efficacy in SCT); (d) show that dietary changes are supported by local people, normative, and beneficial to all (i.e., social influence in DIT); (e) motivate people to take action; and (f) link changes to existing dietary habits. From the homepage, users could access sections presenting basic information on fruits and vegetables; instruction on skills for buying, storing, and preparing fruits and vegetables; ways of including them in their own diet and the family diet (particularly with children); gardening and recipes; and a listing of health resources on the Internet related to fruit and vegetable consumption.

The slow Internet speeds meant that we could not adjust the content to make it more personally relevant. This was unfortunate, because such tailoring enhances intervention effectiveness (Loerzel & Bushy, 2005). Instead, we relied on what Kreuter and McClure (2004) described as a "peripheral approach" to enhancing the effectiveness of health communication by conveying images, photographs, recipes, and other information that might appeal to Valley residents. For example, links took users to information on fruits and vegetables in season locally, and a community directory that listed local organizations that sold fruits and vegetables or supplies for gardening. Photographs of local residents were displayed where appropriate. Local residents were encouraged to submit recipes incorporating fruits and vegetables into common dishes; these recipes were analyzed by a nutritionist; and selected recipes were placed on the website with a picture of the local person who submitted it.

Usability Testing of the Website

Usability of website design is essential for creating online communication that appeals to users, and this was especially important when trying to reach populations with less experience with the Internet, lower education, and low health literacy. We conducted rigorous, iterative usability testing during website production (Zimmerman et al., 2003). We began by exploring users' preferences for the website structure (i.e., site map), and concluded with identification of navigational problems and solutions to minimize them.

The website structure was developed with a card-sorting method (similar to Q-sort and concept mapping) that identified residents' frame of reference and information processing strategies. For this, 43 Native American, Hispanic, and non-Hispanic White adults from the Valley sorted 73 nutrition concepts, and through discussion, developed a website map with section titles. Prototype versions of the website were then tested for usability using protocol analysis, in which participants navigated through the screens while talking aloud and having their reactions recorded. First, 35 adults evaluated a static prototype, containing 27 screens and simple line art and text (written at seventh-grade level). While the feedback was largely positive, several navigational problems were identified. After fixing them, a more operational prototype was evaluated by 31 adults and fewer navigational problems were seen.

In both protocol analyses, adults with more computer experience had less positive reactions to the website. They consistently expected more interactivity and tailoring to personal interests. Unfortunately, telecommunications limitations made both impossible. This remained a persistent shortcoming of the *5 a Day, the Rio Grande Way* website.

Addressing Health Inequities in Internet Access

We initiated two procedures for addressing the inequity in Internet access among Valley residents. These included identifying and promoting public-access sites and making available training on basic computer and Internet skills.

Public-Access Site Assessment

A decade ago, before the Internet was available in most American homes, communities had sites open to the public, where computers connected to the Internet could be used for short periods. Some sites were free, while others charged a small fee. In 2000, we initiated a survey of public-access sites in the Upper Rio Grande Valley (Buller et al., 2001; Medina et al., 2006), assessing the computer and Internet technology hours of operation, costs, and availability of technical support and training. Sixty sites were identified, located in public libraries, community colleges, senior centers, health clinics, local schools, and retail venues (e.g., cybercafés, coffee shops). Managers at 48 sites completed the interview.

Some encouraging results emerged. The large number of public-access sites provided a potential means of providing participants access to the Internet. Also, nearly half of public-access computers had broadband Internet access, ran the most recent Web browsers, and permitted plug-in and cookie technology, which enabled use of the *5 a Day, the Rio Grande Way* website.

On the other hand, several barriers to accessing the Internet and using the website were discovered, suggesting that public-access sites would not address all of the Internet inequities in the Valley. For instance, computers at half of the sites had only 56.6K modem speed and many had limited RAM and small, low-resolution monitors, which would impede website performance. Some sites limited the time spent on a computer, so many adults would be unable to spend long periods on the website. By and large, public-access sites had very little technical support to help people use the Internet.

In a 2003 survey of public-access site users, (Medina et al., 2006), we found notable differences in the populations using sites with free (e.g., at libraries) and paid (e.g., at cybercafés) Internet access. Not surprisingly, sites offering free access attracted more low-income and non-White users, while pay sites attracted more affluent and White residents who were relative newcomers to the Valley. However, free sites provided fewer hours of access than pay sites. Thus, sites with free Internet access were more capable of reaching populations with health and health-care disparities than pay sites, but provided less access overall to these groups. Notably, adults with lower incomes were more satisfied with the public-access sites than more affluent adults.

The assessment of public-access sites found that sites were one solution to reducing Internet inequities within the Valley, and we provided a list of these sites to study participants. Unfortunately, most sites provided far from ideal access, and had little help to actually use the Internet. It was not clear that even if help were available most adults would use it. Non-Hispanic Whites were more likely to take computer classes compared to other ethnic groups. Beyond these problems, many residents had to travel substantial distances to reach one of the public-access sites, as most were located in towns.

Training in Basic Computer and Internet Skills

To address anticipated disparities in computer and Internet skills, and to achieve a diverse sample, we employed a group of local people to identify and recruit Valley residents to the trial. Similar community outreach trainers (COTs) had been used by LaPlaza Telecommunity Foundation in a previous study of online diabetes information. COTs were similar to lay health educators, an intervention used successfully to reach and influence diverse populations, especially those experiencing health and health-care disparities (Buller et al., 1999; Earp et al., 2002; Emmons et al., 2005; Ramirez & McAlister, 1988; Taylor, Serrano, Anderson, & Kendall, 2000). Like lay health educators, LaPlazaTelecommunity

Foundation's COTs reported that they reached users through their personal social networks, organizations (e.g., employers and Women, Infants, and Children (WIC) program centers), and events (e.g., local festivals and farmers' markets) in the community. Unique to our needs, COTs often had to demonstrate the Internet to Valley residents before they could use the diabetes website. COTs provided training both one-on-one and through referral to classes at the Foundation's learning center (Buller et al., 2001).

Community-based training in computer and Internet skills appeared to be essential in closing the gaps in Internet access. To improve upon the COT model, we developed formal training materials, and provided COTs with laptop computers for one-on-one trainings. COTs were provided lists of local public-access sites for study participants who did not have Internet access. COTs helped participants establish e-mail accounts so that they could receive project e-mails. COTs introduced them to the study, and showed them how to locate, log on, and use the *5 A Day, the Rio Grande Way* website (if in the intervention group). Over the entire study, 14 adults were hired and trained to be COTs, and nine worked for more than two months on the project, recruiting 755 adults to the study.

The Success of 5 a Day, the Rio Grande Way

The trial evaluating the *5 A Day, the Rio Grande Way* website occurred in 2002–2004. Participants were recruited on an ongoing basis by COTs, and were enrolled for four months, when 479 (62%) completed the posttest.

Benefits of 5 A Day, the Rio Grande Way Website

The *5 a Day, the Rio Grande Way* website did not appear in general to improve diets (Buller, Woodall, et al., 2008). Adults in the intervention group did report eating more servings of fruits and vegetables on a single-item measure at posttest, but servings of fruits and vegetables assessed by food frequency questions did not improve.

The website's ability to improve diets may have been compromised by variable and mostly infrequent use of it (Buller, Woodall, et al., 2008). Only half of the intervention participants actually logged on (mean = 3.3 logons), and on average, participants spent only 22 minutes using the website. Website exposure was essential for acquiring its benefits: Adults who spent more time using the website reported greater improvement in fruit and vegetable intake. Spending time with sections presenting recipes and information on fruits and vegetables in season and health benefits of eating fruits and vegetables had the strongest association with dietary improvements. Website exposure has emerged in other studies as critical for success (Buller, Borland, et al., 2008; Leslie, Marshall, Owen, & Bauman, 2005; Manwaring et al., 2008; Vernon, 2010).

Evidence of Persisting Inequalities

Unfortunately, there were indications that the website may not have successfully closed gaps in healthy dietary practices of Valley residents. First, participants who were lost to follow-up may have had more health and health-care disparities than the residents who completed the posttest and were evaluated. While this loss was similar between experimental groups, older, White participants were most likely to complete the posttest. By contrast, younger participants with no education beyond high school, and older Native Americans who had used the Internet more frequently, were most likely to dropout (Buller, Woodall, et al., 2008).

Second, analyses of who was most likely to use the *5 a Day, the Rio Grande Way* website suggested that it reached groups less likely to experience health and health-care disparities. Use was highest among older (Spearman Rank $r = 0.36$, $p < .01$), more educated ($r = 0.34$, $p < .01$), and Hispanic adults ($r = 0.24$, $p < .01$), adults in small households($r = -0.23$, $p < .01$) and households without children ($r = -0.20$, $p < .01$), shorter-term residents of the Valley ($r = -0.23$, $p < .01$), adults with more Internet experience (time since last use, $r = 0.25$, $p < .01$; time since first use, $r = 0.23$, $p < .05$; confidence in ability to access Internet on own, $r = 0.17$, $p < .01$), and those already eating healthier diets ($r = 0.12$, $p = .02$), and with more influence over the household diet (shopping/meal planning, $r = 0.14$, $p = 0.01$; foods bought, $r = 0.21$, $p < .01$; foods served, $r = 0.21$, $p < .01$). Likewise, time on the website was highest for divorced adults ($F = 2.86$, $p = 0.0151$), adults who were born outside the Valley ($F = 5.33$, $p = 0.0052$), and Native Hawaiian and Asian adults ($F = 5.32$, $p < 0.0001$). A multivariate regression model found that the strongest predictors of website use were being older ($b = 0.82$, $p < .01$), having longer experience using the

Internet ($b = 3.97$, $p =< .001$), and believing that one ate more fruits and vegetables than peers ($b= 5.38$, $p = .02$). The age effect was moderated by comparison to peers (age x social comparison $b = 0.61$, $p < .01$), such that website use was higher among adults 55 years of age and older when they believed at pretest that they ate more fruits and vegetables than other people (eat less than others $M = 9.2$ min., eat same $M = 23.9$ min., eat more $M = 51.2$ min.), but this social comparison among 25 to 34 year olds was associated with lower use (eat less $M = 12.1$ min., eat same $M = 5.0$ min., eat more $M = 7.6$ min.).

These two trends suggest that we did not reach a large number of the adults in the Valley experiencing inequities. Native Americans, those with lower socioeconomic status, and longer-term residents may have inequities, but at the time of this study may not have had sufficient Internet access, interest, or experience to benefit from the website. Thus, this study may have overestimated the extent to which Internet-based intervention can address disparities in dietary behavior in this rural region, by requiring that all participants access and use the Internet. Some subpopulations still lag in broadband and mobile technology use, so evidence from trials on the newest media technologies may continue to overstate its impact. However, more people are on the Internet, some of the disparities in Internet use have disappeared, and wireless data services may help some groups get online. Thus, technological approaches to addressing health inequities might be best when they rely on online services that are well-established and reach larger segments of society.

Lessons Learned From the *5 a day, the Rio Grande Way* Project

Perhaps the most important lessons we learned from this project pertain to participants' and larger communities' reactions, which were surprisingly consistent. First, despite our initial view that there might be hesitancy in using the Internet for nutrition information, participants were very enthusiastic for this online health information. Second, participants had strong preferences for highly interactive and multimedia Web pages, even though broadband access was very limited. Individuals within rural areas were sophisticated in their expectations about Web content, quickly adopted Internet technology, and possessed the necessary computer and Internet navigation skills, even when the technology was relatively new. Third, there was broad, general agreement among participants and community partners that other Internet health information was both needed and wanted. Fourth, if access to the Internet is made convenient in rural areas, information available on websites will be viewed and utilized. Finally, health information on the Internet may be especially valuable to residents of rural areas, where both health information and services are sparse. The Internet can provide an important leverage point for equality of information access to these Americans.

Back to the Future: The Current Potential for Internet Interventions in Rural America

Undaunted, we have embarked on a new project using the Internet to help mothers and adolescent girls make informed decisions about the human papillomavirus (HPV) vaccine. While our experience on the *5 a Day, the Rio Grande Way* project pointed up challenges to reaching populations with Internet and health disparities, the online landscape has changed dramatically. By 2004, commercial Internet providers had a much larger presence in the Valley; more homes were connected; demand for public-access sites had declined; and participants were skilled in using the Internet (Buller, Woodall, et al., 2008). The second half of the decade witnessed further expansion, including the spread of broadband Internet service. While minority, low-income and rural regions still have the lowest access, at least 60% of these populations use the Internet (Rainie, 2010). Recent federal stimulus funding is addressing the lack of broadband access in rural areas, and the Fast Forward New Mexico project is training rural residents in its productive use. In our new project, we plan to recruit mothers and daughters using online advertisements and off-line promotions, and an online enrollment system, rather than relying on interpersonal outreach.

Another exciting development is the invention of mobile smart phones. Recent surveys suggest that a larger proportion of Hispanics and African Americans go online using smart phones than non-Hispanic Whites (Smith, 2010). Smart phones may be a less expensive means of Internet access than personal computers, and may overcome limitations of landline in rural regions. In another recent study, we provided smoking-cessation support over smart phones using mobile data services to a sample of young adult smokers that included 26% minority participants (6% Hispanic).

However, the results of our study showed that Internet access alone may not solve inequities, unless these populations are motivated to use it. We succeeded in elevating use of the website by sending periodic e-mails, at least for a few days (Woodall et al., 2007). Once again, non-Hispanic Whites, older adults, and those with more Internet experience responded most to these e-mail messages, suggesting that reminders reached more of the same health-conscious adults already using the website who possibly were healthier and experienced fewer disparities (Dutta-Bergman, 2004). Unfortunately, the current fragmentation of the Internet may be creating barriers, even for motivated users, as content and telecommunications providers are exercising more control over who can view their content (Anderson & Wolff, 2010).

Still, the Internet is now reaching large segments of the American population, and its advantages over broadcast and print media, especially its ability to span geographic and social distances, continue to present potential avenues to address health and health-care disparities. Further research is needed to identify the most effective components of Internet interventions for populations experiencing disparities, as well as to identify means of ensuring that the online information reaches populations who can most benefit from it.

Acknowledgments

The research discussed in this chapter was funded by a grant from the National Cancer Institute (CA81864). The contents of this chapter are solely the responsibility of the authors and do not represent the official views of the National Cancer Institute.

Bibliography

The Agency for Healthcare Research and Quality. (2010). *The national healthcare disparities report, 2009*. Washington, DC: Author.

Anderson, C., & Wolff, M. (2010, August 17). The web is dead. Long live the Internet. *WIRED*. Retrieved from http://www.wired.com/magazine/2010/08/ff_webrip/all/1.

Bandura, A. (1986). *Social foundations of thought and action: A social cognitive theory*. Englewood Cliffs, NJ: Prentice Hall.

Boffetta, P., Couto, E., Slimani, N., Jenab, M., Wichmann, J., Ferrari, P., Trichopoulos, D.,...Khaw, K.-T. (2010). Fruit and vegetable intake and overall can-

cer risk in the European Prospective Investigation into Cancer and Nutrition (EPIC). *Journal of the National Cancer Institute, 102*(8), 529–537.
Buller, D. B., Borland, R., Woodall, W. G., Hall, J. R., Hines, J. M., Burris-Woodall, P., Cutter, G. R.,...Saba, L. (2008). Randomized trials on Consider This, a tailored, Internet-delivered smoking prevention program for adolescents. *Health Education & Behavior, 35*(2), 260–281.
Buller, D. B., Morrill, C., Taren, D., Aickin, M., Sennott-Miller, L., Buller, M. K., Larkey, L.,...Wentzel, T. M. (1999). Randomized trial testing the effect of peer education at increasing fruit and vegetable intake. *Journal of the National Cancer Institute, 91*(17), 1491–1500.
Buller, D. B., Woodall, W. G., Zimmerman, D. E., Heimendinger, J., Rogers, E. M., Slater, M. D., Hau, B. A.,...Le Blanc, M. L. (2001). Formative research activities to provide Web-based nutrition education to adults in the Upper Rio Grande Valley. *Family and Community Health, 24*(3), 1–12.
Buller, D. B., Woodall, W. G., Zimmerman, D. E., Slater, M. D., Heimendinger, J., Waters, E., Hines, J. M.,...Cutter, G. R. (2008). Randomized trial on the 5 A Day, the Rio Grande Way website, a Web-based program to improve fruit and vegetable consumption in rural communities. *Journal of Health Communication, 13*(3), 230–249.
Carter, P., Gray, L. J., Troughton, J., Khunti, K., & Davies, M. J. (2010). Fruit and vegetable intake and incidence of type 2 diabetes mellitus: Systematic review and meta-analysis. *BMJ, 341*, c4229.
Committee on Understanding and Eliminating Racial and Ethnic Disparities in Health Care. (2002). *Confronting racial and ehtinic disparities in health care.* Washington, DC: National Academic Press.
Dutta-Bergman, M. J. (2004). Primary sources of health information: Comparisons in the domain of health attitudes, health cognitions, and health behaviors. *Health Communication, 16*(3), 273–288.
Earp, J. A., Eng, E., O'Malley, M. S., Altpeter, M., Rauscher, G., Mayne, L., Mathews, H. F.,...Qaqish, B. (2002). Increasing use of mammography among older, rural African American women: Results from a community trial. *American Journal of Public Health, 92*(4), 646–654.
Emmons, K. M., Puleo, E., Park, E., Gritz, E. R., Butterfield, R. M., Weeks, J. C., Mertens, A., & Li, F. P. (2005). Peer-delivered smoking counseling for childhood cancer survivors increases rate of cessation: The partnership for health study. *Journal of Clinical Oncology, 23*(27), 6516–6523.
Griffiths, F., Lindenmeyer, A., Powell, J., Lowe, P., & Thorogood, M. (2006). Why are health care interventions delivered over the Internet? A systematic review of the published literature. *Journal of Medical Internet Research, 8*(10), e10.
Kagawa-Singer, M., & Kassim-Lakha, S. (2003). A strategy to reduce cross-cultural miscommunication and increase the likelihood of improving health outcomes. *Academic Medicine, 78*(6), 577–587.
Kreps, G. L. (2005). Disseminating relevant health information to underserved audiences: Implications of the Digital Divide Pilot Projects. *Journal of the Medical Library Association, 93*(4), S68–S73.
Kreps, G. L. (2006). Communication and racial inequities in health care. *American Behavioral Scientist, 49*(6), 1–15.

Kreuter, M. W., & McClure, S. M. (2004). The role of culture in health communication. *Annual Review of Public Health, 25*(1), 439–455.

Leslie, E., Marshall, A. L., Owen, N., & Bauman, A. (2005). Engagement and retention of participants in a physical activity website. *Preventive Medicine, 40*(1), 54–59.

Lock, K., Pomerleau, J., Causer, L., Altmann, D. R., & McKee, M. (2005). The global burden of disease attributable to low consumption of fruit and vegetables: Implications for the global strategy on diet. *Bulletin of the World Health Organization, 83*(2), 100–108.

Loerzel, V. W., & Bushy, A. (2005). Interventions that address cancer health disparities in women. *Family and Community Health, 28*(1), 79–89.

Manwaring, J. L., Bryson, S. W., Goldschmidt, A. B., Winzelberg, A. J., Luce, K. H., Cunning, D., Cunning, D.,…Taylor, C. B. (2008). Do adherence variables predict outcome in an online program for the prevention of eating disorders? *Journal of Consulting and Clinical Psychology, 76*(2), 341–346.

Medina, U. E., Rivera, M. A., Rogers, E. M., Woodall, W. G., & Buller, D. B. (2006). Internet access and innovation-diffusion in a National Cancer Institute preventive health education project: Telecenters, cybercafes and sociodemographic impacts on knowledge gaps. *Journal of Public Affairs Education, 12*(2), 213–232.

Miyasaka, K., Suzuki, Y., Sakai, H., & Kondo, Y. (1997). Interactive communication in high-technology home care: Videophones for pediatric ventilatory care. *Pediatrics, 99*(1), e1.

Pomerleau, J., Lock, K., & McKee, M. (2006). The burden of cardiovascular disease and cancer attributable to low fruit and vegetable intake in the European Union: Differences between old and new member states. *Public Health Nutrition, 9*(5), 575–583.

Rainie, L. (2010, January 5). Internet, broadband, and cell phone statistics. *Pew Internet & American Life Project*. Retrieved from http://www.pewinternet.org/Reports/2010/Internet-broadband-and-cell-phone-statistics/Report.aspx.

Ramirez, A. G., & McAlister, A. L. (1988). Mass media campaign—A su salud. *Preventive Medicine, 17*(5), 608–621.

Rogers, E. M. (2003). *Diffusion of innovations* (5th ed.). New York, NY: Free Press.

Slater, M. D., Buller, D. B., Waters, E., Archibeque, M., & LeBlanc, M. (2003). A test of conversational and testimonial messages versus didactic presentations of nutrition information. *Journal of Nutrition Education and Behavior, 35*(5), 255–259.

Smith, A. (2010, July 7). Mobile access 2010. *Pew Internet & American Life Project*. Retrieved from http://www.pewinternet.org/Reports/2010/Mobile-Access-2010/Summary-of-Findings.aspx%20on%20October%2013.

Taylor, T., Serrano, E., Anderson, J., & Kendall, P. (2000). Knowledge, skills, and behavior improvements on peer educators and low-income Hispanic participants after a stage of change-based bilingual nutrition education program. *Journal of Community Health, 25*(3), 241–262.

Thomas, S. B., Fine, M. J., & Ibrahim, S. A. (2004). Health disparities: The importance of culture and health communication. *American Journal of Public Health, 94*(12), 2050.

van't Veer, P., Jansen, M. C., Klerk, M., & Kok, F. J. (2000). Fruits and vegetables in the prevention of cancer and cardiovascular disease. *Public Health Nutrition, 3*(1), 103–107.

Vernon, M. L. (2010). A review of computer-based alcohol problem services designed for the general public. *Journal of Substance Abuse Treatment, 38*(3), 203–211.

Woodall, W. G., Buller, D. B., Saba, L., Zimmerman, D., Waters, E., Hines, J. M., Cutter, G. R., & Starling, R. (2007). Effect of emailed messages on return use of a nutrition education website and subsequent changes in dietary behavior. *Journal of Medical Internet Research, 9*(3), e27.

Zimmerman, D. E., Akerelrea, C. A., Buller, D. B., Hau, B., & LeBlanc, M. (2003). Integrating usability testing into the development of a 5 a Day nutrition website for at-risk populations in the American Southwest. *Journal of Health Psychology, 8*(1), 119–134.

20

Comprehensive Care Center and Computer-Mediated Support Group Approaches to Targeting Health Disparities: Two Case Studies

Kevin B. Wright

Medical researchers, clinicians, government agencies (such as the National Institutes of Health), and social scientists (including communication researchers) have long been interested in innovative approaches to improving health disparities in both the health-care system and in the everyday behaviors of individuals (Baker et al., 2007; Kreps, Viswanath, & Harris, 2002; Neuhauser & Kreps, 2003; Treadwell, Northridge, & Bethea, 2007; Viswanath & Kreuter, 2007). In recent years, the shift to patient-centered approaches to health care in hospitals, clinics, and other health-care organizations, has led to innovative changes, such as environmental restructuring, holistic approaches to health care, and the strengthening of social-support resources among patients (Lefkowitz, 2007; Miller & Fins, 1996; Wright & Frey, 2008).

Moreover, advances in computer technology, and the widespread adoption of the Internet for the fulfillment of information-seeking and social-networking needs has led to an increase in searches for health information, as well as the proliferation of online support groups and communities (Walther & Boyd, 2002; Wright & Bell, 2003). These include traditional bulletin board and chat room support communities, as well as communities within social-networking sites such as Facebook. Communication researchers have begun to examine the implications of these innovations, in terms of their impact on individual, health-related variables (i.e., reduction in physical symptoms, stress, depression, etc.), and more recently, scholars have begun to focus on how such innovations can be used to reduce health disparities among members of underserved populations facing health concerns.

This chapter presents two case studies (drawing from published empirical studies by the author) of how two different innovations: (a) a holistic/community, engagement-based approach to cancer treatment, and (b) online support groups for individuals facing health issues, can be used to target health disparities (and potentially used as interventions). Specifically, the chapter begins with a brief overview of health disparities in the United States, and it examines the role of health-communication research and interventions, in terms of potentially reducing health disparities. Next, it highlights a study of a holistic cancer center, as well as linking it to recent research on the impact of environmental restructuring and holistic health-care approaches to cancer treatment. Then, the chapter focuses on a second empirical study of computer-mediated support groups, as well as research on the use of such groups for reducing health disparities. Finally, the chapter discusses the potential of such innovative approaches to health and health care, and directions for future interventions and research.

Health Disparities in the United States

A number of scholars have defined health disparities as "racial and ethnic differences in the quality of health care that are not due to access-related factors or clinical needs, preferences, and appropriateness of interventions" (Smedley, Stith, & Nelson, 2005, pp. 3–4). Researchers have primarily examined health disparities among racial and ethnic minorities, particularly differences due to cultural, literacy, and linguistic factors (Davis, Williams, Marin, Parker, & Glass, 2002; Hornik & Ramirez, 2006; Neuhauser & Kreps, 2008; Williams, 1999), as well as differences in health behaviors, access to health care and/or health insurance, and health outcomes between socioeconomic groups (Dutta, 2007a, 2007b; Dutta, Bodie, & Basu, 2007; Dutta-Bergman, 2004; Williams & Collins, 2001). However, health disparities can also result from differences in the quality of health care due to other social factors, such as age (Black, Ray, & Markides, 1999); sex (Dutta, 2009); rural versus urban residence (Bierman, 2003; Hill, Weinert, & Cudney, 2006; Weinert & Hill, 2005); access to certain types of media (Kreps et al., 2002; Viswanath & Finnegan, 2002; Wright & Bell, 2003); and the degree of social stigma attached to certain illnesses and health conditions (Brashers, Neidig, & Goldsmith, 2004; Mathieson et al., 1996; Weisgerber, 2004). Researchers have discovered various reasons for health disparities, including discrimination, prejudice, poor provider-patient communication, access to quality health information, and wider health-care system practices (Mayberry, Mili, & Ofili, 2002).

For example, in terms of cancer specifically, cancer affects a large number of racial and/or ethnic groups disproportionately in the US vis-à-vis White Americans. In the US, African American men have the highest death rate for lung, colon, and prostate cancer (U.S. Cancer Statistics Working Group, 2007). African American men and women tend to have higher incident rates for cancer than Whites, and both African American men and women tend to have higher death rates for cancer when compared to White patients (U.S. Cancer Statistics Working Group, 2007). While researchers have found that African Americans differ in terms of knowledge about cancer screening and prevention practices (Kreps, 2008), less is known about behavioral and social differences between African Americans and White Americans in terms of cancer survivorship. Behavioral and social factors may account for more than 50–75% of all cancers and mortality rates (McGinnis & Foege, 1993), and many of these factors continue to pose problems post-cancer diagnosis, and may ultimately impact survival rates. In general, lower levels of education and lower income rates among all racial and/or ethnic groups (including Whites) are also related to higher cancer incident rates and lower survival rates (Freeman, 2004; Kreps, 2006). These and other health disparities have been targeted by a variety of researchers in recent years, including communication scholars (Kreps, 2006; Viswanath, 2005).

Role of Communication in Reducing Health Disparities

While researchers have certainly focused on communication issues surrounding the development, implementation, and assessment of mass-media campaigns targeting health disparities among various racial and ethnic groups (see Beaudoin, Fernandez, Wall, & Farley, 2007; Brodie, Kjellson, Hoff, & Parker, 1999; Hornik & Ramirez, 2006; Kreuter & Haughton, 2006; Wray et al., 2004), relatively few studies have dealt with the impact of communication with a holistic, patient-centered care setting and its health outcomes for people facing health disparities, or the impact of computer-mediated support for individuals living with health disparities. While it is important to identify groups who are at risk for poor health due to sociocultural disparities and the conditions producing these disparities, it is also important for health-communication researchers to identify specific contexts or organizations for gaining access to specific individuals facing health disparities, particularly in terms of developing interventions. For example, locating clinics, hospitals, nonprofit organizations, or even Internet-based communities that serve large numbers of individuals facing health disparities can be a starting point for developing and implementing interventions, particularly if the research is able to gain access to one of these settings.

The following sections present two case studies of published communication research that stemmed from efforts to improve the health of individuals facing health disparities. One study is in the area of comprehensive cancer care for low-income individuals, and the second study deals with computer-mediated support groups for individuals who are stigmatized by living with certain health problems. Before discussing the specific case studies, I will present a brief overview of literature dealing with these communication approaches to health care in general, as well as how they can be used to confront health disparities specifically. In the final section of the chapter, I will examine some of the limitations of these approaches, and provide suggestions for future interventions targeting individuals facing health disparities.

Comprehensive Cancer Treatment, Communicative Involvement, and Health Outcomes

Comprehensive Cancer Treatment

Lefkowitz (2007) described comprehensive cancer treatment as approaches to meeting individual health-care needs through both traditional provider-patient treatments as well as alternative approaches to health care, such as complimentary therapies, support groups, and social activities. Comprehensive cancer care often includes environmental restructuring (i.e., designing the environment of a cancer clinic so that it appears more like a comfortable living space, as opposed to a sterile, clinical environment), the use of complimentary therapies (such as therapeutic massage, guided relaxation exercises, etc.), traditional face-to-face support groups, and social outings for patients (such as group activities, outings, etc.). Such initiatives are part of a larger movement in public-health promotion in which scholars have advocated and tested more holistic or ecological approaches to treating illness (see Franklin & Fins, 1996; Nies & Kershaw, 2002; Spiegel, 1993; Stokols, 1996). This more holistic type of cancer care typically addresses psychosocial health issues in addition to purely biomedical approaches to cancer care. While many low-income cancer patients often lack access to cancer care that focuses on psychosocial issues, some comprehensive care centers (such as the one described in the following research case study) offer low-cost or no-cost treatment to individuals who lack adequate health insurance, in an effort to better serve the community.

However, relatively few studies have examined the impact of comprehensive care center approaches (i.e., such as environmentally restructured diagnosis and treatment centers, support groups, stress-reduction treatment, therapeutic massage, spiritual counseling, etc.) on physical and mental health outcomes for people living with cancer and other health problems, or particularly, how these innovations may impact individuals living with health disparities (see Chin, Walters, Cook, & Huang, 2007; Cooper, Hill, & Powe, 2002 for exceptions). Unfortunately, many individuals facing such disparities do not have access to facilities that offer holistic cancer treatment, due to lack of health insurance or limited insurance options. Yet, given the increased social support, opportunities for stress reduction, and the other features associated with these centers (such as a more comfortable environment and opportunities to interact with people living with similar health issues), they could potentially have a positive impact on people who have traditionally had limited access to this type of health care.

Communicative Involvement

One aspect of comprehensive cancer treatment that is of interest to health-communication scholars is the ways in which center-sponsored activities, such as support groups, social outings for patients, and complementary health-care services (guided relaxation, yoga, etc.) may have health benefits for patients through increased social activity. A number of studies have demonstrated that increased patient communicative involvement in treatment was associated with positive health outcomes, such as increased patient satisfaction with health care, better adherence to treatments, reduced stress, and lower symptom distress (see Brown, Stewart, & Ryan, 2003; Golin, DiMatteo, & Gelberg, 1996; Greenfield, Kaplan,

& Ware, 1985; Nussbaum, Baringer, & Kundrat, 2003; Smith, 2002; Stewart et al., 2003). According to Neuhauser and Kreps (2008), "communication strategies are essential to help people understand ways to prevent and manage cancer" (p. 366). Increased social involvement in cancer prevention, treatment, and survivorship has been linked to positive changes in health behavioral risk factors and increased adherence to cancer treatment (Hiatt & Rimer, 1999; Kreps, 2006).

A Comprehensive Cancer Center Study

To illustrate how the impact of environmental restructuring and communicative involvement within a health-care setting has been assessed, I will discuss the context and findings of a study by Wright and Frey (2008) of a cancer center in Memphis, Tennessee. In terms of health disparities, Memphis ranked in the bottom 10% of the country, in terms of cancer mortality rates at the time the study was conducted. Over 50% of the population of Memphis is African American, and the city has had a long history of problems with racial discrimination, lower incomes among large segments of the African American population, and limited access among African Americans to quality health care.

The cancer center in the Wright and Frey (2008) study was environmentally restructured to resemble a cozy living room-type environment, with comfortable furniture, contemporary décor, and artwork donated by prominent community members. The center also created space for patient interaction, including a large library and reading area, kitchen, Internet access, and space for one-on-one and group meetings between patients. In addition, the center attempted to provide comprehensive cancer care, and patients were encouraged to participate in a variety of center-sponsored activities, in addition to regular appointments with oncologists, for chemotherapy treatments, and for other medical procedures. These activities were predominately social in nature, and included a variety of support groups, complimentary therapies, relaxation therapy, and social-group activities (such as picnics and other outings with fellow cancer center patients). The development of this center was an attempt to provide lower-income individuals in the community with cancer care that addresses psychosocial needs in addition to physical health care. However, the cancer center administrators had not conducted a study to assess the potential impact of the services provided by the center (i.e., support groups, complementary therapies, etc.) on health outcomes, and they asked us to conduct a survey of participants.

Wright and Frey (2008) surveyed 95 cancer patients at the center. Nearly 25% of the sample consisted of low-income African American individuals, while the majority of the sample consisted of lower-income White patients from the surrounding community. The average age of the sample was 59.81. The sample also consisted of individuals who were living with a number of different types of cancer, including breast cancer, lymphoma, colon cancer, lung cancer, and prostate cancer.

The key independent variables in the study were willingness to communicate about health, sources of health information, and generalized perceptions of the cancer environment (e.g., the extent to which the center felt more like an institution, a home, a family, etc.). The outcome measures included perceived health status, perceived helpfulness of the various cancer center activities (i.e., support groups, social outings, etc., and patient satisfaction with the center). In addition, a physical health status measure assessed by a physician was measured as a mediating variable (since physical health status may affect an individual's ability to participate in certain activities that require greater mobility, energy levels, etc.).

Using regression analysis, the authors found that higher willingness to communicate about health predicted greater information-seeking from the various sources of information within the center (i.e., providers, other patients, and staff), increased perceptions of helpfulness of activities, increases in self-perceptions of health status (i.e., coping, overall functioning, etc.), and increased satisfaction with the center. Moreover, perceptions that the center was more like a home and/or family were associated with similar increases in these outcome variables. As expected, physical health status significantly affected the ability of patients to engage in center-related activities and health information-seeking behaviors.

This study provides at least some initial self-reported evidence that environmental restructuring, holistic cancer care, and communicative involvement in care may predict patient satisfaction and perceived physical health status among lower-income and minority cancer patients. However, a patient's willingness to communicate about their health, perceptions of the environment, and physical health status appears to influence the degree to which people are able to participate in activities and seek health information with others.

Yet, these findings may inform future interventions that attempt to use a comprehensive approach to cancer care among individuals facing health disparities. Although the Wright and Frey (2008) study did not find racial differences in terms of willingness to communicate about health, health information-seeking behaviors, and participation in center activities, these variables were negatively associated with age. Future interventions may consider targeting older patients, and finding ways to increase communication and center community involvement among these individuals using a larger sample of lower-income and minority individuals (as well as other groups that face health disparities). In addition, future studies should consider finding ways to increase willingness to communicate about health among populations facing health disparities, and comprehensive cancer treatment centers should consider ways to promote increased involvement among patients with poorer physical health status (perhaps by altering activities so that people with reduced mobility, lower energy levels, etc., can participate more easily).

The next case study involves the issue of stigma related to health issues, and the use of computer-mediated support groups as a means of increasing social support for individuals living with these health issues, and as a potential means for them to reduce social isolation and discrimination due to features of the computer-mediated environment.

Computer-Mediated Social Support and People Facing Health Disparities

In recent years, there has been a considerable amount of research within the social sciences and the health disciplines regarding the potential of the Internet and other computer-mediated technologies to aid individuals with health disparities in finding useful information about their health condition, as well as increased opportunities for social networking, including increasing social-support network members (who are living with the health condition), sharing experiences, and finding behavioral role models within computer-mediated support communities (Walther & Boyd, 2002; Weisgerber, 2004; Wright & Bell, 2003).

Prevalence of CMC Support

In terms of computer-mediated support in particular, over 28% of Internet users participate in support groups online (Pew Internet & American Life Project, 2005). A 2008 study estimated that 1 in 3 Americans used some type of online social network when dealing with health issues (Sarasohn-Kahn, 2008). In addition, Darcy and Dooley (2007) found that individuals coping with eating disorders reported using online eating disorder groups more than 8 hours a week.

Advantages and Disadvantages of Computer-Mediated Social Support

The Internet provides several other advantages over traditional face-to-face support groups for people facing health disparities, including broad reach, 24-hour availability, increased interactivity with others, international participation, multimedia capabilities, and greater anonymity (Cline & Haynes, 2001; Forkner-Dunn, 2003; Neuhauser & Kreps, 2008; O'Conner & Madge, 2004; Wright, 2002, 2009). Previous research has also identified many other advantages of computer-mediated support groups, including:

1. A sense of anonymity, which tends to reduce the social stigma associated with many health conditions, and allows participants to be evaluated on their contributions to the group rather than on their physical appearance or disabilities (Madara, 1997; Wright, 2000),
2. Convenience,
3. Increased opportunities for social comparisons, and
4. Access to in-depth and diverse information about a health condition.

The increased interactivity of computer-mediated support groups, in terms of the ability of participants to exchange information, emotional support, and other health-related messages (Neuhauser & Kreps, 2003; Rice, 2001), may result in messages that are more influential than one-way passive messages (i.e., a health-related, public-service announcement), in terms of influencing health behaviors. However, computer-mediated support groups have also been found to suffer from a variety of problems, such as off-topic discussion, flaming, the use of the groups by privateers (i.e., people who inappropriately use the groups to sell medical products or services to participants), sporadic membership, and other negative aspects, such as deception and cyberstalking (Wright, 2002, 2009; Wright & Bell, 2003).

Computer-Mediated Support and Health Outcomes

Decades of social-support research have linked improved social support to a variety of positive physical and mental health outcomes, such as increased health literacy, reduced stress, increased coping abilities, lower depression levels, greater adherence to treatment, and reduced morbidity and mortality rates (Barrera et al., 2002; Rains & Young, 2009). Communication researchers have contributed to this body of literature by providing a broader understanding of social-support processes, by examining characteristics of supportive messages leading to positive reappraisals and coping behavior (Burleson & MacGeorge, 2002; Goldsmith, 2004), and the relationship between social-support processes and health outcomes among individuals living with a variety of health conditions (Braithwaite, Waldron, & Finn, 1999; Ford, Babrow, & Stohl, 1996; Query & Wright, 2003), and have demonstrated how several key aspects of supportive relationships play critical roles in both psychological and physical health.

A growing number of studies have measured psychological and physical health outcomes related to participation in computer-mediated support groups (Barrera et al., 2002; Rains & Young, 2009; Wright, 2000). There is empirical evidence that these groups provide a wide variety of health benefits for users (such as individuals with cancer, diabetes, and substance-abuse problems), including reduced stress, increased positive coping, increased quality of life, increased self-efficacy in terms of managing one's health problems, reduced depression, and increased physical health benefits (Beaudoin & Tao, 2007; Fogel, Albert, Schnabel, Ditkoff, & Neugut, 2002a, 2002b; Gustafson et al., 2005; Houston, Cooper, & Ford, 2002; Jones et al., 2008; Owen et al., 2005; Rains & Young, 2009; Shaw, Hawkins, McTavish, Pingree, & Gustafson, 2006; Wright, 1999, 2000).

Wright (2000) provided evidence that participation in a computer-mediated support community for older adults predicted reduced stress and increased coping skills. Owen et al. (2005) discovered that participation within an online support group for cancer patients predicted higher quality of life and reduced stress. In addition, in a recent meta-analysis of twenty-eight computer-mediated support-group studies, Rains and Young (2009) found that participation in computer-mediated social-support groups led to increased social support, decreased depression, increased quality of life, and increased self-efficacy in terms of managing health conditions. Finally, two decades of research on the Comprehensive Health Enhancement Support System (CHESS) has demonstrated positive health outcomes for women with breast cancer and people with HIV/AIDS, such as reduced stress, fewer health-care visits and hospitalizations, and fewer incidents of disease symptoms (Gustafson et al. 1999; Gustafson, Julesberg, Stengle, McTavish, & Hawkins, 2001).

Computer-Mediated Support and Health Disparities

According to Barrera et al. (2002), assessing the efficacy of computer-mediated support groups in empirical studies and interventions may ultimately "help efforts to eliminate disparities in health that exist because of inequities in people's access to resources" (p. 638). Donelle and Hoffman-Goetz (2008) found that computer-mediated groups led to increases in social support and health literacy among African American women.

The CHESS projects have been shown to be effective for both low-income, low-education participants as well as higher-income, higher-education level individuals (Gustafson et al., 1993; Gustafson et al., 1999; Gustafson et al., 2001; Gustafson et al., 2005; Shaw et al., 2006). These authors found that African American women used the CHESS groups as frequently as affluent White women. However, African Americans tended to use the groups more for information, as opposed to interactive features, such as discussion groups. It appears, then, that computer-mediated support groups may be well suited for groups facing health disparities, in terms of fulfilling support-seeking needs regarding health-care problems.

However, in general, relatively little is known about how minority groups and other populations facing health disparities use computer-mediated support groups. Buller et al. (2001) found lower Internet ownership and use by Hispanics compared to Whites. Fogel et al. (2002a, 2002b, 2003) found that while African Americans, Hispanics, and Asian Americans tend to use the Internet less than Whites, their Internet use was associated with greater ability to talk with someone about problems and to obtain other types of social support. Researchers have also found that access to traditional, face-to-face support networks (such as close family ties) often deteriorates among members of certain low-income and minority groups due to factors such as unemployment, transience, and substance abuse (Roschelle, 1997). Weinert and Hill (2005) found that rural women (including a high percentage of minorities and individuals from lower socioeconomic backgrounds) using a computer-mediated support intervention had lower levels of depression and higher self-reported management of day-to-day chronic illness symptoms than a control group of similar rural women living with chronic illness. Irizarry, Downing, and West (2002) and Wright (2000) found that computer-mediated support communities were helpful, in terms of aiding isolated older adults facing health concerns in becoming better connected with other individuals with similar circumstances.

A Study of CMC Support Groups as Weak-Tie Support Networks: Implications for Individuals Facing Health Disparities Due to Stigma and Age

Wright, Rains, and Banas (2010) examined characteristics of weak-tie support network preference among members of health-related, computer-mediated support groups. Weak-tie support networks typically consist of individuals who are not interpersonally close, but with whom people interact in a limited way or within highly specific contexts. Computer-mediated support groups have been found to exhibit many of the dimensions (and oftentimes advantages) of weak-tie network support, including lower risk of self-disclosing sensitive health information, increased objectivity regarding health information, greater heterogeneity of health information, and less stringent role obligations (see Wright, 2000, 2002; Wright & Bell, 2003; Wright & Miller, 2010). Given the important role played by computer-mediated support groups in providing access to weak ties, understanding preferences for weak ties (including age, sex, and racial and ethnic differences) may facilitate the development of computer-mediated, support-group interventions.

The sample for the Wright et al. (2010) study included 191 individuals from 40 different health-related support groups (including groups for breast cancer, prostate cancer, HIV/AIDS, and eating disorders). Although the sample included a relatively small number of racial and ethnic minorities (15% of the sample), a large proportion of the sample consisted of older adults (who, as we have seen, often face health disparities). The researchers targeted computer-mediated support groups that varied in terms of the social stigma attached to the health issue focus of the individual groups. For example, HIV/AIDS groups were chosen due to the highly negative social stigma associated with this particular disease (with eating disorders and cancer having a less negative social stigma connected with them). The social stigma associated with illness has been linked to a preference for weak-tie support (Adelman, Parks, & Albrecht, 1987; Wright & Bell, 2003). The researchers contacted moderators for these targeted, online support groups, and asked them for permission to distribute an online survey link to members of the community via e-mail.

The results of the Wright et al. (2010) study indicated that age was negatively associated with a preference for weak-tie support, and that participants not facing a terminal illness were more likely than those facing a terminal illness to prefer weak-tie support. Two factors contained in the weak-tie/strong-tie network support scale were significant predictors of lower perceived stress. These included the objective utility function (e.g., the tendency of weaker ties to give more objective social support than closer ties, such as family members and close friends), and the lower perceived risk dimension of the weak-tie support preference (e.g., the tendency of weaker ties to be removed from a person's primary social network often leads to less risk in terms of disclosing sensitive health information.

The researchers coded for stigmatized health issues, and they found that individuals who were in support groups for health conditions with a more negative stigma attached to the disease or condition, such as HIV/AIDS, alcoholism, bulimia, and mental illness, tended to have higher levels of stress, and they were more likely to have a stronger preference for weak-tie social support. Wright et al. (2010) concluded that the weak-tie functions of the computer-mediated support groups enabled participants to more safely discuss health concerns than with closer ties, due to judgments, risk of self-disclosure, and negative implications of stigmatization associated with these health conditions. The ability of the

Internet to connect people with similar health concerns (despite being geographically dispersed) appears to facilitate weak-tie relationships, motivated by the need to avoid judgment and to safely discuss health concerns in a more objective manner than may be possible with traditional close-tie support-network members.

There are several implications of the findings from this study for developing computer-mediated support interventions for individuals facing health disparities. These implications were shared by the authors with the leadership and members of several of the support groups who participated in the study, in an effort to identify ways in which the leadership could add features or encourage certain practices within their respective online communities that could potentially lead to more favorable outcomes (e.g., increased involvement, greater satisfaction with the community, and positive health outcomes).

First, the negative social stigma attached to some racial and ethnic minority groups (coupled with the negative social stigma associated with certain illnesses and health conditions) may be important variables, in terms of predicting weak-tie support-network preference within computer-mediated support-group interventions. Older individuals appear to prefer stronger ties, particularly when facing a terminal illness, and so age and illness severity may be important variables to assess when deciding whether to recommend a computer-mediated support group or a close friend and/or family support intervention. The findings of the Wright et al. (2010) study demonstrated that weak ties may exhibit more emotional detachment than strong ties, and, consequently, they may be more adept at providing objective feedback about health problems. In other words, weaker ties, such as online support-group members, may serve as a functional alternative for health information and support for individuals facing health disparities who are uncomfortable discussing health issues with stronger ties.

The Potential of These Approaches for Future Interventions and Research for Individuals Facing Health Disparities

The communication approaches to improving health care illustrated in these case studies obviously have a number of theoretical and methodological limitations (consult the discussion sections of each published study for details). However, they do provide a starting point for thinking about how similar approaches could be used in future studies to potentially improve the lives of individuals living with health disparities. Both may offer some promise in terms of reducing health disparities and improving health outcomes.

First, comprehensive approaches to cancer care and other health problems appear to have at least some impact on the psychosocial well-being and physical health behaviors and/or outcomes of patients. Traditionally, these approaches have tended to be applied to health-care settings where patients are more affluent, well-educated, and not a member of a racial and/or ethnic minority group. Future interventions utilizing such approaches should consider the varied social and cultural factors (e.g., including literacy, income, interpersonal communication preferences, media usage, etc.) when targeting such populations, and when making key decisions about environmental restructuring, the appropriateness of various alternative and/or complementary treatment options, and the types of social activities and social-support features that would be ideal for the at-risk population. More preliminary research needs to be conducted on perceptions of environmental restructuring alternatives (e.g., types of designs, physical layout, etc.) among members of specific at-risk groups (perhaps through focus groups or pilot studies). Moreover, researchers would benefit from a more in-depth understanding of preferences for types of health messages, support-group characteristics (e.g., online versus face-to-face, provider-led versus peer support, etc.), and activities from the standpoint of individuals facing different types of health disparities (e.g., by race and/or ethnicity, age, sex, etc.).

In terms of computer-mediated support interventions for people living with health disparities, initial research (i.e., the CHESS studies and others) has demonstrated positive psychosocial and physical health outcomes for members of various racial and/or ethnic minority groups and for older individuals. Given the proliferation of computer-mediated support groups on the Internet, and a shrinking gap in the digital divide (due to lower costs associated with owning a computer and Internet access), it is likely that more members of groups who have traditionally faced health disparities will turn to the Internet for health information and social support. Future research would benefit from capitalizing on what is known about the characteristics of these populations, in terms of their behavioral risk factors, ethnomedical belief systems, new-media usage patterns, and other relevant variables, in terms of designing computer-mediated support interventions that are culturally appropriate. This, combined with increased research on the nature of computer-mediated support, may help researchers and clinicians design online support interventions that are more likely to produce positive health outcomes and reduce health disparities.

Bibliography

Adelman, M. B., Parks, M. R., & Albrecht, T. L. (1987). Beyond close relationships: Support in weak ties. In T. L. Albrecht, M. B. Adelman & Associates (Eds.), *Communicating social support* (pp. 126–147). Newbury Park, CA: Sage.

Baker, D. W., Wolf, M. S., Feinglass, J., Thompson, J. A., Gazmararian, J. A., & Huang, J. (2007). Health literacy and mortality among elderly persons. *Archives of Internal Medicine, 167*, 1503–1509.

Barrera, M., Glasgow, R. E., McKay, H. G., Boles, S. M., & Feil, E. G. (2002). Do Internet-based support interventions change perceptions of social support?: An experimental trial of approaches for supporting diabetes self-management. *American Journal of Community Psychology, 30*, 637–654.

Bierman, A. (2003). Climbing out of our boxes: Advancing women's health for the twenty-first century. *Women's Health Issues, 13*, 201–203.

Brashers, D. E., Neidig, J. L., & Goldsmith, D. J. (2004). Social support and the management of uncertainty for people living with HIV or AIDS. *Health Communication, 16*, 305–331.

Brown, J. B., Stewart, M., Ryan, B. L. (2003). Outcomes of provider-patient interaction. In T. L. Thompson, A. M. Dorsey, K. Miller, & R. Parrott (Eds.), *Handbook of Health Communication* (pp. 141–161). Mahwah, NJ: Lawrence Erlbaum Associates.

Cooper, L. A., Hill, M. N., & Powe, N. R. (2002). Designing and evaluating interventions to eliminate racial and ethnic disparities in health care. *Journal of General Internal Medicine, 17*, 477–486.

Darcy, A., & Dooley, B. (2007). A clinical profile of participants in an online support group. *European Eating Disorders, 15*, 185–195.

Davis, T. C., Williams, M. V., Marin, E., Parker, R. M., & Glass, J. (2002). Health literacy and cancer communication. *CA: A Cancer Journal for Clinicians, 52*, 134–149.

Dutta, M. (2007a). Communicating about culture and health: Theorizing culture-centered and cultural-sensitivity approaches. *Communication Theory, 17*, 304–328.

Dutta, M. (2007b). *Communicating health: A culture-centered approach*. London: Polity Press.

Dutta, M. (2009). Health communication: Trends and future directions. In J. C. Parker & E. Thorson (Eds.), *Health communication in the new media landscape* (pp. 59–92). New York: Springer.

Dutta, M., Bodie, G. D., & Basu, A. (2007). Health disparity and the racial divide among the nation's youth: Internet as an equalizer? In A. Everett (Ed.), *Learning race and ethnicity: Youth and digital media* (pp. 175–197). Cambridge, MA: MIT Press.

Dutta-Bergman, M. (2004). Poverty, structural barriers and health: A Santali narrative of health communication. *Qualitative Health Research, 14,* 1–16.

Fogel, J., Albert, S. M., Schnabel, F., Ditkoff, B. A., & Neugut, A. L. (2002a). Use of the Internet by women with breast cancer. *Journal of Medical Internet Research, 4,* E9.

Fogel, J., Albert, S. M., Schnabel, F., Ditkoff, B. A., & Neugut, A. L. (2002b). Internet use and social support in women with breast cancer. *Health Psychology, 21,* 398–404.

Fogel, J., Albert, S. M., Schnabel, F., Ditkoff, B. A., & Neugut, A. L. (2003). Racial/ethnic differences and potential psychological benefits in use of the Internet by women with breast cancer. *Psychooncology, 12,* 107–117.

Fox, S. (2005). Health information online: Technical report from the Pew Internet and American Life Project. Washington, DC: Pew Research Center.

Freeman, H. P. (2008). Poverty, culture, and social injustice: Determinants of cancer disparities. *CA: A Cancer Journal for Clinicians, 54,* 72–77.

Goldsmith, D. J. (2004). *Communicating social support.* New York: Cambridge University Press.

Greenfield, S., Kaplan, S., & Ware, J. Jr. (1985). Expanding patient involvement in care: Effects on patient outcomes. *Annals of Internal Medicine, 102,* 520–528.

Gustafson, D. H., Hawkins, R., Boberg, E., Pingree, S., Serlin, R. E., Graziano, F., & Chan, C. L. (1999). Impact of a patient-centered, computer-based health information/support system. *American Journal of Preventive Medicine, 16,* 1–9.

Gustafson, D. H., Julesberg, K. E., Stengle, W. A., McTavish, F. M., & Hawkins, R. (2001). Assessing costs and outcomes of providing computer support to underserved women with breast cancer: A work in progress. *The Electronic Journal of Communication, 11,* http://www.cios.org/EJCPUBLIC/011/3/011311.HTML.

Gustafson, D. H., McTavish, F. M., Stengle, W., Ballard, D., Hawkins, R., Shaw, B. R., Jones, E., Julesberg, K., McDowell, H., Wei, C. C., Volrathongchai, K., & Landucci, G. (2005). Use and impact of ehealth system by low-income women with breast cancer. *Journal of Health Communication, 10,* 195–218.

Hiatt, R. A., & Rimer, B. K. (1999). A new strategy for cancer control research. *Cancer Epidemiological Biomarkers Prevention, 8,* 957–964.

Hornik, R. C., & Ramirez, A. S. (2006). Racial/ethic disparities and segmentation in communication campaigns. *American Behavioral Scientist, 49,* 868–884.

Houston, T. K., Cooper, L. A., & Ford, D. E. (2002). Internet support groups for depression: A 1-year prospective cohort study. *American Journal of Psychiatry, 159,* 2062–2068.

Irrizary, C., Downing, A., & West, D. (2002). Promoting modern technology and internet access for under-represented older populations. *Journal of Technology in Human Services, 19,* 13–29.

Jones, M., Luce, K. H., Osborne, M. I., Taylor, K., Cunning, D., Doyle, A. C., et al. (2008). Randomized, controlled trial of an Internet-facilitated intervention for reducing binge eating and overweight in adolescents. *Pediatrics, 121,* 453–462.

Kreps, G. L. (2006). Communication and racial inequities in health care. *American Behavioral Scientist, 49*, 760–774.

Kreps, G. L., Viswanath, K., & Harris, L. M. (2002). Advancing communication as a science: Research opportunities from the federal sector. *Journal of Applied Communication Research, 30*, 369–381.

Lefkowitz, B. (2006). *Community health centers: A movement and the people who made it happen*. New Brunswick, NJ: Rutgers University Press.

Mayberry, R. M., Mili, F., & Ofili, E. (2002). Racial and ethnic differences in access to medical care. In T. A. LaVeist (Ed.), *Race, ethnicity, and health: A public health reader* (pp. 163–197). San Francisco, CA: John Wiley & Sons.

McGinnis, J. M., & Foege, W. H. (1993). Actual causes of death in the United States. *Journal of the American Medical Association, 270*, 2207–2212.

Miller, F. G., & Fins, J. J. (1996). A proposal to restructure hospital care for dying patients. *The New England Journal of Medicine, 334*, 1740–1742.

Monnier, J., Laken, M., & Carter, C. L. (2002). Patient and caregiver interest in internet-based cancer services. *Cancer Practice, 10*, 305–310.

Neuhauser, L., & Kreps, G. L. (2003). Rethinking communication in the e-health era. *Journal of Health Psychology, 8*, 7–22.

Neuhauser, L., & Kreps, G. L. (2008). On-line cancer communication: Meeting the literacy, cultural and linguistic needs of diverse audiences. *Patient Education and Counseling, 71*, 365–377.

Rains, S. A., & Young, V. (2009). A meta-analysis of research on formal computer-mediated support groups: Examining group characteristics and health outcomes. *Human Communication Research, 35*, 309–336.

Roschelle, A. (1997). *No more kin: Race, class, and gender in family networks*. Thousand Oaks, CA: Sage.

Sarasohn-Kahn, J. (2008). *The wisdom of patients: Health care meets online social media*. Available at http://www.chcf.org/documents/chronicdisease/HealthCareSocialMedia.pdf. Accessed August 21, 2008.

Shaw, B. R., Hawkins, R., McTavish, F., Pingree, S., & Gustafson, D. H. (2006). Effects of insightful disclosure within computer mediated support groups on women with breast cancer. *Health Communication, 19*, 133–142.

Smedley, B. D., Stith, A. Y., & Nelson, A. R. (Eds.). (2005). *Unequal treatment: Confronting racial and ethnic disparities in health*. Washington, DC: National Academies Press.

Spiegel, D. (1993). Psychosocial intervention in cancer. *Journal of the National Cancer Institute, 85*, 1198–1205.

Stokols, D. (1996). Translating social ecological theory into guidelines for community health promotion. *American Journal of Health Promotion, 10*, 282–298.

Treadwell, H. M., Northridge, M. E., & Bethea, T. N. (2007). Confronting racism and sexism to improve men's health. *American Journal of Men's Health, 1*, 81–86.

US Cancer Statistics Working Group. (2007). United States cancer statistics: 2004 incident and mortality. Atlanta, GA: US Department of Health and Human Services, Centers for Disease Control and Prevention and National Cancer Institute.

Viswanath, K., & Emmons, K. M. (2006). Message effects and social determinants of health: Its application to cancer disparities. *Journal of Communication, 56*, S238–S264.

Viswanath, K., & Finnegan, J. R., Jr. (2002). Reflections on community health campaigns: Secular trends and the capacity to affect change. In R. C. Hornik (Ed.), *Public health communication: Evidence for behavior change* (pp. 289–312). Mahwah, NJ: Lawrence Erlbaum Associates.

Viswanath, K., & Kreuter, M. W. (2007). Health disparities, communication inequalities, and e-health: A commentary. *American Journal of Preventive Medicine, 32*, S131–S133.

Walther, J. B., & Boyd, S. (2002). Attraction to computer-mediated social support. In C. A. Lin & D. Atkin (Eds.), *Communication technology and society: Audience adoption and uses* (pp. 153–188). Cresskill, NJ: Hampton Press.

Weinert, C., & Hill, W. G. (2005). Rural women with chronic illness: Computer use and skill acquisition. *Women's Health Issues, 15*, 230–236.

Weisgerber, C. (2004). Turning to the Internet for help on sensitive medical topics: A qualitative study of the construction of sleep disorder through online interaction. *Information, Communication, & Society, 7*, 554–574.

Williams, D. R., & Collins, C. (2001). U. S. socioeconomic and racial differences in health. Patterns and explanations. *Annual Review of Sociology, 21*, 349–386.

Wright, K. B. (1999). Computer-mediated support groups: An examination of relationships among social support, perceived stress, and coping strategies. *Communication Quarterly, 47*, 402–414.

Wright, K. B. (2000). Perceptions of on-line support providers: An examination of perceived homophily, source credibility, communication and social support within on-line support groups. *Communication Quarterly, 48*, 44–59.

Wright, K. B. (2002). Social support within an on-line cancer community: An assessment of emotional support, perceptions of advantages and disadvantages, and motives for using the community. *Journal of Applied Communication Research, 30*, 195–209.

Wright, K. B. (2009). Increasing computer-mediated social support. In J. C. Parker & E. Thorson (Eds.), *Health communication in the new media landscape* (pp. 243–265). New York: Springer.

Wright, K. B., & Bell, S. B. (2003). Health-related support groups on the Internet: Linking empirical findings to social support and computer-mediated communication theory. *Journal of Health Psychology, 8*, 37–52.

Wright, K. B., & Frey, L. R. (2008). Communication and care in an acute cancer center: The effects of willingness to communicate about health, healthcare environment perceptions, and health status on information-seeking, participation in care practices, and satisfaction. *Health Communication, 23*, 369–379.

Wright, K. B., & Miller, C. H. (2010). A measure of weak tie/strong tie support network preference. *Communication Monographs, 77*, 502–520.

Wright, K. B., Rains, S., & Banas, J. (2010). Weak-tie support network preference and perceived life stress among participants in health-related, computer-mediated support groups. *Journal of Computer-Mediated Communication, 15*, 606–624.

21

HOW E-HEALTH SYSTEMS CAN HELP THE MEDICALLY UNDERSERVED

Fiona McTavish, Helene McDowell, and David H. Gustafson

The Problem

E-health technologies may provide ways both to improve the health of medically underserved people and the access they have to health care. Although defining the medically underserved can be complex and involve many factors, the most significant characteristic of the population is low socioeconomic status (SES), including income, education, and employment (Reilly, Schiff, & Conway, 1998). Developing technologies that could improve life for this medically underserved population depends on understanding the barriers that now stand in the way of access to quality health care.

Getting health care is complicated for low-SES families (Bade, Evertsen, Smiley, & Banerjee, 2008; Butler, Kim-Godwin, & Fox, 2008; Whitley, Samuels, Wright, & Everhart, 2005). For low-income, rural families, geography, transportation, and childcare are all common barriers to getting medical help. In some very rural areas, primary care is available only on certain days of the week. Acute care must be obtained in distant urgent- or emergency-care facilities. Irregular and distant care makes it hard for low-SES families to establish a trusting relationship with a health-care provider, let alone a medical home (Graham, Logan, & Tomar, 2004). Geography is less of a barrier for the urban poor, but they face similar life complexities as people in rural settings, and getting into the system is often complicated. In addition, many urban, poor African Americans do not trust the health-care system, in part due to infamous biomedical research, such as the Tuskegee experiments. As a result, their lack of desire to establish a relationship with a medical provider may perpetuate poorer health-care outcomes (Matthews, Sellergren, Manfredi, & Williams, 2002; Underwood, 1992). Cost is an obstacle that permeates the other problems of health care for the underserved, including the likes of transportation and childcare costs, health-insurance premiums, copayments, and the cost of prescription drugs. Options for free or reduced-cost health care may be available, but the people who need the help may not know about those options (Ahmed, Lemkau, Nealeigh, & Mann, 2001).

Providing easily accessible, reliable health-care information and support to the underserved poses another significant challenge. This barrier relates mainly to the place of health care in daily life for the underserved. For families struggling to meet the most basic of health and well-being needs (such as having a roof over their heads and food in their stomachs), other health-care needs (i.e., preventive health care, screening) often take lower priority (Cunningham et al., 1999; Gustafson, McTavish, Stengle, Ballard, Jones, et al., 2005). Finding a medical home is not their first priority when basic needs are not met.

Providing health information and communication support can be accomplished via information technology. Unfortunately, racial and ethnic minorities, persons with disabilities, rural populations, and those with low socioeconomic status are significantly less likely than the national average to own a computer, to have Internet access, and to access the Internet at home. This difference is commonly called the digital divide (Chang et al., 2004; Kreps et al., 2004). As technology advances and the Internet becomes integral to health care, closing this gap is essential to improving the health of the medically underserved. Designing information technology systems that help low-income families gain access to quality health information, provide skills-training tools, and offer support when faced with a diagnosis can remove many of the barriers the underserved currently face (geographic distance, access, conveniences, timeliness, childcare, etc.) when dealing with the health-care system.

Technology as a Solution

The pressure to contain health-care costs makes urgent the need to develop more effective ways of providing information, emotional support, decision-making help, and behavior-change assistance for all patients, and especially for the underserved (Shea et al., 2009). People facing a health crisis such as cancer, HIV/AIDS, diabetes, asthma, and addiction often are given information at diagnosis when they are least able to take it in. Often they do not understand the medical terms used and feel intimidated (Davis, Williams, Marin, Parker, & Glass, 2008), and feel a need to make a decision quickly. These problems are greater for the underserved because they often do not have a primary-care physician and have poorer quality care, as well as less formal education (Agency for Healthcare Research and Quality, 2000), lower literacy, and less financial stability. Providing patients and their families with information and support when they have a serious health problem can also be difficult because individuals vary in the amount, type, and timing of the help they want and need (Arora, Rutton, Gustafson, Moser, & Hawkins, 2007; Squiers, Rutten, Treiman, Bright, & Hesse, 2005).

CHESS (Comprehensive Health Enhancement Support System) is a patient-centered, e-health intervention designed to help patients understand their diagnosis and treatment options, become full participants in their health care, and learn and practice communication and lifestyle skills that foster health and well-being.

Figure 21.1. *CHESS Website*

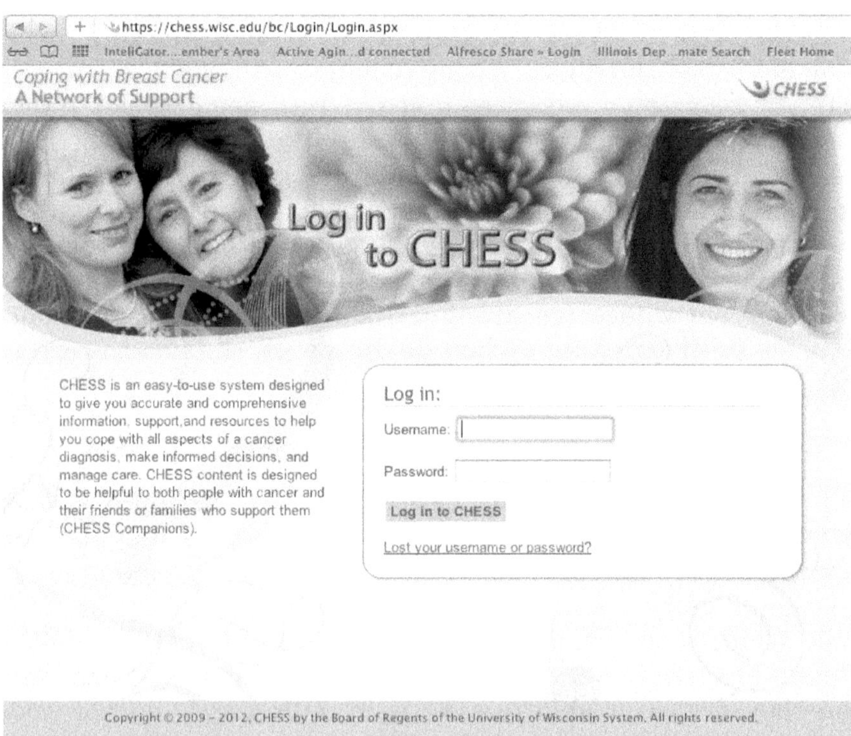

Figure 21.2. CHESS Website

CHESS is based on Self-Determination Theory (SDT), which holds that someone's perceived quality of life is shaped by the degree to which they have three key needs fulfilled:

1. Autonomy—the sense that one's actions and experiences are a conscious choice and not wholly controlled by external or internal forces;
2. Competence—a self-perception of being able to bring about a desired result; and
3. Relatedness—the feeling of connectedness with others, particularly "like others"(Ryan, Patrick, Deci, & Williams, 2008).

The relative importance of these needs varies by individual and situation (Pingree et al., 2010). CHESS has been developed with these three needs in mind. For instance, CHESS modules have tools that allow users to monitor their symptoms, set goals, and track changes, all of which can increase one's autonomy (Baker et al., 2011; Hawkins et al., 2011).

Informational articles, a consumer service, and FAQs are designed to help users understand their disease, know what to expect, and frame questions they might want to ask—all of which increase competency (Arora et al., 2002). To address relatedness, CHESS offers Discussion Groups between patients as well as an "Ask an Expert" service and, in some modules, a direct connection to the patient's clinical team.

Previous research using SDT has shown that when people feel autonomous, they are more likely to feel competent to attain important outcomes. Thus, patients are more likely to develop the skills necessary to manage their health once they freely choose to endorse those behaviors (Williams, McGregor, Zeldman, Freedman, & Deci, 2004), not only improving their health, but their overall quality of life.

CHESS has been used by approximately 5,000 patients and families in randomized clinical trials including HIV/AIDS (n = 741), breast cancer (n = 2,009), prostate cancer (n = 510), lung cancer (n = 556), colon cancer (n = 214), asthma (n = 525), alcohol and other addictions (n = 350), and heart disease (n = 70). An additional 1,500 to 2,000 patients have been given access to CHESS through their health-care provider system.

CHESS modules have been available on various platforms, but always the modules have been designed with noncomputer users in mind, because ease of use is as important as the quality of content. Many CHESS users have had no computer experience before using CHESS, and yet they rate CHESS as very easy to use (Gustafson, McTavish, Stengle, Ballard, Hawkins, et al., 2005; McTavish et al., 1994). Getting information when needed is critical. Our research has found that nearly half of use occurs between 9:00 p.m. and 7:00 a.m., when other services are typically not available (Gustafson et al., 1999).

CHESS may be the most studied of e-health systems; its cancer modules have been the subject of extensive empirical validation as Interactive Cancer Communication Systems (ICCSs). A series of randomized controlled trials has shown that patients use CHESS heavily if it is available, and that patients randomly assigned to a CHESS condition have improved outcomes on information competence, cancer knowledge, social support, and quality of life (Gustafson et al., 2008).

In our studies of the CHESS HIV/AIDS module, patients used CHESS, on average, daily during the 20-week intervention and showed significant improvement in quality of life compared to the control group. Women, minorities, and those with less education used the information and the decision-making and planning tools more than their more educated, White, male counterparts (Pingree et al., 1996). CHESS may be sufficient to overcome preexisting differences in skills and information-use habits due to several factors:

1. The user is highly motivated (i.e., someone diagnosed with cancer, HIV);
2. The system is comprehensive. Information is presented in multiple ways, such as Q/As, Personal Stories, Decision Aids, Discussion Groups, as well as in multiple formats (video, audio, and text). This allows users to get information in the style and format they prefer;
3. Users get the information in the convenience of their own home, reducing barriers such as geographic location and childcare costs;
4. Finally, it is in the user's control. The user controls the pace and timing of information, thereby increasing the likelihood of understanding.

An early pilot study of eight underserved women with breast cancer from inner-city Chicago demonstrated that if given the chance, underserved women were eager to have access to the CHESS breast cancer module. They used CHESS every day, and averaged more than an hour of use per week. They found CHESS easy to use, and valued it greatly, and it appeared to have a positive impact on their quality of life (McTavish et al., 1994).

A study of CHESS with elderly Medicare women with breast cancer in a five-county area around Madison, Wisconsin, found that the elderly used CHESS as much as people in other CHESS studies, and that age, educational level, and familiarity with computers had no effect on use. Consistent with other CHESS studies, the elderly using CHESS had improved quality of life. In addition, we found an increase in cognitive functioning (Gustafson et al., 1998).

A randomized controlled trial funded by the National Library of Medicine studied 246 women with breast cancer receiving care in Madison, Wisconsin; Chicago, Illinois; and Indianapolis, Indiana. Women in the control group received standard care. The data showed that women in the CHESS arm had significantly greater confidence in making major health decisions and in their medical team, felt freer to ask questions, and had improved social support compared to women in the control group. Consistent with other CHESS studies, minority women used CHESS in the same amount as other women, but used it differently. Minority women used the Discussion Group less and the information and decision-making tools more than their Caucasian counterparts (Gustafson et al., 2008).

To further understand the impact of CHESS on underserved people, we conducted a population-based study of low-income women with breast cancer. The women were at or below 250% of the federal poverty level, and lived in rural Wisconsin and urban Detroit. This study, called the Digital Divide Pilot Project (DDPP), loaned computers to participants, and gave them Internet access for four months and training to use CHESS.

The DDPP study found that low-income women logged onto CHESS as often as more affluent women in previous randomized clinical trials, and used CHESS for at least as much time. Again, we found that low-income women used the communication services less than more affluent women did, and used the information and decision-making services more. Both the rural and urban women using CHESS showed significant improvement in quality of life, a finding again consistent with other studies of CHESS. In addition, we found significantly greater participation in health care among the CHESS women (Gustafson, McTavish, Stengle, Ballard, Jones, et al., 2005).

Case Studies

We have repeatedly found that, to be effective, information and support resources such as CHESS must be convenient, timely, nonthreatening, anonymous, and under the user's control. In addition, they must be adaptable to different coping and learning styles (McTavish et al., 1994). These characteristics can only arise from a deep understanding of the users and their continuous feedback. The following case studies demonstrate how CHESS has been used by the underserved, and highlight considerations to weigh in developing e-health systems for the underserved.

Deborah: Empowered to Decide

Allowing the user to control the information they see and for how long is important for everyone, and especially for the underserved, because they tend to have lower rates of literacy (Health Resources and Services Administration, 2011). Having users control the timing and pace of information increases the likelihood of their understanding it. Deborah was an African American woman from inner-city Chicago. She was a single mom with a high school education when she was diagnosed with breast cancer. She had some familiarity with computers, but did not have one in her home. Four months after her diagnosis, she still had not decided whether to have a mastectomy or lumpectomy. Despite her health-care team giving her information and repeatedly encouraging her to schedule surgery, Deborah avoided making a decision. Her health-care providers were concerned that Deborah's not making a decision would have negative effects on her outcome.

A social worker at the hospital where Deborah was diagnosed contacted her to ask if she would take part in a pilot study of the CHESS breast cancer module. After some reluctance, Deborah agreed to participate. In the first week, Deborah spent about one hour in the CHESS Surgery Decision Aid. This tool allowed Deborah to read accurate descriptions of a mastectomy and a lumpectomy, written at a level she could understand. She read about the potential side effects of each surgery, and reviewed criteria that other women used in making their decisions. She determined what criteria were important to her, learned what the research said, and weighed her criteria against the options. In addition to using the Surgery Decision Aid, Deborah spent more than 3.5 hours in Questions and Answers, almost 4 hours reading Personal Stories of other women with breast cancer, and about 2 hours reading Instant Library materials. CHESS gave Deborah information in a variety of formats (a decision-making tool, personal stories that were text-, audio-, or video-based). Deborah controlled how much and what kind of information she accessed, increasing her sense of competence and autonomy. She moved from paralysis about the surgery decision to scheduling it and having it done.

Although Deborah used some of the other communication tools in CHESS such as the Discussion Group, her need for information took precedence over communicating with others. For others, especially the rural underserved, connecting with "like others" is the priority. Because of the smaller population and greater distance to health-care services in rural communities, rural hospitals or clinics often offer support groups for people with all types of cancer, rather than cancer-specific groups, such as a breast cancer or prostate cancer support group. A woman facing breast cancer may not feel comfortable talking about losing a breast in a mixed-gender group, just as a man with prostate cancer might hesitate to talk about impotency as a result of prostate cancer surgery. Internet-based discussion groups provide the option of finding support from similar others, an especially valuable thing for those in rural areas. A CHESS study participant sums it up this way:

> There isn't any support group that is close enough that I want to go to. Because at nighttime, I hate driving and most of them are evening meetings, so I really feel that the computer has its advantage that way, that you don't have to go! You don't have to go out if it's snowing; you don't have to go out if it's dark.... It takes me an hour to get there. That's an hour there, and the meeting is at seven and you won't get out of there until nine or ten and you got another hour to go home!

Computer-mediated support groups help build relatedness—the sense of connecting with others who understand your experience. One woman with breast cancer wrote this:

> With CHESS, you feel like you are connected all the time.... It's wonderful because I can get on at eleven at night or at two in the afternoon, and I can read messages or write a message. It's not like a scheduled support group where it's once a month on Wednesdays and if you can't make it, you just survive. You know, it's like it's just always there, whenever I'm ready for it and that is wonderful.

The connections patients make through discussion groups often last for years. Although CHESS Discussion Groups are designed to be anonymous (participants use code names when writing to discussion-group members), users often choose to reveal their names before long. The Discussion Groups have been intended to run for the length of a clinical trial, but participants have continued to stay in touch long after. Two groups, one from our early CHESS breast cancer studies in the Midwest and another in the Boston area, get together each year even though their CHESS computers were removed more than ten years ago. As one participant wrote,

> Well, if they came and boxed it [the computer] up, I will think that the room will be so empty because to me it's like boxing up 20 neat women. I don't feel or see it as a computer. I see it as my friends.... It really means a lot to me and it is so valuable to my life.

Larissa: Feedback That Improved CHESS.

Larissa's story also demonstrates the power of user-controlled information. Larissa was a single, 44-year-old, African American woman living in Milwaukee. She had one adult son whom she spoke with occasionally via the telephone. At the time she was diagnosed with breast cancer, she shared an apartment with her boyfriend. Shortly after the diagnosis, she had domestic problems, and decided to move in with a female friend. She worked as a semiskilled laborer in light industrial jobs. Larissa found out she had breast cancer after a mammogram at a free clinic. She said she was very scared when she received her diagnosis and did not know where to go. She had no health insurance and, like most people without insurance, she did not have a primary-care physician. Initially, Larissa received some help from the American Cancer Society and a Milwaukee-based program called ABCD (After Breast Cancer Diagnosis). After surgery, Larissa was given access to CHESS as part of a randomized clinical trial. She found the Video Gallery of personal stories of other women with breast cancer particularly helpful. According to Larissa, "It gave me strength to continue my fight against breast cancer and depression. I felt less alone and found their stories encouraging."

Larissa suggested that CHESS needed to provide more practical information. Specifically, she wanted to learn how to apply for Supplemental Security Income (SSI), and how to pay for food and other everyday living expenses while waiting for SSI. Although CHESS had a link to SSI information, we realized we needed to explain the process of applying for SSI, list the required documents, and suggest questions to ask when applying for SSI, SSDI, and other government entitlement programs. We also added information in CHESS about choosing health-care providers and understanding your rights and responsibilities as a patient. In short, feedback from users has made CHESS more relevant and valuable. Getting patients the information and support they needed and wanted allowed them to be more effective partners in their own health care.

Excellent, accessible content is a necessary condition of a good e-health system, but it is not sufficient. The user must also have the skills and ability to use the system. Low-income patients often have little or no computer experience. One common fear among those less familiar with computers is that they will damage the computer if they do not use it precisely as instructed. A well-designed system makes it very difficult to "break" the system. From its inception, CHESS was designed so that someone with no computer experience could easily navigate the system. Building systems that are intuitive and easy to use must be at the front of the design process. Including tutorials for those who prefer more structured learning is also important. In addition, Help Buttons should be tailored to the specific page the user is on (not general tips). As new participants use the system and realize that they cannot "break" it, they are more likely to do further exploration, resulting in increased computer skills (such as navigational and search skills), increased autonomy and competency.

Casey: Afraid to Break the Computer

Casey was a rural, low-income woman in Wisconsin who was 45 years old when she was diagnosed with stage III breast cancer. Casey was a single mother of a teenage daughter and a special-needs son under 10. Casey had no computer experience before she started using CHESS. We loaned her a computer, and trained her to use the CHESS website and do basic Internet searches. Her biggest worry was that she would hit the wrong key and wipe everything off the computer and break it. We stressed during her training that she could not do anything to break the computer unless she threw it across the room. We gave her a toll-free number to call for help. Casey was nervous during training, and did not want to seem stupid. We explained that if using the computer was not intuitive, it was programmed poorly and not a reflection on her. We showed that it is illogical to have to press "start" to turn off a computer, but that is what one must do. With the training, a manual with step-by-step instructions, and the toll-free number, she was ready to tackle the computer.

During the first few weeks of using CHESS, Casey did call the help line to ask questions. During these calls, she said how much better she felt knowing she could go online anytime and look at information about her treatment, ask an expert a question, or post a message to other women in a similar situation. At the end of the four-month study period, Casey asked to keep the CHESS computer until she could afford to buy her own. She kept it for three more months and then got her own. Nine years later, Casey continues to use CHESS, and now acts as a CHESS Discussion Group seed (a survivor who welcomes newly diagnosed women into the group). CHESS worked for Casey by improving her understanding of her disease and treatment (competency), reducing her isolation and increasing her sense of social support (relatedness), and increasing her sense of autonomy.

During our early clinical trials of CHESS, we went into the homes of users (as we did with Casey) to set up the computer and train participants on basic computer skills and on using CHESS as well as doing simple Internet searches. This training typically took approximately one hour. The goal of the training was for the participant to use the keyboard and mouse rather than watch the trainer use CHESS, and do an Internet search. But in-home training was too costly as we considered wider use of CHESS. As a result, we pilot tested shipping laptop computers to participants' homes and having them set up the laptop. All cords were color coded so that the user could quickly understand where each cord plugged into the computer. We also sent a step-by-step CHESS training manual. Then we trained participants over the phone. We concluded that phone training actually gave participants more hands-on experience, more confidence about using the computer, and a better understanding of using the system (increasing autonomy). An unanticipated result of phone training was participants' greater use of CHESS. The cost of training was nearly halved as a result of saving travel time (Gustafson, McTavish, Stengle, Ballard, Hawkins, et al., 2005).

The underserved can be defined in numerous ways—by socioeconomic status, race, gender, and geographic location, among other characteristics—but even among people with the same characteristics, considerable variance exists among individuals. Users who can control the information and support they receive in effect tailor the system to fit their needs. The degree to which the system meets someone's needs influences the person's perceived quality of life, according to Ryan and Deci's self-determination theory (Ryan & Deci, 2000; Deci & Ryan, 2002). Two examples follow of individuals from different underserved populations using CHESS to meet their needs.

Irene: Benefiting From Social Support She Thought She Did Not Need

Irene was a low-income Wisconsin farmwife diagnosed with stage I–II breast cancer when she was 60 years old. This was Irene's second diagnosis of cancer. Irene was the matriarch of her family. She helped run the farm, and, with her grown children, raise the grandchildren. She had some computer experience before she got CHESS, as a result of managing the accounts for the farm. Irene agreed to participate in the DDPP study because she wanted to help others. She said she would not need to use CHESS herself, and did not think it would help her very much. Irene received a laptop on loan from the study, and training in using the CHESS website. During the training, Irene reiterated that she would not use the system much. She took care of others, she said; she did not need to be taken care of herself. Irene began using the CHESS website the first day she was trained. She posted regularly in the Discussion Group, and wrote about her feelings to other women in the group. Irene later told us that she was surprised at how much the Discussion Group helped her through her own treatment. She found that she could not share some of her feelings with her husband and family because she needed to be the caretaker. In CHESS, she was anonymous and could say anything. Nine years later, Irene continues to use the CHESS website. She logs onto CHESS monthly to catch up with her "Chessling friends." Irene used the CHESS Discussion Group almost exclusively, which increased her sense of social support (relatedness) and allowed her to express her emotional concerns without worrying her family (autonomy).

Paris: Arming Herself With Information

Paris was a single, African American mom who had recently lost her job. She was 27 when she was diagnosed with breast cancer. She had had computer experience, but did not have a computer or Internet access in her apartment. Her surgeon referred her to

CHESS because she had requested information. As a young, single mom, she was scared and very worried about her prognosis. Paris received CHESS before she started any treatments. After being trained, she spent most of her time using CHESS for its information services, mostly the Ask an Expert service, where she sought information about her treatments. She also used the FAQs and the Personal Stories. Using CHESS allowed Paris to fulfill her needs for competency and autonomy. She echoed what a Detroit participant said: "Without CHESS, I would have been in the dark and the doctor would have been making all of the decisions. But CHESS armed me with information that made me a part of my treatment." As Paris moved through her treatments, her needs changed. In later weeks, she began using the Discussion Group. Unlike Irene, Paris was a passive user of the Discussion Group, reading but not posting messages.

The examples described in this paper have been of study participants using desktop or laptop computers. With rapid advances in technology, the possibilities available for e-health systems seem endless. CHESS has now been designed in several applications for smartphones, such as one for underserved teens with asthma, one to increase the amount of exercise colon cancer survivors get, and one for people leaving residential treatment for alcohol addiction. In each application, it was important that users have access anytime, anywhere—which is the feature most valuable about smartphones. For instance, A-CHESS (Addiction CHESS) was developed to reduce risky drinking behavior in alcoholics leaving residential treatment. Giving feedback, support, and coping techniques anytime of the day or night is critical. Someone with an urge to drink can get help by pushing the Panic Button on the phone, which sends a text message immediately to a predetermined support system set up by the user. The person could also phone a sponsor, friend, or counselor. While waiting for a response, the phone asks the person several questions. The phone then shows on the screen suggestions of materials to read and actions to take, depending upon the answers to the questions. The important feature is the system's responding quickly to meet the individual's need no matter where the individual is.

Smartphones also put together types of information in a way that makes applications proactive. For example, Survivorship CHESS (S-CHESS) has the goal of increasing the activity level of sedentary colon cancer survivors. Each week the S-CHESS user creates an exercise goal—for example, walking 10 minutes each day of the week. S-CHESS can be programmed to remind the user of having committed to walk each day. Walking routes can be mapped out for the person and a personal trainer can "walk with them," encouraging them to increase their pace, for example, at minutes 5 and 6. The smartphone can automatically enter the time and amount of exercise completed.

The important characteristics of e-health systems remain constant regardless of platform. The systems must be convenient, timely, nonthreatening, anonymous, and under the user's control. Matching the strengths of technology to the needs of the end user is critical to providing effective solutions, not only for the underserved, but for everyone facing a health concern. As one participant said, "CHESS did not change my diagnosis of breast cancer, but it did change how I responded to it."

Bibliography

Agency for Healthcare Research and Quality. *Addressing racial and ethnic disparities in health care—Fact sheet.* AHRQ Publication No. 00-PO41, February 2000. U.S. Department of Health and Human Services. Retrieved from http://www.ahrq.gov/research/disparit.htm

Ahmed, S. M., Lemkau, J. P., Nealeigh, N., & Mann, B. (2001). Barriers to healthcare access in a non-elderly urban poor American population. *Health and Social Care in the Community, 9*(6), 445–453.

Arora, N. K., Johnson, P., Gustafson, D. H., McTavish, F., Hawkins, R. P., & Pingree, S. (2002). Barriers to information access, perceived health competence, and psychosocial health outcomes: Test of a mediation model in a breast cancer sample. *Patient Education and Counseling, 47*(1), 37–46.

Arora, N. K., Rutton, L. J. F., Gustafson, D. H., Moser, R., & Hawkins, R. P. (2007). Perceived helpfulness and impact of social support provided by family, friends, and health care providers to women newly diagnosed with breast cancer. *Psycho-Oncology, 16*(5), 474–486.

Bade, E., Evertsen, J., Smiley, S., & Banerjee, I. (2008). Navigating the health care system: A view from the urban medically underserved. *Wisconsin Medical Journal, 107*(8), 374–379.

Baker, T. B., Hawkins, R., Pingree, S., Roberts, L. J., McDowell, H. E., Shaw, B. R., Serlin, R.,...Gustafson, D. (2011). Optimizing eHealth breast cancer interventions: Which types of eHealth services are effective? *Translational Behavioral Medicine, 1*(1), 134–145.

Butler, C., Kim-Godwin, Y. S., & Fox, J. A. (2008). Exploration of health care concerns of Hispanic women in a rural southeastern North Carolina community. *Online Journal of Rural Nursing and Health Care, 8*(2), 22–32.

Chang, B. L., Bakken, S., Brown, S. S., Houston, T. K., Kreps, G. L., Kukafka, R., Safran, C.,...Stavri, P. Z. (2004). Bridging the digital divide: Reaching vulnerable populations. *Journal of the American Medical Informatics Association, 11*(6), 448–457.

Cunningham, W. E., Andersen, R. M., Katz, M. H., Stein, M. D., Turner, B. J., Crystal, S., Zierler, S.,...Shapiro, M. F. (1999). The impact of competing subsistence needs and barriers on access to medical care for persons with human immunodeficiency virus receiving care in the United States. *Medical Care, 37*(12), 1270–1281.

Davis, T. C., Williams, M. V., Marin, E., Parker, R. M., & Glass, J. (2008). Health literacy and cancer communication. *CA: A Cancer Journal for Clinicians, 52*(3), 134–149.

Deci, E. L., & Ryan, R. M. (2002). *Handbook of self-determination research*. Rochester, NY: University of Rochester Press.

Graham, M. A., Logan, H. L., & Tomar, S. L. (2004). Is trust a predictor of having a dental home? *The Journal of the American Dental Association, 135*(11), 1550–1558.

Gustafson, D. H., Hawkins, R., McTavish, F., Pingree, S., Chen, W. C., Volrathongchai, K., Stengle, W.,...Serlin, R. C. (2008). Internet-based interactive support for cancer patients: Are integrated systems better? *Journal of Communication, 58*(2), 238–257.

Gustafson, D. H., McTavish, F. M., Boberg, E., Owens, B. H., Sherbeck, C., Wise, M., Pingree, S., & Hawkins, R. P. (1999). Empowering patients using computer based support systems. *Quality in Health Care, 8*(1), 49–56.

Gustafson, D. H., McTavish, F. M., Hawkins, R. P., Pingree, S., Arora, N. K., Mendenhall, J., & Simmons, G. E. (1998). Computer support for elderly women with breast cancer. *Journal of the American Medical Association, 280*(15), 1305.

Gustafson, D. H., McTavish, F. M., Stengle, W., Ballard, D., Hawkins, R., Shaw, B. R., Jones, E.,...Landucci, G. (2005). Use and impact of eHealth system by low-income women with breast cancer. *Journal of Health Communication, 10*(1), 195–218.

Gustafson, D. H., McTavish, F. M., Stengle, W., Ballard, D., Jones, E., Julesberg, K., McDowell, H.,...Hawkins, R. (2005). Reducing the digital divide for low-income women with breast cancer: A feasibility study of a population-based intervention. *Journal of Health Communication, 10*(1), 173–193.

Hawkins, R. P., Pingree, S., Baker, T. B., Roberts, L. J., Shaw, B. R., McDowell, H., Serlin, R.,...Salner, A. (2011). Integrating eHealth with human services for breast cancer patients. *Translational Behavioral Medicine, 1*(1), 146–154.

Health Resources and Services Administration. *About Health Literacy* (2011). U.S. Department of Health and Human Services. Retrieved from http://www.hrsa.gov/publichealth/healthliteracy/healthlitabout.html

Kreps, G. L., Gustafson, D. H., Salovey, P., Perocchia, R. S., Wilbright, W., Bright, M. A., Muha, C., & Diamond, C. C. (2004). Using computer technologies to provide relevant cancer information to vulnerable populations: The NCI Digital Divide Pilot Projects. In P. Whitten & D. Cook (Eds.), *Understanding health communication technologies* (pp. 328–336). San Francisco, CA: Jossey-Bass.

Matthews, A. K., Sellergren, S. A., Manfredi, C., & Williams, M. (2002). Factors influencing medical information seeking among African American cancer patients. *Journal of Health Communication, 7*(3), 205–219.

McTavish, F. M., Gustafson, D. H., Owens, B. H., Wise, M., Taylor, J. O., Apantaku, F. M., Berhe, H., & Thorson, B. (1994). CHESS: An interactive computer system for women with breast cancer piloted with an underserved population. *Journal of the American Medical Informatics Association*, (Symposium Supplement), 599–603.

Pingree, S., Hawkins, R., Baker, T., DuBenske, L., Roberts, L. J., & Gustafson, D. H. (2010). The value of theory for enhancing and understanding eHealth interventions. *American Journal of Preventive Medicine, 38*(1), 103–109. PMCID: PMC2826889

Pingree, S., Hawkins, R. P., Gustafson, D. H., Boberg, E., Bricker, E., Wise, M., Berhe, H., & Hsu, E. (1996). Will the disadvantaged ride the information highway? Hopeful answers from a computer-based health crisis system. *Journal of Broadcasting & Electronic Media, 40*(3), 331–353.

Reilly, B. M., Schiff, G., & Conway, T. (1998). Primary care for the medically low-income: Challenges and opportunities. *Disease-A-Month, 44*(7), 320–346.

Ryan, R. M., & Deci, E. L. (2000). Self-determination theory and the facilitation of intrinsic motivation, social development, and well-being. *American Psychologist, 55*(1), 68–78.

Ryan, R. M., Patrick, H., Deci, E. L., & Williams, G. C. (2008). Facilitating health behaviour change and its maintenance: Interventions based on self-determination theory. *The European Health Psychologist, 10*(1), 2–5.

Shea, S., Weinstock, R. S., Teresi, J. A., Palmas, W., Starren, J., Cimino, J. J., et al. (2009). A randomized trial comparing telemedicine case management with usual care in older, ethnically diverse, medically underserved patients with diabetes mellitus: 5 year results of the IDEATel study. *Journal of the American Medical Informatics Association, 16*(4), 446–456.

Squiers, L., Rutten, L. J. F., Treiman, K., Bright, M. A., & Hesse, B. (2005). Cancer patients' information needs across the cancer care continuum: Evidence from the Cancer Information Service. *Journal of Health Communication, 10*(1), 15–34.

Underwood, S. (1992). Cancer risk reduction and early detection behaviors among Black men: Focus on learned helplessness. *Journal of Community Health Nursing, 9*(1), 21–31.

Whitley, E. M., Samuels, B. A., Wright, R. A., & Everhart, R. M. (2005). Identification of barriers to healthcare access for underserved men in Denver. *The Journal of Men's Health & Gender, 2*(4), 421–428.

Williams, G. C., McGregor, H. A., Zeldman, A., Freedman, Z. R., & Deci, E. L. (2004). Testing a self-determination theory process model for promoting glycemic control through diabetes self-management. *Health Psychology, 23*(1), 58–66.

Part IV. Critical Perspectives on Communication and Health Disparities

22

THE BUSINESS OF DOING INTERVENTIONS: A PARADOX FOR THE PARTICIPATORY METHOD

Rebecca de Souza

Nongovernmental organizations (NGOs) play an important role in the health sector of developing countries, and are mandated the grand task of improving global health and eliminating health disparities, especially those related to HIV/AIDS (Green & Matthias, 1997; WHO, 2011). NGOs, also called not-for-profit, citizen, or civil-sector organizations, are defined as organizations not related to either state or business-for-profit sectors (Salamon, 1993). As early as 1978, the WHO Alma-Ata Declaration made it clear that eliminating health disparities was a top priority for governments in partnership with the civil sector; health disparities were described as the "existing gross inequality in the health status of the people" between developed and developing countries as well as within countries (WHO, 1978, p. 1). In India, NGOs have been leading the country's response to HIV/AIDS at the grassroots level, with several hundreds of them working on HIV/AIDS issues at the local, state, and national levels; projects include advocating for the rights of people living with HIV, supporting existing interventions for high-risk populations, such as commercial sex workers and men-who-have-sex-with-men (MSMs), and care for AIDS orphans (World Bank, 2004).

Consistent with the discourse of the "new public health" and mandates set by global health organizations, many NGOs use the language of participation and community empowerment to describe and guide their health programs or interventions (de Souza, 2011; Petersen & Lupton, 1996). The methodologies developed within the community empowerment paradigm of health promotion have been variously referred to as community-based participatory research, participatory action research, and the community empowerment approach (Laverack, 2004). The new paradigm is based on the realization that many of the health and social problems of contemporary times (e.g., HIV/AIDS) are ill-suited to the traditional, "outside expert" approaches to health interventions. In its ideal form, participation is a bottom-up process involving egalitarian processes of decision making, and communities are viewed as "partners," rather than as recipients of the intervention (e.g., Melkote & Steeves, 2001).

Health-communication scholars point out the gap between the rhetoric and reality of participatory programs. Grounded in subaltern and/or postcolonial studies, and the work of Latin American scholars such as Freire (1970), Dutta elaborated the culture-centered approach (CCA) to critique the modernist and Eurocentric bias inherent in programs that claim to be "participatory" (e.g., Dutta & Basnyat, 2006, 2008). Programs fall short because they do not account for the context within which health behaviors arise; structural constraints are ignored, and an ethnocentric view of culture is observed (Airhihenbuwa, 1995; Dutta, 2007; Dutta-Bergman, 2005). For instance, the Radio Communication Project in Nepal claims to use participatory methods, but the project was established within the rubric of family planning, as such messages are centered on improving interpersonal communication between couples, while underlying social, economic, and environmental reasons for larger families are ignored (Dutta, 2008; Dutta & Basnyat, 2006). The CCA calls for the creation of spaces where disenfranchised communities can engage as equal participants in the process of change through dialogue and action.

While NGOs and corresponding interventions have received flak for failing to "hear" communities, in this paper, I locate the discrepancy between rhetoric and the reality of genuine participation against the backdrop of the larger, complex, transnational system that has been set up to "manage" the problem of HIV/AIDS. I argue that the institutionalization of the problem (i.e., HIV/AIDS) as well as the solution to the problem (i.e., the health intervention) exacerbates the complexity, uncertainty, and ethical ambiguities of this model of change. Genuine participatory practice, which involves engaging communities as equals, is limited in a context of multiple stakeholders and million-dollar budgets. While there are many varieties of NGOs, the NGOs referenced in this study refer to local Indian NGOs who work directly with disenfranchised communities at the grassroots level, or are comprised of members of the community, in this instance—people living with HIV/AIDS. This chapter uses examples from my research on NGOs and community-based HIV/AIDS programs in India to identify tensions that arise between NGOs and their stakeholders. The discussion elaborates upon the limitations of participatory interventions to close the global health gap when viewed against the backdrop of economic imperatives of the global "HIV/AIDS industry."

Interventions as "Technologies" of Governance

Following Foucault's (1977) ideas regarding modern systems of governance, the 1990s saw a spate of books demonstrating the increasing institutionalization of the "care" industry, and detailing the technologies (i.e., procedures, policies, methods, discourses, and techniques) through which change "from the top" was stabilized and became status quo. In the context of the United States, as early as 1991, Patton used the phrase, "AIDS service industry," to argue that the disease had led to the development of a new economy operated under the authority of medical and social science (Patton, 1991). More broadly, community organizer, McKnight (1995), urged attention to the economic motives underlying the human-service industry, where paid "experts" got to declare their fellow citizens as deficient, and where conditions of living such as "ageing" and "childhood" became medical, health, and lifestyle problems needing interventions. Poppendieck (1998) examined processes through which emergency food assistance became a stable feature of the economy set up to "manage" the problem of food insecurity. In the context of the "third world," Escobar (1995) showed how policies during the "development era" had become neocolonial mechanisms of control; imperialism occurred by identifying problems to be managed by specific interventions, depoliticizing problems into neutral scientific terms (e.g., poverty indicators), and institutionalizing the technologies to treat stated "problems." More recently, health-communication scholars (e.g., de Souza, Basu, Kim, Basnyat, & Dutta, 2008; Zoller, 2008) have attended to neoliberal technologies of governance perpetuated through the discursive realm by global organizations such as the World Trade Organization (WTO).

It is important to clarify that none of these works undercut the objective reality of social problems (e.g., poverty, hunger, or HIV/AIDS), or the need to solve and/or alleviate the problems, but rather explicate the processes by which real problems and real suffering are recast to suit the agenda of dominant actors. In each instance, the overarching argument is that in the modern system of governance, problems are not solved, only "managed," because "managing a problem" allows for increased movement of capital and labor. Fuelled by the ideology of service and volunteerism, defining problems and coming up with solutions, interventions, and processes (e.g., community-based participatory research (CBPR)) have become a staple feature of the economy.

The "HIV/AIDS Industry"

The prioritization of HIV/AIDS as a global issue of much social and economic consequence, and the careful surveillance of HIV infections world over, have meant that large amounts of money are allocated to the issue of HIV/AIDS. The HIV/AIDS industry is comprised of health-care providers, pharmaceutical companies, researchers, governments, funding agencies, and risk groups who are the named recipients or beneficiaries of the end product of this economy: the health intervention. National, international, government, nongovernment, and private actors are tightly enmeshed in the business of mobilizing HIV/AIDS prevention and support interventions. Money is earmarked for HIV/AIDS interventions; NGOs bid for these funds; funders award money to NGOs and/or government agencies; and NGOs, in collaboration with the government, mobilize interventions amongst disenfranchised communities.

This is especially true in the Indian context, which receives funds from both national and international sources (NACO, 2010). In 1986, the Ministry of Health and Family Welfare formed the National AIDS Control Organization (NACO) to provide leadership to HIV/AIDS control programs in India. NACO receives funding from national budgets and through bilateral and multilateral funders. Phase III of the AIDS-control initiative, for instance, received 2,861 crores (approximately 600 million USD) from direct budgetary support, and 4,148 crores (approximately 900 million USD) from external and/or foreign agencies such as the World Bank; Global Fund for AIDS, TB and Malaria (GFATM); and USAID (NACO, 2010). Large amounts of funds also come from private foundations such as the Bill and Melinda Gates Foundation (BMGF), the Clinton Foundation, and transnational NGOs such as World Vision and Christian Aid (NACO, 2010). In 2009, BMGF announced an increased commitment of $338 million to *Avahan* (BMGF, 2010). The HIV/AIDS industry is thus characterized by large amounts of funding and multiple cross-sector partnerships constituted to utilize this funding.

Data Source

The claims, examples, and excerpts cited in this chapter are drawn from a larger research project involving case studies of five NGOs located in the southern region of India (de Souza, 2007), some of which have been published elsewhere (de Souza, 2009, 2011). Table 22.1 provides a snapshot of each NGO, the intervention with which they were involved, and the particular group that the intervention was centered on. Data collection procedures for the study involved extensive analysis of organizational documents and semistructured, in-depth interviews with participants at each NGO. Drawing on the CCA, the goal was to understand how the term, *empowerment*, was constructed and practiced.

Briefly, I found that HIV/AIDS emerged within a context of poverty, gender inequalities, and stigma and discrimination. Power and empowerment were constructed at the individual, collective, and structural levels, and local NGOs enacted a number of processes (e.g., counseling, collectivization, and capacity building) to mobilize communities, at the heart of which (for some) lay the creation of "communicative spaces," where silenced individuals and groups could engage in dialogue. The study also revealed multiple struggles and dialectics of power amongst NGOs and the various stakeholders they worked with. Economic imperatives at the global level connected actors in a complex web of interdependencies, which shaped the struggles that occurred at a local level between NGOs and their stakeholders. The participatory success of a project depended on the NGO's ability to navigate and maintain relationships with multiple stakeholders, in particular: the funding institutions (local or international) that support the intervention, the government that permits them to work within the borders of the country, and the disenfranchised group or community. The next section of this chapter outlines some of the tensions NGOs faced with each of these primary stakeholders.

Table 22.1. *NGOs and Interventions.*

Name of NGO	Type of Intervention	Target Audience
SPAD	Prevention	Commercial sex workers
MAYA	Medical care; prevention	Commercial sex workers; men who have sex with men (MSMs)
AAZADI	Medical care; sociopsychological support; some advocacy	HIV+ people and children
LPN	Sociopsychological support; advocacy	HIV+ people
SHRUTI	Prevention	Child sex workers, HIV+ children

Tensions With Funders

Since funding is already earmarked for specific interventions (e.g., prevention or support) across the case studies, I found that NGOs do not always have the freedom to identify communities most in need of programs and/or include community voices in the identification of key issues. The particular health agenda (e.g., HIV prevention) is usually already set by national and global imperatives, while epidemiological studies determine the "community" or risk group to be worked with (e.g., sex workers). These directives from above restrict the NGO's autonomy to put resources into communities or issues most in need.

Funders also want to see that specific health goals or outcomes have been met, but tensions arise regarding what counts as success. In the case of HIV/AIDS, funders want to see a decline in HIV infection rates, or that a certain number of women have been integrated into the program. The problem is that it is possible to meet these criteria without actually creating change; it is possible to show that you have done outreach or even provided counseling for 30 women, but this does not mean that meaningful, long-term change has been created for any one of them. Furthermore, not all program successes can be captured within these numeric assessment rubrics. Dhiren, a documentation officer, who works with *SPAD* on an intervention with sex workers, provides an example. The sex workers, as part of the intervention, conduct condom demonstrations; one day, in an attempt to heckle and intimidate the sex worker, a male audience member asked, "What does semen look like?" She answered immediately, "All men know what semen looks like." The act of a sex worker was seen by NGO workers and the women as a sign of empowerment, because in this context, where they are the lowest rung of society, speaking to members of mainstream society was a major step forward. These types of stories are not accounted for in program achievement reports and are hard to convey to funders, who do not have experience in that particular cultural context. So while these are the success stories that the women take pride in and recount amongst themselves, they are not seen as valid outcomes of the intervention itself. The CCA also calls into question the concept of "success" as defined within health interventions, suggesting that we seek out alternative paradigms for evaluating health-education efforts (Dutta & Basnyat, 2008; Dutta-Bergman, 2005).

Disagreements related to program decision making occur because funders may not be familiar with the cultural milieu. *MAYA*'s employees talk about conflicts that occurred with the funding partner, Family Health International (FHI), regarding a medical component of the project. It was decided that the sex workers would be given STI examinations regularly, but a conflict arose about what the term, *regular*, meant. FHI insisted that the women should have health checkups once every month, but Dr. Ramona, the HIV/AIDS specialist argues:

> But, the women here turn around and say if we go every month to the clinic, everyone will think that we are dying of cancer or some very serious disease, because that's the only reason anyone goes regularly to a clinic, so why should we go? There is nothing wrong with us, we'll come once in three months, but we're not going every month.

MAYA was unable to convince the funders about this, and for a while, their field workers had to cajole the women to go to the clinic each month. Here, lack of understanding of the culture led to the development of rubrics that were designed to fail from the outset. Furthermore, genuine participation was compromised, because even when the women described their issues, they were not heard.

Another tension occurred because of the difficulties funders have in recognizing the difference between "capacity building" and simply "giving away things." Capacity building is a process which involves building up the assets and attributes that a community is able to draw upon in order to improve their lives, such as strong leaderships and financial skills (Goodman et al., 1998). The process takes time, patience, and the ability to entrust community members to make their own decisions. *MAYA*'s contention, however, was that funders would rather give money and free services to underserved communities rather than spend time building capacities. For example, following directives from a foreign funding source, the project set up a free beauty parlor for sex workers. Ramona sarcastically referred to it as the "one-stop shop," where the outreach workers could build rapport with the women, increase awareness, distribute condoms, and get their hair done all at the same time. *MAYA* vehemently opposed the parlor, arguing that their goal was to integrate the stigmatized women into the community, not create more segregated structures. Ramona says: "We want them to go to the *regular* beauty parlor, pay Rs. 10 and get your eyebrows done instead of coming to this place for free." To allow for sustainability, one has to utilize and enhance those structures already in place; else, once the project closes, the women are even more vulnerable.

In *MAYA*'s view, funders engage in this behavior because of economic pressures and because of lack of trust in the community. The large amount of money earmarked for specific projects means that there is pressure to spend that money as quickly as possible, and success tends to be viewed in terms of amount of money spent. Ramona says:

> If your budget says 100,000 dollars, they want to know how quickly you can spend the 100,000 dollars. If you spent it, it means you've done good work. Everything is *you* do the work for the community, *you* provide, *you* deliver, it's *all* delivery.

Another reason is that they do not trust the disenfranchised community. A hallmark of the participatory approach is handing control of intervention over from the "experts" to the disenfranchised communities, but participants recall the reaction of representatives from the funding organization (BMGF) when they told them that they would be handing over condom distribution to the sex workers. Ramona says: "There was *total* panic from the funders! What do you mean? How will you guarantee that every woman has got the condoms?" Funders are afraid to let go of project control because they have invested so much. Sara says, "They think 'oh, they are poor, how can they do it'?" Here we see how the lack of trust in the community, as well as the high economic investment in an outcome-driven system, negatively impact genuine participation.

There are more explicit ideological tensions that arise between funders and local NGOs regarding what counts as empowerment. Sara, a social worker from *MAYA*, says that the people from the organization (BMGF) started to say things like, "If you are a sex worker there is nothing wrong with that. You should be empowered and say I am a sex worker and you should be ready to go on a *dharna* (protest) and all that." *MAYA*'s response, per Sara, was this:

> Let us face facts, at *MAYA* we are not *promoting* sex workers. We don't promote *beedi* [cigarettes] rolling for those poor children who later suffer from the dust. We don't promote children working in factories making *pathakas* [firecrackers]. So in the same way we are not talking of *promoting* sex work, but these people [funders] in their excitement of this new thing of working with sex workers got lost. It was you know, identify the sex workers and give them self-esteem building, hug them and say there's nothing wrong in being a sex worker. All that is bullshit! Fine, we are also working with the poor for very very long time. We also have empathy, but let's not treat them as some different animal. They are like you and me. So that was our problem with them initially. We had a problem in trying to convince them.

What Sara is arguing here is for funders to attend to the structural reasons for women being in sex work. Asking a sex worker to be proud of her sex-worker identity ignores the structural reality that went into creating that identity.

While in some parts of India a strong sex workers' rights movement has emerged advocating the three R's: respect of sex workers and their profession, recognizing their professions and their rights, and reliance on their understanding and capability (Jana, Basu, Rotheram-Borus, & Newman, 2004), the issue is a complex one with varying perspectives, even within the activist and sex-worker community in India. One way to parse the differences is by recognizing the difference between sex-worker identities that are an outcome of structural constraint and those that emerge from human agency. A difference must be made between women who choose sex work amidst a range of equally viable options and opportunities, and those who do not. Basu and Dutta (2009) provided a useful analysis of the dialectic tensions between the structure and agency as heard in the narratives of sex workers in the SHIP and New Light's HIV/AIDS project in Kolkata, India; the women are engaged in rational decision-making; they understand their own financial vulnerability, but at the same time take risks, make sacrifices, and fight for 24-hour shelters for their children. However, in this situation, the lack of "culture-centeredness" on the part of external agents means that instead of engaging in a more nuanced understanding of structure and agency, they swing to the extreme pole of cultural relativism. From a Freirian perspective, the risk of adopting this position is that instead of engaging women in a more humanizing process of change, which treats them as equals, and instead of fighting to transform structures that cause oppression, women are *integrated* into the very structure of oppression.

Tensions With Government and State Institutions

NGOs can be shut down at any time and their funding removed for failure to conform to state directives; in fact, NACO recently fired 350 NGOs as part of a massive cleanup of nonperforming partners in the HIV/AIDS prevention and/or support initiatives (Shrivastava, 2008). Because NGOs receive funding from the government and exist "at the pleasure" of the state, it is necessary for them to navigate relationships carefully with state-led institutions.

A large part of the problem, according to NGO actors, is political apathy and denial regarding the extent of the HIV problem in India. According to Nisha, an advocacy officer from *LPN*, cultural misperceptions still exist amongst government officials: "still they [government] have the attitude that India has a morality, we have this culture...they think that HIV is a foreign disease." Local officials also do not see the importance of HIV/AIDS in a context where poverty and malnutrition are rampant; they say "why should we look at HIV, and not other diseases like cancer or TB?" or "No, that is not part of our mandate." Ramona says, "It has taken people in USAID or others to come in and say okay you have a problem, do something about it, but our local people are still in denial." There is a lack of ownership of projects and programs; Ali, an outreach and program manager from *MAYA* explains:

> The government formulates a project, and they take different people to work on it. They come to the project or organization to earn more money, they'll come today, get their money, their good salary, and be well looked after, but after two months or one year, they'll leave. See, they don't have ownership of that project. That is the biggest problem.

Ramona argues that the success of the partnership often depends on whether or not the individual working there at the time is honest and cares about the issue. She says,

> Here in India we find that the partnership largely depends on the individual in the government. If it is a good individual it is very smooth. If it's someone who is not interested or corrupt then no amount of, you know, goodwill on our side or discussions can move things forward, and then we are limited, because most of the major services are, you know, the stronghold of the government, so that is another challenge.

She adds that there is no shortage of resources for HIV, but the problem persists because of vested economic interests. She says:

> Too much money, in fact, we [*MAYA*] feel it's very unfair, because really speaking malnutrition is a hundred times more of a problem than HIV, but no one bothers (laughs) about malnutrition...there's glamour in HIV, you know, right now. Also, then you have to look at it, with HIV the pharmaceutical industries benefit, that's a big issue, we are producing drugs and sending to Brazil, at one tenth the cost that our people are paying for it here, you know, which is quite unfair if you look at it, but, that's the industry.

Here, she is drawing on the notion of the "HIV industrial complex," noting that the problem is not one of lack of resources, but rather the inequitable distribution of resources.

Managing tensions with state institutions is necessary for specific interventions and capacity-building processes to be put into place; this is especially true in the case of sex work, which is an illegal activity. For instance, a few months after *SPAD* mobilized the sex worker intervention, the civic authorities began city-improvement measures, resulting in the destruction of "social marketing outlets." These were small street vendors who had been integrated into the program, and played a crucial role in providing information and condoms to sex workers and clients. The demolition drive resulted in the closing of a number of lodges where sex work took place, so the women moved out of the project area to other places. The NGO had to request the government to end the drive such that the program could continue unabated; they complied, but much of the damage had already been done.

It is particularly challenging for NGOs to maintain relationships when state institutions, procedures, or personnel pose a direct threat to the intervention and community. *SPAD*, for example, found out that police abuse and harassment of sex workers, extortion by lower-ranking police officers, and the misuse of judicial procedures all contributed to the vulnerability of sex workers. Despite being informed of these abuses, the NGO employed less antagonistic strategies to confront the police. In one instance, when a sex worker was wrongly accused and taken to the police station, beaten, and made to pay a fine of Rs. 100, *SPAD* did not take legal action, as it would have compromised rapport-building efforts with the police department; instead a letter was sent to the police commissioner demanding an enquiry. According to the NGO workers, the decision to go to court for a single sex worker was balanced against the larger goals of the program; in their view, they exercised restraint, such that the broader goals of empowerment could be met. When the sex worker community wanted to hold a one-day public conference to voice their grievances, *SPAD* wrote letters to various city officials, including the police commissioner, asking for protection. Permission was granted, and the event was a big success, with political stakeholders and the media present. This might not have happened had *SPAD* taken a different stance in the case of abuse. To their credit, *SPAD* has played an important role in holding sensitizing workshops with the police, and creating spaces where sex workers could confront the police directly about issues faced (see de Souza, 2009).

At the macro-level, NGOs protest national and neoliberal policies through local and international advocacy. Civil society actors agree that only the government can institute long-term programs, not the civil sector, thus it is important to lobby the government for reduced prices on drugs and long-term care and support programs for people with HIV/AIDS. Consistent with the trend towards participatory methodologies, *LPN* has started to train PLHA for the purpose of "positive speaking"; HIV+ people are trained to tell their illness narratives in public, and are invited to speak at events such as the inaugural ceremony of The Global AIDS Week of Action. Nisha, from *LPN*, explains, "through the support group meetings, we will identify people who are well-educated or have a *very* good vocabulary and speaking skills, and then we will train them for speaking." Interestingly, this form of advocacy is also integrated within the structure of the HIV/AIDS industry through funding decisions. The fight against discrimination is no longer a "lonely fight from the margins," but rather the civic engagement of a powerful HIV+ constituency within the mainstream public sphere. Programs are funded for the purpose of advocacy, and people are identified, hired, and trained to advocate for certain issues within the HIV/AIDS arena. *LPN* receives technical and financial support via their national chapter agents such as NACO, USAID, http://www.inpplus.net/partners.htm - 2#2Family Health International, Centers for Disease Control and Prevention, and the United Nations Development Program. *LPN* also received a grant from the International Labor Organization (ILO) to train people to speak on industry and workplace issues. All of this raises questions about how partnerships between the HIV+ community (as organized through the network) and external funding agents (national and international) influence issues on the ground, and what conflicts of interest arise in the process.

Tensions With Communities

Some initial challenges that NGOs face include building relationships with the identified "risk group" and convincing them about the importance of the issue at hand. In the best case scenario, the NGO already has an existing relationship with the target group (e.g., a particular village, sex workers, or street children), but if not, the first step is to build relationships with them. This involves going into the "field" to understand enablers and constraints, and to build rapport and trust with community members. Sara from *MAYA* affirms: "you can't just go out on day one, you know, with condoms in your hand, and say take a condom and STI treatment, there is a whole set of processes involved." First off, NGOs have to convince the target community that HIV/AIDS is important. Ramona says: "I mean, we have to convince them about HIV/AIDS, which is the least priority for them you know, their existence and socioeconomic priorities are much more important to them." When *MAYA* started their intervention with the transgender population, they realized that this group was more vulnerable than female sex workers and much more at risk from HIV, but HIV was still of low priority. Sara says: "So they tell us, 'you don't dare talk to us about

HIV. First, let us come openly and talk to people in the community, let us be accepted, it is better for us to die than to live in this kind of a society…so if HIV is going to kill us—that is no problem." The importance of attending to basic capabilities of life such as food, clothing, and shelter has been well-documented in the health communication literature, as Dutta writes:

> Health decisions might be located in the capability of community members to gain access to some of the primary resources of life, such as food, clothing, and shelter. In the face of the absence of these basic resources, engaging in higher order health behaviors such as getting mammograms, not smoking cigarettes, or having safe sex might seem irrelevant. (Dutta-Bergman, 2005, p. 109)

However, even when communities prioritize their own risks and speak out about what is important to them, they are often not heard because of top-down directives.

In this larger context of funding opportunities, an interesting tension observed is between *SPAD* and the sex-worker community regarding the use or "abuse" of power. NGOs build capacities of the disenfranchised community, who may then use their newly learned skills to bargain with the NGO. *SPAD* refers to a time when they dealt with such a situation. The women realized that *SPAD* as the mediator between the funders and the community was dependent on the sex-worker community for the success of the intervention, and so they threatened to go to the funders directly. Sukumar, the secretary of *SPAD*, describes this change:

> See, actually, the sex workers in the initial stage, they are only listeners. They listen to you and whatever you say they will do also. They will say yes. But when they understand the situation and they realize who you are, then they start to question. That is the beginning of their understanding of the situation, and beyond that, they start threatening you also…they start to saying *areee*, what you are trying to do is nonsense, we know that you are trying to showcase us to get funds. In these events we begin to think that they are overempowered.

Dhiren insists that "There is nothing like overempowerment, maybe overstrengthening and definitely abuse of empowerment… but that is a different case." *SPAD* was caught off guard in this situation; the staff was forced to reckon with the dialectical nature of power. The women were thinking critically about their situations, and using their collective power to garner resources, thereby highlighting the relative nature of power. There were other, less explicit instances in the data where the power of the community could be seen. For instance, *MAYA* found itself pleading with the women to go to the clinics each month, such that they could meet the project target. Ramona says,

> No, but the guideline says this is how it should be done, if you don't do it every month, that means you're not performing, the pressure is on our staff, so our staff will say, "please, you go there monthly for *my* sake"... (laughs) literally.

Thus, not only do NGOs bring funding and legitimacy to communities, but community participation legitimizes the NGO in the eyes of the funders, and the interdependencies make for dynamic power relations. As Labonte so eloquently put it: "Empowerment is not static. It is a fascinating dynamic of power given and taken all at once, a dialectical dance between consensus and conflict, professional expertise and lay wisdom, hierarchic institutions and community circles" (as quoted in Bernstein et al., 1994, p. 285).

While there are power struggles that emerge between the NGO and the community, there are also power struggles that emerge *within* the disenfranchised community, and NGOs are forced to address these tensions for the success of the intervention. *SPAD* dealt with one such situation. Tensions arose within the sex-worker group because some members wanted more authority. They wanted their peer group to resemble the organizational structure of *SPAD*; they wanted a staff, a secretary, and a president. The female staff members at *SPAD* who are older and more experienced were called "Ma'am" as a sign of respect, so the women also demanded to be called "Ma'am" by their peers, instead of the traditional "sister." The staff at *SPAD* also noticed that the women started to prioritize the financial functions of the group; ignoring other group functions such as community building. Thomas, the President of *SPAD*, said:

> They spend most of the time managing the money, rather than looking into the program, looking into the community, and relating to the community. It has gone to the extent that, when another sex worker walks into the office, they abuse them, scold them, and send them away. They say we don't have time. We are very busy (laughs).

These behaviors proved to be a serious threat to community solidarity; the alienated group members started to question the activities of the power holders, and threatened to leave the group. *SPAD* eventually called a meeting, where everyone was given a chance to express themselves. It got quite heated at one point, with personal allegations being made against the peer-group leaders, but the group eventually devised a solution that involved decentralizing various functions. This incident illuminates that even disenfranchised communities, who organize on the principle of equal participation, can turn into oppressive structures reinforcing power imbalances because of personal agendas and opportunistic behaviors.

At a larger scale as well, communities are not immune to economic and political pressures; even within the HIV+ community, issues lean in the direction of the most dominant groups. The discourse of community participation has resulted in increased incentives for HIV+ people to work in prevention and support programs. NGOs and government agencies actively seek out HIV+ people to be part of the interventions. HIV+ people work as health workers, peer educators, and positive speakers, providing much benefit to their clients and to themselves personally. People who may not have been able to find lucrative careers, either because of HIV/AIDS stigma and/or because they lack skills necessary to get jobs in mainstream society, are now being trained to be a part of this industry. But there is a tendency for those that are already relatively powerful to take advantage of the opportunities offered by participation (Labonte & Laverack, 2001). In my conversations with actors outside of *LPN*, I noted much finger-pointing about corruption and the misuse of funds within the national and local HIV+ network chapters, to the extent that a common dictum amongst them was the phrase, "HIV is not a qualification," that entitles one to resources, but rather true empowerment lies in good citizenship defined as self-reliance and giving back to the community (de Souza, 2011).

Health Interventions: A Trickle-Down Technology

Managing relationships with stakeholders each step of the way is challenging, but necessary for the success of health interventions. NGOs invest time and money in managing these relationships, and where resources exist, they create unique positions, such as "documentation officer" and "grant writer," to enable them to communicate in the language of funding agents or to talk to local officials. It is through these endeavors that NGOs are able to organize disenfranchised communities, create spaces where they can voice their issues, provide medical care and support for sick patients, and advocate on behalf of disenfranchised groups. That said, taking a step back (as even many NGO employees do), one sees the looming, billion-dollar HIV/AIDS service industry. It is easy to observe how the influx of funds has created multiple economic opportunities for the government, NGO, and private sector: the creation of global and state agencies to oversee HIV/AIDS initiatives, the creation of funding institutions to distribute funds, the creation of grant-writing institutes to apply for these funds, and unprecedented job growth in the pharmaceutical sector. So, in the end, it seems as though the intervention becomes what McKnight (1995) called a "counter-productive service technology," similar to the kind where a low-income mother is given $1 in income and $1.50 in medical care and/or services, when it is perfectly clear that the single greatest cause of her ill-health is her low income; in a modern, increasingly neoliberal world, the economic opportunities created by providing her with medical care and services motivates this paradoxical allotment of funds.

In a large industry such as this, it is almost impossible to ascertain where moral and ethical boundaries are crossed. In my study, I found sensitive and pragmatic people committed to service at the NGOs; many were aware of the critiques against them, and wary of external agents, including researchers such as myself. In fact, when I was first starting out this work, inspired by the qualitative ethic, I ended my letter to gain entry into *MAYA* by saying "I really do believe this study will make an important contribution to HIV/AIDS initiatives in India." Across time zones, I received a response from the Director three hours later. It said,

> Dear Rebecca, Your request is about the 50th we have received…all claim to be of help…but in fact few have come up with any new insights. May be, like the good ol' US of A we in India should start charging for time…and giving the proceeds to a vulnerable person…any suggestions??

Besides being a slap in the face, this response forced me to rethink some of my assumptions—that the NGOs would be delighted to talk to me, that a Western academic institution would increase my credibility, and that being Indian would increase their trust. Academics engaged in social research are very much implicated in the business of intervening; there is no shortage of "successful" health researchers, who are not much more than million-dollar grant administrators. McKnight (1995) and Poppendieck (1998) noted that the service sector is not necessarily conspiratorial, but involves well-intentioned groups of people who have bought into the belief that they were involved in good works. I am sure that if I interviewed individuals from the funding agencies, I would also find folk, both conservatives and die-hard liberals, committed to solving problems in the developing world. They may be lacking in cultural savvy, and they may be ethnocentric and unaware, but they may be well-intentioned, none the less. It is the presence of well-intentioned people that points to the importance of reflexivity in our work and recognizing the larger institutional mechanisms that have evolved to mobilize interventions. To put it bluntly, just because we are in the business of "doing good," it does not mean that we are actually doing good, especially since "good" is always defined by those within the industry.

Limitations of Participatory Health Interventions

In some circles, CBPR is referred to as a "social movement," because it uses the discourse of rights, social inequities and disparities, and citizen participation to discuss health and illness. However, it is unrealistic to expect that health interventions could bring about changes to the social order or reduce health disparities for everyone and for the long term. What they can bring about are "trickle-down" changes. Power imbalances across and within each stakeholder group mean that at every level, geospatial hierarchies continue to persist, and benefits flow in the direction of those who already have access to social and material power. NGOs are in an especially difficult situation; on the one hand, they have to protect communities against the threats posed by powerful institutions, but on the other hand, they have to tread lightly, since they are accountable to funding agents and government stakeholders. At the level of the community, there are also imbalances. A small proportion of funding has "trickled down" to high-profile risk groups in the industry (e.g., sex workers), and those individuals with existing social networks, education, and skills have been able to take advantage of the lucrative opportunities that the industry provides. However, there are many who have not benefitted and remain beyond the reach of this global industry.

There remains an underclass of people who exist in the very margins of society beyond the scope of the transnational HIV/AIDS industry. Grounding politics in one identity (e.g., sex worker, transgender, and HIV+ person) tends to obscure oppression that arises along other, more fundamental forms of structural violence. For instance, Kaveri, from *AAZADI*, described the case of an HIV-negative woman who is married to an HIV+ man, but cannot protect herself from HIV/AIDS because he rapes her. In this instance, the woman knows everything she needs to know in terms of the public-health agenda, but she is still unable to defend herself from being infected with HIV/AIDS; the woman exists at the periphery of all existing institutions—economic, health, and social. In the case of children as well, participatory methodologies are meaningless, because children cannot be treated as "partners" in any intervention, especially sexual ones. In most cases, NGOs working with children are forced to use harm-reduction strategies. Aruna, the founder and program manager at *SHRUTI*, provides an example:

> We were doing this street play once of safe sex—I remember this very clearly…one small little boy came to me and said "you give me also that whatever you are giving to the others." So I told him, "See, this is not for you, you should not use it. This is for the adults." He says, "what is that, *Nirodh* (condom brand)? What do you think, I don't have private organs?" It sounded so disgusting! In other words, he has seen all these things.

She continues, "See, being an NGO, we have a very very sensitive role to play and we have to be nonjudgmental." The staff knows that they are providing temporary solutions that ignore the elephant stomping in the room. Gayathri, a children's counselor, provides another example:

> We had one child like that. She was a sex worker about 15 years old. She was stinking of alcohol when we got here. She did not even have underwear, just some dirty dress. We tried to rehabilitate her…she stayed for some time, but after that she said she wanted men.

The stories of these nameless people point to the long-term effects of "structural violence"—large-scale social forces, such as political violence, poverty, and other social inequalities rooted in historical and economic processes that sculpt the distribution and outcome of disease and illness (Farmer, 2005). In these instances, health interventions, even those that encourage genuine participation, are grossly inadequate, because they leave open the question of who is *their* "community" and what does participation entail for *them*? Despite the money poured into global health problems, interventions are small-scale projects that take place in the long history of social stratification. They are successful at removing some barriers to health, such as lack of information or access to medical care in *some* areas for *some* communities, for a certain period of time, but large-scale problems of poverty and violence against women and children (for example), located and engendered historically, are not easily ameliorated through small-scale interventions. Political and economic reform, involving a redistribution of wealth, resources, and social power at the global and local scale, is needed to eliminate power inequitie—a key predictor of health disparities.

Perhaps we can imagine that ten years later the little girl who slipped through the cracks of *SHRUTI's* intervention finally gets caught in the intervention mobilized by *SPAD*. She is standing on a street corner awaiting clients, when she is approached by a peer who invites her to participate in the sex-worker program. She hears from the funders, the NGOs, the health experts, and peers about the importance of using condoms, of joining together with women like her, of speaking out against stigma, and of taking pride in her identity as a sex worker. After some rational consideration, she becomes a "partner" in a multimillion transnational conglomerate, which will provide her (at best) with health information, free condoms, a savings account, and some new friends. This is a step in the right direction, but someone should ask her if she, as the final "customer" of this industry, is satisfied with her product? Does she know that her package of benefits comes with a multimillion-dollar price tag? Does she know how many hundreds of people have been involved in "helping" her? And also, should we ask her if she can think of better ways in which we could spend this money on her?

Co-optation of the Participatory Method

Today, the terms *participation* and *empowerment* are buzzwords co-opted in the field of global health and health disparities. The participatory method is used to solicit millions of dollars in grant funding; surprising, because the method was born and developed in some of the most impoverished parts of the world, with relatively few resources. Freire, Escobar, and other Latin American scholars protest the co-optation of this sacred principle of participation and dialogue, arguing that it was "easily co-opted by the established system and rendered ineffective or counter-productive" (Escobar, 1999, p. 326). Even when communities speak about what is important to them, they are not heard, and participation is rendered as mere tokenism. The voices of marginalized communities can rarely compete with the logic of the neoliberal agenda, its paid experts, and the system devised to govern the problem. For instance, NGOs may recognize that the goals of the intervention conflict with the needs of the people, but in this competitive climate of grant-seeking, they must either force the intervention on the community, compromise on issues and/or methodology, or risk losing funding.

Especially in this climate, there is a need to recognize the difference between participation which is actually allowed to bring about transformation in the social order, and participation which cannot bring about change because it is locked within a system of governance that profits from its own existence. From a Freirian lens (Freire, 1970), participation implies that all stakeholders are changed by the process of dialogue and communication. Change "at the bottom" involves disenfranchised individuals and communities becoming conscientized, and recognizing the sociopolitical context within which they live; change at the top involves people in positions and institutions of power coming to realize the role they play in perpetuating the status quo. As practitioners, we have to intervene in the dominant modes of "thinking and doing" to bring about both these types of change. We must educate ourselves and others about the nuances of human agency and structural constraint. We have to liberate ourselves and the communities that we work with from the authoritative hold of science, not simply counting everything, but finding out what actually counts to people. Our interventions must develop in dialogue with the way people live, rather than force people to live in ways that fit our theories, models, and rubrics. Culturally meaningful health communication entails creating a dialogic framework that allows community voices to be heard: "These voices present opportunities for social change by challenging the dominant articulations of social reality and by suggesting alternative interpretations" (Basu & Dutta, 2009, p. 108; Dutta & de Souza, 2008). In keeping with the spirit of social reform that has historically rooted the participatory paradigm, there is a need to destroy the causes which nourish false charity, recognizing that disparities between rich and poor, healthy and sick will not be fixed merely through well-intentioned, billion-dollar interventions, but through understanding the causes of injustice, and recognizing the counterproductive nature of the service machinery that has evolved to "manage" health disparities.

Bibliography

Airhihenbuwa, C. (1995). *Health and culture: Beyond the Western paradigm.* Thousand Oaks, CA: Sage.

Basu, A., & Dutta, M. J. (2009). Sex workers and HIV/AIDS: Analyzing participatory culture-centered health communication strategies. *Human Communication Research, 35*(1), 86–114.

Bernstein, E., Wallerstein, N., Braithwaite, R., Gutierrez, L., Labonte, R., & Zimmerman, M. (1994). Empowerment forum: A dialogue between guest editorial board members. *Health Education Quarterly, 21*(3), 281–294.

Bill & Melinda Gates Foundation (BMGF). (2010). Avahan: India AIDS initiative. Retrieved from http://www.gatesfoundation.org/avahan/Pages/overview.aspx.

de Souza, R. (2007). *NGOs and empowerment: Creating communicative spaces in the realm of HIV/AIDS in India* (Doctoral dissertation). Purdue University, West Lafayette, IN. Retrieved from Dissertations & Theses @ CIC Institutions database. (Publication No. AAT 3291160).

de Souza, R. (2009). Creating 'communicative spaces': A case of NGO community organizing for HIV/AIDS prevention. *Health Communication, 24*(8), 692–702.

de Souza, R. (2011). Local perspectives on empowerment and responsibility in the new public health. *Health Communication, 26*(1), 25–36.

de Souza, R., Basu, A., Kim, I., Basnyat, I., & Dutta, M. (2008). The paradox of 'fair trade': The influence of neoliberal trade agreements on food security and health. In H. Zoller & M. Dutta (Eds), *Emerging perspectives in health communication: Interpretive, critical and cultural approaches* (pp. 411–430). Mahwah, NJ: Lawrence Erlbaum.

Dutta, M. J. (2007). Communicating about culture and health: Theorizing culture-centered and cultural sensitivity approaches. *Communication Theory, 17*(3), 304–328.

Dutta, M. J. (2008). Participatory communication in entertainment-education: A critical analysis. *Communication for Development and Social Change, 2*(1), 53–72.

Dutta, M. J., & Basnyat, I. (2006). The Radio Communication Project in Nepal: A culture-centered approach to participation. *Health Education and Behavior, 35*(6), 459–460.

Dutta, M. J., & Basnyat, I. (2008). The case of the Radio Communication Project in Nepal: A culture-centered rejoinder. *Health Education and Behavior, 35*(4), 459–460.

Dutta, M. J., & de Souza, R. (2008). The past, present, and future of health development campaigns: Reflexivity and the critical-cultural approach. *Health Communication, 23*(4), 326–339.

Dutta-Bergman, M. J. (2005). Theory and practice in health communication campaigns: A critical interrogation. *Health Communication, 18*(2), 103–122.

Escobar, A. (1995). *Encountering development: The making and unmaking of the Third World*. Princeton, NJ: Princeton University Press.

Escobar, A. (1999). Discourse and power in development: Michel Foucault and the relevance of his work to the Third World. In T. L. Jacobson & J. Servaes (Eds.), *Theoretical approaches to participatory communication* (pp. 309–335). Cresskill, NJ: Hampton.

Farmer, P. (2005). *Pathologies of power: Health, human rights, and the new war on the poor*. Berkeley: University of California Press.

Foucault, M. (1977). *Discipline and punish: The birth of the prison*. London, UK: Allen Lane.

Freire, P. (1970). *Pedagogy of the oppressed*. New York, NY: Seabury Press.

Goodman, R. M., Speers, M. A., McLeroy, K., Fawcett, S., Kegler, M., Parker, E., Smith, S. R.,…Wallerstein, N. (1998). Identifying and defining the dimensions of community capacity to provide a basis for measurement. *Health Education & Behavior, 25*(3), 258–278.

Green, A., & Matthias, A. (1997). *Non-governmental organizations and health in developing countries*. New York, NY: St. Martin's Press.

Jana, S., Basu, I., Rotheram-Borus, M. J., & Newman, P. A. (2004). The Sonagachi project: A sustainable community intervention program. *AIDS Education & Prevention, 16*(5), 405–414.

Labonte, R., & Laverack, G. (2001). Capacity building in health promotion. Part 1: For whom? And for what purpose? *Critical Public Health, 11*(2), 111–127.

Laverack, G. (2004). *Health promotion practice: Power and empowerment*. Thousand Oaks, CA: Sage.

Macedo, D. (2002). Introduction to 30th anniversary edition. In P. Freire, *Pedagogy of the Oppressed*. New York, NY: Continuum.

McKnight, J. (1995). *The careless society: Community and its counterfeits*. New York, NY: Basic Books.

Melkote, S., & Steeves, L. (2001). *Communication for development in the Third World: Theory and practice for empowerment*. Thousand Oaks, CA: Sage.

National AIDS Control Organization (NACO). (2010). Funds and expenditures. Retrieved from http://www.nacoonline.org/About_NACO/Funds_and_Expenditures/.

Patton, C. (1991). *Inventing AIDS*. London, UK: Routledge.

Petersen, A., & Lupton, D. (1996). *The new public health: Health and self in the age of risk*. Thousand Oaks, CA: Sage.

Poppendieck, J. (1998). *Sweet charity? Emergency food and the end of entitlement*. New York, NY: Penguin.

Salamon, L. M. (1993). *The global associational revolution: The rise of the third sector on the world scene* (Occasional paper No. 15). Baltimore, MD: Institute for Policy Studies, Johns Hopkins University.

Shrivastava, B. (2008, January 30). 350 NGOs sacked in mega NACO clean-up. Retrieved from http://www.livemint.com/2008/01/29235754/350-NGOs-sacked-in-mega-Naco-c.html

World Bank. (2004). *Report: Preventing HIV/AIDS in India: World Bank.* Retrieved from http://go.worldbank.org/FKLVQCCF40.

World Health Organization (WHO). (1978). Declaration of Alma-Ata: International conference on primary health care, Alma-Ata, USSR, 6–12 September 1978. Retrieved from http://www.who.int/publications/almaata_declaration_en.pdf.

World Health Organization (WHO). (2011). Civil society initiative (CSI): NGOs and health. Retrieved from http://www.who.int/civilsociety/health/en/.

Zoller, H. M. (2008). Technologies of neoliberal governmentality: The discursive influence of global economic policies on public health. In H. Zoller & M. Dutta (Eds.), *Emerging perspectives in health communication: Meaning, culture, and power* (pp. 390–410). New York, NY: Routledge.

23

Culture-Centered Deconstructions: In-depth Interviews With HIV Planners Targeting Tribals in Orissa, India

Lalatendu Acharya and Mohan J. Dutta

The historical marginalization of the indigenous tribes of India continues to be carried out in contemporary development policies and programs (Dutta-Bergman, 2004a, 2004b). Health is articulated within this broad paradigm of development, and the health-care inequalities faced by the indigenous tribes of India are documented across a wide variety of health-care contexts, ranging from access to preventive resources and services to access to hospitals, health services, and treatment options. Of growing relevance is the documentation of the high risks of HIV among tribal populations, exacerbated by the globalization policies of the state, the structural adjustment programs, the uprooting of tribal populations from their homes, and the development of migratory patterns, with many members of the tribal communities migrating to work sites in search of a living (Dutta, 2008). It is against this backdrop that HIV programs have been developed in the state of Orissa that focus on the many indigenous tribes of Orissa as the targets of their interventions ("Orissa Vision 2010," 2003). How, then, do indigenous tribes emerge in discourses of HIV programming?

In this chapter, we seek to interrogate the discourses of the HIV interventions targeting tribals against the backdrop of health-care disparities, attending to the symbolic representations of the target population that circulate in mainstream campaign discourses, and engaging theoretically with the implications of these representations. Through in-depth interviews conducted with HIV program planners in the state of Orissa, we simultaneously coconstruct and deconstruct the signifiers of representation that occupy the articulations of campaign planners. The in-depth interviews with program planners constitute the mainstream discourses of campaigns, and the deconstruction of mainstream campaign discourses creates entry points for interrogating the presence/absence of the subaltern sectors in these discourses.

We situate our project on the theoretical framework of the culture-centered approach, paying attention to the goals of the approach in creating participatory spaces for addressing health-care disparities by first examining the erasures and silences that are written into the dominant frameworks of health communication (Dutta, 2008; Dutta-Bergman, 2004a, 2004b). The interpretive lens of the culture-centered approach is situated within the broader commitments of Subaltern Studies theory to examine the erasure of subaltern voices in narratives of development and health communication. The culture-centered approach fundamentally argues that the structural disparities in health are perpetuated through the inequalities in the terrains of symbolic representations and through the erasures of subaltern agency in mainstream discourses (Dutta, 2008, 2009). Therefore, the objective of this chapter is to critically evaluate the narratives constructed by campaign planners, examining these narratives for the representations they utilize to talk about the subaltern target audience of the intervention. We wrap up the chapter with discussions of the implications of such constructions, particularly in light of the material inequalities that are faced by the subaltern sectors.

Theoretical Framework: Culture-Centered Approach

The theoretical framework of the culture-centered approach puts forth a deconstructive element in the theorization of existing bodies of knowledge, questioning the representations that constitute these bodies of knowledge, and engaging critically with the erasures that are mapped out through the discourses of mainstream knowledge (Dutta, 2008, 2009; Dutta-Bergman, 2004a, 2004b; Basu & Dutta, 2009). Questions of erasure and silencing become particularly relevant when we examine discourses of marginalization, attending to the communicative erasures of certain communities from dominant discursive spaces, and engaging critically with the implications of these erasures in terms of material inequalities (Dutta, 2008). The critical engagement in the culture-centered approach is driven by a continuous attention to the silences, fixities, and erasures that are achieved through dominant discourse about subalterns.

The culture-centered approach puts forth the fundamental argument that the material marginalization of disenfranchised communities is carried out, constituted, and reinforced symbolically, meaning that the erasure of voices of the subaltern sectors is directly tied to material violence carried out on the subaltern sectors through a variety of practices framed under the logic of Enlightenment (Dutta, 2008). Therefore, through deconstruction, the structures underlying dominant discourses of health become explicit, drawing attention to the marginalizing practices situated within these structures, and creating openings for critically interrogating these dominant structures for the ways in which they go about erasing subaltern agency and fixing subaltern agency as a target-audience segment to be worked over through messages of Enlightenment. Therefore, methodologically, the deconstructive turn in the culture-centered approach examines the discourses of those very programs of development that are targeted toward subaltern communities, evaluating these discourses in the context of the silences, erasures, and perpetuation of disenfranchisement. It is then on the foundations of these deconstructions that spaces are created for listening to the voices of subaltern communities, and for centering subaltern knowledge claims as entry points for envisioning health.

Indigenous Tribes in Orissa

Context

Indigenous people make up almost 400 million of the world's population, which constitutes around a third of the 900 million extremely poor, rural people of the world (UNDESA, 2010). India has the second largest concentration of tribal population in the world (Census of India, 2001). According to the Census of India report (2001), the Indian tribes constitute around 8.2% of the nation's total population, constituting nearly 84.3 million. There are 635 different tribes in India, located in five major tribal belts across the country, with seven Indian states accounting for more than 75% of the tribal population. The main concentration of tribal people is the central tribal belt in the middle part of India and in the northeastern states. The tribal communities of India belong to different ethnolingual groups, profess diverse faiths, and are at varied and/or different levels of economic, educational, and cultural development (Planning Commission of India, 2008).

Orissa is a state on the eastern coast of India and has a 36.8 million (Census of India, 2001) population, and ranks high in poverty and vulnerability of its people. Orissa has a total tribal population of around 7 million, which constitutes around 19% of the state's total population. It is the third highest tribal population in India, accounting for11% of India's tribal population (Planning Commission of India, 2003). More than 50% of the tribal population in Orissa is found in the undivided Koraput, Sundargarh, and Mayurbhanj districts of Orissa. The southern district of Koraput is home to more than 25% of Orissa's total tribal population. The Indian constitution lists 62 different types of scheduled tribes in Orissa, most of whom live in designated scheduled areas. Out of these tribes, the Indian government identifies 13 tribes as "primitive" tribes, based on the government's view of the tribes' cultural and technological status (Planning Commission of India, 2003). The tribal population is predominantly rural, with close to 94% living in villages (Integrated Social and Environmental Assessment Study, 2008), situated in remote and hilly lands of the state. Reports note that this has been the traditional dwelling of many of the tribes for centuries, and also that many of them are driven there after losing their lands on fertile plains to the nontribal and various government and private projects (Planning Commission of India, 2003). The tribal historically depended on gathering forest produce, shifting cultivation, hunting, and farming in valley lands for their survival. But now, with the industrialization, land alienation of tribes, displacements, government policies regarding forest preservation, and lack of sufficient markets for forest produce, there is an increasing shift toward taking up jobs as daily laborers, private household servants, temporary workers in government development schemes, vehicle and/or machinery operators, and unskilled migrant workers.

The Planning Commission of India further noted that the Orissa tribal population is amongst the poorest in the state and is routinely exploited by dominant social, cultural structures. Due to their history of vulnerability, exploitation, and prolonged marginalization, the indigenous tribes of Orissa have a comparatively poor socioeconomic status (Integrated Social and Environmental Assessment Study, 2008). It is reflected in the following indicators from the Government of India Census (Census of India, 2001). The overall literacy rate of the scheduled tribes in Orissa is 37.4%, as compared to the state average of 63.08%. The maternal, child, and infant mortality rates are very high amongst the tribes in Orissa.

Per the "Orissa Vision 2010—A Health Strategy" document (2003) developed by the Government of Orissa, the health of the tribal population in Orissa is worse than the national average, with an infant mortality rate of 84.2, and under-five-years-old mortality rate of 126.6 per 1,000 live births. There are some tribal pockets where the infant mortality rate is as high as 200 per 1,000 live births (Planning Commission of India, 2003). Similarly, the tribal children aged 12–23 months who received basic vaccinations are much lower than the rest of the population, and have a much higher incidence of anemia and overall malnutrition. Of the tribal children under three years of age 56.7% are underweight (National Family Health Survey (NFHS)-III, 2006). The survey further found that approximately 69% of the tribal women between 15–49 years of age are anemic, and 46.6% of the women have a body mass index lower than normal (NFHS-III, 2006). Similarly, a report by the Planning Commission of India (2008) noted that the health status of the scheduled caste and scheduled tribes are far worse than any other sectors in the society. Access to care is limited for the tribal, and since most of the tribal population lives in remote rural areas, barely 18% had deliveries in a health facility, compared to 51% among other communities (Planning Commission of India, 2008). An example would be Koraput district, where the institutional delivery amongst the tribal populace is less than 15% (National Rural Health Mission (NRHM) Orissa Report, 2009). The tribal population of Koraput, like many other tribal populations elsewhere, is marginalized, lives in remote areas with little access, and suffers from many endemic diseases such as malaria, HIV/AIDS, tuberculosis, diarrhea, and sickle cell anemia (Planning Commission of India, 2008).

The tribal communities are highly vulnerable to various health problems, especially, communicable diseases including HIV/AIDS, due to their poor health infrastructure, high levels of poverty, and ignorance (Naik et al., 2005). The National Family Health Survey-III (NFHS-III), conducted in 2006, reported that only 32.6% of tribal women and 56% of tribal men in Orissa have ever heard of HIV/AIDS. When we look at the percentages that have a comprehensive knowledge about HIV/AIDS,[1] we find that it is only 3.3% of tribal women and 11.5% of tribal men in Orissa (NFHS-III, 2006). If we look at the percentage of total rural men and women in Orissa that have ever tested for HIV/AIDS prior to NFHS-III, only 0.3% of rural women and 1.4% of rural men reported having done so. These figures paint the serious health situation prevalent in the tribes of Orissa and their vulnerability to HIV/AIDS. This vulnerability is further enhanced by the increasing encroachment of the dominant society on the tribal spaces.

A social assessment study commissioned by the National AIDS Control Organization (NACO, 2008) arrived at a conclusion that certain tribal groups in India have relatively lenient attitudes toward premarital sexual activity, have permissive attitudes, and hence are more vulnerable to risks of STIs and HIV/AIDS infections. A similar conclusion was noted by Naik et al. (2005) in their article: "Lack of awareness, permissiveness of tribal societies for premarital or extra-marital sexual relationships, and sexual mixing patterns predispose these communities to HIV/AIDS and STD infections." We came across several such articulations when we discussed and perused research documenting the prevalence and spread of HIV/AIDS amongst tribal populations. Most of the utterances located the problem in the unsafe sexual practices of the tribal culture, in their permissive cultural practices, and their attitude toward casual sex. This prompted us as researchers to critically engage with the understanding of the program planners of the problematic of HIV/AIDS, and its construction with respect to the tribal.

Sampling

The researchers conducted one-to-one, in-depth interviews with 21 participants; representative of the group of the HIV/AIDS program planners and implementers in Koraput, a predominantly tribal district of Orissa. The district of Koraput, with 5.38% of Orissa's area, has a population of 1.18 million, of which more than 50% are tribal and scattered widely in villages and hamlets (Census of India, 2001). The respondent list consisted of two program managers of the district government health department; two program managers of international nongovernmental organizations; two program officers from a United Nations organization; two heads of leading, local NGOs working with HIV/AIDS issues; 11 NGO workers, each with more than 5 years of experience working with HIV/AIDS campaigns and interventions; and two medical doctors working with HIV/AIDS prevention and treatment programs in Koraput. The researchers also reviewed some available, published documents on HIV/AIDS in Koraput in order to gain a sense of the field, and to supplement the discourses that emerged through the in-depth interviews.

Sampling Strategy

The sample was constructed from respondents who could be of any gender, any class, any ethnic background, any health status, and above 18 years of age; the sole inclusion criterion was that they should belong to the group of HIV/AIDS program planners and implementers working with tribal populations in Koraput. This group of people was defined to consist of government workers, nongovernment workers, international agency program officers, local NGOs, and key health workers working in the Koraput district. Participants were involved in either planning or implementation of the HIV/AIDS programs among the tribal population in Koraput. To cover all key people in the district, the researchers followed a purposeful sampling strategy, in which they selected the key individual respondents and the sites for study (Hatch, 2002). Flexibility and room was maintained for adding to the sample, depending on suggestions from the field and emergent opportunities as defined in opportunistic sampling (Creswell, 2007). All interviews were recorded with the permission of the respondents after the research purpose was explained and they had signed on an IRB Informed Consent Form. The researchers continued the investigation until they reached a point of theoretical sufficiency (Charmaz, 2008). The interviews varied between 25 minutes to an hour in length, and were conducted in the field at different places, which were decided by the participants based on accessibility and convenience. The 21 interviews resulted in 80 pages of single-spaced text for further analysis. The researchers' field notes and reflexive journaling yielded an additional 30 pages of data.

Data Analysis

A constructivist, grounded theory approach was employed, which adopts the grounded theory guidelines and methods but does not subscribe to the objectivist, positivist assumptions inherent in its earlier formulations (Charmaz, 2008). As Charmaz (2008) wrote,

> The critical stance in social justice in combination with the analytic focus of grounded theory broadens and sharpens the scope of inquiry. Such ef-

forts locate subjective and collective experience in larger structures and increase understanding of how these structures work. (p.508)

The transcripts were read once and then in several passes to discover the codes and, subsequently, the overarching themes. The initial passes resulted in 293 open codes. Then the data was examined in detail and memos written which linked the field experiences and learning when doing the interviews and discussions in the field. The memos threw significant insights and questions on the way the tribal is positioned in the articulations of the participants and in their interactions.

With the memos, we started focused coding, which meant using the most significant and/or frequent earlier codes to sift through the large amounts of data and take decisions about which initial codes made the most analytic sense (Charmaz, 2008). We started axial coding, where the objective was to relate categories with subcategories. As Charmaz (2008) mentioned, "Axial coding relates categories to subcategories, specifies the properties and dimensions of a category, and reassembles the data you have fractured during the initial coding to give coherence to the emerging analysis" (p. 60). Axial coding, according to Strauss and Corbin (1998), brings back the data together again in a coherent whole. From the open codes, we reassembled 23 axial codes and categories, and went back to the data again to do selective coding, and arrived at a theoretical integration. At this stage, the relationships among the distinct categories were established at a more abstract level (Strauss & Corbin, 2007), and enabled us to conceptualize and pull together the different meanings into overarching themes.

Discourses of Marginalization: Paradoxes of Intervention

Our discursive engagement with the campaign planners threw light on the postcolonial contexts of intervention planning, demonstrating the ways in which intervention discourses continue to carry out the modernist depictions of the tribal as the primitive, removing the fundamental capacity of tribal community members to engage in interpretations, meaning making, and actions. The marginalization of tribal communities of India is played out through discourses of development and progress that continue to cast tribal people as primitive members of a community that is set back in time and space, and in need of the mantras of Enlightenment. The discourses of HIV-intervention planners continue to carry out these narratives, and in doing so, further continue to constitute the tribal as a homogeneous entity at the margins of contemporary health care.

Primitive Indigenous Cultures and Modern Interventions

The discourses of HIV-intervention planners and field workers operate through the construction of the tribal communities in terms of lack of agency, set backward both in terms of time and space. Worth noting in the discourses of HIV programmers and planners is the interpenetration of time and space in portraying a community in need, situated back in time amidst a space of underdevelopment. Communities are constructed as primitive, and it is this precise construction that offers the framework of problem definition underlying the articulation of the rationale for the health intervention. Consider the following articulation by one of the participants, who is an HIV program planner in the area:

> Koraput is fully tribal and people depend on forests and mountains. Agricultural lands are few, so they depend on towns, urban centers and markets. Here education is very less and people migrate out to Khurda, Bangalore, Raipur. Before they did not have much knowledge about HIV AIDS and so outside they were maintaining relationships outside and then coming back and starting relationships here. Also, in tribal culture, sex is free. During festivals it's totally free.

Note the economic logic of development that works into the constructions of health interventions; the forests, mountains, and agricultural lands are juxtaposed in the backdrop of the towns, urban centers, and markets. Tribal behavior is set in the context of the lack of education and the absence of knowledge among tribal communities. The depiction of the outside contacts of the community is constituted in terms of the lack of education and awareness within the community, and not in terms of the structural factors underlying tribal displacement, migration, and structural marginalization. Culture emerges in HIV program planner discourse as a marker of the bizarre and the mysterious, serving, therefore, as the site of the intervention, and continuing to reify the colonial logic of development and progress.

Worth noting in program-planner discourse is the representation of the community, in terms of the absence of knowledge, and the simultaneous absence of articulations of social, economic, and political structures surrounding the lived experiences of tribal communities in Orissa and throughout several parts of India. Also worth noting is the discourse of tribal culture, based on dominant understandings of tribal culture, where sex is constructed as "free." Emphasized in the cultural discourse is the primitivization of culture, tied to the articulation of cultural logics in the context of problematized tribal behaviors, as seen through the lens of the campaign practitioners. Also consider the following articulation that carries out the depiction of the tribal as primitive:

> They have a tradition also. In their village they do not have many restrictions on sex. In tribal culture there are many traditions...during festivals, say, during Chaitra Parba,[2] Jatra,[3] there is not much restriction on sex between boys and girls in villages...everyone knows...between tribals. Say a boy and girl like each other...the boy takes the girl and goes off...the girl comes back and again goes out with others.

Note the emphasis on the tradition. Tradition offers an interpretive frame for locating the etiology of HIV/AIDS in the tribal communities of Koraput, Orissa. The depiction of tradition in tribal villages fundamentally serves as a window into tribal life within the modernist logics of intervention discourse. The description of the tribal festivals emerges as a window into tribal life, once again constituting the logic of development embedded in HIV-campaign discourse.

Awareness and Explanation

It is against this backdrop of the portrayal of the tribal culture as primitive that intervention planners emphasize the lack of education in tribal communities:

> Education is a big problem here. If we focus on education, we can have success...now also there are a lot of uneducated tribals. This is a tribal dominated district...and they are illiterate. If we do not focus on the major group...so they will automatically be aware. They will gain knowledge, they will know their rights and they think about themselves, their village, their community.

Worth emphasizing is the depiction of the uneducated tribal in the discourse. Education is established as the problem configuration under which, then, HIV/AIDS programs are framed. Tribals are constructed as uneducated, being evaluated within the Enlightenment framework of development, progress, education, and health. Literacy and education offer the frames for understanding the problems of the tribals, and therefore the interventions are framed in terms of educating and empowering the community so that the community becomes aware of its rights, having been touched with the mantra of Enlightenment.

Worth noting here is the hierarchical flow in the depiction of knowledge, where knowledge flows from the intervention planners to the community, and gaining knowledge is tied to the logic of coming into consciousness and becoming aware of one's rights. The creation of consciousness within the community, then, is seen as the task of intervention planners, who are charged with the tasks of making the tribal communities aware of their rights, teaching them to think about themselves, their villages, and their communities. HIV/AIDS programming, therefore, is framed in terms of awareness and education of the tribal. Consider the following excerpt:

> Suppose I put it this way. I would say that awareness has increased but the behavior change which should be the result of the awareness of the program, that has not been achieved till now. As it is to say that out of sight is out of mind...so we keep on educating people about awareness. People's awareness has increased to some extent, they are aware. But they are not concerned and behavioral change has not happened, that way more awareness is needed.

The problem of HIV/AIDS is framed in terms of a lack of awareness, and therefore, as the need for education. Worth noting here is the paradox of the awareness-driven approach that continues to circulate in the articulations of program planners. Although, in this instance, the program planner notes that awareness has increased in the area and has not really manifested in the form of behavior change (therefore, depicting the gap between awareness and behavior), his suggestion is that more awareness is needed. The acknowledgment of the limits of the awareness-based framework is then tied to the very logics of the framework, ultimately resulting in recommendations for increasing awareness efforts.

Here is another example that depicts the framework of awareness running through HIV/AIDS program planner discourse in Koraput:

> Then, there is the lack of awareness which is very important. We are trying to increase awareness. We are promoting condoms, but they know condoms are available, but they do not know where they are available and how to use it. They do not have the behavioral practice that if they use condoms they will benefit from it. They are not interested. Lack of interest to use condoms is another thing. And even in many of the internal belts, condom is not reaching them. This is essential, as HIV/AIDS is not happening to all people, HIV/AIDS is affecting illiterate people…unawareness, they are least concerned about safe sexual practices…also there is a lot of lack of access here.

In this discourse, there is an explicit emphasis on illiteracy as the root of HIV/AIDS. It is precisely on this logic that the enterprise of HIV/AIDS-campaign planning is established. Making tribal community members literate through the provision of knowledge is seen as the way to reduce the risks of HIV in the tribal communities of Koraput.

Community Participation

The awareness-based framework of the HIV program planners is intertwined with ideas of participation, framing local community participation as a method for diffusing the health intervention in the local communities. Here, community participation is framed as a tool for carrying out the awareness-based goals of the intervention, creating greater awareness in the target communities by offering increasing access for campaign planners and field staff. Consider, for instance, the following narrative of HIV program planning:

> We make community-based groups, village institutions, federations…farmers federation…and when we are making those groups we are making inclusive groups, we do not discriminate…we do not see whether disabled patients are there or positive people but we try to make the groups inclusive and try for maximum inclusion. So whenever we get opportunity we interact about health and education and at different times we make drug distribution centers where people can get immediate medicines when required like during epidemics or in risky periods like rainy season…everybody is ready and in villages they distribute.

The group settings offer spaces for carrying out the intervention programs. It is in the collective context of the community-based groups that the program planners and field workers interact with the local community in order to distribute information and education materials. Also, once set up, the community groups become resources for distributing intervention materials in the local, rural communities. Community participation therefore becomes a conduit, offering a space of access to the local village and to the wider community, emerging as a space for carrying out the objectives of the intervention.

The location of community participation within the broader framework of program awareness is further evident in the following excerpt:

> Awareness is a very broad term. In awareness levels whatever tools we use that may be applicable to a certain pocket of the society....And for tribal areas we have to make tribal volunteers...taking the people from their society and training them, those who will be peer educators and such and they will only explain it to them.

The role of the peer educators in the framework of awareness is constituted in terms of carrying out the information to the community, primarily conceived of as a communication channel. Worth noting here is the paradox of participation in HIV/AIDS program planner discourse, where the participation of community members is constituted in their role as communication channels, almost perceived in the form of vessels for carrying out information to the remote locales of tribal communities that are otherwise difficult to reach.

Discussion

Our culture-centered interrogation of the discourses of the HIV/AIDS campaign planners in the tribal regions of Koraput point toward the depictions of tribal culture that continue to circulate in dominant discourses of HIV/AIDS about tribal people. In these narratives of tribal health, tribal culture is constituted as a barrier, depicted in the form of cultural categories and practices as perceived by the campaign planners. Our discursive engagement with the perceptions of the campaign designers draws attention to the portrayal of the tribal culture as backward and primitive. Health is conceptualized in the context of lack of knowledge tied in to the cultural practices, and therefore, the intervention is conceptualized as an awareness and/or information program that seeks to bring about education about HIV/AIDS in the tribal population.

The culture-centered approach equips program planners and intervention developers working with marginalized populations with a critical lens that interrogates the marginalizing nature of dominant intervention discourses. It is by turning this lens toward the processes of knowledge creation that the culture-centered approach creates spaces for listening to the voices of the subaltern communities that have historically been erased by the mainstream processes and markers of knowledge production. For health communication interventions seeking to address health-care disparities, critical deconstructions such as the one carried out in this project create openings for continuously interrogating the silences that are created through the languages, attitudes, and programs that are reflected in interventions. To the extent that the subaltern sectors of the globe continue to be scripted as passive and without agency, dominant health programs continue to reproduce the structural marginalization of such subaltern communities by denying them dignity and the right to have a say in things that matter to them.

We would specifically like to emphasize the examination of paradoxes that emerge through culture-centered deconstructions. For instance, the discourses of program planners continue to perpetuate the construction of the tribal as pathology, and therefore erase tribal agency, even as they articulate strategies of community participation (see, for example, Dutta & Basnyat, 2008a, 2008b). The paradox of participation lies in the contradictory impulses of participatory processes, as framed within the language of empowerment in dominant program-planning discourses. Even as programs discuss processes of empowerment, they do so by fundamentally denying the capacity of local communities to participate in making choices that are meaningful to them. Even as program planners discuss processes of empowerment, their articulations of empowerment are constrained within narrowly defined logics of using local communities to diffuse top-down messages, and continue to reify paternalistic notions of local communities and their practices, thus fundamentally silencing these communities and minimizing the possibilities for their participation.

Culture-Centered Praxis

We end this chapter with a call for further work in the development of reflexive methodologies for policy makers and program planners, critically interrogating the ways in which their narratives of their target audiences continue to perpetuate the material inequalities with which they seek to work. Of particular importance is the need for developing reflexive tools that attend to the dichotomies in projects of empowerment that continue to perpetuate the logics of marginalization within objectives of empowerment. Worth noting in this culture-centered deconstruction of campaign-planner discourses are the tensions that imbue enlightenment-based principles of health programming and intervention development that seek to mitigate health inequalities and disparities.

To the extent that campaign discourses continue to fix their target audiences as passive recipients, and marginalize them through the ways in which they capture these audiences as homogeneous masses frozen in antiquity, they continue to perpetuate the fundamental inequalities in health by denying local communities entry points for dialogue and for having a voice in determining their health problems and corresponding health solutions. Furthermore, the perpetuation of colonial tropes within the so-called programs of empowerment continue to locate power in the outside sender of messages, therefore working toward disenfranchising local communities by erasing local possibilities for participation and enactment of agency.

Reflexive tools for campaign planners developed on the frameworks of the culture-centered approach would continually need to question the positions of power that are being served through discourse, the erasures that are played out through the perpetuation of the Enlightenment logic, and the possibilities of working against these erasures by creating entry points for listening to subaltern communities through the fundamental acknowledgement of the capacity of local communities to develop solutions that are meaningful to their lived experiences. Reflexivity therefore involves the continuous interrogation of the privileges that are embodied in the positions of power occupied by intervention planners, campaign staff, and evaluators. A turn toward listening would begin with the fundamental questioning of the foundations of those very knowledge claims that constitute the core of campaign knowledge and campaign practice. It is through this very interrogation of the core, for instance, that possibilities are opened up for critically engaging with the meaningfulness of theories through conjectures and refutations, connected with the willingness to reject those theories that do not seem to fit into the lived experiences and shared narratives of local communities.

Notes

1. Respondents with comprehensive knowledge of HIV/AIDS say that the use of a condom for every act of sexual intercourse, and having just one uninfected, faithful partner can reduce the chance of getting HIV/AIDS; say that a healthy-looking person can have HIV/AIDS; and reject the two most common perceptions, namely that HIV/AIDS can be transmitted by mosquito bites and by sharing food.
2. *Chaitra Parba* is the most important festival for the tribal population of Koraput, and is observed in the months of March–April, when whole villages go into festive mood for almost a month.
3. *Jatra* is the local word for *theatre*.

Bibliography

Basu, A., & Dutta, M. J. (2009). Sex workers and HIV/AIDS: Analyzing participatory culture-centered health communication strategies. *Human Communication Research, 35*(1), 86–114.

Census of India. (2001). States at a glance. Retrieved from http://www.census india.gov.in/Census_Data_2001/ /States_at_glance/State_Links/21_ori.pdf (no longer accessible).

Charmaz, K. (2008). *Constructing grounded theory: A practical guide through qualitative analysis*. Newbury Park, CA: Sage.

Creswell, J. W. (2007). *Qualitative inquiry and research design: Choosing among five approaches* (2nd ed.). Thousand Oaks, CA: Sage.

Dutta, M. J. (2008). *Communicating health: A culture-centered approach*. London, UK: Polity Press.

Dutta, M. J. (2009). On Spivak: Theorizing resistance—Applying Gayatri Chakravorty Spivak in public relations. In Ø. Ihlen, B. van Ruler, & M. Fredriksson (Eds.), *Public relations and social theory: Key figures and concepts* (pp. 278–300). New York, NY: Routledge.

Dutta, M. J., & Basnyat, I. (2008a). The case of the Radio Communication Project in Nepal: A culture-centered rejoinder. *Health Education and Behavior, 35*(4), 459–460.

Dutta, M. J., & Basnyat, I. (2008b). Interrogating the Radio Communication Project in Nepal : The participatory framing of colonization. In H. M. Zoller & M. J. Dutta (Eds.), *Emerging perspectives in health communication: Meaning, culture, and power* (pp. 247–265). New York, NY: Routledge.

Dutta-Bergman, M. J. (2004a). The unheard voices of Santalis: Communicating about health from the margins of India. *Communication Theory, 14*(3), 237–263.

Dutta-Bergman, M. J. (2004b). Poverty, structural barriers and health: A Santali narrative of health communication. *Qualitative Health Research, 14*(8), 1107—1122.

Hatch, J. A. (2002). *Doing qualitative research in education settings*. Albany: State University of New York Press.

Human Development Report. (2004). *Human development report*. Bhubaneswar, Orissa, India: Planning and Coordination Department, Government of Orissa.

Integrated social and environmental assessment study. (2008). Retrieved from http://www.orissapanchayat.gov.in/English/download/Final%20Report%20Tribal%20Assessment.pdf.

Naik, E., Karpur, A., Taylor, R., Ramaswami, B., Ramachandra, S., Balasubramaniam, B., Galwankar, S.,...Salihu, H. M. (2005). Rural Indian tribal communities: An emerging high-risk group for HIV/AIDS. *BMC International Health and Human Rights*, *5*(1). Retrieved from http://www.biomedcentral.com/1472-698X/5/1 (no longer accessible).

National AIDS Control Organization (NACO). (2007). *Social assessment report on HIV/AIDS amongst tribal people in India*. Delhi, India: Author.

National AIDS Control Organization (NACO). (2008). *Social assessment of HIV/AIDS among tribal people in India—Report prepared by AC Nielsen—ORG-MARG*. New Delhi, India: National AIDS Control Organization, Ministry of Health and Family Welfare, Government of India.

National Family Health Survey (NFHS)-III. (2006). Key indicators from National Family Health Survey-III, India by wealth index and caste/tribe. Retrieved from http://www.nfhsindia.org/pdf/IN_WICT.pdf (no longer accessible).

National Rural Health Mission (NRHM) Orissa Report. (2009). Retrieved from http://mohfw.nic.in/nrhm/Documents/High_Focus_Reports/Orissa_Report.pdf (no longer accessible).

Orissa Vision 2010—A Health Strategy. (2003). *Orissa state integrated health policy: Strategies and action points*. Bhubaneswar, Orissa, India: Health and Family Welfare Department, Government of Orissa.

Planning Commission of India. (2003). An overview of the scheduled tribes of Orissa. Delhi, India: Author. Retrieved from http://planningcommission.nic.in/plans/stateplan/sdr_orissa/sdr_orich15.pdf.

Planning Commission of India. (2008). *Expert group report on development challenges in extremist affected areas*. Delhi, India: Author.

Strauss, A. L., & Corbin, J. M. (1998). *Basics of qualitative research: Grounded theory procedures and techniques* (2nd ed.). Newbury Park, CA: Sage.

UNDESA. (2010). *The state of the world's indigenous peoples*. New York, NY: Department of Economic and Social Affairs, Division of Social Policy and Development, Secretariat of the Permanent Forum on Indigenous Issues, United Nations.

24

THE GOOD-NEIGHBOR CAMPAIGN: A COMMUNICATION INTERVENTION TO REDUCE ENVIRONMENTAL HEALTH DISPARITIES

Heather Zoller and Lisa Melancon

The environmental health and environmental justice movements seek recognition of the health effects of pollution and the health disparities that result from the often discriminatory siting of industrial toxins (Brown, 2007; Bullard & Wright, 1990; Gibbs, 2002). This form of health activism among community members, advocacy organizations, scholars, and scientists is a significant means of resisting and transforming the political roots of health disparities by targeting industrial practices (Zoller, 2009). Both public-health scholars and practitioners are paying more attention to the use of communication campaigns to change health-threatening corporate behavior (Freudenberg, Bradley, & Serrano, 2009). In this chapter, we describe the good-neighbor campaign as a community-based communication intervention that seeks to reduce health disparities resulting from exposure to environmental toxins (Heiman, 1997; Lewis, 1997). Groups across the US, Canada (http://www.goodneighbourcampaign.ca /about_gnc), and Europe (Global Community Monitor, 2006) have used the good-neighbor model to work directly with corporations in their neighborhoods to reduce or eliminate their exposure to harmful chemicals. Among the multiple communication processes involved in health activism (Zoller, 2005), this chapter focuses on methods of organizing and campaign tactics, providing details of two cases.

We describe two campaigns by the community-based organizer, Ohio Citizen Action (OCA)—the Eramet campaign in Marietta, OH, and the Lanxess campaign in Addyston, OH. The Eramet (a metallurgical manufacturing company) campaign achieved a commitment from company executives to invest $150 million to improve the plant's environmental and operational performance. The combined furnace and baghouse projects alone are expected to reduce emissions by 54% in the No. 1 furnace, and provide an overall 20% emissions reduction for the entire facility (Cooley, 2008). The good-neighbor campaign targeting Lanxess (a producer of plastic resins that may be associated with cancer and other health risks), resulted in corporate investments of well over two million dollars to reduce leaks and malfunctions, and a Notice of Violation Findings and Orders (Klepal, 2005; Klepal & Smith Amos, 2005) by the U.S. EPA that led to a settlement agreement predicted to reduce 360 tons of butadiene emissions, 1 ton of acrylonitrile, and an additional 59 tons in hazardous air pollution per year under the Clean Air Act (U.S. EPA, 2009).

After a brief description of environmental hazards as a significant source of health disparities, we detail the steps involved in a good-neighbor campaign. We discuss the common communication challenges in this process, and the resources that citizens can use to overcome them. We follow this with specific discussion of the two OCA campaigns.

Environmental Pollution and/or Toxins as a Source of Health Disparities

Health disparities are differences that occur by gender, race and ethnicity, education level, income level, disability, and geographic location. Some health disparities are unavoidable, such as health problems that are related to a person's genetic structure (Association of Public Health Association, 2004), but most can be attributed to social causes and effects (Frank, Mustard, & Fraser, 1994). Policy-makers and government organizations have prioritized the health disparities as a key area for concern. For example, Healthy People 2020 includes specific initiatives to eliminate health disparities.

As a specific subarea, environmental health disparities are the result of a complex consideration of social and economic disparities (such as income level, race, literacy rates), combined with environmental exposures tied to a specific geographic region (Comacho, 1998; Morello-Frosch & Lopez, 2006; Payne-Sturges & Gee, 2006). Although the effect is often difficult to measure, toxins in the air and soil contribute to birth defects, developmental delays, asthma, cancer, and other illnesses (Brown, 2007). These pollutants often can be directly connected to a local industry. Researchers and policy-makers have made persuasive cases that many environmental health disparities affect racial and ethnic minorities and the poor at a higher frequency (Institute of Medicine, 1999; Pamies &

Nsiah-Kumi, 2009, p. 18; O'Neill, et al., 2003), and that pollution-siting decisions often target these groups explicitly (Bullard & Wright, 1990). The Commission for Racial Justice of the United Church of Christ reported in 1987 that race was the most significant variable associated with the siting of hazardous waste facilities: "Most industries seek to locate in communities that are economically depressed, politically disenfranchised, and therefore subject to economic blackmail" (Head & Leon-Guerrero, 1997, p. 323).

To reduce environmental health disparities requires a comprehensive and multilayered approach that considers both social factors and environmental hazards. Researchers have proposed various theoretical approaches to reduce environmental health disparities (Gee & Payne-Sturges, 2004; Hornberg & Pauli, 2007; Payne-Sturges, Zenick, Wells, & Sanders, 2006). Because of the undue burden that many poor and minority communities face, academics, policy-makers, and public-health practitioners alike cannot overlook the power and potential of an engaged community working toward reducing environmental health disparities in their own communities. One method for engagement is activism through the good-neighbor campaign.

The Good-Neighbor Campaign as a Communication Intervention

The good-neighbor campaign is an innovative model for grassroots environmental activism. The organizing model, attributed to Sanford Lewis, promotes the development of win-win relationships among neighbors, workers, and a corporate polluter (Heiman, 1997; Lewis, 1997). The model encourages cooperation and relationship building, offering corporations the opportunity to bolster their reputation by improving their environmental performance (Den Hond & De Bakker, 2007). Despite this approach, most companies do not engage willingly in negotiations, but "can be pressured during contract negotiations, at times of license renewal, when an accident has already occurred leading people to question the operation of the plant, and through the threat of regulation or product liability suits" (Heiman, 1997, p. 637). Neighbors may apply pressure by building awareness, organizing neighbors, gaining media exposure, or involving regulators (Ryder, 2006). Often, companies will not engage in *binding* agreements with neighbor groups, unless the community can affect production or profitability through threats to reputation or regulatory permits (Della Porta & Diani, 1999; Heiman, 1997). Good-neighbor campaigns empower local residents by encouraging them to talk with their neighbors, meet directly with management, learn to perform their own chemical sampling, and build coalitions with other community groups.

Communication Processes in Organizing a Good-Neighbor Campaign

As a grassroots activism tool, the good-neighbor campaign is a communication intervention aimed at changing corporate behavior. As we describe below, the campaign gives neighbors a model for communicating with corporate managers, regulators, and community members to collaboratively find ways to reduce or eliminate toxic exposures and other problems. Campaigns involve multiple methods, but generally, the goal is to achieve a serious commitment from management to solve the company's environmental problems (a Good-Neighbor Agreement). Although we acknowledge that every community-based intervention unfolds somewhat uniquely, we describe below communication processes involved in conducting most campaigns. The information provided here is based primarily on the Good-Neighbor Campaign Handbook produced by OCA (Ryder, 2006).

OCA is an environmental advocacy nonprofit that provides the tools and/or support to help communities focus their concerns and organize citizens to take action. OCA promotes citizen involvement toward the goal of reducing environmental harms, and toward environmentally protective policies at the political level. The organization usually becomes involved in communities when local residents request assistance with efforts to reduce pollution in their neighborhood. Initially, much of the communication work involves organizing neighbors. Rachael Belz recommends starting with a core group of at least five dedicated neighbors and ideally growing that number to several hundred. OCA builds awareness, and promotes neighbor participation by knocking on doors to survey neighbors about problems they experience resulting from their corporate neighbor. These "walk and talks," along with flyers and posters, encourage residents to become a member of the neighbor organization or to attend organizing meetings.

The good-neighbor model encourages residents to meet directly with management to ask questions and clearly state their desired outcomes, following the "Getting to Yes" method recommended by the Harvard Negotiation Project (Fisher & Ury, 1981). OCA (Ryder, 2006) suggests that neighbors speak with management as equals (for instance, calling for a meeting rather than requesting one). Ryder (2006) reminded readers that managers are people, too, and that neighbors should first make them aware of the problems, and, second, appeal to their conscience. When pointing out the problems, neighbors should avoid calling for specific changes, such as installing a filter or reducing emissions by 75%. Discussing specific community interests, such as "I want clean air so my child does not have asthma attacks," allows for creative solutions that may achieve even better results and simultaneously improve the company's efficiency. OCA's handbook recommends that neighbors remain calm when they communicate with managers, regulators, and others involved in the campaign, because angry people are easy to dismiss.

Because most managers and corporate polluters will initially not cooperate, neighbors should be prepared to raise the stakes. One of OCA's most significant influence tactics is the letter-writing campaign. OCA hires staff to campaign throughout the state of Ohio. Staff members knock on doors and tell residents about what is happening in the current campaign. They provide residents with brief talking points, and ask them to write even a short letter to management asking the company to cooperate with the neighbors (and include drawings from children, if possible). Residents tape the letters outside their door and canvassers pick them up within a few hours. Ryder (2006) estimated that half to three-quarters of those who agree actually produce the letter. Management begins to feel pressure when thousands of letters begin arriving in the mail. If managers do not respond, OCA will send letters to the board of directors. Most neighbors do not have the resources for a statewide canvass on their own, but they can conduct letter-writing campaigns in their communities by knocking on doors, talking with neighbors about the problems, and asking them to spend five minutes expressing their thoughts about the problems the company is causing, right away—before they forget about it.

Neighbors should be creative in raising their campaign's profile. For instance, community members may hang white flags outside, which communicate how dirty the air is when the flags quickly become gray or black. Residents may initiate a petition calling for a particular step that will forward the campaign, such as removing a plant manager or CEO who is not cooperating. Speaking with the press can raise a campaign's profile, as long as campaigners have significant news to share.

Communication Challenges and Resources for Addressing Them

Good-neighbor campaigns are not easy. Corporations would prefer that neighbors not interfere with their operations, and many residents become hostile because they see environmental campaigns as a threat to the town's economic base or to their own jobs. Campaigns can go on for years before achieving acceptable results. In this section, we discuss some of the major communication challenges involved in good-neighbor interventions and resources for managing these challenges.

First, even though neighbors have direct experience with problems caused by a plant—odors, watering eyes, chemical smells, loud trucks, alarm bells going off during the night—they also need to understand what the target company produces and how it produces it. Residents need information about the plant's financial standing, particularly because many companies will claim they do not have enough money to invest in improvements. At this stage, campaigners need to conduct research into these issues so that they are better informed and therefore less likely to be dismissed by the company. Campaigners can begin with the company's website (if at least one member of the group has Internet access or there are library resources). Librarians can help residents search press accounts of the company. Information about the plant's emissions is available through the Toxics Release Inventory (TRI) (http://www.epa.gov/tri/). Neighbors can request that state or city environmental and health regulators provide them with permit descriptions and other public information.

Neighbors should seek input from similar companies, particularly those that have implemented solutions to the problems neighbors are experiencing. Neighbors may keep a log of their experiences, including chemical smells, noises, or other problems caused by the company. Pooling these logs helps neighbors to understand what chemicals are released and when, as well as how the plant operates (e.g., when supply trucks arrive).

Second, neighbors struggle with the technical language used by the company, regulators, or scientists. OCA's campaign organizers tell neighbors to insist that the people they encounter speak in clear language, and continue to ask questions until they are answered satisfactorily. Residents should seek out neighbors who work in similar industries so that they can help explain some of the technical processes. Neighbors can also reach out to scientists or researchers who are willing to help them interpret scientific data and health research.

Third, campaigners face barriers when they seek to understand the relationship between toxic exposures and health problems. Proving causal links is very difficult because humans are exposed to multiple potential agents of disease, preventing isolation of causes. Moreover, we lack understanding of the toxic effects (and interaction effects) of thousands of chemicals, and research methods are sometimes biased against establishing positive results (Brown, 1992; Wing, 2005). To avoid this difficulty, neighbors may call for companies to reduce exposure based on more immediate physical effects such as bad smells and watering eyes, or nuisances such as dirt and grime on cars that plants are responsible for addressing. Additionally, the environmental health and justice movements developed innovative methods that empower citizens to understand their exposure without a lot of money or expertise. For instance, neighbors may form a "bucket brigade" to measure air quality. Neighbors can sample air with inexpensive home-made canisters containing a vacuum pump at the moment they smell chemicals, rather than wait hours or days for environmental agencies to investigate. The canisters are U.S. EPA-approved, and take legitimate samples that can be analyzed by labs scientifically. Effective campaigns should continue organizing activities to maintain momentum if they explore partnerships with scientists and regulators that may provide scientific data.

Fourth, good-neighbor campaigns have to address corporate issue management tactics of co-optation, denial, and delay. Management may attempt to co-opt outspoken neighbors by inviting them to join the company's public-advisory group, where management largely controls the agenda. This keeps activists busy and out of the public spotlight. Companies frequently deny exposure, or that exposures are harmful. Managers also delay solutions by slowly investigating alternatives, or waiting for scientific research into alternatives. OCA generally recommends against joining the public-advisory group, at least initially. Neighbors should take the initiative and avoid responding to each new move on the company's part. Using multiple methods maintains momentum by encouraging flexibility. Campaigners may need to change tactics or influence targets. If one method is not working, move to another. Neighbors may decide to shift from dealing with plant managers to CEOs, or from CEOs to the board of directors. In addition, management often builds support by suggesting that good-neighbor campaigns are a threat to a town's economic base. It is important for residents to understand that investments to improve environmental performance can increase corporate efficiency. Indeed, such investments often result in modernizing plants that otherwise would have been abandoned by the company (Kazis & Grossman, 1991; Mann, 1993).

Fifth, another potential problem is that these campaigns may be lengthy, from a year or two upward to ten years, and neighbors can suffer from fatigue. It is very important, in the opening stages of the campaign, to discuss the time commitment that may be involved. The key group of community organizers will bear the brunt of maintaining momentum, which means that the leaders of a campaign need to be the most committed. A potential solution is to create one or more subcampaigns that focus on smaller, winnable goals that show campaigners that real progress is being made. For instance, Ryder (2006) described a subcampaign in Middletown, Ohio, pressing AK Steel to build a fence keeping children out of a polluted creek. Campaigns are also more sustainable when neighbors support one another and make sure that organizing meetings and events are fun. Even small victories, such as reaching a certain number of letters submitted to the polluting company, should be celebrated. By marking internal milestones, campaigners can remain focused and motivated on the larger task at hand.

Cases

Marietta

Many Appalachian communities bear an undue burden of environmental health disparities from industries (Hendryx, 2011; Haynes et al., 2010; Hendryx, 2011). Marietta, Ohio is a relatively isolated Appalachian community with a median family income of $36,042, where 13.6% of families are below the poverty level (U.S. Census Bureau, 2000). As part of the Mid-Ohio River Valley, the town is also home to an industrial corridor with more than 20 industrial plants along the river, including Eramet, a ferromanganese refinery. Until recently, it was the only ferromanganese plant in the United States, releasing thousands of pounds of manganese (Mn) into the Marietta airshed (U.S. E.P.A. TRI Database, 2009).

In 2000, a group of community members began sharing their experiences about the pervasive odors in their community. From 2000–2006, this group of "Stink Friends" tracked odors that were emanating from the Eramet refinery at all hours. One particular community member kept a "Stink Diary" in which she logged every occurrence of smells and the effects they had on her and her family. Community members also collected dust swipes and samples, used Tedlar bags for air sampling, and personally paid for analysis of these materials. Neighbors contacted government agencies and officials at all levels, and sent letters to the then CEO of Eramet, who refused to meet with them or acknowledge their concerns.

Distressed by the poor air quality and concerned for the potential health effects associated with their exposure, the "Stink Friends" called a community meeting on the subject of air pollution and Eramet, which was identified as the main company polluter. Approximately 50 people attended the meeting in March, 2006. Most of the attendees complained of the extreme odors from the plant, a metallic taste in the air, and the visible, brown pollution coming from the facility. Many in attendance also expressed frustration with the local company, the Ohio EPA, and the Ohio Department of Health. They felt that it was up to the citizens themselves to do something about the problem. At this meeting, the "Stink Friends" became a nonprofit, citizen-action group, Neighbors for Clean Air (NCA), and they officially partnered with OCA to launch a good-neighbor campaign.

Following OCA's template of strategic steps, NCA members wrote letters and encouraged their friends and neighbors to write letters to the company polluter. In four months, the community had written 3,000 letters. Other actions undertaken by NCA's good-neighbor campaign included door-to-door canvassing to convey information and to increase the letter-writing campaign, researching files at the Ohio EPA, and printing and placing over 200 yard signs that read, "Eramet, let's clear the air." From September, 2006 to February 2007, NCA walked the picket line with locked out workers, and held a food drive for workers and their families over Thanksgiving.

During this same period, NCA was working to complete a "Citizens' Audit." As recommended by OCA, an audit helps to compile all the research and information that community members have gathered, secures credibility that the community has done the necessary research and legwork, gives clarity to the goals of the campaign, and acts as a ready resource that can be offered to reporters. NCA's citizens' audit included a recommendation:

> We commend Eramet for the reductions of emissions they have made thus far in Marietta. We believe that further significant reductions are necessary and feasible. We recommend that the decision makers at Eramet Marietta work with citizens, agencies and the Eramet corporate headquarters to make the changes necessary to be a good neighbor. (NCA, 2006, p. 1)

NCA and OCA continued the letter-writing campaign. OCA asked Ohio citizens to write letters to the corporate headquarters in France, and an OCA member hand-delivered a set of letters to the corporate headquarters in Paris, France. NCA members sent letters to the editors of local papers to maintain interest in the company; they canvassed neighborhoods surrounding the plant; they organized "walk and talk" events downtown; and they had a presence at local fairs and events, such as Earth Day.

In Marietta, neighbors used citizen science by measuring air-pollution levels using portable gas analyzers, and collected "swipe" samples to be sent off and analyzed. This type of participation can be key to instigating community-based change.

Two interdependent goals have always been present for the campaign in Marietta: to reduce the amount of air emissions, particularly manganese, from local industry, and to understand the health effects, if any, of the continued exposure to emissions. These goals were realized at the June 2006 press conference organized to release and present the Citizen's Audit. There, NCA met Dr. Erin Haynes[i] from the University of Cincinnati, who had been invited by OCA organizers. In the conversations that ensued, Dr. Haynes was moved by the NCA's stories and their frustrations.

She committed to trying to find answers to address their concerns about the potential health effects associated with their exposure.

The year 2008 was a watershed year. The management of Eramet finally agreed to meet with NCA. Conversations took place over a series of months in the first part of the year. In April 2008, Dr. Haynes was awarded a $2.6 million grant from the National Institutes of Environmental Health Science, which is known in the local community as CARES (Communities Actively Researching Exposure Study). In September, 2008, after 28 months and over 44,000 letters, the pressure portion of the good-neighbor campaign was officially halted (Shawver, 2008). Ongoing follow-up suggests that the good-neighbor campaign was successful in addressing both goals of NCA. First, it was able to achieve positive results to reduce the amount of air emissions. The good-neighbor campaign resulted in a $150 million commitment by Eramet to replace the most problem-plagued furnace with a new state-of-the-art model, use odor-abatement technology, and construct a new baghouse to prevent particulate stack emissions from their largest furnace by 54% (Cooley, 2008).

Second, the good-neighbor campaign achieved the goal of finding a way to get answers concerning the health effects of air pollution. Local resident and NCA member, Dr. Richard Wittberg, always believed that "science can give the answers" to the community's concerns about air pollution (http://www.eh.uc.edu/cares/study/profiles/erin.html). The bidirectional partnership that NCA formed with the University of Cincinnati and Marietta College is premised on the research framework known as community-based participatory research (CBPR) (Minkler & Wallerstein, 1997, 2003). CBPR is changing the scientific research paradigm so that scientific research is done *with* rather than *on* a community. The bidirectional partnership has worked, in large part, because of the continued commitment of NCA. At the time of writing, the scientific research study is still recruiting participants, but it has exceeded the anticipated enrollment numbers in its first year. This case illustrates how university-based CBPR can aid health activists by giving them relevant, understandable, and timely data to understand how toxic exposure affects their health. NCA followed OCA's template for conducting a good-neighbor campaign, which was a key attribute to the campaign's success.

The Lanxess Campaign

In Addyston, OCA initiated a good-neighbor campaign in 2004 targeting the Bayer Chemical plant in Addyston, Ohio. (The plant spun off into Lanxess Corporation in 2004, and was absorbed by the English INEOS Group in 2006). The plant used acrylonitrile, butadiene, and styrene (ABS) to make plastic resins. The state of Ohio does not regulate these chemicals, so the company operated under a county permit agreement that approved certain levels of air and water emissions. The organizer, Sara (a pseudonym used in Heather's research), chose Lanxess because TRI data showed increased chemical emissions and plant malfunctions above permitted levels. In addition, a community canvass showed that neighbors had concerns about odors, health effects, and the safety of children in the preschool-through-first-grade school directly across from the plant.

Addyston is a historically White, working-class neighborhood on the Ohio River, with an increasing number of minority and poor residents, surrounded by several middle-class communities. Sara organized concerned citizens in the Westside Action Group (WAG) to achieve decreased plant accidents and pollution. Despite worries from minority residents who lived near the plant, the roughly 40 men and women who participated were mostly White residents with a mix of working- and middle-class backgrounds. The core 12 participants were women.

Neighbors were surprised that managers agreed to meet with Sara, which she credited to the statewide letter-writing campaign that sent hundreds of letters to the plant manager. In response, managers arranged multiple meetings to answer questions about the plant. They spoke openly about their products and production processes, but avoided health questions and denied that neighbors were exposed to plant chemicals.

WAG members chose not to participate in the company's public-advisory group, but met directly with management to discuss emissions. The meetings were long, and managers used technical language to describe changes they were implementing to reduce smells. Neighbors talked with managers and the press about their frustration with smells, headaches, and nosebleeds that they associated with emissions. They expressed fears about the high incidence of cancer in the area and the possible health effects for children and pregnant women. During this time, three chemical releases over permitted levels in October, November, and December of 2004 led to local press coverage of the campaign. As a result, the mayor and a local state representative created the Addyston Task Force to address the controversy. The company denied that the exposure could lead to health problems.

The good-neighbor campaign used citizen science to promote neighbor participation in discussions about chemical exposure. Sara encouraged neighbors to keep an Odor Log to track when the company was emitting chemicals or noxious odors, and call local and state regulatory agencies with odor complaints. WAG members also learned from OCA and activists from Texas's "cancer alley" to do "bucket" sampling and use handheld Cerex monitors. These methods found consistent chemical readings above levels permitted in states with ABS standards.

In addition to citizen science, the WAG developed a "community standard" for the plant's performance, going beyond permit requirements to demand zero air and water emissions. Concerned residents also began a petition calling on the German CEO to replace the plant manager. A neighbor traveling to Germany hand-delivered a letter to the CEO asking for plant improvements. By April, Ohio citizens sent over 1,300 citizen letters to Lanxess, and the WAG mailed their petition to remove the plant manager to German headquarters with 500 signatures. Some neighbors dropped out as the smells began to fade in response to sewage projects, but the core WAG members grew more confident, demanding "answers at meetings and not slides," and noting that "We can demand things—they need us, it is expensive to move."

The WAG also criticized the local regulatory agency for its failure to adequately regulate the company and monitor exposure. After tough neighbor questioning during a town meeting focused on possible health effects, particularly for children, the environmental agency began ambient air monitoring every six days, and eventually installed a real-time monitor.

On July 15, 2005, the company announced a reorganization that included the retirement of the Plant Manager and the removal of the Head of North American Operations. The company publicly attributed the change to a lack of plant profits (Klepal, 2005). Yet Lanxess also announced a $1.5 million investment in manufacturing-process controls, as requested by OCA. After continued organizing throughout the summer, Lanxess announced an additional $1 million investment in September. OCA posted on their website on September 26, 2005:

> In a direct response to Ohio Citizen Action's good neighbor campaign, Lanxess Plant Manager...announced on Thursday that Lanxess will invest $1 million to reduce butadiene emissions that go into the air. They also will seek to reduce accidents and call on external experts to evaluate its performance.

Sara announced that she would end the pressure phase of the good-neighbor campaign, satisfied with the commitments the company made to its neighbors. The campaign moved to the cooperation and implementation phase. Some WAG members worked through the task force to monitor changes and request health-effects research.

In this case, neighbors did not build alliances with environmental scientists or epidemiologists, having found no local experts on ABS. Two environmental health researchers who volunteered their time to talk with residents acknowledged the potential risks from the plant, but provided conflicting advice about the possibility of establishing health effects. Instead, an environmental agency legitimized neighbor concerns.

On December 5, 2005, the Ohio EPA released its first report of the Air Toxic Study with a press release, and issued a Notice of Violation Findings and Orders to Lanxess. The plant was found to be in violation of its Emulsion Polymerization permit and classified as a "public nuisance," because carcinogens were measured off-site. The report suggested that continued exposure (20–30 years) to the chemicals would raise the risk of cancer by an additional 5 cases out of 10,000 in the area, which they characterized as exceeding acceptable levels by 6 times, and noted that children may be at greater risk (Klepal & Smith Amos, 2005).

WAG members felt that the report legitimized their concerns. The company and plant supporters disputed the findings, but managers worked with Ohio EPA to reduce emissions. Thus, the regulatory agency helped to negotiate additional environmental upgrades and monitored implementation. During this process, the plant dedicated an additional million dollars in environmental upgrades ("Plant going beyond compliance," 2005).

At the same time, though, the report (along with pressure from the task force) spurred a health study of cancer in Addyston. The study was designed and conducted by scientists at the county health department without participation from neighbors. After long delays resulting from methodological questions, the 2006 report found a higher incidence of lung, colon, bronchus, and rectum cancer than expected. However, the report stated, "but it is impossible to determine the cause of each case of cancer due to the multitude of risk factors, and the interaction of these factors, that play a role in cancer development" (*Cancer Incidence Among Residents of Addyston Village, Hamilton County, Ohio*, 2006, p. 4). Despite the stated impossibility of determining causes, the report did describe risk factors such as tobacco smoking, eating habits, exposure to radon and asbestos, and family history. The study did not recommend reducing plant emissions, but did promote improving exercise and eating habits as a way to improve health.

Lanxess managers touted the report as proof of the plant's safety. Neighbors privately questioned the study's methods and findings, but residents did not continue to pursue evidence about health effects. Several WAG members and an OCA leader sit on the company's public-advisory group (PAG) in order to continue to monitor emission reductions (U.S. EPA, 2009), but there was little ongoing discussion about environmental illness, particularly illnesses such as asthma and children's developmental disorders that were not a part of the health study. Neighbor organizing achieved significant plant investments in reducing neighbor exposure to ABS. Yet, because it was not central to the initial goal of the campaign, and due to uncertainty regarding health effects, residents' questions about environmental illness were not resolved.

Conclusion

Good-neighbor campaigns can empower communities, and "empowered communities may be able to protect themselves from the instruction of new hazards and eliminate extant ones" (Gee & Payne-Sturges, 2004, p. 1649). Health activism through community involvement is a key factor in the elimination of environmental health disparities, and increasing dialogue among health communication researchers, activists, and practitioners builds our understanding of the everyday experiences and challenges of participatory methods of health and social change (Camacho, Yep, Gomez, & Velez, 2008; Dutta, 2007; Ford & Yep, 2003; McLean, 1997; Zoller, 2009).

The Marietta campaign was trying to address two interdependent goals. The first, reducing air emissions, is a standard goal of good-neighbor campaigns, while the second, trying to get a scientific answer about the health effects, if any, of air pollution in the area, is not typically a goal of good-neighbor campaigns. By all accounts, at the time of the writing of this chapter, the Marietta campaign has been a success, but it was not without its challenges. Keep in mind that on the one hand, Marietta's first attempt at action was not successful, in part because it was not as strategic as the campaign started in 2006. The Addyston campaign, on the other hand, led to important upgrades to the facility, but mixed results regarding the community's concern about health effects. The two case studies offer a lesson for other communities and communication scholars who may work with them: it does take a strategic plan and an intense commitment from the community to achieve reductions in chemical exposure and an increase in knowledge about possible health effects.

The cases differ in their relationship to science. As other researchers have found (see Israel et al., 2010), CBPR can play an important role in eliminating health disparities. As Addyston showed, traditional science can both help and hinder neighbors, and campaigns can proceed without definitive proof of illness. Yet, the Eramet case suggests that good-neighbor campaigns involving complex environmental and health issues have much to gain from CBPR partnerships that enable community groups to get the science they need to enact long-term change, when those relationships focus on the needs and perspectives of neighbors.

These cases also raise important questions about how we measure success. Good-neighbor campaigns are used to address a diverse set of issues within communities with diverse sets of problems. Many parts of localized campaigns are difficult, if not impossible, to replicate in other areas. Overrelying on "measurable outcomes" of health improvements to judge the success of campaigns can lead us to overlook the power of grassroots activist campaigns that cannot carry out health-effects assessment work. This may involve measuring reductions in toxic exposure, rather than measuring reductions in illness. Additionally, the ability to measure health outcomes resulting from reduced toxic exposure is constrained by resource availability, lack of scientific knowledge about toxins (particularly low dosage and interactive effects), and the inherent uncertainties involved in epidemiological research.

At the same time, researchers and environmental health advocates do need to continue to innovate tools for improved campaign communication, including finding systematic ways to translate epidemiology into user-friendly language and distribution. For instance, OCA leaders, Melissa English and Rachael Belz, noted the need for "citizen ready" information, a clearinghouse for health-science information, available local health data (such as asthma rates in a particular area) that would be laypeople friendly. Additionally, communication researchers can advocate for improved models of dialogue about risks that involve neighbors as legitimate stakeholders and knowledgeable health citizens.

Finally, campaigns that focus on a single company are important, but they may feel isolated or disconnected from larger issues of environmental health caused by industry across the country. Such campaigns may be confused with NIMBYism (not in my backyard). However, environmental health and justice activists have created national and international networks where citizens can compare tactics and share innovative solutions. For instance, the Louisiana Environmental Action Network (http://www.leanweb.org/) organizes local environmental activist groups into a statewide network. The Environmental Health Coalition promotes environmental justice in San Diego, addressing disparities in toxic exposures in the Latino community in the US, as well as the toxic effects of *maquiladoras* in the border areas near Tijuana (http://www.environmentalhealth.org/About_EHC/index.html). The Blue-Green Alliance (http://www. bluegreenalliance.org/) brings labor and environmental activist groups together to build a more environmentally sustainable economy. Many of these groups work together on larger political issues such as the regulation of toxins, right-to-know legislation, safe workplaces, and the promotion of the precautionary principle—an important policy tool that shifts the burden of proof for proving chemicals safe to the chemical producer rather than on citizens who must prove chemicals to be hazardous before they are regulated (Brown, 2007).

Researchers and practitioners have much to contribute in helping us to understand the relationship between local campaigns and this larger social movement. However, what we do know is that an engaged community using a good-neighbor campaign as a communication intervention, has the potential to disrupt existing power structures and positively affect change in communities experiencing health disparities. Open, ethical, and effective communication, including inquiry, dialogue, debate, persuasion, and protest, is key to the success of health social movements seeking to reduce environmental illness.

Note

1. Dr. Erin Haynes holds a doctorate in public health, and is an advocate of community-based research. Dr. Haynes's primary research interest is to examine the effect of low-level toxicant exposures on neurobehavioral outcomes. Specific research interests include evaluating biological and health outcomes in children resulting from exposure to multiple metals, such as lead and manganese, evaluating gene-environment interactions that may influence the neurobehavioral effect of metal exposure, and increasing public knowledge of environmental toxicants that threaten public health. She is currently working on an NIEHS-funded community-based participatory research R01 to study the neurobehavioral effects of metal exposure in two rural Appalachian communities: Marietta and Cambridge, Ohio.

Bibliography

Association of Public Health Association (APHA). (2004). *Eliminating health disparities: Communities moving from statistics to solutions toolkit.* Washington, DC: Author.

Brown, P. (1992). Popular epidemiology and toxic waste contamination: Lay and professional ways of knowing. *Journal of Health & Social Behavior, 33*(3), 267–281.

Brown, P. (2007). *Toxic exposures: Contested illnesses and the environmental health movement.* New York, NY: Columbia University Press.

Bullard, R. D., & Wright, B. H. (1990). The quest for environmental equity: Mobilizing the African-American community for social change. *Society and Natural Resources, 3*(4), 301–311.

Camacho, D. E. (1998). The environmental justice movement: A political framework. In D. E. Camacho (Ed.), *Environmental injustices, political struggles: Race, class, and the environment* (pp. 11–30). Durham, NC: Duke University Press.

Camacho, A. O., Yep, G. A., Gomez, P. Y., & Velez, E. (2008). El poder y la fuerza de la pasión: Toward a model of HIV/AIDS education and service delivery from the 'bottom-up.' In H. M. Zoller & M. J. Dutta (Eds.), *Emerging perspectives in health communication: Meaning, culture, and power* (pp. 224–246). New York, NY: Routledge.

Cancer incidence among residents of Addyston Village, Hamilton County, Ohio. (2006). Ohio Department of Health, Hamilton County General Health District.

Cooley, P. (2008, March 29). Eramet plans $20M upgrade. *Marietta Times*. Retrieved from http://www.mariettatimes.com/page/content.detail/id/502173/Eramet-plans—20M-upgrade.html?nav=5002.

Della Porta, D., & Diani, M. (1999). *Social movements: An introduction*. Malden, MA: Blackwell.

Den Hond, F., & De Bakker, F. (2007). Ideologically motivated activism: How activist groups influence corporate social change activities. *Academy of Management Review, 32*(3), 901–924.

Dutta, M. J. (2007). Communicating about culture and health: Theorizing culture-centered and cultural sensitivity approaches. *Communication Theory, 17*(3), 304–328.

Fisher, R., & Ury, W. (1981). *Getting to yes: Negotiating agreement without giving in*. Boston, MA: Houghton Mifflin.

Ford, L. A., & Yep, G. A. (2003). Working along the margins: Developing community-based strategies for communicating about health within marginalized groups. In T. L. Thompson, A. M. Dorsey, K. I. Miller, & R. Parrott (Eds.), *Handbook of health communication* (pp. 241–262). Mahwah, NJ: Lawrence Erlbaum.

Frank, J., Mustard, F., & Fraser, J. (1994). The determinants of health from a historical perspective. *Daedalus, 123*(4), 1–20.

Freudenberg, N., Bradley, S. P., & Serrano, M. (2009). Public health campaigns to change industry practices that damage health: An analysis of 12 case studies. *Health Education and Behavior, 36*(2), 230–249.

Gee, G. C., & Payne-Sturges, D. C. (2004). Environmental health disparities: A framework integrating psychosocial and environmental concepts. *Environmental Health Perspectives, 112*(17), 1645–1653.

Gibbs, L. (2002). Citizen activism for environmental health: The growth of a powerful new grassroots health movement. *The Annals of the American Academy of Political and Social Science, 584*(1), 97–109.

Global Community Monitor. (2006, September 14). Good neighbor campaign launched against Europe's dirtiest refinery. Retrieved from http://www.gcmonitor.org/article.php?id=443.

Haynes, E. N., Heckel, P., Ryan, P., Roda, S., Leung, Y. K., Sebastian, K. (2010). Environmental manganese exposure in residents living near a ferromanganese refinery in southeast Ohio: A pilot study. *Neurotoxicology, 31*(5), 468–474.

Head, L., & Leon-Guerrero, M. (1997). Fighting environmental racism. In C. Levenstein & J. Wooding (Eds.), *Work, health, and environment: Old problems, new solutions* (pp. 320–328). New York, NY: Guilford Press.

Heiman, M. (1997). Community attempts at sustainable development through corporate accountability. *Journal of Environmental Planning and Management, 40*(5), 631–643.

Hendryx, M. (2011) Poverty and mortality disparities in central Appalachia: Mountaintop mining and environmental justice. *Journal of Health Disparities Research and Practice, 4* (3), 44–53.

Hornberg, C., & Pauli, A. (2007). Child poverty and environmental justice. *International Journal of Hygiene and Environmental Health, 210*(5), 571–580.

Institute of Medicine. (1999). *Toward environmental justice: Research, education, and health policy needs*. Washington, DC: Institute of Medicine, Committee on Environmental Justice, Health Sciences Policy Program, Health Sciences Section.

Israel, B.A., Coombe, C., Cheezum, R. R., McGranaghan, R.J., Schulz, A.J., Lichtenstein, R., Reyes, A. G., Clement, J., & Burris, A. (2010). Community-based participatory research: A capacity-building approach for policy advocacy aimed at eliminating health disparities. *American Journal of Public Health, 100*(11), 2094–2102.

Kazis, R., & Grossman, R. L. (1991). *Fear at work: Job blackmail, labor, and the environment*. Philadelphia, PA: New Society.

Klepal, D. (2005, July 15). Addyston plant is revamping. *The Enquirer*. Retrieved from http://news.cincinnati.com/article/20050715/NEWS01/507150414/Addyston-plant-revamping.

Klepal, D., & Smith Amos, D. (2005, December 6). Cancer risk closes Addyston school. *The Enquirer*. Retrieved from http://news.cincinnati.com/article/20051206/NEWS01/512060355/Cancer-risk-closes-Addyston-school.

Lewis, S. J. (1997). Citizens as regulators of local polluters and toxic users. In C. Levenstein & J. Wooding (Eds.), *Work, health, and environment* (pp. 309–312). New York, NY: Guilford Press.

Mann, E. (1993). Labor's environmental agenda in the new corporate climate. In R. Hofrichter (Ed.), *Toxic struggles: The theory and practice of environment justice* (pp. 179–185). Philadelphia, PA: New Society.

McLean, S. (1997). A communication analysis of community mobilization on the Warm Springs Indian Reservation. *Journal of Health Communication, 2*(2), 113–125.

Minkler, M., & Wallerstein, N. (1997). Improving health through community organization and community building: A health education perspective. In M. Minkler (Ed.), *Community organizing and community building for health* (pp. 204–232). New Brunswick, NJ: Rutgers University Press.

Minkler, M., & Wallerstein,N. (Eds.). (2003).*Community based participatory research for health*. San Francisco, CA: Jossey-Bass.

Morello-Frosch, R., & Lopez, R. (2006). The riskscape and the color line: Examining the role of segregation in environmental health disparities. *Environmental Research, 102*(2), 181–196.

Neighbors for Clean Air and Ohio Citizen Action. (2006). *Citizens' Audit of Eramet*. Marietta: Marietta, Ohio.

O'Neill, M.S., Jerrett, M., Kawachi, I., Levy J. I., Cohen, A. J., Gouveia, N., Wilkinson, P., Fletcher, T., Cifuentes, L., Schwartz, J. (2003). Health, wealth, and air pollution: *Advancing theory and methods. Environmental Health Perspectives, 111*(16), 1861–1870.

Pamies, R. J., & Nsiah-Kumi, P. A. (2009). Addressing health disparities in the 21st century. In S. Kosoko-Lasaki, C. T. Cook, & R. L. O'Brien (Eds.), *Cultural proficiency in addressing health disparities* (pp. 1–36). Sudbury, MA: Jones and Bartlett.

Payne-Sturges, D., & Gee, G. C. (2006). National environmental health measures for minority and low-income populations: Tracking social disparities in environmental health. *Environmental Research, 102*(2), 154–171.

Payne-Sturges, D., Zenick, H., Wells, C., & Sanders, W. (2006). We cannot do it alone: Building a multi-systems approach for assessing and eliminating environmental health disparities. *Environmental Research, 102*(2), 141–145.

Plant going beyond compliance. (2005). *Insight*, 4.

Ryder, P. (Ed.). (2006). *Good neighbor campaign handbook: How to win*. Lincoln, NE: iUniverse.

Shawver, S. (2008) Eramet praised for $150M proposal. *The Marietta Times*. Retrieved from http://www.mariettatimes.com/page/content.detail/id/506582.html?nav=5002.

United States Census Bureau. (2000). American fact finder. http://factfinder.census.gov/home/saff/main.html?_lang=en.

US EPA. (2009). Reports. Toxic Release Inventory Explorer. Retrieved from www.epa.gov/triexplorer/.

U.S. EPA. (2009, August 6). Ineos ABS USA/Lanxess Clean Air Act settlement information sheet. Retrieved from http://www.epa.gov/compliance/resources/cases/civil/caa/ineos-lanxess-infosht.html.

Wing, S. (2005). Environmental justice, science, and public health. *Environmental Health Perspectives, 113*, 54–63. Retrieved from http://www.ehponline.org/docs/2005/7900/7900.html.

Zoller, H. M. (2005). Health activism: Communication theory and action for social change. *Communication Theory, 15*(4), 341–364.

Zoller, H. M. (2009). Narratives of corporate change: Public participation through environmental health activism, stakeholder dialogue, and regulation. In M. Dutta & L. Harter (Eds.), *Communicating for social impact: Engaging theory, research, and pedagogy* (pp. 91–114). Cresskill, NJ: Hampton Press.

25

Relational Tensions in Academic-Community Partnerships in the Culture-Centered Approach (CCA): Negotiating Communication in Creating Spaces for Voices

Mohan J. Dutta, Sydney Dillard, Rati Kumar, Shaunak Sastry, Christina Jones, Agaptus Anaele, U. Dutta, William Collins, Titilayo Okoror, and Calvin Roberson

In this chapter, we engage with participatory reflections on a project that uses the culture-centered approach (CCA) (Airhihenbuwa, 1995; Dutta, 2008; Dutta & Basu, 2008) to develop culturally based, comparative effectiveness research summary guides (CERSGs) in two African American communities in two different counties of a Midwestern state. Funded by the Agency for Healthcare Research and Quality (AHRQ), this project works with two Midwestern counties that report large health disparities for African Americans, as well as large health disparities in the realm of cardiovascular disease. As a critique of dominant models of health communication, CCA seeks to foreground the voices of community members and their lived experiences and localized meanings of health, in demonstrating how dominant models of health communication have led to the silencing and erasure of such voices (Dutta-Bergman, 2004a, 2004b; Lupton, 1994). The goals of the CCA processes in this project are to create culturally centered CERSGs on heart disease that build on the participatory capacity of local communities to develop locally meaningful solutions, more specifically, to develop health-information capacities in the local communities. The CCA processes are driven toward transferring the spaces of power from the academic-community organization partnership to the more flexible, permeable elements of the community.

Connecting with the works of Barge (2004), Barge and Shockley-Zalabak (2008), Cheney (2008), Simpson and Seibold (2008), and the notion of "engaged scholarship," we explore the conceptualizations of the culture-centered approach as a model of participatory approach to research that represents a commitment to engagement with community members, the research practice, and the theoretical rooting of CCA within the discipline of Communication. Further, we demonstrate that by emphasizing what Barge (2004) called "reflexive practice," CCA represents a fecund marriage of its theoretical lineage in critical scholarship and intellectual critique with the realities, uncertainties, and complexities of academic-community partnerships. The data for this paper come from the journal entries of the research team, as they engage with the community advisory board and the community in the initial stages of the grant. Finally, we present a grounded theory analysis of the data from the journal entries in order to articulate reflexive entry points on communicative processes described in CCA work.

CBPR and CCA: Participatory Approaches

A growing number of communication scholars have expressed concern that the dominant approach to health communication focuses primarily on individual behavior change (Airhihenbuwa, 1995; Basu & Dutta, 2007; Dutta, 2008; Lupton, 1994). Embedded within this approach is cognitive or behavioral reasoning that looks to individual attitudes, beliefs, and behaviors to solve vital health problems (Basu & Dutta, 2007; Dutta & Zoller, 2008; Dutta-Bergman, 2004a, 2004b; Lupton, 1994; Yehya & Dutta, 2010). In employing this logic, the leading course of action to improve health disparities has been to strategically develop messages targeted at optimizing effectiveness, thus ultimately aiming to shape health decisions of marginalized populations (Dutta, 2008).

Health experts have begun to question the gaps between knowledge production in health-communication research and the translation of conventional research into interventions or transformative policies in favor of disadvantaged populations (AHRQ, 2009; Airhihenbuwa, 1995; Lupton, 1994; Viswanathan et al., 2004). This has led to the development of multiple theoretical frameworks that call for more participatory involvement in the entire health-communication research process. Thus, the following similar, yet fundamentally different orientations have emerged: (a) Community-Based Participatory Research (CBPR), and (b) the Culture-Centered Approach (CCA).

Theorizing the Purpose of Participation

"Outside expert" research and interventionist approaches lack the ability to obtain sustainable, positive change, given the complexities of health problems and the corresponding location of power in the hands of academics at every point in the research process (Dutta, 2007; Lupton, 1994; Minkler & Wallerstein, 2003; Minkler, 2005). Thus, the concept of increased community participation has emerged as fundamentally crucial in health-communication processes. Though now viewed as essential to the entire communication process, the conceptualization of participation in the development of CBPR and CCA has spurred further dialogue that calls for more clearly identifiable definitions of such key terms. More importantly, the larger issue under scrutiny is the conceived and actual function of community participation as it informs research.

CBPR Participation as a Means to an End

As defined by Israel, Schulz, Parker, and Becker (1998), centralizing participation in health-communication research sets forth principles of CBPR based on the following assumptions: (a) genuine partnership through colearning (academic and community partners learning from each other); (b) research efforts that include capacity building (in addition to conducting the research, there is a commitment to training community members in research); (c) provision of knowledge that benefits all partners; and (d) involvement of long-term commitments to effectively reduce disparities. The way in which participation is centralized in the implementation of CBPR is one that uncovers lay knowledge by enhancing researchers' understanding of sensitive health issues particular to disadvantaged populations. It is through cultural sensitivity and humility that studies based in CBPR principles investigate unexplored contributors to health and social problems. This important information is often explored through local community needs assessments and formative survey research. For instance, the implementation of CBPR in Lee, Krause, and Goetchius (2003) led to a 69% response rate among a largely immigrant, and highly marginalized, hotel-room-cleaning population, thus yielding culturally sensitive data revealing the needs of this group. CBPR's ultimate contribution to the health field is its ability to develop culturally appropriate measurement instruments built upon the participation and responses of community members, therefore making projects more effective and efficient (Goodman et al.,1998; Viswanathan et al., 2004; Wallerstein & Bernstein, 1994).

Despite CBPR's logistical grounding in community participation, questions of purpose have also emerged (Dutta & Basnyat, 2008a, 2008b). Health-communication scholars with ontological commitments to the tenets of critical theory and subaltern studies continue to question the ways in which such orientations claim exclusion from the status quo. In centralizing participation as a means to an end, structural and cultural contexts are far removed from discussions of health issues within their local context. Dutta (2007) maintained that

> Efforts of health promotion are typically based on the universal logic of scientific rationality, drawn upon individualistic assumptions about the constitution of health risks and hence are ignorant of cultural context, and are unresponsive to the social-cultural-economic contexts within which health experiences are located. (p. 305)

Though participation is placed at the forefront of CBPR, its system of reasoning still stems from that which dominates most health-communication research.

CCA Participation Under Interrogation

The culture-centered approach (Airhihenbuwa, 1995; Dutta, 2008; Dutta & Basu, 2008; Dutta-Bergman, 2004a, 2004b; Ford & Yep, 2003) foregrounds the voices and lived experiences of cultural members in seeking to establish how traditional approaches to health-communication campaigns have contributed to the erasure of voices of marginalized communities, even in working to improve the health outcomes of such communities (Dutta, 2008). The notions of culture, structure, and agency represent three concepts central to the CCA. It is only through the interactions of these that a discursive opening can be found to "provide opportunities for co-constructing the voices of those who have traditionally been silenced" (Dutta, 2008, p. 5). In opposition to CBPR's usage of participation as a means to an end, the CCA takes participation and investigates its meaning within the process of seeking structural change while simultaneously finding relevance in the local narratives of health within a given community. Therefore, participation emerges in the CCA as meta-discourse about participatory processes. The CCA places emphasis on the need for investigating alternative entry points to interpreting health, and the ways in which ideas of health are understood in addition to those that are scientifically, clinically, or biomedically supported (Dutta, 2007, 2008; Dutta & Zoller, 2008). The CCA also centralizes participatory approaches as a pathway to understanding health meanings, and in doing so, continually turns the lens on the dominant structures. Constantly revisiting processes of reflexivity and performativity, it engages with the active, meaning-making processes through continual interaction with individuals as agentic entities, while interrogating the structural barriers that enable and constrain them in their everyday lives. Reflexivity, in the CCA, is ultimately connected to the project of emancipation that is embedded within it. This emancipatory impulse within the CCA is the legacy of its theoretical roots: in critical theory, postcolonial theory, and the subaltern studies literature. We now proceed to locate within the CCA some of the core tensions inherent in these traditions.

CCA and Critical Theory: Structure, Knowledge, and Power

Culture-centered theories of health communication owe significant debts to the various strands of critical theory. In their interrogation of the nexus between knowledge and power, culture-centered theories reflect this core tension of Marxist critical theory. The relationship between knowledge and power has been the subject of many strands of critical theory, most of which are traced back to the work of Marx and Engels (Artz, Macek, & Cloud, 2006; Marx, Engels, & Tucker, 1978). CCA takes from these traditions a commitment to interrogate the interplay of knowledge production and power within the realm of global health communication, and is interested in exploring how conditions of economic marginalization are related to patterns of knowledge production within a health context (Dutta, 2008). For instance, in culture-centered work conducted amongst commercial sex workers in Kolkata, India, by Basu and Dutta (2009), it was found that the dominant knowledge claims about sex workers as passive and ignorant about matters of sexual health were debunked, and instead, the voices of these sex workers emerged as those of "experts," given that their recommendations and understandings of what stood in the way of sex workers enacting effective sexual prevention behaviors was grounded materially within their lived experiences.

CCA and Subaltern Studies: Voice, Erasure, and Agency

Originally emerging from the discipline of history, the subaltern studies collective represented scholars that were interested in the politics of historiography—or the writing of history (Guha, 1982). Subaltern studies scholarship begins with the notion that historiography is complicit in the erasure and silencing of voices that exist at the fringes of societies. Within a contemporary, health-communication context, the legacy of subaltern studies gives CCA the tools by which to analyze the current erasure of large populations of individuals from the construction of history and knowledge. For instance, Dutta-Bergman's (2004a, 2004b) work with the Santalis, a tribe that lives in the eastern part of India, explicated how the health choices made by the Santalis were largely contingent on what they earned everyday, given that most of them were daily wage laborers and needed to secure food for themselves and their children on an everyday basis. The author argued, then, that health experiences in marginalized, subaltern sectors must engage with issues of the structural barriers that impede access to health care and effective health options (Dutta-Bergman, 2004a). The concept of agency, as theorized in CCA, emerges in very direct ways from the subaltern studies literature, inasmuch as it attends to the individual enabling acts that subaltern communities engage in, in order to challenge the structural barriers that affect their daily lives. It is therefore with this emphasis on creating avenues for the participation of local communities in processes of change that the CCA turns the reflexive lens on the research process, continually interrogating the meanings that emerge in the negotiation of academic-community partnerships. Therefore, in engaging with processes of change in CCA, we seek to examine the following question: How does participation get conceptualized in the academic-community partnership?

Method

Data

The data for this chapter are derived from the reflexive journal entries maintained by research team members through the life cycle of the project. This is accomplished through the initial setting up of a community advisory board, which codesigns and coconstructs the intensive, in-depth interviews and focus groups to be conducted with community members. These in-depth interviews and focus groups become the foundations for organizing community coalitions, which further become avenues for working with the advisory board and the academic-community partnership in recruiting community members for communication capacity-building workshops, where the communication capacities of the community are built by community members through the development of communication strategies and tactical solutions.

Throughout the entire period of the project, the research team—comprised of one principle investigator (PI), three coinvestigators, one community leader from the partner community organization, and two community organizers—is required to maintain weekly reflexive entries as part of the process. These entries are maintained on a blog, giving the different project team members the opportunity to respond to each other's blogs, thus building a dialogue along the lines of CCA. This chapter is built on our analysis of the phase I of the blog entries maintained by the research team between August, 2010 and March, 2011. The blog entries resulted in 98 pages of blogging.

Data Analysis

We conducted a grounded theory analysis of the research team blogs in order to develop insights into the participatory processes of CCA as articulated through the lens of the research team. Given the culture-centered theoretical lens applied in this project, we utilized a coconstructive grounded theory approach for data analysis, that emphasized the involvement of the participants in the meaning-making process. The data were analyzed using open, axial, and selective coding, as suggested by Charmaz (2000). Open coding identified the concepts as were explicit from the interviewees' responses. In the next step of axial coding, commonalities were taken from the open coding and related categories were formed. Finally, these categories were bound together to form theoretical integration (Strauss & Corbin, 1990).

Negotiating Communicative Spaces

The two themes that emerged from the data analysis and constitute the key methodological questions for academic-community partnerships driven by the CCA are (a) heterogeneity of participatory directions, and (b) reflexivity as process. Our results depict the contextually rooted, continually contested nature of academic-community partnerships, played out amidst the competing agendas of various academic and community stakeholders in the participatory processes of the CCA.

Heterogeneity of Participatory Directions

Throughout the processes of engaging with the advisory boards, conducting the in-depth interviews, and articulating the directions of the project, the research team noted the heterogeneity of responses articulated by different participants from the communities in the two counties where we have been conducting our work. These multiple responses point toward the multiplicity of interpretive frames within the community, and the competing agendas, goals, and strategies articulated within these competing frames. Local meanings of heart health, comparative effectiveness research (CER), clinical decisions, medical practice, and experiences of physician-patient interactions are articulated within often contradictory understandings. At times, the locally situated agendas of change converged on some common threads. At other times, these agendas differed dramatically from each other. Even in instances where community members pointed toward a specific problem configuration, their notions of addressing the problem configuration differed dramatically. Here is an example of a tension in meaning-making noted by Latoya:

> I've been thinking about several facets of the interview. How do we negotiate between conflicts in information? If one participant says one thing and another says something different or in opposition to the former, how do we as researchers negotiate these tensions. For instance, I noticed at last night's debriefing session, Ahmed noted how some participants noted that having a community advocate at the hospital would help the CERSG information be clearer and digested more easily. However, I remember another participant stating that this would not be a good idea as waiting to access doctors is already a job in itself. After bringing it up to the advisory board, it seemed they also agreed that having a heart health advocate will place a further burden on patients as well as the grant and community financially. So here we have conflicting ideas both from interview's participants and advisory board members. I wonder whose voice gets heard. Is it our job to choose whose voice is heard or should we try to make every voice heard? Also if our goals are to have all the voices of the community heard, then what are we to do when it comes time for the practical applications of these thoughts?

The contested nature of participation is foregrounded in this reflection. The subjective meaning articulated by Latoya through her interactions with a participant does not match up with the subjective meaning articulated by Ahmed through his interactions with some of the participants. Furthermore, the advisory board, which represents community members as well as staff members of the health coalition we have been working with, also had competing understandings of the role of the heart-health advocate and the costs that would be associated with having heart-health advocates at hospitals and clinics. It is against this backdrop, then, that the culture-centered processes of the project critically engage with the participatory frameworks of the project, raising questions about whose voice gets heard amidst these competing tensions, and what happens when such tensions arise. Of particular interest are the questions raised regarding the development of practical applications (such as having a heart-health advocate to explain the CERSGs) by working through the competing interpretations by different members of the communities, and by different stakeholder groups who are involved in the processes. For the research group, engaging in participatory dialogues with community members is constituted within working through the different and competing perspectives about meanings of health, heart health, experiences of interactions with physicians, and experiences in using evidence for making health decisions.

Along these lines, Ahmed noted the tensions in the politics of representation, as participatory spaces are turned into spaces for developing policies and specific interventions that are rooted in the voices of communities:

> The politics of representation is perhaps the most difficult question to engage in with CCA work that is mainly concerned with making structural transformations. That communication is central to this structural transformation becomes apparent in the interviews. It is also through the politics of representation that CCA takes individual voices as legitimate entry points for theory building. However, the aggregation of individual stories as representative articulations of communities is the fundamental discursive shift that is embodied in the move from individual stories to policy articulations. Authenticity of the scholarship is negotiated precisely in the midst of this shift; from the individual to the community, at once recognizing the points of solidarity in articulating a community-based politics, and simultaneously continually acknowledging the fractures in the politics of representation that moves from individuals to communities. It is also amidst these tensions that I am

> continually drawn to acknowledge the tensions in my desires to respond, to speak for, to become the voice of legitimacy for the community. This also is the key challenge: how do I lend my expert structures and the legitimacy that comes with these structures in solidarity with the voices in the community such that we speak together in working out a politics of change, a politics that envisions a just, equitable world.

The question that is raised by Ahmed here ties to the discursive move from individual subjectivity of the participant voices as narrated through the in-depth interviews to the act of aggregation in representing the voices of the community in the development of interventions and policies. When policy- and intervention-based decisions are made by the advisory group-research group partnership, such decisions are made amidst extrapolations of interpretations from individual stories to collective data at the aggregate level. Even as Ahmed spells out a politics of solidarity, inherent in this politics is the vital move from the individual stories shared by the community participants to specific solutions from the participant stories that are picked by the partnership of the research team and the advisory board, and are highlighted as entry points of change. In seeking to speak together with the communities in the context of cocreating a just and equitable world, Ahmed notes the tensions in working out what counts as the voices of the communities, privileging certain themes, and in that very process erasing and/or omitting other themes. The same thread plays out in the data transcription and analytic phase of the project, as the academic team works through the 150 hours of in-depth interview data to draw out specific themes at the aggregate level from the individual-level dialogues with community members.

Noting this tension in the participatory direction of the project as it moved from the individual subjectivities to an aggregated entry point for a politics of change, Vikram noted in his posting, titled "Aphorisms from the alcoves of a church":

> Sitting in a recessed office at the back of a small church, our pastor-participant spoke about his struggles. Against the overwhelmed, unbothered medical system; against the systematic bureaucracies in the institutions around him. City Hall. The church. Corrupt officials; nepotistic pastors. A leaking roof; an aching back. Outside, the snow ravaged his rapidly decaying city: unplowed pavements, foreclosed houses; his struggle was his city's struggle—at the wrong end of a wantonly greedy system.

> Harnessing the power of youth. He spoke passionately about the need to bring the youth back. Told us about his effort to start a youth center next to the church. A basketball court. Not as a way to 'exercise regularly to reduce heart disease risk', but more fundamentally, to build a youth-based community. To show kids a way out of the stress. The stress from living in the hood; from being poor and uninsured. To 'turn them around', as he said. Can there be a stronger example of health as negotiating structural barriers?
>
> ...He spoke about how he and his brother would talk about the medications they had each been prescribed, and between them, figuring out what was best for them; going and asking their doctors for more information, better medication, less side effects. They compared medications with each other, figured out what was common to their symptoms, keep challenging uncaring doctors only too eager to push pills on to poor patients. The brothers used a 'cultural barrier'; a family history of heart disease into an agentic resource; as a comparison group. Even a small fissure, like the n of 1 becoming an n of 2, can start the ball rolling, encouraging individuals to challenge the structures around them. What a story.

Worth noting in Vikram's reflection of the interview with one of the first participants in the in-depth interviewing element of the project is the articulation of the movement from the subjectivity of individual participation to the coconstruction of a collective narrative that may be mobilized in working toward change. Vikram points out that the story at the level of n equals 1 creates an entry point for building a politics of change constituted on the foundations of broader, community-specific data that are then used in working toward efforts of social change. As the in-depth interview engaged with the participant, it emerged as a site for storying the participant's understanding of his own life and his own relationship with his brother. It is through this relationship that the participant talked about making many of his health decisions, comparing notes with his brother, and arriving at solutions through the comparison of notes. Vikram notes how the brothers in this story re-storied the narrative of family history to create an alternative entry point where the relationship as brothers became a source for comparing between various treatment options. For Vikram, the movement from n equals 1 to n equals 2 constitutes the politics of change in CCA.

Contrary to the articulations of moving from the individual realms of subjectivity to the politics of aggregation that work in supporting each other and in creating an entry point for change, the research team also noted the contradictions that emerge from the differences between localized articulations by community members as opposed to the articulations by the advisory board members, who are also from within the community. Arpita points out the following:

> Throughout the CCA process there is constant discussion and awareness of the top down approach. As far as the research team is concerned, there have certainly been reflections on the recurring instances which arose during the advisory board meetings, where we as the research team would have to curb the urge to resort to a top down interaction in keeping with the CCA philosophy.
>
> However, the same was not always the case during the feedback by the Advisory Board as regards what would and would not be a viable approach to take with the community members themselves. Now, as we move into the next level of interactions, it would certainly be a demonstration of how we have to be careful about the possibility of taking a top down approach, even when we get feedback from the people working in the community, if we do discover any significant differences between what the community members see as sustainable and workable solutions to certain heart health message communication practices, on the ground, in comparison to the feedback given by the board members, who, being in interaction with the community would still not necessarily be part of the community....It certainly puts in perspective the need to recruit not a convenient sample, but one that truly gives us data from a different perspective that maybe triangulates our prior data or takes us in a different direction that we may not have explored with the board.

Worth noting in Arpita's articulation is her perception of the gap between the voices of the advisory board and the voices of community members. Even as she notes the reflexivity of the research team in interrogating the top-down processes in order to create spaces for the advisory board, she also notes the power embodied in the advisory board and the top-down decisions that were being made by the advisory board, as it often took the position of being the voice of the community. In taking the position as representatives of the community, the advisory board was making certain decisions, and Arpita interrogates the top-down nature of these decisions as they relate to the politics of representation embedded in the notion of serving as a voice of the community. This is where Arpita also talks about the need for reflexivity at the level of the advisory board, particularly in instances where the board members took up the voice of the community to say things such as, "community members would not understand this or that." She further engages the question about the possible course of action when the articulations of community members do not really match with the articulations of the advisory board. Furthermore, worth noting in Arpita's construction is the emphasis of the culture-centered approach on difference, pointing toward the need for recruiting community members in the next phase of the project (in-depth interviews) that provide alternative entry points, points of enunciation that attend to the dynamic and complex nature of meaning-making within communities. Highlighting this construction of difference as an entry point to change, Samiran similarly notes the following:

> Lastly, given the reaction of the community members interviewed for the Gospel tour video, I am looking forward to seeing how many folks from the community grasp and imbibe the notion of the culture centered approach while analyzing these heart health messages for their community. This seems to be a fantastic opportunity to observe the fine balance between the materiality of working within a community, and simultaneously drawing out the possibilities of a solution based on listening and participation, which may not always be the most "sensible" approach initially, in the eyes of the community members themselves.

The voices of heart health at the gospel tour was a video-voice project that was undertaken by the research team at the request of an advisory group member, who headed the social equities initiative of the American Heart Association (AHA). The AHA historically conducts the gospel tour as a method for raising awareness about heart health through African American churches. During the video narratives with community members at the tour, the research team came across localized articulations by community members that did not match with the articulations of the advisory board. For example, some members of the advisory board in one of the interviews raised questions about the question, "What does heart health mean to you?" saying that the question would not work. Yet, when the question, "What does heart health mean to you?" was placed to community members at the gospel tour, they utilized the question to construct complex narratives of heart health. The research team would have scrapped this question had it not been for an alternative data point that had emerged through the prior interviews at the gospel tour, where community members responded positively and in an involved way to the question,

"What does heart health mean to you?" It is against this backdrop, then, that the research team continually worked back and forth about the methodological questions of decision-making, and the location of decision-making in the hands of the advisory board members as representatives of the community vis-à-vis placing decisions in the hands of individual participants from the community. At what points do acts of listening to advisory-board members as representatives of the community work in difference with acts of listening to other groups and subgroups within the community?

Reflexivity as Process

Throughout our entries, the research team continually struggled with what it means to engage authentically with the two communities we were working with, and particularly as it relates to the decisions that were being taken by the research team through negotiations with the advisory board. The meaning of the term, *authenticity*, became a term for contestation, not only in terms of what would constitute authenticity, but also in terms of how does one act authentically in relationship to the community partners and advisory-board members. For example, building up to February 14, Angela suggested that we send out Valentine's Day cards to community members to show that we care about them, and as a way to build a connection that takes the relationship beyond the impersonal domain of research. This, for Angela, was the way that the academic partner could demonstrate its care for the community partner and advisory-board members. She recalled how her mother always shared this story about this one research project that her mother participated in and would recall even after many years, because she received a card in the mail from the project director. To this suggestion, Vikram responded:

> As I traverse through my personal struggles with CCA, with the meaning of critical pedagogy, I have some haunting questions. As an 'owner' of the method in a rather pragmatic sense, I benefit from it—I get published, I get taken seriously, I am 'listened' to in academic contexts, and so on. And at the same time, this ownership also demands of me an accountability; a responsibility to the standards that I proclaim. In that sense, this moment of 'ownership' is a daunting one; it means different things to me to be a student of CCA than it does to practice it. Perhaps this is just a displacement of my soon-to-be-on-the-market anxieties, but

questions of 'What commitments does culture-centered scholarship entail?' keep me occupied much more lately.

Why do I say all this? This is because the line between what constitutes solidarity and the touchy-feelyness of cultural sensitivity is always watermarked. Ergo, there needs to be a constant 'holding up to the sun' process; a reality check, a vigilant self-examination. The reason CCA can never be—dare I say must never be—a feel good exercise is precisely because the latter is the defining characteristic of its impostor: the 'ethics of care' as exemplified by culturally-sensitive global health communication.

What does it mean, then, for us to send an 'American Heart Month' celebration card to our community as a token of our appreciation? To clarify, this is not meant to be an ad hominem attack on either the person or the idea of sending cards out to people. Rather, this is an important point to deliberate because this is precisely the moment where the impostor puts on his mask. We;ve read all along that putting one's privileges on the table is an essential part of the CCA process. But what does that entail? What are its nuts and bolts?

I've spoken in earlier posts about my struggles to establish trust within our research community. As an outsider, I've learnt the hard way what it means to own up to one's privilege. I've found, to my great surprise, that it is not enough to belong to a group of trustworthy outsiders; the road to developing trust is always a solitary one. (I say this with profound humility—I don't necessarily think I can take as assumed the trust that the advisory board and our interview participants have in me). It is a process that takes time, reflection, swallowing a heck of a lot of one's own ego and notions of expertise. People who know me closely know how much I've thought about this process, and the journey that I have made even as I have come to know this community. I'm sure colleagues of mine who've been on this project with me, and who've joined this project after me, are, just like me, currently going through this journey, or will arrive at this moment in the future—this process always reminds me of standing at the immigration counter at US airports; and the relief that comes from getting your papers stamped.

Maybe this is too personal an impulse to be communicated, but I feel that sending 'American Heart Month' cards, well-meaning as the effort is, dilutes that moment and reverses that process in so many ways. I mean, as far as special months and occasions go, I don't think our efforts should point out that February is 'Black History month', and so we're doing that little bit extra for you this month. I wonder what it means to send an 'American Heart Month' card to an individual whose doctor is not bothered to even listen to what she has to say about her symptoms. Wouldn't it remind her of all the infrastructural differences between people who can actively seek out screenings during heart month, and those

> who have to constantly change doctors so as to find one that actually gives a hoot? We're seeing from our interviews that dominant understandings of black communities are that they are passive, lazy, unbothered, ill-informed people. We're hearing their stories of actively struggling, making choices within what they have, coming together to help each other. Let's listen to these active voices harder.

What gets foregrounded in Vikram's articulation is his struggle with the development of authentic relationships, the development of trust with individual participants, the advisory board, and the community partners. He notes the politics of authenticity outlined in CCA that seeks to engage with listening as a way to transform existing power inequities. At the same time, he notes the tensions that emerge from the academic politics of CCA work, and the need to define and demarcate boundaries in order to etch out a space for the CCA work in the academic marketplace. Vikram locates the relationship of authenticity with community members amidst this question of authenticity that goes back and forth between a politics of change and a politics of bureaucratic functioning that operates upon the needs of the marketplace.

For Vikram, the sending out of the Valentine's Day cards to the community members amounts to the superficial touchy-feelyness of cultural sensitivity, an inauthentic move that superficially attempts to craft a relationship of solidarity without really listening to the active voices of community members and the stories of their struggles in securing access to health. Without attending to the structural issues of access, sending out cards are reminders of the inequities that constitute the inaccess to health. Vikram feels that carving out the relationship of authenticity with community members is one of sincere listening, one that attends to the struggles of inequities and injustices that are articulated by the community members, bringing these struggles to the forefront. Whereas Angela thought that sending out the cards would serve as a marker of building an authentic relationship with community members, and as an indicator of a relationship that aspired to move beyond the traditional relationship of researcher and researched, Vikram resisted the idea based on the notion that a card is superficial and does not really create the scope for building a relationship of authenticity that is sensitized to the inequities written into the differential positions of power occupied by the research team and the community partners.

Latoya, herself an African American who grew up in a context very similar to one of the communities in this project, responds to Vikram's notion of authenticity through the acknowledgement of the struggles articulated by community members:

> Every time someone says the African American community members said this or that, I become physically tense and defensive as if any and everything someone says about the African American community applies or represents my thoughts, goals, and experiences. I know that everyone has different experiences and understandings of their life, but it is quite difficult for me to suppress these feelings. Notice in *Vikram's* blog earlier, he states (about the voices of African Americans interviewed so far) that "We're hearing their stories of actively struggling, making choices within what they have, coming together to help each other." My gut response was, "oh my goodness, here we go again with the rhetoric that still somehow addresses us as struggling, always trying to do better, to be better...simply always struggling." Now I know that *Vikram* never meant it in this way, to further marginalize the group as "always struggling" or which some would say, never reaching a level of contentment or solace, but it's so difficult not to respond to this. It has been my experience that the dominant understanding of the African American experience is one in which we are portrayed as lazy (and all the other adjectives *Vikram* pointed out), but another way that African Americans have been talked about has been from the perspective of a group of people always struggling against the structures that bound them, especially in media.
>
> This perspective has also been a more acceptable view as it doesn't demonize African Americans and their perceived unwillingness to carry their own weight in the name of rugged individualism. If you think about it, this rhetoric has been around since slavery, similar to other forms of depictions such as the lazy African American. Let's think about it, if you've ever taken an African American studies course or any course that talks about slavery, you will learn how African Americans used old Negro spiritual songs to aid in getting the message out about fleeing plantations. Additionally, you found out that these songs helped in the struggle of African Americans, allowing them an outlet to bear the daily hardships of their lives. Moving in time, you'll also see many discussions of African Americans during the civil rights movement, when the "struggle" turns towards their fight for equal rights. I'm not saying that is it wrong to state that people are actively struggling with the structures as we all do it every day, but I just hope that we as researchers are careful that we do not further marginalize the group by reintroducing another interpretation of the African American experience, even if it may be true. Note here I do think it is true that African Americans in our interviews spoke about actively struggling and making "do" with what they have, but I just want to bring up that even in doing this, in claiming this ex-

perience, we walk a fine line of placing another dominant understanding of African Americans in to the dialogic spaces of heart health discussions. One which has been used before and one which may quickly become trite.

What Vikram sees as the focal point for a politics of authenticity rooted in the everyday struggles of the African American community members who share with us the stories of their pains and struggles in the context of securing access to health resources, appears as a politics of aggregation based on patronizing stereotypes to Latoya. To Vikram, the reality of the in-depth interviews was articulated through the stories of pain and suffering in the context of defining meanings of health. Vikram arrived at his narratives through the processes of coconstructions in the in-depth interviews. Latoya, however, notes the problem in moving from the individual stories of pain and suffering to aggregate responses that sought to represent the narratives of the African American community in terms of struggles. Note here the tension Latoya feels as herself being a member of the community, making a discursive move to take ownership of her own identity in the context of the project and in the context of her relationship with the community. To Latoya's response, Ahmed notes during a one-on-one conversation with Latoya:

> You make a really good point about the importance of being careful about how and how far we are extrapolating the stories that emerge from the in-depth interviews. You encourage each of us to be mindful of how far we extrapolate? In this process of extrapolation though *Latoya*, the question also worth asking is: How do you make the transition from being the researcher to talking about the community as "we," based on the assumption that you are a member of the community? So my interest is also in this personal transition for you, from being a researcher to being one of the community. So now you live in one of these communities, so you can claim perhaps that you are a resident and hence a community partner. And yet, you are also a member of the research team, the one who is sitting amidst the privileges of the dominant structure and getting to set up the rules of the game. So what does this mean in terms of your own responses when you hear these stories?

Reflexivity is not only embedded in the questioning of one's role and position within the academic-community partnership, but also in the questioning of the specific questions that are asked in the in-depth interviews, the reflections on these questions as the interviews unfold, and the flexible adjustment of questions based on the articulations of community members. Vikram had the following to say about observing an in-depth interview conducted by Ahmed:

> When he asked him what health meant to him, he spoke from his rich experience....He was detailed, confident and clear in his responses. However, when we asked him a 'fact' question; what appeared to be a typological question; a 'how many kinds of...are there' question, he felt distinctly uncomfortable. It became a moment where we 'experts' were unwittingly testing his knowledge of heart disease. When the question was skillfully rephrased as a 'meaning' question, back to his normal eloquence.
>
> So much of CCA hinges on little choices we make. Rhetorical, nonverbal, gestural, lexical. Ahmed, I saw you get that moment of discomfort and react to that. I wonder if you were consciously aware of that; and remember that now. How much vigilance is hyper vigilance?

Vikram's articulation of the in-depth interviewing process brings attention to the constant vigilance at the dialogic moments of the in-depth interviews, where listening to the stories of participants also continually calls for "working on the self" for the interviewer, being aware of the gaps and fissures in the questions she/he asks, the foreclosures that are brought about by particular phrasing of questions, and the need to continually revisit the questions at the very sites of the in-depth interviews. Worth noting here is the gap between the in-depth interview protocol developed by the academic-community partnership and the advisory board, and the phrasing of the question that creates an entry point for a conversation. Whereas Vikram observed that asking the fact-based question about the different types of heart disease led the interviewee to feel uncomfortable, the rephrasing of the question into a question of meaning made him feel at ease again. Vikram discusses these as the little choices we make, and yet these little choices we make are situated amidst the continual reflexivity of the research team throughout the processes of the in-depth interviews.

Conclusion

The two key themes that emerge from our engagement with the reflexive processes that the research team was going through are constituted around the contingent and contradictory nature of communities, and the relationships between community and academic partners. The academic-community partnership in our culture-centered model is situated amidst multilayered relationships with the community, advisory board, and the minority-health coalition. At each of these interfaces of communication between the academic entity and the community constituents, tensions arise in the articulation and representation of needs, and in the interrogations of the terrains of representation. Emphasizing the Subaltern Studies roots of CCA, our reflexive entries continually engage with the erasures in our participatory processes, interrogating the ways in which these processes, at multiple interfaces—from the academic-organization partnership, to the partnership with the advisory board, to the partnership with community members—are continually constituted at entry points of erasure. Even as the participatory processes of CCA seek to engage with local community voices, they are intrinsically tied to the erasures of these voices through specific elements of the methodologies. It is in the context of such erasures that our reflexive journal entries point toward the "work on the self" through which CCA seeks to create entry points for listening to voices of community members from the margins of dominant structures.

Bibliography

Agency for Healthcare Research and Quality (AHRQ). (2009). *National health disparities report 2008*. Rockville, MD: Author.

Airhihenbuwa, C. O. (1995). *Health and culture: Beyond the Western paradigm*. Thousand Oaks, CA: Sage.

Airhihenbuwa, C. O., & Obregon, R. (2000). A critical assessment of theories/models used in health communication for HIV/AIDS. *Journal of Health Communication, 5*(1), 5–15.

Artz, L., Macek, S., & Cloud, D. L. (2006). *Marxism and communication studies: The point is to change it*. New York, NY: Peter Lang.

Barge, K. J. (2004). Reflexivity and managerial practice. *Communication Monographs, 71*(1), 70–96.

Barge, J. K., & Shockley-Zalabak, P.(2008). Engaged scholarship and the creation of useful organizational knowledge. *Journal of Applied Communication Research, 36*(3), 251–265.

Basu, A., & Dutta, M. J. (2007). Centralizing context and culture in the co-construction of health: Localizing and vocalizing health meanings in rural India. *Health Communication, 21*(2), 187–196.

Basu, A., & Dutta, M. J. (2009). Sex workers and HIV/AIDS: Analyzing participatory culture-centered health communication strategies. *Human Communication Research, 35*(1), 86–114.

Beverly, J. (2003). Testimonio, subalternity and narrative authority. In N. K. Denzin & Y. S. Lincoln (Eds.), *Strategies of qualitative inquiry* (p. 22). Thousand Oaks, CA: Sage.

Cheney, G. (2008). Encountering the ethics of engaged scholarship. *Journal of Applied Communication Research, 36*(3), 281–288.

Dutta, M. J. (2007). Communicating about culture and health: Theorizing culture-centered and cultural sensitivity approaches. *Communication Theory, 17*(3), 304–328.

Dutta, M. J. (2008). *Communicating health: A culture-centered approach*. London, UK: Polity Press.

Dutta, M. J., & Basnyat, I. (2008a). Interrogating the Radio Communication Project in Nepal: The participatory framing of colonization. In H. M. Zoller & M. J. Dutta (Eds.), *Emerging perspectives in health communication: Interpretive, critical and cultural approaches* (pp. 247–265). Mahwah, NJ: Lawrence Erlbaum.

Dutta, M. J., & Basnyat, I. (2008b). The Radio Communication Project in Nepal: A culture centered approach to participation. *Journal of Health Education and Behavior, 35*(4), 442–454.

Dutta, M. J., & Basu, A. (2008). Meanings of health: Interrogating structure and culture. *Health Communication, 23*(6), 560–572.

Dutta, M. J., & Zoller, H. (2008). Introduction. In H. M. Zoller & M. J. Dutta (Eds.), *Emerging perspectives in health communication: Interpretive, critical and cultural approaches* (pp. 30–38). Mahwah, NJ: Lawrence Erlbaum.

Dutta-Bergman, M. J. (2004a). Poverty, structural barriers and health: A Santali narrative of health communication. *Qualitative Health Research, 14*(8), 1107–1122.

Dutta-Bergman, M. J. (2004b). The unheard voices of Santalis: Communicating about health from the margins of India. *Communication Theory, 14*(3), 237–263.

Ford, L. A., & Yep, G. A. (2003). Working along the margins: Developing community-based strategies for communicating about health with marginalized groups. In T. L. Thompson, A. M. Dorsey, K. I. Miller, & R. Parrot (Eds.), *Handbook of health communication* (pp. 241–261). Mahwah, NJ: Lawrence Erlbaum.

Goodman, R. M., Speers, M. A., McLeroy, K., Fawcett, S., Kegler, M., Parker, E, Smith, S. R.,...Wallerstein, N. (1998). Identifying and defining the dimensions of community capacity to provide a basis for measurement. *Health Education & Behavior, 25*(3), 258–278.

Guha, R. (1982). *Subaltern studies: Writings on South Asian history and society*. New York, NY: Oxford University Press.

Guha, R. (1987). On some aspects of the historiography of colonial India. In R. Guha & G. C. Spivak (Eds.), *Selected subaltern studies* (pp. 37–44). New York, NY: Oxford University Press.

Israel, B. A, Schulz, A. J., Parker, E. A., & Becker, A. B. (1998). Review of community-based research: Assessing partnership approaches to improve public health. *Annual Review in Public Health, 19*, 173–202.

Lee, P., Krause, N., & Goetchius, C. (2003). Participatory action research with hotel room cleaners: From collaborative study to the bargaining table. In M. Minkler & N. Wallerstein (Eds.), *Community based participatory research for health* (pp. 390–404). San Francisco, CA: Jossey-Bass.

Lupton, D. (1994). *Medicine as culture: Illness, disease and the body in Western societies*. Thousand Oaks, CA: Sage.

Marx, K., Engels, F., & Tucker, R. C. (1978). *The Marx-Engels reader* (2nd ed.). New York, NY: Norton.

Minkler, M. (2005). Community-based research partnerships: Challenges and opportunities. *Journal of Urban Health, 82*(2), ii3–ii12.

Minkler, M., Breckwich, V., Warner, J. R., Steussey, H., & Facente, S. (2006). Sowing the seeds of sustainable change: A community based participatory research partnership for health promotion in Indiana, USA and its aftermath. *Health Promotion International, 21*(4), 293–300.

Minkler, M., & Wallerstein, N. (Eds.). (2003). *Community based participatory research for health*. San Francisco, CA: Jossey-Bass.

Sastry, S. J., & Dutta, M. J. (2011). Postcolonial constructions of HIV/AIDS: Meaning, culture, structure. *Health Communication, 26*(5), 437–449.

Shome, R., & Hegde, R. (2002). Postcolonial approaches to communication: Charting the terrain, engaging the intersections. *Communication Theory, 12*(3), 249–270.

Simpson, J. L. & Seibold, D. R. (2008). Practical engagements and co-created research. *Journal of Applied Communication Research, 36*(3), 266–280.

Spivak, G. C. (1987). Subaltern studies: Deconstructing historiography. In R. Guha & G. C. Spivak (Eds.), *Selected subaltern studies* (pp. 3–32). New York, NY: Oxford University Press.

Spivak, G. C. (1998). *A critique of postcolonial reason: Toward a history of the vanishing present*. Cambridge, MA: Harvard University Press.

Strauss, A. L., & Corbin, J. M. (1990). *Basics of qualitative research: Grounded theory procedures and techniques*. Newbury Park, CA: Sage.

Viswanathan, M., Ammerman, A., Eng, E., Garlehner, G., Lohr, K. N., Griffith, D., Rhodes, S.,...Whitener, L. (2004). *Community-based participatory research: Assessing the evidence*. Rockville, MD: Agency for Healthcare Research and Quality (AHRQ).

Wallerstein, N., & Bernstein, E. (1994). Introduction to community empowerment, participatory education, and health. *Health Education Quarterly, 21*(2), 141–148.

Yehya, N., & Dutta, M. (2010). Health, religion, and meaning: A culture-centered study of Druze women. *Health Communication Research, 20*(6), 845–858.

Voices of Hunger: A Culture-Centered Approach to Addressing Food Insecurity

Mohan J. Dutta, Christina Jones, Abigail Borron, Agaptus Anaele, Haijuan Gao, and Sirisha Kandukuri

Food insecurity has increasingly emerged as a core health issue in the US, reflecting similar patterns of disparities in inaccess to food globally. Food insecurity explains the condition where "the availability of nutritionally adequate and safe foods or the ability to acquire acceptable foods in socially acceptable ways [e.g., without resorting to emergency food supplies, scavenging, stealing, or other coping strategies] is limited or uncertain" (Seligman, Davis, Schillinger, & Wolf, 2010, p. 1227). In 1995, based on a benchmark developed by the U.S. Department of Agriculture Food and Nutrition Service (FNS), the hunger crisis was severe in the US. Even in the midst of the economic boom, the analysis revealed that an estimated 11.9% of U.S. households (35 million persons) were food insecure at the time of the survey's administration, with 4.1% of households (6.9 million adults and 4.3 million children) reporting recurring food insecurity (Carlson, Andrews, & Bickel, 1999). Recent government data obtained in 2008 indicated growing numbers; of all households in the United States, at least 14.6% (17.1 million households) were food insecure at some point during the year (Banks, Marmot, Oldfield, & Smith, 2006). Moreover, 5.7% (6.7 million households) had, at some point, very low food security, characterized by disruptions in eating patterns and reductions in food intake of one or more household members, from the inability to afford enough food. These disruptions are even more common in households with children under 18 (6.6% of such households, or 2.6 million, have very low food security) (Nord, Andrews, & Carlson, 2009).

Against the backdrop of food insecurity and hunger throughout the US, there are national and federally funded programs pledged to reduce these phenomena. However, such expert interventions lack consideration of localized experiences of food insecurity and hunger (Chilton & Rose, 2009). It is in this context that the need to understand the meanings of hunger among food-insecure persons becomes crucial. The Culture-Centered Approach (CCA) (Dutta, 2008), applied in the context of food insecurity, suggests that program planners and policy-makers need to fundamentally listen to marginalized communities in the design of health and food interventions, with the ultimate goal of handing over the ropes of decision-making into the hands of communities. In the context of food insecurity, CCA deconstructs the missing links in contemporary food-assistance programs in the United States, and offers constructive ways on how to meaningfully engage with marginalized populations that are the recipients of the food assistance programs by foregrounding the very voices of communities at the margins through participatory programs. In this chapter, we coconstruct narratives of hunger with food-insecure persons in one mid-sized Midwest community, outlining a CCA project that resulted in the formation of a "Voices of Hunger" coalition.

The Context: Domestic Food-Assistance Programs

Numerous public policies and programs exist to ameliorate and prevent poverty-related food insecurity. There are a number of federal food-assistance programs overseen by the government. Approximately half of food-insecure households (55%) participate in one or more of the three largest federal food- and nutrition-assistance programs: the Supplemental Nutrition Assistance Program (SNAP, formerly called the Food Stamp Program); the Special Supplemental Nutrition Program for Women, Infants, and Children (the WIC Program); and the National School Lunch Program (NSLP).

Nationally, when policy funding fails to adequately address the hunger issue, there are a number of organizations providing emergency food assistance to the food insecure. Many of these organizations operate in conjunction with the largest, domestic hunger-relief charity in the United States, Feeding America (FA). Formed through the merging of America's Second Harvest and Food Chain (two of the nation's largest food-rescue programs) in 2001, Feeding America is a national network that serves an estimated 37 million different people annually. Outside of FA, about 5,300 emergency kitchens and 32,700 food pantries exist as part of The Emergency Food Assistance Program (TEFAP), which provide more than 173 million meals and distribute around 3 billion pounds of food per year (Ohls & Saleem-Ismail, 2002). As such, based primarily in localized contexts, the TEFAP makes extensive use of volunteers.

Emergency kitchens and food pantries are two of the most commonly assessed organizations of the TEFAP. Nearly two-thirds of emergency kitchens nationwide are operated by faith-based organizations, and most have been operating for longer than five years. Almost half of emergency kitchens serve meals only on weekdays, and another 10% only serve meals on weekends. The typical food pantry distributes food just two days a week; 16% are open less than once a week; and on a day when a pantry is open, the typical pantry remains open for just 3–4 hours. Typically, however, pantries allow households to obtain food once or twice a month. While some pantries allow households to select their own food, most pantries apportion the food available to the needy.

Both emergency kitchens and food pantries rely heavily on food banks as the "wholesalers" of the TEFAP, obtaining food in bulk amounts and distributing it to local providers. Most food banks are private, nonprofit, and nonreligious organizations. On average, the typical food bank services 17 emergency kitchens, 96 food pantries, three other food banks, and 12 shelters, in addition to numerous other charitable agencies (Ohls & Saleem-Ismail, 2002). Most food banks receive their food from other food banks or wholesalers and retailers, including both food that could have been sold and "salvage" food considered safe but not salable due to mislabeling or packaging defects. Some food banks are also able to obtain food from USDA commodity programs. Within the context of our culture-centered project, the centralized location soliciting, collecting, and redistributing food to those in need is that of Food Finders Food Bank, Inc. Food Finders Food Bank in Tippecanoe County, Indiana currently serves a number of surrounding counties, both suburban and rural, in the Northwest-Central region of Indiana. Specifically, food is donated from farmers, individual businesses (grocers and/or restaurants), large manufacturers or processors, the USDA, personalized local donations, and larger emergency food-relief organizations (Feeding America) to the local food bank, Food Finders. This organization then redistributes the food to local, nonprofit, food-bank member agencies, such as food pantries, soup kitchens, and shelters for the needy. It is at these locations where those experiencing food insecurity are able to gain access to food, whether that be by receiving food products to be taken home or by receiving actual cooked food to be eaten at a shelter or kitchen.

According to a 2010 survey conducted by the Feeding America organization, representing 101 food agencies served by Food Finders, and 231 clients of those agencies, most emergency food agencies in the Food Finders service area are faith-based organizations (Feeding America, 2010). The average length of operation of these service providers is 18 years, with almost half having served clients for 11–30 years. In this service area, 36% of food pantries, 22% of kitchens, and 11% of shelters reported turning clients away during the previous year, with the most common reason being lack of food.

A Culture-Centered Approach to Food Insecurity

Based on the argument that the structural marginalization of the food insecure is connected to their communicative marginalization, CCA seeks to foster spaces for participation of those at the margins, with the notion that their voices define the parameters of participation as well as the specific solutions to be developed. The experiences of the food insecure are situated amidst the broader structural constraints that determine the inaccess to food services, and therefore, the participation of the food insecure in processes of meaning-making creates entry points for disrupting the structural invisibility of the food insecure, as well as for developing program and policy parameters that are guided by the needs of the food insecure (Dutta, 2008, 2011; Dutta-Bergman, 2004). Noting that communicative erasure lies at the heart of the perpetuation of structural erasures, CCA seeks to cocreate avenues for achieving structural transformations through the creation of participatory spaces for communities at the margins. Because structural inequities are simultaneously marked by the absence of communicative resources of participation for the subaltern sectors, the emphasis of fostering spaces for listening to the voices from the margins disrupts the taken-for-granted logics and assumptions of mainstream structures. The process of fostering communicative spaces for participation begins with the development of relationships of solidarity with those at the margins, fostering spaces of meaning-making that privilege their localized understandings of problems and corresponding solutions. In-depth interviews offer initial frameworks for the cocreation of meanings accompanied by focus groups where community members come together in iterative cycles to coconstruct their meanings of problems and solutions, to analyze in collaboration with the academic partner the key themes that emerge from the interviews, and to develop research- and performance-based narratives for communicating problem-and-solution configurations with key stakeholders with the goals of addressing social change.

Figure 26.1. A Diagrammatic Representation of the CCA Process

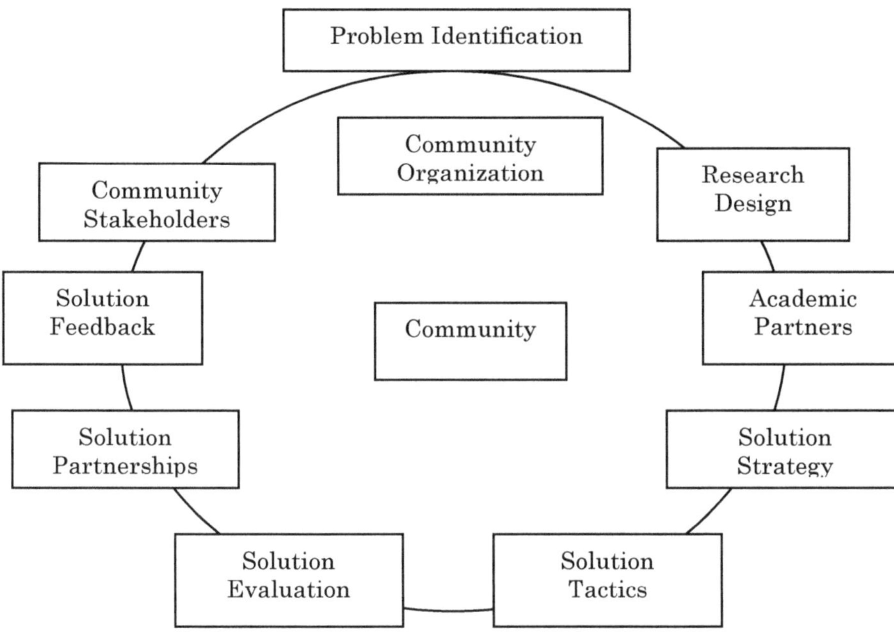

Structure, culture, and agency offer the theoretical framework for fostering entry points for local meanings. Structures refer to systems of organizing of resources that prevent and enable access to resources; culture refers to the local contexts and experiences within which meanings of hunger are constituted and negotiated; agency refers to the capacity of cultural members to enact their choices, and to participate actively in negotiating the structures within which they find themselves. It is through this interplay of structure, culture, and agency that CCA fosters spaces for developing contextually and culturally viable solutions in addressing the problem of food insecurity. As stated earlier, food programs have worked diligently to serve the hungry in communities across the country. While Feeding America conducted a national study that interviewed more than 61,000 clients served by local food programs, and surveys of 37,000 feeding agencies (Feeding America, 2010), critical components of the narratives from the clients of these services, as well as the services themselves, were left untouched as new policies were enacted and program developments took place. The missing critical components include the epistemologies of the cultures involved—whether it is from the local, marginalized level or the dominant agencies organizing and distributing needed resources. In order to tap into these critical components of the narratives, in-depth cultural studies are needed that can provide the necessary focus on local and global worlds of knowing (Delvecchio Good, 1995).

Listening to Voices of the Hungry

First, the culture-centered approach provides an opportunity to consider the client narratives through coparticipation on behalf of the researcher. The result is a more comprehensive understanding that communication about hunger revolves around the negotiation of shared meanings embedded in socially constructed identities, relationships, social norms, and structures (Dutta, 2008). The result is the recognition that these identities, relationships, social norms, and structures are mutually interdependent. Through coparticipation, ongoing conversations with members of the marginalized communities attempt to privilege the narratives. By doing this, emerged narratives lend themselves to the opportunity to truly question how knowledge is articulated, and discover how this knowledge is often located (or absent) in the dominant structures and social systems, which are commonly far removed (even unavailable), not understood, or even accepted by the marginalized communities.

By revealing the local meanings within a dynamic cultural setting, CCA equips the researcher to recognize the common forms of exploitation that can and often do take place within the dominant structures that are actively engaged in addressing the needs and desires of the marginalized audience. The voices of the marginalized draw attention to these exploitations and inequities that are embedded within the mainstream structures of organizing. Exploitative practices on behalf of the dominant agency or organization are often seen in the area of public health (Darby & Svoboda, 2007; Delvecchio Good, 1995; Dutta-Bergman, 2004; Hahn & Kleinman, 1983; Sharp, 2000), where health has become a commodity by becoming less affordable and tangible for those that these agencies and organizations are meant to serve; this happens through a variety of bureaucratic processes that are often far removed from the localized needs of communities. Culture-centered processes of social change therefore are fundamentally directed at undoing these inequities in the politics of recognition and representation within discursive spaces and processes by turning to subaltern communities at the margins as entry points for dialogue.

Centering Conversations With Stakeholders: The Opportunity for Policy Change

Through the processes of listening to the voices from the margins, the goal of CCA is to foster entry points for communication with the mainstream. Through CCA, a researcher must move in and out of nearly polar opposite worldviews in order to allow marginalized narratives (Worldview 2) to effectively inform the dominant worldviews that are embedded within the mainstream structures and constitute the organizing processes of these mainstream structures (Worldview 1). To enter worldview 2, as a researcher, means that there is an attempt to achieve Verstehen—to develop, perhaps an empathetic understanding (i.e., what does it mean to be an individual experiencing food insecurity from that individual's perspective?). But where Weber/Verstehm stops and CCA picks up is in how to effectively reenter worldview 1, not through the researcher's voice (as the researcher), but with the voices of the marginalized community in solidarity. They are the ones speaking; they are the ones who ultimately offer a form of Verstehen to the dominant structure by finding spaces for recognition and representation within dominant discursive processes and structures.

Culture-Centered Participatory Processes

Our Participation

In order to gain a general understanding of the broader context of food insecurity, we volunteered at the food bank in the West Lafayette area several times, and participated in multiple food pantries and mobile pantries as well. We conducted in-depth interviews with the staff members of the organization to gain an understanding of the structure of the organization, procurement of and distribution of the food, the donors, transportation facilities, and other organizational matters. We also volunteered twice to help organize and send out food from the food bank, so that the team could get first-hand experience as it related to the process of food gathering, sorting and distribution. Also, we volunteered at two of the local pantries where the food was distributed.

Interviews. Interviews served as the primary method for centering the meanings of food insecurity, as means for foregrounding the stories of the food insecure in the discursive space. We placed recruitment posters at food pantries throughout the community. Individuals interested in participating in the in-depth interview contacted our team members directly to schedule an interview at a time and place that was convenient for them. Venues for the in-depth interviews included public libraries, university libraries, and participants' homes. We conducted semistructured, open-ended approach for the purposes of encouraging the participants to lead the direction of the discussion. Each interview lasted between 60 minutes and 120 minutes, depending on the willingness of the participant. A total of 17 in-depth interviews were conducted through the first phase of the project, offering entry points for foregrounding the localized meanings of food insecurity.

Focus groups and photo voice. From the broader group of interview participants, focus-group participants were recruited for making sense of the interviews and for developing action steps. Three longitudinal focus group discussions were conducted, accompanied by photo-voice sessions, where participants received cameras, recorded pictures, and then discussed them in depth. The first focus group primarily discussed the key problems and issues as identified by community members, as well as the solutions as seen through their worldviews. The focus groups also emerged as sites for conducting analyses of transcribed interview data. This phase was composed of a series of three focus groups that incorporated a photo-voice component for the purposes of additional data collection. Photo-voice is a method in which the participants use photographs to display meanings, opinions, ideas, and values. It is a process through which personal experiences and knowledge are portrayed visually. These aspects of the participants' lives might otherwise be hard to express in words (Wang & Burris, 1997).

During this initial focus-group meeting, participants were provided disposable cameras, and asked to capture photographs they felt would describe meanings of hunger, health, and food accessibility and insecurity, in addition to other issues that they felt were significant in understanding the whole process of the food distribution, and the opportunities and risks that they face. One week later, the participants joined together for the second focus group, at which point they turned in their spent cameras. The duration of the second focus group was spent brainstorming and prioritizing issues they felt were relevant and important in addressing food insecurity in the community. During the third focus group, participants shared and discussed their printed photos, and then determined which photos best described the issues they face, as they relate to hunger, food insecurity, and accessibility, connecting their photos with the stories they shared about their experiences of hunger, and grounding their narratives in the backdrop of the broader themes that emerged from the interviews.

Photo exhibit and dialogue. The photographs were further put on display in the School of Visual and Performing Arts at Purdue University, for a more formal gathering of the Food Finders Organization, local food pantries and volunteers, the media, and the participants in the project to discuss these issues in detail. The photo exhibit resulted in opportunities for the food insecure to share their stories with key policy-makers and program planners in the community, as well as to have their voices and stories highlighted in the local media.

Voices of Hunger: Problems and Solutions

In the collaborations between the food-insecure community members and the academic team, six key themes emerged as the in-depth interviews and focus-group data were systematically analyzed. Through iterative cycles of discussions about what emerged during the analysis, food-insecure community members identified three core issues to address in their photo-voice exhibit, as they narrated the stories from the images they captured to mobilize for change.

The Reality of Food Insecurity

That food insecurity could happen to anyone, and that it has a human face were the key threads in messages that were most important to participants. Community members recalled the average lifestyles and jobs they enjoyed before getting into the situation where they were completely out of financial resources. The reality of hunger for them therefore lies in how suddenly it becomes an issue, because of job loss and because of the economic inability to pay for food. The following quotation is useful in drawing attention to the reality and magnitude of food insecurity:

> Umm, I don't know, but, I don't know, but I think it is also hard in America here because umm it just doesn't seem like it's a real problem. But I guess it is, I guess there are other people in this situation but I've always in past times thought how can people ever get to a point where there is no food, where there is no food in their house, how can they get to that point? And because there is so much help out there, there is so much help out in America. But for me, umm, I don't receive assistance. So, umm, and because of income I could not receive assistance. So I feel like especially with the economy the way it is now umm it's become a real thing to me (beginning to cry, sniffle). So, umm, anyway, trying not to get too emotional here. (Participant # 1)

As evident in the quotation, the participant, like many other Americans, was previously unaware of the magnitude of the problem of hunger and food insecurity until he/she was affected.

The reality of the problem of hunger and food insecurity is further evident in the following narrative:

> The emotions of having to go through something like this. It is definitely emotionally harder. Umm, I don't know, it just kind of makes you wake up and realize I guess what you take for granted. I mean, you know, before when food was getting a little bit low, oh I can just go to the food pantry, you know, and do whatever. Umm, do something like that. Umm, but now it's not, we're really hitting hard these past couple of days waiting for the next pay check to come in. Umm, and so during that time there has been no money. Nothing in the bank accounts, nothing. And it's not just food we're lacking, it's the gas. How are we going to get to our jobs if we hardly have any gas in our vehicles to get to our jobs? (Participant # 1)

Based on these realities of their lived experiences, for the participants in our project, it was important to center the everydayness of food insecurity. That it could happen to anyone was a message that they wanted to communicate to others, in order to foster spaces of understanding and also in order to resist the stigmatization tied to dominant structures of political and economic organizing.

Overcoming Social Stigmas

Participants narrated stories of stigma and shame associated with begging for food, and in their analysis of the data, sought to generate greater awareness and information in order to fight the stigma. Consider, for instance, the following narrative of Sam:

> But stuck in a rough spot, but anyway, back to the experience with food pantries. As far as going there, it's really embarrassing. I try not to let myself get embarrassed but it is, it is embarrassing going. Umm, I don't know, the people are always nice when I go. They seem very, like they want to help. I mean I realize they are volunteers from churches and stuff like that but they are always nice. They give you, I don't know, sometimes I'm just like I'm in a position where I'm begging for food, I hate being in that position in the first place.

There is visible indication of stigma associated with going to the food pantries to source food. Another participant painted the picture thus:

> My kids wouldn't go with me. Oh, not now, never. These kids nowadays, they won't even use coupons. If you go to Goodwill, even if you buy Abercrombie and Fitch, they better not know where it came from. My kids would absolutely die than to go into a food pantry and eat. So, I did these things because, to survive. Me and Andy, and by golly they ate the food but they don't want to know where it came from. Today's generations are just, or at least all the kids that I've ran in to, young ones, consider it second-hand food. I'm not no, do you not like my fried chicken? Green beans are green beans! But yeah, the kids, oh they all loved the food, don't get me wrong, but they just didn't want any part of it. My kids anyways. So, I mean, that was just my house. So I just went and did what I had to do, and cooked the food. And they didn't ask questions. (Participant # 4)

What becomes evident here is the fact that the participant's children refuse to go with their parents to the pantries because of the social stigma and loss of dignity associated with going to the pantries in a Midwest community.

Quality of Food

Participants articulated multiple perspectives about quality of food, but their primary concern was regarding the expiry date of the food that was served to them. They appreciated receiving fresh food, and often referred to the partnerships with the farmers as an option they enjoyed. They also talked about the fact that they did not like the stale food that was sometimes given out. They stressed the importance of storing meats properly, and making sure that steps are taken to ensure that the food is both presentable as well as consumable. The following excerpt from a coconstructive journey with Sara conveys in detail the dissatisfaction of the food insecure with the quality of food they receive from the pantries:

> I guess my biggest complaint would be the day-old food, the stale food. I mean it's kind of, two things about it, is one it's, you know a relief to get the food and then you're all excited about it, but then you're quickly disappointed because there's something wrong with it and you can't eat it. That would be one complaint I have. Then the other one would be, oh I just lost it, I had two thoughts on it. Oh, it's kind of a, I can't think of a better term, it's kind of like a slap in the face. Oh you're begging for food so we'll just give her leftovers. So those two things are the things I dislike the most about it. It's almost like ridicule.

The quotation presents a legitimate concern about the mixture of frustration and disappointment that greets the food insecure, especially when they visit the food pantries with high expectations to receive food like every other human being only to be handed expired foods. Another participant, Jonathan, further explains the challenge of quality of food in the following quotation:

> The quality of food is low rated. It's the food people don't want. It's off-brand names. Its food that's, you're not getting the quality of food that everyone else is getting when you go to these food pantries. It's always the minimum things. You never get any high-squality food out of there.

Participants discussed that food pantries do not really care about what they need or want, noting that the distribution of food and the decisions tied to it are made in top-down structures. Due to power differentials and structural configurations in the ownership and operations of the food pantries and food banks in the US, the food insecure are not involved in the ownership and running of food pantries in the US. In such circumstances, food subsidy therefore falls within the logic of broader, top-down interventions, with an emphasis on the number of foods distributed, rather than quality, as markers of effectiveness.

Logistical Issues: Quantity, Transportation, Information

Related to food quality is the quantity of food and frequency at which such foods are given out at the food pantries. The participants in our project noted that the quantity of food served at any one pantry was typically not enough to fulfill their needs. Therefore, they had to make trips to multiple pantries in order to make sure that their food needs were met. Articulating the degree of humiliation associated with waiting for food at the pantries, one of our participants painted the picture this way:

> That's basically it, the gas money; you have to wait outside in line sometimes 25 minutes in January, when it's 10 degrees outside. And the third one, that's why we are going to more and more food pantries, we used to get the same amount of food by going to say, 6 food pantries. But now we need to go to 12, because the food they are passing out is about half, maybe 60% of what they used to pass out, but it has gone down towards half. (Participant # 9)

According to the participants, due to the drop in the quantity of food distributed by the pantries, the food insecure are forced to queue in lines in frozen weather for up to 25 minutes waiting for food.

Another participant explains the challenges associated with the quantity in the following excerpt:

> So, but yeah, the access. There's just, there are some good ones but you're only allowed to go once a month. What good is it to go once a month? They give you one bag of food for once a month and it's like (pauses). Now, I don't look a gift horse in the mouth, and I appreciate every single bag that I get, but, I don't understand how one bag is going to help a family survive because...you know, once a month. And I understand that there's, you're trying to feed a lot of people.

Participants also discussed the difficulties they faced in getting to the food pantries. In this context, they discussed the importance of setting up public-transportation facilities in order to transport them to the food pantries. These key issues discussed and identified by the participants as the issues they sought to have addressed emerged as the narrative frameworks for their photo exhibits. In an iterative process, the images that the participants collected tied back to their stories, the key themes that emerged through their analysis of the interviews, and the key points they wanted to express as a collective.

Project Impact

The impact of CCA projects is tied to the possibilities for listening to the voices of those at the margins that are fostered by the coconstructive processes of CCA; the measurement of impact therefore is tied to (a) the communicative opportunities that are fostered through CCA processes; and (b) the material transformations that are brought about in inequitable structures through the participation of the disenfranchised in mainstream, discursive spaces. As depicted in our media story that appeared in the local newspaper, *Journal and Courier*, as well as the story that appeared on the local television and radio stations, the voices of the hungry found entry points into public discourses and public sites in the local community, not simply as individualized subjects written into stories as subjects of interventions, but as collective formations that strategically utilized the mainstream, discursive platforms in sharing their stories and in drawing attention to the possibilities of structural transformation. As noted by Jenna, one of the participants in the project:

> That our voice matters is something we learned through this process. Getting to talk to Katy Bunder (the director of Food Finders) was really important because it gave us a chance to share our problems and issues with her. Never have we been able to do this. Now if we can come together as a coalition and keep pushing for this, we can make a lot of changes happen.

As noted in the narrative of Jenna, the "Voices of Hunger" project created spaces for the articulation of problems by community members who experience food insecurity. Fostering these spaces also resulted in the Food Finders staff, as well as food pantries, noting their greater learning about the key issues that are faced by the food insecure, including issues of quality of food as well as the issues of inadequacy of food supplies for community members. What was being served in the food pantries emerged as a key point of discussion, and resulted in the sensitization of the food organizations in the community about the quality and quantity of food being distributed by the pantries (a material intervention). Most importantly, the formation of an ongoing coalition named the "Voices of Hunger" coalition emerged as an avenue for foregrounding the narratives and experiences of the food insecure in the context of food policies and programs in the local community.

Discussion

In conclusion, this chapter documents a culturally centered "Voices of Hunger" project that seeks to address the issue of hunger locally through the fostering of spaces for listening to the voices of the hungry. The central argument in this CCA project builds on the notion that fostering spaces for listening creates entry points for social change and structural transformation. When those who have been materially disenfranchised participate in discursive spaces and processes in the mainstream, they offer entry points, avenues, and frameworks for conceptualizing problems, and for coming up with solutions as understood through their interpretive frames and localized frameworks of meaning-making. CCA rests on the notion that these interpretive frames are the most meaningful because they are built on the active participation of local communities at the margins in the definition of solutions, and in the development of specific strategies and tactics for implementing the solutions that are configured by community members.

Therefore, problem configurations and solutions are driven by local ownership of collaborative processes, resource allocations, and coconstructions among community members in conversations with research team members. Community-academic partnerships become ways for distributing the sites of power, and for engaging disenfranchised communities in shaping the configurations of research, and in participating collaboratively in analyzing the data, making sense of the field data, and drawing conclusions that may then be utilized toward the goals of structural transformations. In the case of our project, the local participation of the food insecure in the processes of identifying problems and developing solutions drew attention to the issues of stigma and social perception; quality of food supplies; and logistical issues such as issues of transportation, quantity of food received, and information about the operating hours and accessibility of food pantries. The opportunities for the food insecure to come together to brainstorm about solutions also fostered openings for collaborations between the academic and community partners in brainstorming solutions directed at addressing the structures around food provision. The contribution of CCA in this project lies in the formation of the local networks of solidarity among the food insecure through the development of the hunger coalition.

Bibliography

Banks, J., Marmot, M., Oldfield, Z., & Smith, J. P. (2006). Disease and disadvantage in the United States and in England. *Journal of the American Medical Association, 295*(17), 2037–2045.

Carlson, S. J., Andrews, M. S., & Bickel, G. W. (1999). Measuring food insecurity and hunger in the United States: Development of a national benchmark measure and prevalence estimate. *Journal of Nutrition, 129*(2S), 510S–516S.

Chilton, M., & Rose, D. (2009). A rights-based approach to food insecurity in the US. *American Journal of Public Health, 99*(7), 1203–1211.

Cook, J. T. (2002). Clinical implications of household food security: Definitions, monitoring, and policy. *Nutrition in Clinical Care, 5*(4), 152–167.

Darby, R., & Svoboda, J. S. (2007). A rose by any other name? Rethinking the similarities and differences between male and female genital cutting. *Medical Anthropology Quarterly, 21*(3), 301–323.

Delvecchio Good, M. (1995). Cultural studies of biomedicine: An agenda for research. *Social Science Medicine, 41*(4), 461–473.

Don't cut SNAP to pay for other priorities. (2011). *Frac.org.* Retrieved from http://frac.org/leg-act-center/updates-on-snapfood-stamp-cuts/.

Dutta, M. J. (2008). *Communicating health: A culture-centered approach.* London, UK: Polity Press.

Dutta, M. J. (2011). Agriculture and food; Global inequalities. In M. J. Dutta, *Communicating social change: Structure, culture, and agency* (pp. 96–122). New York, NY: Routledge.

Dutta-Bergman, M. J. (2004). Poverty, structural barriers, and health: A Santali narrative of health communication. *Qualitative Health Research, 14*(8), 1107–1122.

Feeding America. (2010). *Hunger in America 2010: National report.* Chicago, IL: Author.

Hahn, R. A., & Kleinman, A. (1983). Biomedical practice and anthropological theory: Frameworks and directions. *Annual Review of Anthropology, 12*(1), 305–333.

Mayer, J. (1990). Hunger and undernutrition in the United States. *Journal of Nutrition, 120*(8), 919–923.

Nord, M., Andrews, M., & Carlson, S. (2009). *Household food security in the United States, 2008.* U.S. Department of Agriculture, Food and Nutrition Service. Economic Research Report No. 83 (ERS-83). Retrieved from http://ers.usda.gov/Briefing/FoodSecurity (no longer accessible).

Ohls, J., & Saleem-Ismail, F. (2002). *The emergency food assistance system: Findings from the provider survey.* Washington, DC: U.S. Department of Agriculture.

Orshanski, M. (1965). Counting the poor: Another look at the poverty profile. *Social Security Bulletin, 28,* 3.
Powell, M., & Bao, Y. (2009). Food prices, access to food outlets, and child weight. *Economics and Human Biology, 7*(1), 64–72.
Prakash, G. (1994). Subaltern studies as postcolonial criticism. *The American Historical Review, 99*(5), 1475–1490.
Seligman, H. K., Davis, T. C., Schillinger, D., & Wolf, M. S. (2010). Food insecurity is associated with hypoglycemia and poor diabetes self-management in a low-income sample with diabetes. *Journal of Health Care for the Poor and Underserved, 21*(4), 1227–1233.
Sharp, L. A. (2000). The commodification of the body and its parts. *Annual Review of Anthropology, 29,* 287–328.
Turrell, G., Hewitt, B., Patterson, C., Oldenburg, B., & Gould, T. (2002). Socioeconomic differences in food purchasing behaviour and suggested implications for diet-related health promotion. *Journal of Human Nutrition and Dietetics, 15*(5), 355–364.
Wang, C., & Burris, M. A. (1997). Photovoice: Concept, methodology, and use for participatory needs assessment. *Health Education & Behavior, 24*(3), 369 – 387.
The WIC program: Background, trends, and economic issues (Economic Research Report #73). (2009). Washington, DC: U.S. Department of Agriculture.
Zoller, H. (2008). Technologies of neoliberal governmentality: The discursive influence of global economic policies on public health. In H. M. Zoller & M. J. Dutta (Eds.), *Emerging perspectives in health communication: Meaning, culture, and power* (pp. 390–410). New York, NY: Routledge.

27

Communication as Health Activism: The Case of the Campaign to Stop the Gardasil Clinical Trials in India

Mohan J. Dutta

Increasingly, health-communication scholars have established the importance of exploring the structural loci of health inequalities, drawing attention to the inequities that are constituted amidst globalization processes and the political economy of neoliberalism (Dutta, 2008; Zoller & Dutta, 2008). The focus of health-communication processes in the context of challenging the structural forces of neoliberal governance has emphasized grassroots, resistive processes in the form of health activism (Coburn, 2000, 2004; Dutta & Zoller, 2008). Similarly, in public health, scholars have called for situating the articulations of health inequalities within the broader frameworks of health policies under neoliberal configurations. This suggests the importance of activism and advocacy that are directed toward bringing about transformations in policies that fundamentally produce inequalities in health by foregrounding the agendas of transnational corporations (TNCs) and simultaneously erasing the interests of subaltern communities that exist at the margins of contemporary neoliberal politics (Coburn, 2004). Of specific emphasis in projects of health activism are the structural injustices perpetrated by the contemporary political-economic order, and therefore, the emphasis is on challenging these structural injustices by fundamentally bringing into question the underlying, taken-for-granted assumptions of the political economic order (Dutta, 2008; Zoller, 2008, 2009; Zoller & Dutta, 2008).

Therefore, the relevance of health activism as intervention becomes evident in the context of challenging the political and economic configurations of neoliberalism that create and sustain unequal points of access to basic health-care services and treatment options, and perpetuate the exploitation of the poor in the form of specific health policies and interventions in the global arena. Activism therefore becomes an entry point to constituting and sustaining structural changes, with the central emphasis on structural transformation. What, then, are the communicative entry points to resisting the structural marginalization of the poor played out through the mainstream programs of health policies, health programs, and health interventions? It is precisely within this space of transformative politics that health activists participate in creating alternative symbols and frameworks of meaning that challenge the structures of neoliberalism, critically engage with the unequal practices under neoliberalism, and offer entry points for material transformations through the realms of the symbolic.

Of particular import in the context of global health disparities are the uses and exploitations of subaltern populations in the form of clinical trials in resource-poor sectors. Such clinical trials are carried out with the limited structures of monitoring and evaluation in global spaces. They are carried out with limited spaces of subaltern participation in neoliberal platforms of civil society and knowledge production where policies and programs are planned and evaluated. They are conducted without adherence to the processes of accountability that are set up in the domains of biomedicine, thus violating the human rights of the subaltern sectors in the Third World, who emerge in the discursive and material spaces of global biomedicine as profitable subjects of the clinical-trials industry (Prasad, 2009). The Third World subject gets situated within the constellation of the clinical-trials industry as a profitable body to be recruited, and the Third World emerges as a site for supplying bodies and patients for carrying out the clinical trials, often following processes that would not hold up to the global standards that are circulated in mainstream communities of health in the West and/or North. It is precisely against this backdrop, then, that global processes of health activism in the Third World raise questions about accountability, sharing of knowledge, and articulation of transparent processes that can be evaluated both locally as well as globally (Sarojini & Shenoy, 2010).

In this chapter, I will examine the communicative processes of organizing and representation that constitute the local activist movements in India that sought to ban the Merck-manufactured Gardasil clinical trials that were approved by the Ethical Committee and Advisory Groups at state and center levels in India, were funded by the Bill & Melinda Gates Foundation, and were carried out by the global NGO, PATH, in collaboration with the local government in the Indian states of Andhra Pradesh and Gujarat. Gardasil is a drug developed by Merck that is being positioned as vaccination against the human papillomavirus (HPV), the virus that causes certain strands of cervical cancer. Fairly soon after the Gardasil clinical trials were launched in India by PATH, women's activist groups such as SEVA and SAHELI started protesting against the clinical trials locally—based on the articulation that informed consent was not being ethically secured from the participants—and sending field workers to the sites to observe how the clinical trials were being conducted, and how consent of the project participants was being secured.

The activist processes gained momentum when six girls who were enrolled in the trial died, resulting in a wide range of resistive, communicative strategies that involved the subaltern groups in the villages of Andhra Pradesh and Gujarat who were being directly affected by the clinical trials. The representations of the deaths were constituted amidst questions raised about the causes of the deaths and the multiple contested stories about the deaths, with the activists arguing that the deaths were caused by the vaccines, and asking for reports of more systematic monitoring of the girls by the providers (Sarojini & Shenoy, 2010). In August 2010, the central government in New Delhi issued a halt on the Gardasil clinical trials. What, then, were the communicative processes that were mobilized by the activists in their communication with stakeholders at the state and federal levels, in order to create entry points for structural transformations? What were the processes of organizing through which the activist groups came to articulate their strategies for change, and came to share a voice as a collective? What communication strategies were utilized in order to create spaces of social change and structural transformation?

The discussion of health activism in the context of global clinical trials introduces the frames of legitimacy and accountability for monitoring the practices of transnational capital operating in the domains of biomedicine, exploring the linkages between the interests of capital, biomedical knowledge, and nongovernmental organizations (NGOs). Health activism opens up avenues for disrupting the hegemony of biomedical knowledge claims, situating discussions of risks and efficacy among political, economic, social, and cultural processes, and engaging the global with the local. The legitimacy of biomedicine that has historically been placed in the realm of the sacred through the articulation of the scientific discourse of progress is ruptured through the participation of subaltern voices that have historically been rendered silent through such projects driven by the Enlightenment logic. The participation of activist groups in processes of change articulates alternative rationalities, and puts forth alternative frameworks for the evaluation of programs carried out by transnational hegemony, the collaborations of the state, TNCs, nongovernmental organizations (NGOs), and policy-making bodies.

Clinical Trials, Transnational Capital, and Accountability

The agendas of transnational capital and clinical trials are intertwined, as clinical trials serve as the predominant mechanism for the emergence of pharmaceutical-manufactured medicines and treatments into the markets of profit and economic gains in global capitalism (Prasad, 2009). As medical products and technologies navigate through the gatekeeping functions of regulatory agencies in the global arena, they do so in the context of the predominant function of clinical trials in the processes of getting a product or technology approved through the regulatory spaces of the market. Clinical trials are established within certain frameworks of evaluative processes, which then are utilized by federal agencies in the West (such as the Food and Drug Administration) to determine whether a medical product or service ought to be launched into the global market, based on evaluations of efficacy, safety, and side effects. Therefore, established within the domain of risks and potential benefits, clinical trials operate in global markets on the basis of the recruitment of participating subjects, in order to create the sufficient, necessary data that would be essential in the regulatory approval of the product and the launching of the product into the market. The globalization of the clinical-trials industry, therefore, is based on the effectiveness of the industry in capturing clinical-trial markets across borders and in securing participants in clinical-trials processes.

The globalization of market economies has resulted in the global shifting of the risks and benefits in the contexts of clinical trials, played out within the purviews of power, control, and political economic access to the dominant structures of health. The global control of the clinical-trials industry is based on its effectiveness in recruiting cheap sources of clinical labor, i.e., participant pools for clinical trials. In a nutshell, therefore, the Third World emerges in the global arena as an economically viable source of clinical trials for products that are then targeted toward the consumer economies of the North. On one hand, these clinical trials then become income-generating mechanisms for national economies in the South, and at the same time, they become vast resources for drawing out human subjects for the consumer economies of neoliberal hegemony.

Inequities and health-care disparities are played out in the inequities in the distribution of risks and in the benefits that are obtained from the information gleaned from the clinical trials. Therefore, whereas the poor in the global South becomes the site for the displacement of the risks, those with economic access and access to resources in global market economies become the benefactors of the products and technologies so developed. It is against this backdrop of the positioning of the clinical trials in the global South that questions arise in the domain of human-rights violations of the standards of accountability in the recruitment and remuneration of participants from the subaltern sectors of the global South in clinical trials, in the processes of securing informed consent, in the communication of risks and benefits, and in the evaluation of transparency in the processes of conducting the clinical trials. It is with these demands for accountability and transparency that activist groups in the global South raise demands on the nation-state and on international legislative bodies to hold global TNCs and their knowledge-producing mechanisms accountable.

Culture-Centered Processes of Grassroots Health Activism

The culture-centered approach begins with the essential theorization of the large-scale material and symbolic inequalities that exist in the realms of health, putting forth the argument that communicative inaccess in the realms of the symbolic are intertwined with material inaccess (Dutta, 2008). In other words, the communities and groups that have historically been erased from discourse also exist at the material margins of political economies of knowledge production. The structural marginalization of the poor is played out in policy discourses where the poor are constructed as passive and without agency, and therefore decisions are continually made on their behalf by policy-makers, program planners, and academics who are mostly out-of-touch with the lived experiences of the poor. The economic marginalization of the subaltern sectors is fundamentally situated in this erasure of subaltern voices and in the negation of the subaltern capacity to participate in processes of change (Dutta, 2008). Simultaneously, the subaltern emerges into the spaces of contemporary neoliberal economics as a profitable resource; the subaltern knowledge base becomes an exploitable resource for biopiracy, and the subaltern body becomes the site for carrying out neoliberal interventions (Dutta & Pal, 2010).

Therefore, the culture-centered approach draws largely from the body of scholarship in Subaltern Studies to continually interrogate the ways in which the agency of local communities is co-opted and configured within the agendas of dominant structures in neoliberalism (Dutta, 2006, 2008). Deconstructing the co-optation of subaltern agency creates entry points for interrogating the consolidation of global power in the hands of transnational hegemony. Also, a culture-centered critique attends to the processes of knowledge production and the ways in which these processes articulate entry points for expertise, which simultaneously erase the participatory capacity of local communities, and configure these communities as sites of interventions (Basu & Dutta, 2009; Dutta, 2006, 2008). Local communities, when configured into the discourses of knowledge, emerge as passive subjects who must be subjects of interventions carried out by dominant structures (Dutta & Basnyat, 2008a, 2008b).

Therefore, with an emphasis on theorizing the discursive erasure of the subaltern sectors of the globe from dominant discursive spaces of health, the politics of the culture-centered approach is situated amidst the cocreation of spaces where subaltern voices are heard through participatory processes, and entry points are created for coconstructing narratives of transformation with locally based processes of resistance to the dominant structures of health (Basu & Dutta, 2008, 2009; Dutta, 2006, 2008; Dutta-Bergman, 2004a, 2004b). First and foremost, the culture-centered approach draws attention to the historical materiality of the politics of change; paying attention to these possibilities of structural transformations, it therefore creates openings for listening to subaltern voices through journeys of solidarity and listening (Dutta, 2008). The emphasis of the culture-centered approach is on centering subaltern agency as an entry point for change, in listening to the stories of subaltern participation and agency in processes of transformative politics. In the context of health, culture-centered stories of change that engage meaningfully with subaltern contexts create entry points for alternative rationalities that challenge the taken-for-granted assumptions in the dominant structures of knowledge production, and render impure the dominant terrains of truth claims through the presence of subaltern voices (Dutta & Pal, 2010).

The Case: PATH Carries Out Gardasil Clinical Trials in Rural Regions of India

In July and August 2009, the nongovernmental organization (NGO), Program for Appropriate Technology in Health (PATH), with funding from the Bill & Melinda Gates Foundation, and in collaboration with the Indian Council of Medical Research (ICMR) and the state governments of Gujarat and Andhra Pradesh, launched a demonstration project for vaccination against cervical cancer in the Khamman district of Andhra Pradesh and the Vadodara district of Gujarat (Sarojini & Shenoy, 2010). The vaccine administered in Andhra Pradesh was Gardasil, manufactured by Merck Sharpe & Dohme (India) Pharmaceuticals Private Limited (MSD), the Indian subsidiary of U.S.-based pharmaceutical company, Merck & Co. Inc. Since the early days of the implementation of the demonstration project, women's health activists from groups such as Sama, Jan Swasthya Abhiyan, and Anthra raised questions about the processes utilized in the demonstration project. They visited the Bhadrachalam site of the project in Andhra Pradesh, and this visit became the entry point for local-national activist organizing that drew attention to the violation of ethical procedures in the implementation of the project. Furthermore, four girls died following the vaccination, and although PATH and local authorities noted that the causes of the deaths were not because of the vaccines, the women's health activist groups raised questions about the lack of appropriate monitoring and systematic follow-up of these deaths. The women's groups started organizing by writing several petitions to the ministry of health and family welfare, demanding an immediate halt to the demonstration projects. On April 29, 2010, ICMR acknowledged that ethical guidelines had been violated, and the Government of India temporarily suspended the trials. As an example of health activism, the local organizing of Sama, Jan Swasthya Abhiyan, and Anthra was successful in impacting specific policies in the context of the Gardasil clinical trials. In this chapter, we will critically engage with the discourses of activist organizing in this particular, localized context of the Gardasil clinical trials in India to create entry points for discussing the communicative processes in activist organizing in health care.

Communicative Processes in Health Activism

The Sama: Women's Resource for Health activist group communicated its resistance through the organizing of meetings with key stakeholders, by engaging the medical community, by drafting letters to the state and federal governments, by running traditional media campaigns in mainstream media, and by articulating its presence and activities on a blog that drew a great deal of participation both locally as well as globally. Because the blog presented archival details of the communication materials utilized in the activist campaign, this chapter mostly focuses on analyzing the blog entries made by Sama, along with media reports on the activist organizing processes, to offer contextual information. On World Health Day on April 8, 2010, the activist group issued a memorandum to the Health Minister of India, signed by key feminist activists and political leaders in India, demanding that the clinical trials be stopped, additional evaluation be provided, and the information about the clinical trials be made transparently available to the activist groups. I will also draw my analysis from this letter and the corresponding blogs, attending to the communicative processes that are presented in the letter.

Reframing Communication and Information Capacities

Essential to the challenges raised against the Gardasil clinical trials was the fundamental interrogation of the frames of risks, benefits, and side effects that were communicated by the dominant configurations of the pharmaceutical industry, the global NGO, and the local governments. That the clinical trials would inherently benefit the girls from the resource-poor settings who were going through the clinical trials was interrogated in the discourses of resistance, drawing on articulations of materiality based on lived experience, and asking for empirical evidence. The interrogation of the dominant framings of risks, benefits, and side effects created entry points for challenging the veil of opacity that marked the operation of clinical trials and the related campaigns on Gardasil, thus demanding greater transparency, accountability, and communication of truth.

Here is an excerpt from the letter written by women's resource groups and activists to the Minister of Health on World Health Day, demanding additional information about the processes utilized in the trials and their approvals:

> Although, the nomenclature suggests otherwise, the secrecy regarding the non-transparent selection criteria of the area, the girls and the aim of the project has raised undeniable fears that this is a Phase IV clinical trial being carried out under the guise of a 'demonstration project.' Moreover, there is an inexcusable lack of clarity with regards to the role and accountability of international agencies such as PATH and Bill and Melinda Gates Foundation that appear to be funding this project. It is beyond doubt the worst case of human rights violation, where young healthy girls have to die for being part of a state-endorsed experiment, initiated by a profit making private company. (http://samawomenshealth.wordpress.com/2010/04/08/memorandum-to-the-health-minister-on-world-health-day-opposing-hpv-vaccinations/)

The emphasis placed by the activists is on pressing for greater transparency, and on greater availability of information regarding the selection criteria that were utilized for the purposes of picking the areas, the girls, and the overall aims of the project. The frame articulated here operates on the basis of raising tensions between the frame offered by mainstream articulations and the reality of the lived experiences at the margins.

Furthermore, in the absence of information and transparency, the activists suggest that the project is a phase IV clinical trial, as opposed to being a demonstration project, as framed by public discourses about the project. The naming and categorization of the project by the activists resists the ontology placed by the dominant configurations of Merck, PATH, Bill & Melinda Gates Foundation, as well as the Indian government. The demand for local information capacities that provide transparent details about the processes involved and the accountability of international agencies such as PATH and the Bill & Melinda Gates Foundation is positioned in terms of local entry points for holding accountable national and global structures and processes. It is also worth noting that the discursive constructions of the activists frame the clinical trials as violations of human rights, attributing causality of the death of the girls to the complicit relationship between the state and the profit-making pharmaceutical industry, and thus opposing the logic of progress and development that are written into the dominant frames of the clinical trials in India. The framing of the clinical trials as human-rights violations, as opposed to markers of development, is crucial to the organizing processes of the health activists, who demand greater points of accountability and transparency in the system. Here is another excerpt that positions local articulations directly in opposition to the claims made by the status quo:

> Consent and information, reserved for only a few, is a farce and was based on provision of wrong information. Many of the girls were told that the vaccine would prevent ***uterine cancer*** and would provide life-long protection. This information is factually incorrect and provides the recipient with a false sense of security against the "dreaded" disease. (http://samawomenshealth.wordpress.com/2010/04/08/memorandum-to-the-health-minister-on-world-health-day-opposing-hpv-vaccinations/)

In this instance, the alternative frame offered by the activists points toward the articulation that the processes of consent and information sharing, as evidenced in the dominant structures of clinical trials in India, is farcical, and is based on the delivery of incorrect information, precisely within the manipulative goals of the clinical-trials industry to recruit subjects for the projects, based on manipulation of information about risks and safety.

Based on locally narrated experiences, the information provided by the dominant structures is countered. For example, the excerpt points out that the girls were misled about the effects of the vaccine (that it would prevent uterine cancer and would provide lifelong protection) based on information that is factually incorrect. Here, the local articulations of facts question the articulations made by the NGO in its recruitment efforts, pointing out that many of the pieces of information were not based on biomedical evidence. Digging into the very same evidence base, the activists point out that Gardasil does not actually provide the kind of security (lifelong protection) that is claimed in the recruitment efforts. Consider the following articulation:

> The literature circulated in the project makes outright false statements about its safety, efficacy and duration of effectiveness. The girls and their parents have been told through the project documents that the vaccine will give life long immunity, has no side effects other than minor ones like fever and rash and will not affect future fertility of the young

girls. (http://samawomenshealth.wordpress.com/2010/04/08/ press- conference-on-hpv-vaccines/)

Once again, the local voices of the activists directly point to the false statements made by the literature circulated in the project about safety, efficacy, and side effects. Questions are also raised about the ways in which the girls and their families were communicated to about the efficacy, side effects, and risks of the vaccines. Once again, the biomedical framing of truth being situated within the dominant structures of biomedicine is reframed utilizing the precise markers of biomedicine that are utilized as the bases of making truth claims. Questions of process are raised on the foundations of the very structures of legitimacy that constitute the market logic of the clinical-trials industry in India.

In the context of the Third World location of subaltern subjects in the backdrop of projects of clinical trials, the asking for evidence points toward an attempt in shifting the directions of power, particularly in terms of who should have access to knowledge and in what ways the knowledge should be distributed in the local communities. Local demands for information and/or knowledge at local sites resists the mainstream constructions of biomedicine that typically operate on the basis of the idea that local communities do not really have a say in matters related to their health because they do not really have the capacity to engage with the languages and processes of biomedicine in the dominant structures (Dutta, 2008). The health activists' demanded greater information, and greater communication of how this information is shared at the local level in the securing of informed consent, with the goal of creating information capacities that are meaningful to the lived experiences of the local communities. They demanded that the clinical-trials industry clearly communicate the processes utilized in the trials, with the call for the setting up of local information capacities in helping the local communities make informed decisions about their participation or the participation of their children in the clinical trials.

Resisting Structures Through Witnessing

One of the primary strategies used by the activists in the context of their interrogation of the clinical-trials processes was their presence at the sites where the experiments were being conducted and the vaccines were being administered, and their interrogation of the processes through which informed consent was secured in the context of the administration of the vaccines. The dominant rationale articulated by the government-NGO-pharmaceutical industry linkage was interrogated by the methodological questions raised by the activists. Here is an excerpt from the Sama press conference held in April, 2010:

> As per the extensive documentation by a fact finding carried out by local groups in Khammam district in Andhra Pradesh (where 14000 girls in the age group of 10–14 years have been vaccinated with the three doses of HPV vaccine), the adverse reactions faced by over 120 girls include epileptic seizures, severe stomach ache, head aches and mood swings.... The matter has further gained grave urgency with the deaths of four tribal girls in the area. Moreover, the very basis of the project is in direct violation of both the Governments own Ethical Guidelines of Research on Human Subjects and the rights of children as secured by national law as well as international conventions that India is a signatory to. (http://samawomenshealth.wordpress.com/2010/04/08/memorandum-to-the-health-minister-on-world-health-day-opposing-hpv-vaccinations/)

The extensive documentation by the fact-finding mission constituted of local groups shifts the methodological tools of finding facts into the hands of the community. The local groups consider the evidence base, the articulations of the dominant structure, and situate the articulations in the backdrop of the claims made by the dominant structure. The processes of informed consent for securing the participation of the girls is questioned, as well as the adverse reactions faced by 120 girls, that include epileptic seizures, severe stomach aches, headaches, and mood disorders.

The claims of biomedicine were disrupted through organizing at the local sites of biomedical governance. The very presence of activists at the sites as evaluators and observers of processes in the realm of biomedicine disrupted the status-quo articulations of the processes of securing consent and implementing the clinical trials. The methodological expertise situated in the hands of the experts was turned into the expertise base in the hands of the community. The health activists questioned the claims made by experts about the informed-consent processes by themselves observing the processes and how they were being implemented. Here is an excerpt from a blog entry on the Sama website that describes the communicative processes of participation in the form of witnessing:

> During March 27–30, 2010, a team of women's and health activists visited Bhadrachalam *mandal*, one of the three *mandals* of Khammam district where the 'demonstration project' was undertaken to understand the ground reality; in particular, to look at the nature and procedures of taking consent and providing information to the girls and their parents, and the availability of the health infrastructure required to support cancer screening and prevention.

The presence of the team of women's and health activists at the Bhadrachalam mandal resisted the dominant configurations of biomedicine in the specific context of how clinical trials are conducted. The expertise-based space of the sacred in biomedical clinical trials is disrupted by the presence of the activists as witnesses and as evaluators of the processes. The activists were present to understand and evaluate the ground reality.

Worth noting here is the emphasis on shifting the frame of evaluation into the hands of the activists, who were doing the monitoring, observation, and evaluation of the processes that were carried out in the trials. The positioning of the activists as the experts and evaluators of truth is critical to disrupting the expert-driven spaces of biomedical knowledge and praxis, and is transformative in its relocation of the evaluative capacity of the project in the hands of activist publics, who demand more process-based information and information about how the ethical guidelines established in policy documents were carried out. The women's groups and health activists in the project evaluated the processes of securing consent, the provision of information to the girls as well as their parents, and the availability of basic health-care infrastructure in the communities to support cancer screening and prevention. It was on this basis of direct presence at the very sites of intervention that the activists countered the intervention-based model of biomedicine that operates on the basis of the top-down expertise of biomedical actors. The articulation of alternative entry points to structural injustices is achieved through the reconfiguration of the very structures of knowledge production within which claims of scientific rationality are made.

Interrogating "Truth" Claims

The very act of witnessing in the local contexts introduces into the discursive space locally situated subaltern voices that interrogate the "truth" claims made by the dominant structure. The very basis of the truth claims juggles between the articulation of the local rationality and the simultaneous juxtaposition of global criteria. The local accounts then provide insights into the global processes and the "truth" claims that are circulated in these processes. Consider the following articulation from the press release:

> The girls, 10–14 years old, belonging to poor families, were enrolled in a study being carried out jointly by PATH (an International NGO), Indian Council of Medical Research and the respective state governments funded by Bill and Melinda Gates Foundation. The objective of this two year study is to look into acceptability and service delivery issues of Gardasil, marketed in India by MSD Pharmaceuticals Pvt. Ltd, being misleadingly promoted as a preventive for cervical cancer. (http://samawomenshealth.wordpress.com/2010/04/08/press-conference-on-hpv-vaccines/)

Note here, the framing of the poverty of the families that the girls belonged to. As we will see later, it is this very context of poverty that becomes the entry point for interrogating the narratives of ethical procedures in recruitment. Also note that the local narratives point out that the marketing agendas of evaluating acceptability and service-delivery issues of Gardasil were instead framed as preventive for cervical cancer (in order to push Gardasil into the country). It is precisely, then, against this backdrop that the truth claims regarding the purposes of the vaccines, the trials, and their effectiveness are brought into question.

> When deaths started getting reported local groups in Andhra Pradesh were alarmed by and carried out a fact finding to discover that no consent was taken from parents and the girls and their families have been left uncared for. Post mortem reports were also not easily accessible, and in cases of death cover up was the general dictum. (http://samawomenshealth.wordpress.com/2010/04/08/press-conference-on-hpv-vaccines/)

Worth noting in this narrative are interrogations of the processes that were used in securing consent from the parents. Unlike the articulations made within dominant structures about the sanctity of processes, the local activists point out that no consent was taken from parents, and that the girls and their families have been left uncared for. Furthermore, the activists question the epistemology of the dominant structure (in the form of the postmortem), pointing out that the real causes of death are often covered up.

The interrogation of the truth claims is further played out in the form of the following questions raised by the activists:

> Why are poor girls and their families being actively misled? Who is liable for the debilitating effects of the vaccine and who will medically look after these girls and pay compensation for the damages suffered by these people? (http://samawomenshealth.wordpress.com/2010/04/08/press-conference-on-hpv-vaccines/)

Here, the activists interrogate the reason for misleading the girls and their families. The question of liability is raised, as well as the onus of looking after the girls because of the damages experienced by them. The legal framework introduced into the discursive space raises questions about responsibility for the treatment of the girls who have suffered from debilitating effects of the vaccine. Similarly, the document raises the question of compensating the girls for the damages they suffered.

> How has the government embarked on this study of giving three injections to the girls when it is also planning a massive multi-centric dose determination study to see if two doses will suffice? (http://samawomens health.wordpress.com/2010/04/08/press-conference-on-hpv-vaccines/)

The specific processes and underlying knowledge configurations are brought into question, throwing light on the divergence between the current state of knowledge and the intervention proposed by the government, thus drawing attention to the risks that the girls were exposed to in the absence of adequate knowledge to support the interventions. One of the key elements in the discourse is the emphasis on the idea of risks placed on the girls and their families in the absence of an adequate knowledge base. Attention is also drawn to the inadequacy of the regulatory mechanisms:

> How has the Drugs Controller General granted approval to the vaccine without proper research in India? For a drug to be administered to children, it has to go through stages of clinical trial, including Phase 3 adult clinical trials. So far with Gardasil only one trial has been carried out with just a small sample of 110 girls which has followed them up for just one month after the completion of vaccination and that too only to look at the immune response post vaccination. The vaccine has also been approved for adult women till 27 years of age without doing any trials with them at all. (http://samawomenshealth.wordpress.com/2010/04/08 /press-conference-on-hpv-vaccines/)

Here once again process-related questions are presented in the context of the regulatory mechanisms, presenting the notion that appropriate regulatory mechanisms were not followed. For example, the activists raise the question regarding the importance of Phase 3 adult clinical trials to be carried out before a drug is administered to children; they point out that these processes were not carried out.

The knowledge of the processes creates an entry point for the activists to draw attention to the discrepancy between the processes outlined in the regulatory mechanisms and the processes that were actually followed. Therefore, the process-based claims to truth become entry points for organizing discourse. Finally, the activists also raise the question regarding the economic accessibility of the vaccines in a subaltern context with limited financial resources. They note:

> Why are these studies being carried out when at various times the concerned government officials have gone on record to say that it is not feasible for the vaccine to be introduced in the Indian Public Health system given its costs? Is it not then using these poor girls as guinea pigs for a vaccine which can be used only in the private market by well to do families? Why is a two year study being carried out with no future guarantee even for the subjects of this study to keep them protected with boosters when they actually get married by when the effect of this vaccination will wear off? The effect of the vaccine seems to wear off after 4–5 years, and will require periodic booster injections to retain claimed effectiveness. (http://samawomenshealth.wordpress.com/2010/04/08/press-conference- n-hpv-vaccines/)

The principles of the clinical trials in using subaltern subjects are interrogated amidst questions of accessibility of the vaccines among the subaltern sectors of India. The market logic of the clinical-trials industry is juxtaposed in the backdrop of the ethics of public health, pointing to the ethics in utilizing girls from poor communities in the studies when the vaccines are expensive, and therefore, beyond the reach of the girls and their families. The information that the effects of the vaccines will wear off in 4–5 years and therefore will require booster injections to retain claimed effectiveness is juxtaposed with the costs of the vaccines on the private market, thus leading to the discursive constructions of the girls as guinea pigs for a vaccine that is to be used in the commercial market for economic profits. Once again, the process-based "truth" claims of the project are interrogated vis-à-vis the framing of health as human right. The public health and ethical contexts of clinical trials are juxtaposed in the backdrop of the market economics of the clinical-trials industry in India.

Discussion

The culture-centered processes of local organizing directed at transforming the unhealthy policies at national and global levels draw attention to the relevance of examining health activism as a health-communication intervention. The global reach of neoliberal policies has resulted in the increasing reach of global pharmaceuticals in developing markets, usurping indigenous knowledge under patent acts, and finding exploitable human subject pools in the global South as sources of cheap labor in the clinical-trials industry, particularly in the materially marginalized and disadvantaged spaces of the global South. It is in this landscape of material marginalization and inaccessibility to health-information infrastructures in subaltern contexts that the nexus of pharmaceuticals, NGOs, and national and state governments play out the politics of power and control through the co-optation and manipulation of information (Dutta, 2008). Culture-centered projects of social change in health care draw attention to the locally situated politics of change in the subaltern sectors of the globe that seek to bring about changes in the material structures of global oppression and exploitation. A culture-centered reading of health-care disparities therefore calls for communicative processes of social change and structural transformation that are rooted in the participatory capacities and voices of local communities.

As noted in the articulations of the health activists in the organizing against the Gardasil trials in India, culture-centered processes begin with the acknowledgement of local agency as an entry point to developing problem configurations and working on corresponding solutions that are meaningful to the lived experiences of communities. Therefore, in grassroots organizing of local health activists, social change is accomplished through the emphasis on building local structural information capacities. In the biopolitics of neoliberalism, entry into discursive spaces for making truth is legitimated on the basis of access to truth and the processes of knowledge creation through which activists can generate claims about truth. Therefore, in local organizing in health, the emphasis of activist politics is on transparency and on creating structures of information at the local level that render the processes of biomedicine visible to local communities. The health activists continually emphasize transparency, and the politics of activism in the global South continually works to rupture those structures of expertise and knowledge claims-making that enjoy their hegemony precisely because of the points of inaccess they create. Therefore, activist organize works with one of the specific goals of rendering health-care structures and processes transparent and visible to local communities so that judgments and evaluations can be made. Embedded in the culture-centered processes of health activism is the opening up of the very communicative infrastructures and processes that keep hidden the processes of scientific-technical decision-making in the global marketplace of biomedical power.

The articulation of transparency is situated amidst simultaneous re-presentations of truth claims based on local experiences and local narratives. The truth claims made by the dominant structures of biomedicine (in this case, PATH, Bill & Melinda Gates Foundation, Merck, state and federal governments) are interrupted by the truth claims made by the activists based on their acts of witnessing. The embodiment of activist organizing at the sites of implementation of the Gardasil vaccinations becomes communicative processes for disrupting the status quo, by creating alternative rationalities about the processes that were used in securing informed consent, and by interrogating the actual implementation of the recruitment processes in the Gardasil trials. The local presence at the site of claims-making becomes an entry point to change through the creation of a space for interrogating the processes embodied in dominant structures. Culture-centered processes of social change draw attention to the capacity of local organizing in disrupting oppressive structures through the presence of local agency at sites of generating knowledge claims. Such local agency on the one hand engages with possibilities of alternative stories that disrupt the mainstream stories narrated by the structure; simultaneously, they refer back to the processes and legitimating narratives of the structure to draw attention to the paradoxes, hypocrisies, and failures written into the structures.

Finally, the local emerges as the site for making truth claims based on questions of social justice, equity, and health as human right. As noted previously, these truth claims on the one hand draw upon the language of the very structures they seek to critique; on the other hand, the engagement with the discourses of the dominant structures becomes entry points for raising questions about health, human rights, social justice, and equity in the backdrop of the politics of market economics underlying the clinical-trials industry. In analyzing the communicative processes in the local activism of the women's health groups, it is worthwhile to draw attention to the dialectics between the local and the global that continues to resonate through organizing for change. The universal standards of health as a human right offer the foundations for raising locally situated questions about the processes used in the Gardasil demonstration project; simultaneously, the locally articulated questions of justice draw upon references to universal standards for human subject protection and securing of informed consent in the context of clinical trials.

The science of clinical trials is disrupted with local narratives that draw upon the very standards articulated in the language of science to draw attention to alternative narratives of the clinical trials story. This continual dialogue and departure between the local and the global elucidates the contradictory and complementary processes of organizing in the politics of social change in health care, which on the one hand seeks to establish entry points for local voices in global platforms, and on the other hand, utilizes the language of universal rationality of social justice and scientific processes in doing so. The Enlightenment rationality embedded in the global clinical-trials industry is rendered impure through attention drawn to the various processes and ways in which human subjects are manipulated and information is misrepresented to recruit bodies of labor for the industry. The communicative processes in this case study of organizing for social change point toward the continual negotiation of the terrains of the local and the global within spaces of legitimacy to make locally situated truth claims of social justice that are negotiated within national-global terrains of transnational hegemony (Dutta & Pal, 2010).

Bibliography

Basu, A., & Dutta, M. J. (2008). Participatory change in a campaign led by sex workers: Connecting resistance to action-oriented agency. *Qualitative Health Research, 18*(1), 106–119.

Basu, A., & Dutta, M. J. (2009). Sex workers and HIV/AIDS: Analyzing participatory culture-centered health communication strategies. *Human Communication Research, 35*(1), 86–114.

Coburn, D. (2000). Income inequality, social cohesion and the health status of populations: The role of neoliberalism. *Social Science & Medicine, 51*(1), 135–146.

Coburn, D. (2004). Beyond the income inequality hypothesis: Class, neoliberalism, and health inequalities. *Social Science & Medicine, 58*(1), 41–46.

Dutta, M. J. (2006). Theoretical approaches to entertainment education campaigns: A subaltern critique. *Health Communication, 20*(3), 221–231.

Dutta, M. J. (2008). *Communicating health: A culture-centered approach.* Malden, MA: Polity Press.

Dutta, M. J., & Basnyat, I. (2008a). The case of the Radio Communication Project in Nepal: A culture-centered rejoinder. *Health Education and Behavior, 35*(4), 459–460.

Dutta, M. J., & Basnyat, I. (2008b). The Radio Communication Project in Nepal: A culture-centered approach to participation. *Health Education and Behavior, 35*(4), 442–454.

Dutta, M. J., & de Souza, R. (2008). The past, present, and future of health development campaigns: Reflexivity and the critical-cultural approach. *Health Communication, 23*(4), 326–339.

Dutta, M. J., & Pal, M. (2010). Dialog theory in marginalized settings: A subaltern studies approach. *Communication Theory, 20*(4), 363–386.

Dutta, M. J., & Zoller, H. (2008). Theoretical foundations: Interpretive, critical, and cultural approaches to health communication. In H. M. Zoller & M. J. Dutta (Eds.), *Emerging perspectives in health communication: Interpretive, critical and cultural approaches* (pp. 1–28). Mahwah, NJ: Lawrence Erlbaum.

Dutta-Bergman, M. (2004a). Poverty, structural barriers and health: A Santali narrative of health communication. *Qualitative Health Research, 14*(8), 1107–1122.

Dutta-Bergman, M. (2004b). The unheard voices of Santalis: Communicating about health from the margins of India. *Communication Theory, 14*(3), 237–263.

Prasad, A. (2009). Capitalizing disease: Biopolitics of drug trials in India. *Theory, Culture, & Society, 26*(5), 1–29.

Sarojini, N. B., & Shenoy, A. (2010, September 12). Gardasil research targets girls from vulnerable communities. *The Scavenger.* Retrieved from http://www.thescavenger.net/health/gardasil-research-targets-girls-from-vulnerable-communities-21675.html (no longer accessible).

Zoller, H. M. (2008). Technologies of neoliberal governmentality: The discursive influence of global economic policies on public health. In H. M. Zoller & M. J. Dutta (Eds.), *Emerging perspectives in health communication: Meaning, culture, and power* (pp. 390–410). Mahwah, NJ: Lawrence Erlbaum.

Zoller, H. M. (2009). Narratives of corporate change: Public participation through environmental health activism, stakeholder dialogue, and regulation. In L. Harter, M. J. Dutta, & C. Cole (Eds.), *Communicating for social impact: Engaging communication theory, research, and pedagogy* (pp. 221–242). Cresskill, NJ: Hampton Press.

Zoller, H. M., & Dutta, M. J. (2008). Afterword: Emerging agendas in health communication and the challenge of multiple perspectives. In H. M. Zoller & M. J. Dutta (Eds.), *Emerging perspectives in health communication: Interpretive, critical and cultural approaches* (pp. 449–463). Mahwah, NJ: Lawrence Erlbaum.

28

Voices of Resistance: The Niyamgiri Movement of the Dongria Kondh to Stop Bauxite Mining

Mohan J. Dutta

Critical health communication scholars studying health disparities locate inequalities amid questions of distributions of power, and interrogate the ways in which the positions of power carry out the disenfranchisement of the poorest sectors of society (Dutta, 2008a, 2011; Dutta & Zoller, 2008). In this context, a Subaltern Studies reading of inequities attends to the communicative erasures of subaltern voices and the ways in which these erasures are tied to the material disenfranchisements of the poorest sectors of societies. Building on the Subaltern Studies framework, the culture-centered approach attends to the inequities in the discursive sites of recognition and representation, drawing attention to the ways in which discourses are mobilized to disenfranchise the poorest sectors of the globe, and suggesting coconstructive strategies for resistance that seek to transform the structural inequities written into global policies and programs (Basu & Dutta, 2008a, 2008b, 2009; Dutta, 2008a, 2011, in press-a, in press-b; Dutta-Bergman, 2004a, 2004b, 2005). Particularly trenchant is the critique of neoliberal policies offered by critical health scholars, who note that the large-scale promotion of the principles of privatization, trade liberalization, industrialization, and minimization of social welfare through structural adjustment programs (SAPs) imposed by international financial institutions (IFIs) such as the World Bank (WB) and the International Monetary Fund (IMF) have led to increasing inequities globally and to the increasing material marginalization of the poor (Dutta, 2008a, 2011; Dutta & Pal, 2010, 2011; Millen & Holtz, 2000; Millen, Irwin, & Kim, 2000).

At the heart of this work is the examination of the links between health outcomes and economic policies of development and growth, especially as they relate to the health of the margins (Millen & Holtz, 2000; Millen, Irwin, & Kim, 2000; Navarro, 1999). For instance, economic policies of industrialization accomplished through the establishment of special economic zones (SEZs) with minimal labor and environmental regulations create conditions of exploitations of workers, and have tremendous impact on health outcomes of workers (Dutta, 2011; Millen & Holtz, 2000; Millen, Irwin, & Kim, 2000). Similarly, industrialization projects built in the global South under the framework of SAPs often threaten the livelihoods of local indigenous communities, threaten the environment, and displace communities from their homes, all serving as pathways for poorer health outcomes. The increasing use of violence to displace indigenous communities in order to create openings for industrial projects is another instance of health threat for the poor. Similarly, the large-scale migrations and economic insecurities that result from projects of land grab under neoliberal development result in economic vulnerabilities of subaltern communities, also therefore increasing the health risks attached to the struggles of securing a livelihood.

What, then, are the processes of resistance through which the structures of inequity that have fundamentally silenced subaltern voices are resisted? What are the communicative processes through which the inequities written into structures of neoliberal governance are resisted? In this chapter, we will attend to the grassroots performance of struggles of resistance in the Kondh community residing in the Niyamgiri Hills of Orissa, an Eastern Indian state, where the community faces threats of displacement by a multinational mining and distillery project. By paying attention to the processes of resistance at the grassroots, it is my hope that we will learn about the ways in which subaltern communities seek to impact policy by taking over sites of recognition and representation. Also, attending to issues of displacement and land acquisition foregrounds some of the key health inequities under global neoliberal governance that typically go unchallenged in the mainstream literature in health communication.

Save Niyamgiri Movement

The Save Niyamgiri Movement is a localized, indigenous movement that is led by the indigenous tribe of the Dongria Kondh residing in the Niyamgiri Hills of Orissa, a resource-rich region that has emerged on the national landscape of contemporary India as a site for projects of mining. The Niyamgiri Hills, the specific site for the activism represented in this chapter, are rich in bauxite and were contracted out to the British aluminum giant, Vedanta, for bauxite mining and for the running of a refinery. The site of struggle at Niyamgiri is constituted amidst the resistance of the Dongria Kondh community against their displacement from the hills where they live and that they consider as being sacred to their livelihood. The movement, on the one hand, is organized through local, indigenous leadership, and on the other hand, leverages local-global networks of activism to foster spaces of pressure and change in policy circles nationally and globally. The goal of this chapter is to depict the communicative strategies that are mobilized to exert local-global pressures on policy circuits to bring about transformations in the inequitable policies of neoliberalism that achieve their hegemony through the very disenfranchisement of the subaltern sectors from spaces of recognition and representation that dictate and arbitrate the policies of land acquisition. In this sense, the chapter depicts the agentic capacity of subaltern communities in participating in collective processes of transformation that fundamentally challenge the very logics of erasure that constitute mainstream structures.

By way of background, mining projects such as the one in Niyamgiri are typified by wide discrepancies between the promises made to tribal community members as the projects are built, and the actual opportunities and compensations offered to tribal community members once the projects have been built. Promises of employment are often offered as enticements for the acquisition of land, although in multiple instances, such promises are not carried out in reality. The promises of employment to tribal people that have often been offered as strategic tools to carry out the displacement projects have not been met, and even in instances where they have been met, have resulted in the employment of tribal community members as unskilled labor in the mines, living in high-risk conditions, and barely being able to make ends meet. This background picture of exploitation of tribal communities through false promises offers a frame for understanding the Save Niyamgiri Movement amid the everyday skepticisms of tribal community members toward displacement-based programs of mining that are typically framed within the narrative of development. As supported by earlier work with the culture-centered approach, the collective resistance of indigenous communities to neoliberal projects of development is often based on the interrogation of the very taken-for-granted claims of development that are made by the displacement programs.

In 1997, amid the large-scale liberalization of India, the Orissa authorities had signed a contract with Sterlite India, a subsidiary of the UK-based multinational, Vedanta, for the Niyamgiri mining project (Amnesty International, 2010). In direct violation of Schedule V of the Indian constitution that protects tribal land, the local people living in the communities were not consulted in the decision-making process. Under Schedule V, local authorities are required to consult local people through *Gram Sabhas* and *Gram Panchayats* before acquiring land in Schedule V areas, thus meant as a mechanism for protecting indigenous rights over land. The recognition and representation of affected communities lies at the center of policy frameworks on land acquisition; yet, in this instance, affected community members were not adequately informed or involved in the processes of decision-making. The Kalahandi District Collector's office sent out land-acquisition notices to landowners in June 2002, noting that the District Administration was going to compulsorily acquire land for the refinery project; people that lost their land would be adequately compensated, and people that lost both their home and their land would be compensated as well as resettled. These letters were not sent to landless laborers whose livelihoods were going to be affected by the acquisition as well. People who had complaints were asked to register their complaints by June 22, 2002, and the public meetings were held on June 26, 2002, thus offering a small window for the voicing of opinions. Within two weeks, the land acquisition had started. In September, 2004, additional land was acquired by the District Administration and then transferred to the refinery. Throughout the process, then, the material marginalization of the indigenous communities living in the Niyamgiri Hills was carried out through their communicative marginalization from the discursive spaces of consultation and decision-making.

In 2006, Sterlite, a subsidiary of Vedanta Aluminium, built a refinery in Lanjigarh at the bottom of the Niyamgiri Hills of Orissa in India. The building of the refinery resulted in widespread human-rights violations, as well as environmental pollution in the community. Amidst the rising protests from local community members, in 2007, the parent company, Vedanta Aluminium, applied for expansion of the refinery in the area. The refinery had been built on land that was earlier used by the indigenous communities in the area for farming, and was compulsorily acquired from tribal communities in 2002 and 2004, displacing 118 families and forcing approximately 1,200 families to sell their farmlands to the refinery. It is against this backdrop of the building of the refinery that the local tribal communities of the Dongria Kondh started organizing protests, participating in public demonstrations, and raising their voices of resistance in organized collectives. It is my hope that listening to the voices of resistance in this case study expands the scope for how we think about the role of communication interventions in addressing health disparities.

Participation, Recognition, and Representation as Resistance

First, the participation of the subalterns as leaders in the Save Niyamgiri Movement challenges the key, top-down frameworks of neoliberal development, where experts offer the mantras of modernization embodied in principles of industrialization, urbanization, and land redevelopment as solutions to underdevelopment and poverty (Dutta, 2011). The expertise of the experts as enunciators of knowledge is interrogated by the indigenous tribe through its direct participation at sites of discourse, and through its struggles to lay claims on truth formations and on configurations of knowledge. The representation of indigenous voices at alternative sites of knowledge production seek to displace the neoliberal forms of decision-making that privilege expert knowledge in order to further disenfranchise indigenous communities (see, for example, Dutta, 2011 for an extensive argument about the role of communication in perpetuating disenfranchisement under the neoliberal configuration). In almost all of these development projects under neoliberalism, the politics of development has been carried out through the marginalization of tribal communities from discursive spaces, without tribal participation in decisions of resource management and resource utilization. The participation of the Dongria Kondh, therefore, in the struggles of resistance reoccupies the discursive spaces and sites from which the tribal communities have been erased. The act of transforming inequities lies in the very act of taking over the structures of representation through subaltern participation.

At the center of the protests was the articulation of the violations of local, national, and international laws that mandate tribal consultations, participation, and informed consent in projects that impact tribal communities. In other words, access to the structures of communication itself was the subject of struggle. India's PESA legislation enacted in 1996 mandates that prior recommendations through *Gram Sabhas* or *Panchayats* involving tribal community members at the appropriate level shall be conducted before granting mining leases in Scheduled Areas (Amnesty International, 2010). Between 2002 and 2009, throughout the entire period where decisions were made, the tribal community members were not consulted. Consultation meetings and public hearings that were held regarding the mining project and the distillery were held in spaces that were difficult to reach. Publicity of the meetings and consultations was circulated in newspapers that were published in Oriya and English, and available in the cities. No efforts were made on behalf of the state government or the mining company officials to engage the local tribal community members in the discussions, and most of the tribal community members from the affected areas had not even heard of the consultations and public meetings.

In a report on the processes utilized in securing the land and setting up the mining and distillery projects, Amnesty International (2010) noted that officials of the District Administration and company representatives were present at the consultations, but no one from the Dongria Kondh communities that were affected by the mine site. The consent of the Dongria Kondh community members was not sought; also, they were not involved in processes of decision-making regarding the acquisition of the land and the setting up of the mining and the distillery projects. Noting this absence of villagers from discursive spaces of decision-making, N.S., a Dongria Kondh man, shared with the Amnesty researchers, "Please write to Vedanta Resources and ask them to go talk to the Dongria Kondh." Along similar lines, another Dongria Kondh participant, S.M., noted, "Our message to the company and Sarkar [the government] is simple. We will sit together, us Dongria people, and decide directly" (Amnesty International, 2010, p. 35). Resistance, therefore, is embodied in the very demand for recognition and representation.

In meetings with villagers in the village council meetings for land acquisition, villagers were misinformed that they would each receive Indian Rupees 100,000 for each acre of land, in addition to jobs provided by the company for every family member who sold land. The dream sown among the indigenous community members was that the area would get supplies of electricity and water, and would soon turn into a Bombay or Dubai. According to the official records, the district administration assured the villagers that the proposed refinery would be directly beneficial to the local village and to the entire country, as well as generate job opportunities for the unemployed men and women. Misinformation and manipulation were two key strategies that were deployed by Vedanta public relations and by government officials, and these became the targets of collective organizing. Consider the articulations in the resistive voices in the Amnesty International (2010) report:

> At no stage, they said, were they told that the refinery processes involved risks of substantial pollution. "We were not told that they will make alumina powder and send it elsewhere," said S. and K., two women, who attended two of the village council meetings: "We were later shocked to discover so many trucks bringing the bauxite and taking the powder away. We felt deceived, as we were not told that everything would be done here. The officials did not share in the *gram sabha* meeting or elsewhere that there would be so much dust, chimney smoke, noise, that our river would become dirty. We had never seen a refinery so had no experience or information on what life could be like staying so close to it." (p. 41).

Central to the voices of resistance was the articulation of the misinformation or absence of information in the community before the acquisition of the land and the building of the refinery. Pointing to their lack of access to structural resources of information, community members noted the manipulative strategies utilized by the administration to deceive them. They also noted within this context the environmental pollution caused by the refinery, and the consequences of the environmental pollution for the families living in the area. Furthermore, the communities living on top of the Niyamgiri Hills had no representation in the village council meetings, and this became a key point of contention in the resistance. Accompanying these manipulative strategies, the government used violence to displace the Dongria Kondh from their spaces of living: "Once the village officer and the police came here. They did not give any notice to take away the land. Forcefully they built the road and the pipeline" (http://www.youtube.com/watch?-

v=VJt59wbNI6s). Here is another excerpt: "Our land on which we were farming was taken away forcefully by the company" (http://www.youtube.com/watch- ?v=VJt59wbNI6s). A common thread that runs in this backdrop is the desire for the subaltern community of the Dongria Kondh to be represented and recognized as decision-making agents within the dominant structures of decision-making.

Offering Alternative Principles of Organizing

The discursive erasure of the Dongria Kondh from the modernist platforms of development was essential to circulating a monolithic narrative that foregrounded economic growth. An alternative aesthetic of development celebrated the beauty of the Niyamgiri Hills, as opposed to the objectification of the hills as spaces for mining (http://www.youtube.com/watch?v=VJt59wbNI6s):

> Today I saw Niyamgiri mountains
> And my eyes have become holy
> The mountain looks so beautiful
> And the nightingale is singing
> Since ages, deer and antelopes and bears
> Have been living here.

The sacredness of the Niyamigiri Hills depicted in the song resists the framing of the hills as sites of bauxite mining. The Niyamgiri Hills constitute the economic, cultural, and social fabric of the Dongria Kondh, offering the material and symbolic resources that make up their livelihood. Another protestor echoes the theme of the sacredness of the Niyamgiri Hills. "Yes Niyamgiri is our God. We live in the mountains and survive. We don't have any land on which we can produce and live. We are dependent on the mountain. We won't leave Niyamgiri." T he valuing of the sacred over the logic of economic cost-profit fosters an alternative principle for organizing.

Similarly, laying claim to the historic ownership of the land by the *adivasi* (*ethnic* and/or *tribal*) community, a protestor shares, "We have been living here for generations, how can the government now just say that it is their land and decide to allow mining without talking to us?" (Amnesty International, 2010, p. 3). Establishing claim to the land that has been their home for generations, the Dongria Kondh voices interrogate the frame of development offered by the government that usurps tribal land. The question of ownership is foregrounded to narrate the rights of indigenous communities living on the land for generations. The mining project gets articulated within a narrative of rights, establishing the rights of indigenous communities to the land on which they have lived for generations. The indigenous ownership of land through generations of living on the land is juxtaposed against modernist notions of ownership of land.

Further noting an alternative framework of organizing, these voices of resistance discuss the unhealthy effects of the mining operations understood through the lived experiences of the Dongria Kondh living in the region. The framing of development as economic growth is ruptured by the stories of the health effects of the distillery. Here is the voice of a community member who discusses the likely effects of mining operations on Niyamgiri: "we will not get any water even if we dig deep. It is only because of Niyamgiri that we get water." Here is another voice that documents the increasing adverse health effects of the distillery: "When we bathe, the skin itches. When we drink water, we get sores in our mouth" (http://www.youtube.com/watch?v=VJt59wbNI6s). Voices of local community members from the area, and their stories of suffering, are recorded through mobile phones and uploaded on YouTube and Facebook, physically offering evidence of the adverse health effects brought about by the emissions from the distillery (see, for example http://www.youtube.com/watch?v=l_s6HhDAuAY). Videos, photographs, and audio recordings document the effect of the Vedanta operations in the region, and disrupt the monolithic narrative of mining as a source of economic growth. Videos point to the red mud spill and document the toxic hazards, and then circulate the evidence through websites and social-media groups.

The resistance of the local community of the Dongria Kondh is expressed in the form of the Niyamgiri SurakshaSamiti that becomes a space for discussion of alternative logics. Here is the voice of LadoSikaka, a local leader (http://www.youtube.com/watch?v=ipHmVee_uXw&list=UUw67nB2cjNXlJQOJeeF2Zsw&index=5&feature=plcp):

> Vedanta has troubled us greatly. It is fighting with us for Niyamgiri. The company people try to lure us with schools and roads in our villages. They cannot trick us like that. We will do whatever it takes to save Niyamgiri. We have a right over the mountain. If they take Niyamgiri our

> world will be destroyed. We won't give up Niyamgiri for any price. Can they give us the wealth of Niyamgiri in return? Niyamgiri is not a pile of money. That mountain is our life. We won't flinch if you take our flesh but we won't tolerate Niyamgiri being dug up. They have bought Niyamgiri from the Govt. but it doesn't belong to the Govt. It belongs to the tribals...makes me so angry I get tempted to use the axe. We are not afraid of them because it is for Niyamgiri. They might outnumber us but every one of us will fight. We are not afraid of monsters like Vedanta.

Evident in the narrative is the resolve of the Dongria Kondh to stand up in resistance, and to resist the various means of control exercised by the company that operates hand in hand with the state and the district. The voice of LadoSikaka challenges the dominant structures of articulation by foregrounding the right of the Dongria Kondh to Niyamgiri. It is against this backdrop that the leader expresses the will of his community not to give up Niyamgiri at any price. The narrative of resistance is expressed in the conviction to not leave Niyamgiri and in the resolve to fight.

The dominant forces of development that have been deployed in the region are presented as the backdrop of the grassroots efforts of organizing (Dutta, 2011). Throughout 2009, the Niyamgiri SurakshaSamiti organized protests, and the images of these protests were shared through online resources (http://www.youtube.com/watch?v=Dm8ljBHv61Y&feature=autoplay&list=UUw67nB2cjNXlJQOJeeF2Zsw&lf=plcp&playnext=2). Videos of the protests, shared through various social media, depicted community members marching; voicing slogans such as, "We won't be afraid for the sake of mother earth," and "We won't tolerate devastation in the name of development"; and singing songs and performing dances in groups. Songs such as "Gaonchorabnahin" (we won't leave our village) sung by local performers depict the conviction of the Dongria Kondh to not leave their land, laying out their collective rights to the land. Penned by an *adivasi* movement leader, bhagwaanmajhi, the imagery of the song depicts the lush, green Niyamgiri mountains, the plants and the trees, the mines and the aftermaths of the mining operations, utilizing visuals to document the impact of the bauxite mining operations. Set to an *adivasi* tune and interwoven with *adivasi* art and animation depicting the different aspects of *adivasi* life in relationship to the mining operations, the lyrics narrate the resolve of the Dongria Kondh (http://www.youtube.com/watch?- v=Q49PnMl3HH8&feature=related):

We will not leave our village
Nor our forests
Nor our mother earth
We will not give up our fight!
They built dams...
Drowned villages and...
Built factories
They cut down forests
Dug out mines
And built sanctuaries
...Without water, land, and forest, where do we go?
Oh God of development, pray tell us, how to save our lives?

In the song, the mainstream narrative of development is challenged, portrayed as a perpetrator of oppressions on subaltern communities, and not as an emancipator, as depicted in the mainstream. Specific development projects are referred to in order to document the oppressions that are perpetrated by these projects. Similarly, video films with performances including songs, dances, poems, marches, etc. on different aspects of the mining operations and tribal resistance are widely circulated through social media. Also, reports are circulated across various outlets to raise awareness. For instance, in a report written by Das and Padel (2010), BhagabanMajhi, the leader of the Kashipur movement, is noted as saying:

> *Agya, unnotiboilekono?* (Sir, what do you mean by development?) Is it development to displace people? The people, for whom development is meant, should reap benefits. After them, the succeeding generations should reap benefits. That is development. It should not be merely to cater to the greed of a few officials. To destroy the millions of years old mountains is not development. (http://www.foilvedanta.org/articles/ battles-over-bauxite-in-east-india-the-khondalite-mountains-of-khondistan/)

Performances on streets share the stories of the community and its experiences of loss. Protest participants carry axes as symbolic representations of resistance. Here is a statement made by a leader at the site of the protest march, "If they come with guns, we will pick our arrows, spears, and axes." Here is another leader: "We won't allow mining and we will remain united.... We won't be afraid of death to save Niyamgiri. Die we may we won't leave Niyamgiri."

Local-National-Global Networks of Solidarity

The communicative strength of local organizing is complemented by online activist networks as well as national and global networks. The organizing of the protests against the top-down structures of development is shared in the weaving together of the movement with similar other movements against mega-development projects, providing a broader backdrop for resisting the industrialization projects carried out in the name of development. Solidarity across individual sites against projects of mega-development weave together the resistive elements of the movement into broader, interconnected narratives that question the taken-for-granted assumptions of development. Scholarly articles documenting the human-rights violations, environmental hazards, and displacements of communities from their sources of livelihood offer entry points for disrupting the dominant structures of development planning (see, for example, Das & Padel, 2010).

Tribal communities from the Niyamgiri area participated in concerted efforts of resistance at various sites. For instance, in May 2008, hundreds of Dongria Kondh traveled to the capital of Orissa, Bhubaneswar, to voice their resistance to the mining project to be set up on their sacred mountain. The protest march was covered across several activist outlets, and generated global-national-local networks of reporting, reaching a global audience. Here is a report excerpt from Survival International (2008), a global NGO that works on issues of indigenous rights:

> Survival has launched a campaign targeting Vedanta, and is urging shareholders, including major British companies Standard Life, Barclays Bank, Abbey National and HSBC, as well as Middlesbrough and Wolverhampton Councils, to disinvest unless Vedanta abandons its plans.
>
> Survival's director Stephen Corry said today, 'If Vedanta goes ahead with this mine, the Dongria Kondh will be destroyed. They cannot survive as a people without their land. The Norwegian government has already sold its shares in Vedanta, and other investors should follow suit, or face boycotts over their human rights record.' (http://www.survivalinternational.org/news/3294)

The offline voices of protest organized by tribal community members found their way into online discursive spaces through media reports, reports organized by local and national NGOs, and reports created by global NGOs to organize a global campaign against Vedanta's mining project in Niyamgiri. The resistive narratives voiced at the local sites are picked up and renarrated by NGOs and activists at other sites of protest both offline and online, drawing global attention to the unethical and coercive practices of Vedanta that severely impacted the environment as well as the local, cultural practices of the Dongria Kondh.

Resistance is constituted in the disruption of those spaces of dominant discourses in mainstream sites nationally and globally that serve the strategic functions of Vedanta. For instance, when in 2009, the World Environment Foundation announced that its Golden Peacock Award for best Environmental Management practices was to be awarded to Vedanta Alumina Limited, over twenty activists representing environment and social-action groups organized a protest at the opening ceremony of the "Global Convention for Climate Change" being held at Palampur Agricultural University. The activists carried banners stating, "Stop greenwashing corporate crimes," and "Stop selling climate change," utilizing the space of the opening ceremony to disrupt the environmentally conscious image of Vedanta by narrating the story of the Vedanta mining project and the refinery plant. Documenting the oppressions carried out by Vedanta on tribal communities, the displacement of tribal community members, the displacement of local livelihood, large-scale environmental abuse, health hazards, and the environmental effects of the distillery, the activists took over the microphone to note the hypocrisy of the awards that conferred the prestige of being environmentally conscious on a corporation that was destroying the environment in a local community in Orissa. Distributing reports on the environmental and human-rights abuses of Vedanta, the activists also noted that awards such as the Golden Peacock Award serve as public-relations strategies for corporations such as Vedanta to deflect attention away from their environmentally disruptive practices, and to counter efforts of regulating the environmental abuses caused by companies such as Vedanta.

The discursive presence of the voices of the Dongria Kondh at public sites of discussion disrupt the hegemony of the dominant structures that have traditionally narrated frameworks of development. The Belamba Public Hearing witnessed a concerted effort by the Dongria Kondh to resist the discursive erasures of the Dongria Kondh from the platforms of decision-making and from the discursive articulations of policy-making. Kondh voices foregrounded the struggles of the community and the effects of the mining and refinery operations on the community. The local sites of activism are connected to the global platforms of activist politics through new-media platforms such as Web pages, bulletin boards, and discussion groups; through social media such as Facebook; and through audiovisual tools such as Youtube. For example, the Facebook page of the Save Niyamgiri project seeks to organize public opinion and support through the connections and networking opportunities it offers across local and global spaces (http://www.facebook.com/#!/group.php?gid=31785088220). The project has a Facebook presence through which it shares important news directly from the site, compiles a variety of media stories, as well as connects with videos that narrate through the voices of the tribal people the experiences they face in the backdrop of the mining project, the threats to their life posed by the London-based Vedanta Corporation, the violence enacted by the corporation, and the state-sponsored police violence on tribal protests (http://www.facebook.com/#!/group.php?gid=31785088220). The Facebook page of the Save Niyamgiri campaign emerges as a site for sharing news and information about the movement, and for providing access to images and videos taken from field sites as communities participated in their efforts of resistance.

Videos of protests and police tortures from local spaces of protest emerged into the networks of solidarity through Facebook. Similarly, videos linking to documentaries on the resistance, videos of performances, and videos utilized by Vedanta for its public-relations practices fostered networked sites of resistance. Websites such as *Foil Vedanta* (http://www.foilvedanta.org/articles/battles-over-bauxite-in-east-india-the-khondalite-mountains-of-khondistan/) offer information resources, environmental reports, and links to articles, photographs, and steps for action. The local-global solidarity networks of the movement emerged through the involvement of organizations such as Amnesty International and Survival International (http://www.survivalinternational.org-/tribes/dongria) in documenting the corrupt practices of Vedanta, and in developing campaigns in London and at other global sites that carried the message of the Save Niyamgiri Movement. Campaigns involved online petitions, resources for letter writing and communicating with key stakeholders, and videos for raising awareness. The local-national-global organizing of the Save Niyamgiri Movement was instrumental in pressuring the Indian government to refuse clearance to Vedanta (http://www.survival international.org/tribes/dongria); along these very lines, the global solidarity networks pressured the Church of England and the Norwegian Government and investment firm, Martin Currie, to withdraw their investments from Vedanta Resources. Celebrating the story of the movement, the Survival International website combines a number of videos with press releases and petitions, juxtaposing the voices of the tribal community members with information on specific campaigns organized globally against Vedanta.

Discussion

The Save Niyamgiri Movement is an example of a culturally centered health intervention that seeks to transform the inequitable structures of neoliberal governance by rupturing the inequities written into the organizing of discursive spaces of recognition and representation. The agency of subaltern resistance is enacted in the struggles of representation through which the Dongria Kondh challenge the mainstream frameworks of development imposed by the nexus of the Vedanta Corporation and the state. The participation of Dongria Kondh community voices in protest songs, marches, slogans, dances, and speeches is circulated in offline and online spaces, seeking to influence key stakeholders through direct action on the field as well as through circulation in new-media channels and amid local-global activist networks. The leveraging of local-global activist networks creates entry points of solidarity for the intervention, seeking to leverage the possibilities of impact at multiple local, national, and global sites through which neoliberal policies can be transformed (Dutta, 2011). The exploitative capacity of neoliberal governance is enacted in the disenfranchisement of subaltern communities through the very uses of the language of democracy, participation, community, and dialogue. As depicted in the example of the Niyamgiri movement, the transformative capacity of the culture-centered approach lies in skeptically turning these very concepts on their head, and in leveraging local community participation as a site for enactment of subaltern agency to disrupt the oppressive policies carried out by mainstream structures of neoliberalism.

Bibliography

Amnesty International. (2010). *Don't mine us out of existence: Bauxite mine and refinery devastate lives in India.* London, UK: Author.

Basu, A., & Dutta, M. J. (2008a). Participatory change in a campaign led by sex workers: Connecting resistance to action-oriented agency. *Qualitative Health Research, 18*(1), 106–119.

Basu, A., & Dutta, M. J. (2008b). The relationship between health information seeking and community participation: The roles of health information orientation and efficacy. *Health Communication, 23*(91), 70–79.

Basu, A., & Dutta, M. J. (2009). Sex workers and HIV/AIDS: Analyzing participatory culture-centered health communication strategies. *Human Communication Research, 35*(1), 86–114.

Das, S., & Padel, F. (2010). Battles over bauxite in East India: The Khondalite mountains of Khondistan. Retrieved from http://www.foilvedanta.org/articles/battles-over-bauxite-in-east-india-the-khondalite-mountains-of-khondistan/.

Dutta, M. J. (2006). Theoretical approaches to entertainment education campaigns: A subaltern critique. *Health Communication, 20*(3), 221–231.

Dutta, M. J. (2007).Communicating about culture and health: Theorizing culture-centered and cultural sensitivity approaches. *Communication Theory, 17*(3), 304–328.

Dutta, M. J. (2008a). *Communicating health: A culture-centered approach.* London, UK: Polity Press.

Dutta, M. J. (2008b). A critical response to Storey and Jacobson: The co-optive possibilities of participatory discourse. *Communication for Development and Social Change: A Global Journal, 2*, 81–90.

Dutta, M. J. (2008c). Participatory communication in entertainment education: A critical analysis. *Communication for Development and Social Change: A Global Journal, 2*, 53–72.

Dutta, M. J. (2009). On Spivak: Theorizing resistance—Applying Gayatri Chakravorty Spivak in public relations. In Ø. Ihlen, B. van Ruler, & M. Fredriksson, (Eds.), *Public relations and social theory: Key figures and concepts* (p. 278–300). New York, NY: Routledge.

Dutta, M. J. (2011).*Communicating social change: Culture, structure, agency.* New York, NY: Routledge.

Dutta, M. J. (in press-a). A culturally centered approach to communication for social change. In S. Melkote (Ed.), *Development communication and directed change: A reappraisal of theories and practices.* AMIC.

Dutta, M. J. (in press-b). A culture-centered critique of global public relations. In K. Sriramesh (Ed.), *Cultural theories of public relations.* Routledge.

Dutta, M. J., & Basu, A. (2008). Meanings of health: Interrogating structure and culture. *Health Communication, 23*(6), 560–572.

Dutta, M. J., & Pal, M. (2010). Dialog theory in marginalized settings: A subaltern studies approach. *Communication Theory, 20*(4), 363–386.

Dutta, M. J., & Pal, M. (2011). Public relations in a global context: Postcolonial thoughts. In N. Bardhan & K. Weaver (Eds.), *Public relations in global cultural contexts* (pp. 195–225). New York, NY: Routledge.

Dutta, M. J., & Zoller, H. (2008). Theoretical foundations: Interpretive, critical, and cultural approaches to health communication. In H. M. Zoller & M. J. Dutta (Eds.), *Emerging perspectives in health communication: Interpretive, critical and cultural approaches* (pp. 1–28). Mahwah, NJ: Lawrence Erlbaum.

Dutta-Bergman, M. (2004a). Poverty, structural barriers and health: A Santali narrative of health communication. *Qualitative Health Research, 14*(8), 1107–1122.

Dutta-Bergman, M. (2004b). The unheard voices of Santalis: Communicating about health from the margins of India. *Communication Theory, 14*(3), 237–263.

Dutta-Bergman, M. (2005). Theory and practice in health communication campaigns: A critical interrogation. *Health Communication, 18*(2), 103–112.

Millen, J. V., & Holtz, T. (2000). Dying for growth, Part I: Transnational corporations and the health of the poor. In J. Y. Kim, J. V. Millen, A. Irwin, & J. Gershman (Eds.), *Dying for growth: Global inequality and the health of the poor* (pp. 177–223). Monroe, ME: Common Courage Press.

Millen, J. V., Irwin, A., & Kim, J. Y. (2000). Introduction: What is growing? Who is dying? In J. Y. Kim, J. V. Millen, A. Irwin, & J. Gershman (Eds.), *Dying for growth: Global inequality and the health of the poor* (pp. 3–10). Monroe, ME: Common Courage Press.

Navarro, V. (1999). Health and equity in the world in the era of 'globalization.' *International Journal of Health Services, 29*(2), 215–226.

Survival International. (2008, May 8). Tribe stages mass protest against British company Vedanta. Retrieved from http://www.survivalinternational.org/news/3294.

Zoller, H. (2005). Health activism: Communication theory and action for social change. *Communication Theory, 15*(4), 341–364.

Zoller, H. (2006). Suitcases and swimsuits: On the future of organizational communication. *Management Communication Quarterly, 19*(4), 661–666.

Zoller, H., & Dutta, M. J. (2008a). Afterword: Emerging agendas in health communication and the challenge of multiple perspectives. In H. Zoller & M. J. Dutta (Eds.), *Emerging perspectives in health communication: Meaning, culture, and power* (pp. 449–463). Mahwah, NJ: Lawrence Erlbaum.

Zoller, H., & Dutta, M. J. (2008b). Introduction: Communication and health policy. In H. Zoller & M. J. Dutta (Eds.), *Emerging perspectives in health communication: Meaning, culture, and power* (pp. 358–364). Mahwah, NJ: Lawrence Erlbaum.

29

EPILOGUE: QUESTIONS AND DEBATES IN ADDRESSING HEALTH DISPARITIES

Mohan J. Dutta and Gary L. Kreps

As we wrap up this edited volume comprised of what we see as key works in the communication literature that are directed at addressing health disparities, we point toward what we see as some of the emerging trends as well as the debates that are foregrounded in the literature presented in these pages. We attend to these debates and trends as a way of outlining frameworks for moving ahead, both in terms of engaging the theorizing of health disparities, as well as in the development of methodologies and applications for specific communication interventions that are directed toward addressing health disparities in local community-specific contexts. That health disparities are ethically unacceptable underlies our approach toward health interventions that frames the issue within the broader umbrella of communication for social justice (Dutta, 2008, 2011, 2012; Kreps, 2006).

While we initially sought to offer a global scope with the book, reflecting our limited positions of engagement as scholars located in the US, we are aware of the limited geographic contexts that emerge in our chapters. Perhaps, ironically, reflecting the broader biases underlying communicative inequities globally and the inequities in the politics of representation and recognition (Dutta, 2008, 2011, 2012), a large number of our chapters are U.S.-centric, voicing the articulations of U.S.-based scholars. Whereas we have tried to correct this bias by including scholarship engaging the contexts of Israel, Ghana, Kenya, and India, there remains much scope for further engagement of diverse voices in these discursive spaces, particularly in drawing upon scholarship on health equities in Cuba and Scandinavian countries (Chomsky, 2000). Similarly, in our U.S.-based work, many of the voices presented here within our book are the voices of the dominant mainstream, therefore suggesting the need for additional work that engages the margins within the dominant spaces of the global North.

However, in justification of our overrepresentation of U.S.-based examples in the book, the US tends to be a particularly problematic site for health disparities, perhaps because of the tremendous diversity and high levels of immigration within this complex "melting pot" country. Due to the high levels of racial, ethnic, international, religious, educational, and socioeconomic diversity in the US, there are a plethora of inequities in health outcomes evident in this nation (Kreps, 2006). In addition to the widespread problems of prejudice and discrimination within the US that can lead to health disparities, the for-profit economic model for the delivery of health care in the US also leads to disparities in health outcomes for those who are uninsured (or underinsured), for those who are poor, for the elderly, for sexual minority groups (gay, lesbian, and transgender people), and for many sex workers. The US is a hotbed for health disparities, and needs communication interventions to improve health outcomes for vulnerable populations.

Questions and Debates Addressing Health Disparities

The literature on the underlying causes for health disparities points toward an emergent complexity that documents the complex interplay of the structural, community-specific, and lifestyle-based contexts of health disparities. Responding to these complexities at multiple levels and in multiple frameworks for understanding the lived experiences of inequities, the communication interventions documented in this book offer a polymorphic framework for engaging health disparities, responding to health disparities at the level of the individual, at the level of the community, through wider uses of societal-level mass media, and through the politics of activism that is directed at bringing about structural transformations in policies as well as programs. Whereas many of the interventions described in this book are coconstructive, other analyses presented in the book document the existing communicative disparities within mainstream structures that produce the conditions for health disparities. In this section, we note some of the key themes that emerge in the array of work represented in this book.

Levels of Analysis

Health-communication interventions addressing health disparities vary in their level of analysis of the problem of disparities and the level at which the communication interventions are developed. Whereas a large number of projects are focused on the level of individual lifestyles, individual communication skills (both of the patient as well as of the provider), others are focused on the level of the relationship (such as projects on addressing disparities by improving the relationship between the provider and the patient); yet others are focused on addressing disparities at the level of the community, by building community resources and by addressing the structural barriers that are experienced in community settings. Furthermore, other health interventions are developed at the level of policy-making and programming, seeking to transform political, societal, and economic structures that perpetuate inequities. Projects of health activism are directed at transforming the oppressions and exploitations that are written into hegemonic formulations of policies and programs. The continual dialogues among these different levels of analysis offer opportunities for engaging the complex interplays of various factors that contribute to health disparities, and also document the debates between the divergent paradigms of health disparities work.

Methodological Choices

Reflecting the diversity in the level of analysis, we also observe a wide variety in the methodological choices that are taken up by scholars in documenting the effects and effectiveness of communication interventions. Whereas in some instances we do see the emphasis on the traditional, quantitative measures of effectiveness, the chapters in this book also move beyond these traditional measures to document individual cases of policy change; analyses of media reports; in-depth interviews and focus groups with community members, policy-makers, and program planners; postintervention interviews with community members and policy-makers; ethnographically "rich" descriptions of contexts, etc. What these diverse arrays of methodologies depict is the growing trend in health-disparities scholarship toward opening up the discursive spaces of knowledge structures for engaging with, innovating on, and creatively articulating diverse methodologies for communicating the value of communication interventions, engaging in thoughtful debates about what it means to measure the success of an intervention, and situating these debates in the context of the meaningfulness of the interventions in the lives of community members. Ultimately, the question confronting scholars working on developing interventions addressing health disparities relates to the continual development of adequate tools for measuring the impact of the communication interventions being developed, the continual dialogue about how best to measure the success of interventions in meaningful terms, as well as the creation of adequate strategies for fostering learning cycles through which these communication interventions can continually be improved. Rather than clinging to a methodological dogmatism that is wedded to the blind commitment to a certain narrow set of methodological choices for evaluation because that is what has been traditionally done, the question of methodological choices highlighted in this book suggests the need for additional thinking and conversations about how best to measure the success of communication interventions situated within specific localized contexts.

Paradigmatic Diversity

In the various chapters represented in the pages of the book, we also note the extensive philosophical differences in theorizing the underlying causes of health disparities and the appropriate strategies for countering these disparities. Whereas some of the work presented in the pages of this book depict the role of technologies as enablers in addressing disparities, other work documents the limits of a technologically deterministic approach; yet other writings suggest that the underlying causes of disparities are rooted in structures, and that for change to take place, these underlying structures need to be transformed. Whereas in some of the projects there is an emphasis on understanding culture as a barrier to effective health-behavior change, other work takes this concept of culture as barrier to task, and instead suggests the importance of looking at culture as an enabler. We believe that this paradigmatic diversity reflected in this subarea of communication interventions addressing health disparities offers opportunities for continued conversations, directed toward accomplishing the pragmatic goal of addressing the various sites and sources of health disparities. The debates and dialogues open up continued opportunities for building the study of communication interventions directed toward addressing health disparities.

Questions of Change

That existing forms of inequities need to be changed is a common thread that runs across the various chapters that are presented in the book. However, how to go about addressing this change varies widely. Much of the change suggested in the literature on health interventions occurs at the individual level, and therefore is situated amidst the existing structural patterns of distribution of resources. Other change-driven work focuses on addressing the communicative inequities that are at play in the domain of health inequities. Yet other change-driven work is focused on transforming the structural inequities by explicitly challenging the structures through direct action, activism, and processes of change. We see these differences in theorizing about the locus of change as vital to the debates and conversations that could possibly be opened up by this area of health-disparities interventions. Furthermore, engaging these differences fosters opportunities for conversations about the implicit values that are carried out by health interventions, thus rendering visible the underlying values and agendas that often remain hidden within mainstream structures of knowledge production.

Questions of Democratizing Knowledge

That inequities are fostered through the large-scale differentials in the distribution of knowledge is at the heart of communication interventions that are directed at addressing health inequities. In this line of research, the focus is on building local information processing and information participation capacities in communities that are directed at bringing about structural transformations both through individualized interactions in physician-patient relationships and also through participation of community members in processes that hold the structures accountable in the context of the distribution of health resources, the opportunities for local participation, and the openings for dialogue. For instance, the development of communication interventions that focus on building culturally specific and community-based resources driven by comparative effectiveness research (CER) documents the role of participation at multiple levels of health-care decision-making in addressing health disparities. Similarly, scholarship on innumeracy and literacy building can be seen from the standpoint of building information capacities in disenfranchised communities.

Within the broader backdrop of health-care reform in the US, and the democratizing of scientific knowledge production processes globally, we see a growing need for health-communication scholarship that is directed at fostering participatory spaces of decision-making rooted in disenfranchised communities, rather than using the traditional persuasion-based frameworks that often hide behind claims of scientific expertise to render decision-making processes opaque to communities at the margins.

Directions for Future Work

Based on our observations about some of the key trends in communication scholarship that is directed at addressing health disparities, in this section we suggest some of the directions for future work as we see them. First and foremost, we see the need for more work that theoretically seeks to understand the ways in which communication works to reinforce and reproduce disenfranchisement, and then connect the theorizing of this work to develop forms of interventions that seek to transform these underlying forms of communicative disenfranchisement. This calls for increasing conversations among the different paradigms of doing health-communication work in productive ways, constituted within the broader agenda of addressing health disparities. For instance, critical-cultural health-communication work examines the underlying power differentials and differentials in distribution of resources (Dutta, 2008, 2009, 2010, 2011; Zoller, 2005). Engaging the theoretical insights of this work with the work on development of interventions can be productively directed toward transforming policies and programs that seek to disrupt these underlying causes of health disparities (Kreps, 2006). The PBS series on unnatural causes is an example of a communication-based intervention that seeks to address the underlying causes of health disparities by creating awareness about these causes (http://www.unnatural causes.org/). Other application- and methodology-driven works that seek to build a bridge between the various approaches to conducting health-disparities scholarship would provide vital entry points for collaboration.

There is also a greater need for multimethod, multilevel approaches to the development of communication interventions addressing health disparities at multiple sites. Although many of the chapters highlighted in this section do indeed present a wide array of methodologies, we look forward to seeing more work being done that actually connects these conceptual frameworks, and offers complex treatments of the underlying dimensions of health disparities. What is perhaps most important methodologically in health-disparities work is the process of reflexivity, the continual interrogation of the positions of privilege embodied in the methods we apply, and the critical questioning of the ways in which this privilege carries out the agendas of the status quo. It is through reflexivity that health-disparities work can come to critically examine the agendas of dominant structures, carefully interrogate the co-optive possibilities in the scholarly work we conduct, and foster avenues for collaboration with disenfranchised communities so that structural oppressions may be resisted and transformed. For instance, when working on and working with health campaigns in the broader framework of health disparities, health-disparities research also needs to challenge the values embodied in these campaigns and carefully examine whether these values are actually positioned in changing the existing configurations of inequities and disparities.

Although we have highlighted some projects that discuss the role of communication activism and communication for social change in addressing health disparities, we also see the need for more work that is directed at addressing unhealthy policies that foster inequities and create structural contexts for health disparities. Because health disparities are largely situated amid inequitable structures, it is important to envision what role communication scholars might play in generating greater awareness about the underlying causes of health disparities, and fostering entry points for interventions that seek to bring about transformations in unhealthy policies that breed and reproduce inequities.

We also see the need for descriptive work that documents health-care efforts in international contexts that have been particularly effective in addressing health disparities. For instance, in Cuba, the overall health outcomes of the population are strong, and the disparities in health outcomes are limited through vital state-driven policies that are specifically focused on equities (Chomsky, 2000). What are the communicative processes through which health policies and programs are developed that foster opportunities for minimizing health disparities and creating health equities? Similarly, as pointed out by one of our reviewers, excellent health interventions in the Scandinavian countries directed at health disparities emerge from within the context of state-based efforts at regulating health services and technologies with specific attention to health disparities. As we move ahead, it is our hope that this edited book fosters spaces for additional debates and dialogues, especially as these debates and dialogues relate to listening to the voices of the global margins, and engaging in conversations about building just and equitable social, political, and economic systems globally.

In conclusion, then, we see this edited volume as an excellent starting point for conversations and collaborations that work toward fostering a world where inequities are mitigated by combinations of locally situated, mediated, face-to-face, and community-based interventions that address the multiple levels of health disparities, ranging from the individual to the relational to the group to the community to national and global contexts.

Bibliography

Airhihenbuwa, C. O., & Dutta, M. J. (2012). New perspectives on global health communication: Affirming spaces for rights, equity, and voices. In R. Obregon & S. Waisbord (Eds.), *The handbook of global health communication* (pp. 34–51). Chichester, UK: Wiley-Blackwell.

Chomsky, A. (2000). The threat of a good example: Health and revolution in Cuba. In J. Y. Kim, J. V. Millen, A. Irwin, & J. Gershman (Eds.), *Dying for growth: Global inequality and the health of the poor* (pp. 331–357). Monroe, ME: Common Courage Press.

Dutta, M. J. (2008). *Communicating health: A culture-centered approach.* London, UK: Polity Press.

Dutta, M. J. (2009). On Spivak: Theorizing resistance—Applying Gayatri Chakravorty Spivak in public relations. In Ø. Ihlen, B. van Ruler, & M. Fredriksson (Eds.), *Public relations and social theory: Key figures and concepts* (pp. 278–300). New York, NY: Routledge.

Dutta, M. J. (2010). The critical cultural turn in health communication: Reflexivity, solidarity, and praxis. *Health Communication, 25*(6–7), 534–539.

Dutta, M. J. (2011). *Communicating social change: Culture, structure, agency.* New York, NY: Routledge.

Dutta, M. J. (2012). *Voices of resistance.* West Lafayette, IN: Purdue University Press.

Kreps, G. L. (2006). Communication and racial inequities in health care. *American Behavioral Scientist, 49*(6), 760–774.

Zoller, H. (2005). Health activism: Communication theory and action for social change. *Communication Theory, 15*(4), 341–364.

INDEX

Health Communication

Gary L. Kreps, Series Editor

This series examines the powerful influences of human and mediated communication in delivering care and promoting health.

Books analyze the ways that strategic communication humanizes and increases access to quality care as well as examining the use of communication to encourage proactive health promotion. The books describe strategies for addressing major health issues, such as reducing health disparities, minimizing health risks, responding to health crises, encouraging early detection and care, facilitating informed health decisionmaking, promoting coordination within and across health teams, overcoming health literacy challenges, designing responsive health information technologies, and delivering sensitive end-of-life care.

All books in the series are grounded in broad evidence-based scholarship and are vivid, compelling, and accessible to broad audiences of scholars, students, professionals, and laypersons.

For additional information about this series or for the submission of manuscripts, please contact:

Gary L. Kreps
University Distinguished Professor and Chair, Department of Communication
Director, Center for Health and Risk Communication
George Mason University Science & Technology 2, Suite 230, MS 3D6
Fairfax, VA 22030-4444
gkreps@gmu.edu

To order other books in this series, please contact our Customer Service Department:

(800) 770-LANG (within the U.S.)
(212) 647-7706 (outside the U.S.)
(212) 647-7707 FAX

Or browse online by series:
www.peterlang.com

www.ingramcontent.com/pod-product-compliance
Ingram Content Group UK Ltd.
Pitfield, Milton Keynes, MK11 3LW, UK
UKHW021835100926
8556IPUK00004B/356